Advances in Bearing Lubrication and Thermodynamics 2023

Advances in Bearing Lubrication and Thermodynamics 2023

Editors

Ke Yan
Bin Fang
Bei Yan
Fei Chen

MDPI

Basel • Beijing • Wuhan • Barcelona • Belgrade • Novi Sad • Cluj • Manchester

Editors

Ke Yan
Xi'an Jiaotong University
Xi'an
China

Bin Fang
Xi'an Jiaotong University
Xi'an
China

Bei Yan
Chang'an University
Xi'an
China

Fei Chen
Xi'an Jiaotong University
Xi'an
China

Editorial Office
MDPI
St. Alban-Anlage 66
4052 Basel, Switzerland

This is a reprint of articles from the Special Issue published online in the open access journal *Lubricants* (ISSN 2075-4442) (available at: https://www.mdpi.com/journal/lubricants/special_issues/2PKIP1GH3L).

For citation purposes, cite each article independently as indicated on the article page online and as indicated below:

Lastname, A.A.; Lastname, B.B. Article Title. *Journal Name* **Year**, *Volume Number*, Page Range.

ISBN 978-3-7258-1129-8 (Hbk)
ISBN 978-3-7258-1130-4 (PDF)
doi.org/10.3390/books978-3-7258-1130-4

Contents

lubricants

Article

Research on the Nonlinear Stiffness Characteristics of Double-Row Angular Contact Ball Bearings under Different Working Conditions

Bin Fang [1,2,*], Jinhua Zhang [1,2], Jun Hong [1,2] and Ke Yan [1,2,*]

1 Key Laboratory of Education Ministry for Modern Design and Rotor-Bearing System, Xi'an Jiaotong University, Xi'an 710049, China
2 School of Mechanical Engineering, Xi'an Jiaotong University, Xi'an 710049, China
* Correspondence: binfang@mail.xjtu.edu.cn (B.F.); yanke@mail.xjtu.edu.cn (K.Y.)

Abstract: To study the variation rules of nonlinear stiffness of double-row angular contact ball bearings (DR-ACBB), this paper proposed a general mathematic model for DR-ACBB under three different configurations based on the improved quasi-static model of ball bearings, an explicit expression stiffness matrix of DR-ACBB is analytically derived, and a double-layer nested iterative algorithm based on the Newton–Raphson method is designed to realize the efficient solution of the proposed model. Then, the effects of the preload, speeds, and loads on the nonlinear stiffness variations of DR-ACBB under different arrangements are comparatively analyzed. The results show that DR-ACBB under the DB and DF configurations have the same variation rule in axial and radial stiffness; that is, a nonlinear soft-spring stiffness characteristic (i.e., the stiffness decreases with the external load) within the low-speed range and light load condition, and a nonlinear hard spring stiffness characteristic (i.e., the stiffness increases with the external load) within the high-speed range or heavy load condition.

Keywords: double-row angular contact ball bearing; nested iterative algorithm; nonlinear stiffness; soft/hard spring stiffness characteristic

Citation: Fang, B.; Zhang, J.; Hong, J.; Yan, K. Research on the Nonlinear Stiffness Characteristics of Double-Row Angular Contact Ball Bearings under Different Working Conditions. *Lubricants* **2023**, *11*, 44. https://doi.org/10.3390/lubricants11020044

Received: 28 December 2022
Revised: 22 January 2023
Accepted: 26 January 2023
Published: 28 January 2023

1. Introduction

As the core functional component of various rotating machinery systems, double-row angular contact ball bearings (DR-ACBB) have been widely used in rail transportation, CNC machine tools, and aerospace and other fields. Compared with traditional single-row angular contact ball bearings, DR-ACBB show better rigidity and carrying capacity. Due to the strong nonlinear coupling between the loads and deformations of single-row angular contact ball bearings (SR-ACBB), the stiffness characteristics of DR-ACBB cannot be obtained by the linear superposition of two single ball bearings. Unfortunately, studies on DR-ACBB are relatively few, and only a small number of static simplified models have been reported in recent years.

Bercea [1] proposed a unified model for different types of double-row rolling bearings under a face-to-face arrangement; however, a study on the stiffness formulation of double-row rolling bearings was not involved. Gunduz and Singh [2,3] conducted a comprehensive study on the mechanical modeling and stiffness matrix calculation for static DR-ACBB, and the effects of the preload and arrangement forms are detailed, discussed, and analyzed. On this basis, Dick [4] presented a simplified mathematical model to predict the load distribution, stiffness variations, and vibration responses for DR-ACBB with raceway defects of varying length, depth, and surface roughness. Xu et al. [5] further studied the influence of ring angular misalignments on the load distribution and stiffness coefficients for DR-ACBB. Above all, not only the effects of ball inertia forces have been ignored—the relationship between DR-ACBB and SR-ACBB has not been explicitly determined.

In order to accurately predict the mechanical and stiffness characteristics of DR-ACBB at different speed ranges, the influence of the centrifugal forces and gyroscopic moments of internal balls need to be fully considered. Furthermore, according to the structure features of DR-ACBB, the mechanics and motion states of the SR-ACBB should be determined first. Different from the study on DR-ACBB, large numbers of the modeling methods and characteristics analyses of SR-ACBB have been widely reported [6–18]. Among them, due to its high computational efficiency and accuracy, the improved quasi-static model based on the no-macro-sliding hypothesis has been widely used in load distribution calculations and stiffness characteristic analyses [10,13].

As the basic aspect of the dynamic modeling and characteristic analysis of bearing and rotor systems, research on the stiffness matrix of ball bearing has been paid attention to for a long time [19–34]. Especially in recent years, multi-degree-of-freedom stiffness matrix calculations based on analytical differential [24–28] and numerical difference [29–31] methods have become a trend in this field. Compared with the numerical method, the analytical method has higher calculation efficiency, stability, and accuracy. However, all the above studies were only conducted for SR-ACBB, mainly focusing on stiffness softening behaviors; that is, the stiffness of ball bearings decrease as their rotating speed increases. However, an analysis of the nonlinear stiffness characteristics of ball bearings is rarer [32]. Considering the significant difference between the dynamic response of nonlinear and linear systems, it is impossible to accurately predict the dynamic characteristics of the bearing and rotor support system based on linear stiffness assumptions. Furthermore, DR-ACBB can have different configurations, which may lead to more abundant nonlinear stiffness characteristics for DR-ACBB under different load conditions and speed ranges. Therefore, it is of great significance to systematically and comprehensively study the nonlinear stiffness characteristics of DR-ACBB.

This paper presents a comprehensive study on the nonlinear stiffness characteristics of DR-ACBB under three different arrangements. First, based on the improved quasi-static model of SR-ACBB, the load and displacement relationships between DR-ACBB and two SR-ACBB are given by the vector transformation under multi-coordinate systems. Then, the analytical derivation of the stiffness matrix for DR-ACBB under different arrangements is presented, and a double-layer nested iterative algorithm based on Newton–Raphson is designed for the high-efficiency solution of the proposed model. At last, the influences of the rotating speed, preload, and external force on the nonlinear stiffness characteristics of DR-ACBB under different arrangements are comparatively analyzed.

2. Theoretical Analysis

The three typical configuration forms of DR-ACBB are shown in Figure 1: (i) back-to-back (DB) arrangement; (ii) face-to-face (DF) arrangement; and (iii) tandem (DT) arrangement. To build a universal model of DR-ACBB in the above different configurations, the following assumptions are given to simplify the complexity:

(1) Two SR-ACBBs have the same structural parameters, that is, the raceway contact curvature radii r_i and r_o, the raceway contact diameters d_i and d_o, and the ball diameter and number D and Z, respectively;
(2) The influence of the lubrication and cage is not considered;
(3) The outer raceway is fixed while the inner ring moves and rotates with the central shaft.

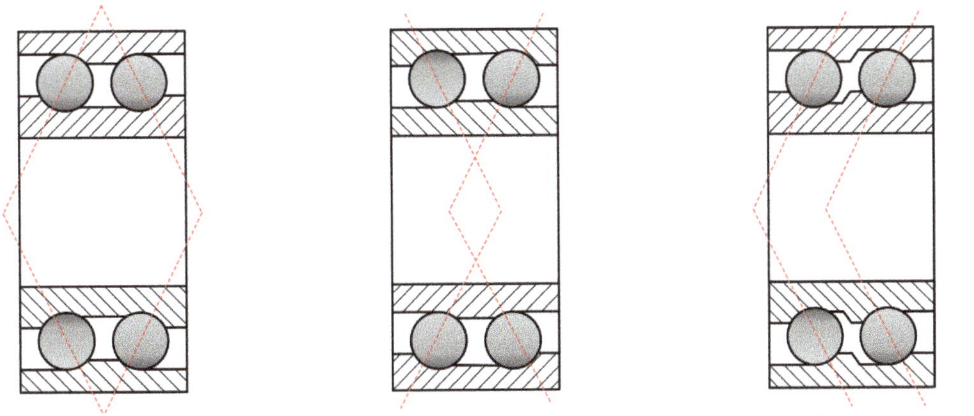

Figure 1. Three configurations of DR-ACBB: (**a**) DB arrangement; (**b**) DF arrangement; (**c**) DT arrangement.

In order to further improve the supporting rigidity and operation accuracy, DR-ACBB is always designed with a split inner raceway, which provides a lock nut or spacer convenient for the axial preloading [3,18]. Therefore, compared with SR-ACBB, DR-ACBB have two additional geometry parameters as presented in Figure 2: the center distance between two rows of rolling elements (d_c) and the axial clearance of the split inner ring (δ_p). During the installation process of double-row ball bearings, the axial clearance δ_p needs to be eliminated by axial locking force to achieve effective pre-tightening.

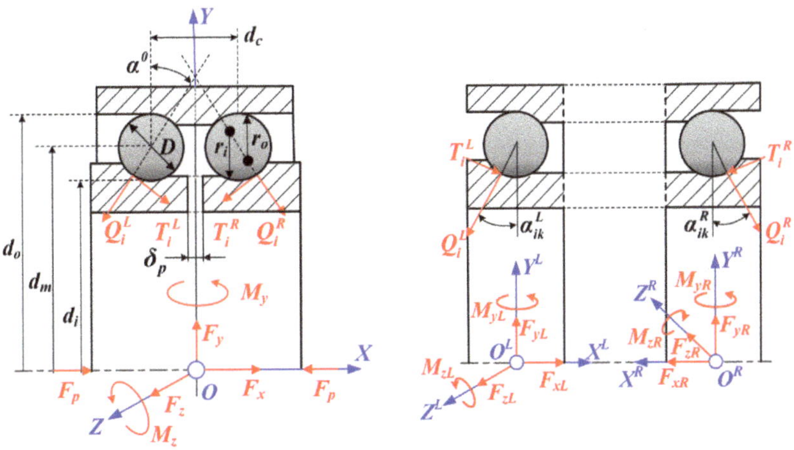

Figure 2. The coordinate systems and force state of DR-ACBB under DB arrangement: (**a**) the diagram of DR-ACBB; (**b**) the diagram of the single-side ACBB.

2.1. Mechanics Analysis of DR-ACBB

As shown in Figure 2, the DR-ACBB under the DB configuration are chosen as an example to illustrate the detailed modeling process of DR-ACBB under variation operating conditions. At first, it is assumed that DR-ACBB are subjected to arbitrary combined loads at their geometric center, and then two type of coordinate systems are established to aid the derivation the equilibrium equations of DR-ACBB: (i) the global coordinate system of DR-ACBB: $O\text{-}XYZ$; (ii) the local coordinate systems of the single-row ACBB: $O^L\text{-}X^LY^LZ^L$ and $O^R\text{-}X^RY^RZ^R$. Furthermore, the Y_L and Y_R axes, respectively, pass through the ball centers of the left and right side of the ACBB. On this basis, the mechanical equations of the moving inner raceway for DR-ACBB under a DB arrangement can be given as:

$$\left\{\begin{array}{l}
F_x = \sum\limits_{k=1}^{Z}\left[\left(Q_{ik}^L \sin\alpha_{ik}^L - T_{ik}^L \cos\alpha_{ik}^L\right) - \left(Q_{ik}^R \sin\alpha_{ik}^R - T_{ik}^R \cos\alpha_{ik}^R\right)\right] \\
F_y = \sum\limits_{k=1}^{Z}\left[\left(Q_{ik}^L \cos\alpha_{ik}^L + T_{ik}^L \sin\alpha_{ik}^L\right) + \left(Q_{ik}^R \cos\alpha_{ik}^R + T_{ik}^R \sin\alpha_{ik}^R\right)\right]\cos\psi_k \\
F_z = \sum\limits_{k=1}^{Z}\left[\left(Q_{ik}^L \cos\alpha_{ik}^L + T_{ik}^L \sin\alpha_{ik}^L\right) - \left(Q_{ik}^R \cos\alpha_{ik}^R + T_{ik}^R \sin\alpha_{ik}^R\right)\right]\sin\psi_k \\
M_y = \sum\limits_{k=1}^{Z}\left\{\begin{array}{l}\left[\frac{d_m - D\cos\alpha_{ik}^L}{2}\left(Q_{ik}^L \sin\alpha_{ik}^L - T_{ik}^L \cos\alpha_{ik}^L\right) + \frac{d_m - D\cos\alpha_{ik}^R}{2}\left(Q_{ik}^R \sin\alpha_{ik}^R - T_{ik}^R \cos\alpha_{ik}^R\right)\right] + \ldots\ldots \\ \left[\frac{d_c + D\sin\alpha_{ik}^L}{2}\left(Q_{ik}^L \cos\alpha_{ik}^L + T_{ik}^L \sin\alpha_{ik}^L\right) + \frac{d_c + D\sin\alpha_{ik}^R}{2}\left(Q_{ik}^R \cos\alpha_{ik}^R + T_{ik}^R \sin\alpha_{ik}^R\right)\right]\end{array}\right\}\sin\psi_k \\
M_z = -\sum\limits_{k=1}^{Z}\left\{\begin{array}{l}\left[\frac{d_m - D\cos\alpha_{ik}^L}{2}\left(Q_{ik}^L \sin\alpha_{ik}^L - T_{ik}^L \cos\alpha_{ik}^L\right) - \frac{d_m - D\cos\alpha_{ik}^R}{2}\left(Q_{ik}^R \sin\alpha_{ik}^R - T_{ik}^R \cos\alpha_{ik}^R\right)\right] + \ldots\ldots \\ \left[\frac{-d_c + D\sin\alpha_{ik}^L}{2}\left(Q_{ik}^L \cos\alpha_{ik}^L + T_{ik}^L \sin\alpha_{ik}^L\right) + \frac{d_c - D\sin\alpha_{ik}^R}{2}\left(Q_{ik}^R \cos\alpha_{ik}^R + T_{ik}^R \sin\alpha_{ik}^R\right)\right]\end{array}\right\}\cos\psi_k
\end{array}\right. \tag{1}$$

where Q and T are the ball raceway's normal contact load and tangential friction load, α and ψ are the contact angle and position angle of ball, and d_m and D are the pitch diameter and ball diameter, respectively. Furthermore, in this paper, unless otherwise stated, the subscripts i and o, respectively, indicate the parameters used for inner and outer ball raceway action analysis, and the subscript k indicates the parameters used for the k_{th} ball. The superscripts L and R, respectively, denote the parameters used for modeling the left-side and right-side DR-ACBB.

To simplify the above mechanical equations of DR-ACBB, the mechanical equations of two SR-ACBBs need to be determined first. As shown in Figure 2b, it is assumed that the restoring force vectors generated by two SR-ACBBs at the local coordinate center points O_L and O_R are written as $\mathbf{F_L} = \{F_{xL}, F_{yL}, F_{zL}, M_{yL}, M_{zL}\}$ and $\mathbf{F_R} = \{F_{xR}, F_{yR}, F_{zR}, M_{yR}, M_{zR}\}$, respectively. Then, the expressions of the mechanical equations of two SR-ACBBs are presented in Equations (2)–(5):

$$\left\{\begin{array}{l}
F_{xL} = \sum_1^Z F_{ak}^L \\
F_{yL} = \sum_1^Z F_{rk}^L \cos\psi_k \\
F_{zL} = \sum_1^Z F_{rk}^L sin\psi_k \\
M_{yL} = \sum_1^Z \left(\frac{d_m - D\cos\alpha_{ik}^L}{2}F_{ak}^L + \frac{D\sin\alpha_{ik}^L}{2}F_{rk}^L\right)\sin\psi_k \\
M_{zL} = -\sum_1^Z \left(\frac{d_m - D\cos\alpha_{ik}^L}{2}F_{ak}^L + \frac{D\sin\alpha_{ik}^L}{2}F_{rk}^L\right)\cos\psi_k
\end{array}\right. \tag{2}$$

with

$$\left\{\begin{array}{l}
F_{ak}^L = Q_{ik}^L \sin\alpha_{ik}^L - T_{ik}^L \cos\alpha_{ik}^L \\
F_{rk}^L = Q_{ik}^L \cos\alpha_{ik}^L + T_{ik}^L \sin\alpha_{ik}^L
\end{array}\right. \tag{3}$$

and

$$\left\{\begin{array}{l}
F_{xR} = \sum_1^Z F_{ak}^R \\
F_{yR} = \sum_1^Z F_{rk}^R \cos\psi_k \\
F_{zR} = \sum_1^Z F_{rk}^R sin\psi_k \\
M_{yR} = \sum_1^Z \left(\frac{d_m - D\cos\alpha_{ik}^R}{2}F_{ak}^R + \frac{D\sin\alpha_{ik}^R}{2}F_{rk}^R\right)\sin\psi_k \\
M_{zR} = -\sum_1^Z \left(\frac{d_m - D\cos\alpha_{ik}^R}{2}F_{ak}^R + \frac{D\sin\alpha_{ik}^R}{2}F_{rk}^R\right)\cos\psi_k
\end{array}\right. \tag{4}$$

with

$$\left\{\begin{array}{l}
F_{ak}^R = Q_{ik}^R \sin\alpha_{ik}^R - T_{ik}^R \cos\alpha_{ik}^R \\
F_{rk}^R = Q_{ik}^R \cos\alpha_{ik}^R + T_{ik}^R \sin\alpha_{ik}^R
\end{array}\right. \tag{5}$$

Substituting Equations (2) and (4) into Equation (1), then one can obtain:

$$\begin{cases} F_x = F_x^L - F_x^R \\ F_y = F_y^L + F_y^R \\ F_z = F_z^L - F_z^R \\ M_y = M_y^L + M_y^R + \frac{d_c}{2}\left(F_z^L + F_z^R\right) \\ M_z = M_z^L - M_z^R - \frac{d_c}{2}\left(F_y^L - F_y^R\right) \end{cases} \tag{6}$$

The above Equation (6) can be further rewritten to the following matrix operation:

$$\mathbf{F}^T = \mathbf{N}_1(\frac{d_c}{2}) \times \mathbf{F}_{\mathbf{L}}^T + \mathbf{N}_2(\frac{d_c}{2}) \times \mathbf{F}_{\mathbf{R}}^T \tag{7}$$

with

$$\mathbf{N}_1(\frac{d_c}{2}) = \begin{bmatrix} 1 & 0 & 0 & 0 & 0 \\ 0 & 1 & 0 & 0 & 0 \\ 0 & 0 & 1 & 0 & 0 \\ 0 & 0 & \frac{d_c}{2} & 1 & 0 \\ 0 & -\frac{d_c}{2} & 0 & 0 & 1 \end{bmatrix}, \mathbf{N}_2(\frac{d_c}{2}) = \begin{bmatrix} -1 & 0 & 0 & 0 & 0 \\ 0 & 1 & 0 & 0 & 0 \\ 0 & 0 & -1 & 0 & 0 \\ 0 & 0 & \frac{d_c}{2} & 1 & 0 \\ 0 & \frac{d_c}{2} & 0 & 0 & -1 \end{bmatrix} \tag{8}$$

Furthermore, the displacement vectors ($\mathbf{d_L}$ and $\mathbf{d_R}$) of two SR-ACBBs at the local coordinate center points O_L and O_R can be calculated by the relative displacement vector of DR-ACBB at the global coordinate center points O ($\mathbf{d} = \{\delta_x,\delta_y,\delta_z,\theta_y,\theta_z\}$):

$$\mathbf{d}_{\mathbf{L}}^T = \begin{bmatrix} \delta_x^L \\ \delta_y^L \\ \delta_z^L \\ \theta_y^L \\ \theta_z^L \end{bmatrix} = \begin{bmatrix} 0.5\delta_P + \delta_x \\ \delta_y - \frac{d_c}{2}\theta_z \\ \delta_z + \frac{d_c}{2}\theta_y \\ \theta_y \\ \theta_z \end{bmatrix} = \mathbf{N}_1^T(\frac{d_c}{2}) \times \mathbf{d} + \begin{bmatrix} 0.5\delta_P \\ 0 \\ 0 \\ 0 \\ 0 \end{bmatrix} \tag{9}$$

$$\mathbf{d}_{\mathbf{R}}^T = \begin{bmatrix} \delta_x^R \\ \delta_y^R \\ \delta_z^R \\ \theta_y^R \\ \theta_z^R \end{bmatrix} = \begin{bmatrix} 0.5\delta_P - \delta_x \\ \delta_y + \frac{d_c}{2}\theta_z \\ -\delta_z + \frac{d_c}{2}\theta_y \\ \theta_y \\ -\theta_z \end{bmatrix} = \mathbf{N}_2^T(\frac{d_c}{2}) \times \mathbf{d} + \begin{bmatrix} 0.5\delta_P \\ 0 \\ 0 \\ 0 \\ 0 \end{bmatrix} \tag{10}$$

Based on the above equations, it can be found that the relative displacement vectors of two SR-ACBBs at the local coordinate center points O_L and O_R can be determined by the matrix transformation of the relative displacement vector of DR-ACBB. Furthermore, the same initial preload displacement ($\delta_p/2$) is also considered in the relative displacement vector calculation of two SR-ACBB.

Similarly, the ring mechanical equations of DR-ACBB under DF and DT arrangements can also been given in a simplified matrix formation, as follows:

i. DF arrangement:

$$\mathbf{F}^T = \mathbf{N}_3(\frac{d_c}{2}) \times \mathbf{F}_{\mathbf{L}}^T + \mathbf{N}_4(\frac{d_c}{2}) \times \mathbf{F}_{\mathbf{R}}^T \tag{11}$$

$$\mathbf{d}_{\mathbf{L}}^T = \begin{bmatrix} \delta_x^L \\ \delta_y^L \\ \delta_z^L \\ \theta_y^L \\ \theta_z^L \end{bmatrix} = \begin{bmatrix} 0.5\delta_p - \delta_x \\ \delta_y - \frac{d_c}{2}\theta_z \\ -\delta_z - \frac{d_c}{2}\theta_y \\ \theta_y \\ -\theta_z \end{bmatrix} = \mathbf{N}_3^T(\frac{d_c}{2}) \times \mathbf{d}^T + \begin{bmatrix} 0.5\delta_p \\ 0 \\ 0 \\ 0 \\ 0 \end{bmatrix} \tag{12}$$

$$\mathbf{d}_{\mathbf{R}}^{T}=\begin{bmatrix}\delta_x^R\\ \delta_y^R\\ \delta_z^R\\ \theta_y^R\\ \theta_z^R\end{bmatrix}=\begin{bmatrix}0.5\delta_p+\delta_x\\ \delta_y+\frac{d_c}{2}\theta_z\\ \delta_z-\frac{d_c}{2}\theta_y\\ \theta_y\\ \theta_z\end{bmatrix}=\mathbf{N}_4^T(\frac{d_c}{2})\times\mathbf{d}^T+\begin{bmatrix}0.5\delta_p\\ 0\\ 0\\ 0\\ 0\end{bmatrix} \quad (13)$$

ii. DT arrangement:

$$\mathbf{F}^T=\mathbf{N}_1(\frac{d_c}{2})\times\mathbf{F}_{\mathbf{L}}^T+\mathbf{N}_4(\frac{d_c}{2})\times\mathbf{F}_{\mathbf{R}}^T \quad (14)$$

$$\mathbf{d}_{\mathbf{L}}^{T}=\begin{bmatrix}\delta_x^L\\ \delta_y^L\\ \delta_z^L\\ \theta_y^L\\ \theta_z^L\end{bmatrix}=\begin{bmatrix}0.5\delta_p+\delta_x\\ \delta_y-\frac{d_c}{2}\theta_z\\ \delta_z+\frac{d_c}{2}\theta_y\\ \theta_y\\ \theta_z\end{bmatrix}=\mathbf{N}_1^T(\frac{d_c}{2})\times\mathbf{d}^T+\begin{bmatrix}0.5\delta_p\\ 0\\ 0\\ 0\\ 0\end{bmatrix} \quad (15)$$

$$\mathbf{d}_{\mathbf{R}}^{T}=\begin{bmatrix}\delta_x^L\\ \delta_y^L\\ \delta_z^L\\ \theta_y^L\\ \theta_z^L\end{bmatrix}=\begin{bmatrix}0.5\delta_p+\delta_x\\ \delta_y+\frac{d_c}{2}\theta_z\\ \delta_z-\frac{d_c}{2}\theta_y\\ \theta_y\\ \theta_z\end{bmatrix}=\mathbf{N}_4^T(\frac{d_c}{2})\times\mathbf{d}^T+\begin{bmatrix}0.5\delta_p\\ 0\\ 0\\ 0\\ 0\end{bmatrix} \quad (16)$$

where the transfer matrices $\mathbf{N}_3$ and $\mathbf{N}_4$ used in the above expressions are written as:

$$\mathbf{N}_3(\frac{d_c}{2})=\begin{bmatrix}-1 & 0 & 0 & 0 & 0\\ 0 & 1 & 0 & 0 & 0\\ 0 & 0 & -1 & 0 & 0\\ 0 & 0 & -\frac{d_c}{2} & 1 & 0\\ 0 & -\frac{d_c}{2} & 0 & 0 & -1\end{bmatrix},\ \mathbf{N}_4(\frac{d_c}{2})=\begin{bmatrix}1 & 0 & 0 & 0 & 0\\ 0 & 1 & 0 & 0 & 0\\ 0 & 0 & 1 & 0 & 0\\ 0 & 0 & -\frac{d_c}{2} & 1 & 0\\ 0 & \frac{d_c}{2} & 0 & 0 & 1\end{bmatrix} \quad (17)$$

Then, to further determine the ball raceway's normal contact loads and tangential friction loads, the force analysis of the local balls in DR-ACBB needs to be conducted. As shown in Figure 3, taking the k_{th} ball inside the left-side ACBB to build the local mechanical equations [9]:

$$\begin{cases}Q_{ik}^L\sin\alpha_{ik}^L-T_{ik}^L\cos\alpha_{ik}^L-Q_{ok}^L\sin\alpha_{ok}^L+T_{ok}^L\cos\alpha_{ok}^L=0\\ Q_{ik}^L\cos\alpha_{ik}^L+T_{ik}^L\sin\alpha_{ik}^L-Q_{ok}^L\cos\alpha_{ok}^L-T_{ok}^L\sin\alpha_{ok}^L+F_{ck}^L=0\end{cases} \quad (18)$$

where F_c is the ball centrifugal force, and then the ball raceway's normal contact load Q and tangential friction load T can be further given as [13]:

$$\begin{cases}Q_{ik}^L=K_{ik}^L(\delta_{ik}^L)^{\frac{3}{2}}\\ Q_{ok}^L=K_{ok}^L(\delta_{ok}^L)^{\frac{3}{2}}\end{cases},\ \begin{cases}T_i^L=\frac{2Q_{ik}^L}{Q_{ik}^L+Q_{ik}^L}\frac{M_{gk}^L}{D}\\ T_o^L=\frac{2Q_{ok}}{Q_{ik}^L+Q_{ok}^L}\frac{M_{gk}^L}{D}\end{cases} \quad (19)$$

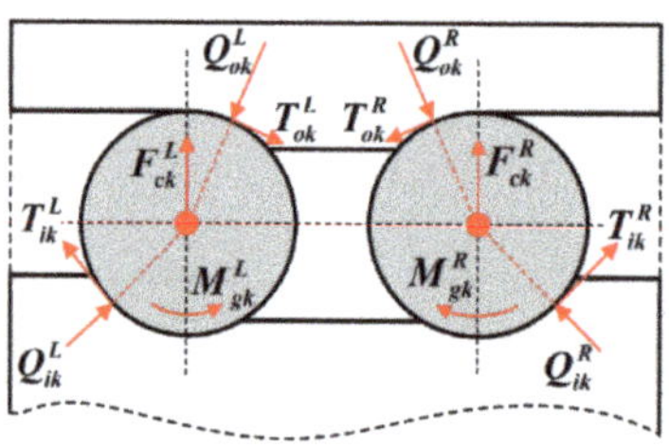

Figure 3. The force analysis of local ball for DR-ACBB under DB arrangement.

In addition, the geometric position change of the k_{th} ball inside the left-side ACBB for DR-ACBB under a DB arrangement is shown in Figure 4, and the initial positions of the inner-ring's curvature center and ball center are changed from points and O^L_{bk} to points $\dot{O}^L_{ik}$ and $\dot{O}^L_{bk}$, respectively. Therefore, the ball raceway's normal deformation is given as:

$$\begin{cases} \delta^L_{ik} = \left[(A^L_{1k} - X^L_{1k})^2 + (A^L_{2k} - X^L_{2k})^2\right]^{1/2} - (r_i - 0.5D) \\ \delta^L_{ok} = \left[(X^L_{1k})^2 + (X^L_{2k})^2\right]^{1/2} - (r_o - 0.5D) \end{cases} \tag{20}$$

with

$$\begin{cases} A^L_{1k} = (r_i + r_o - D)\sin\alpha^0 + \delta^L_x + 0.5d_m \sin\psi_k \theta^L_y - 0.5d_m \cos\psi_k \theta^L_z \\ A^L_{2k} = (r_i + r_o - D)\cos\alpha^0 + \cos\psi_k \delta^L_y + \sin\psi_k \delta^L_z \end{cases} \tag{21}$$

where α^0 is the static initial contact angle of DR-ACBB without preload.

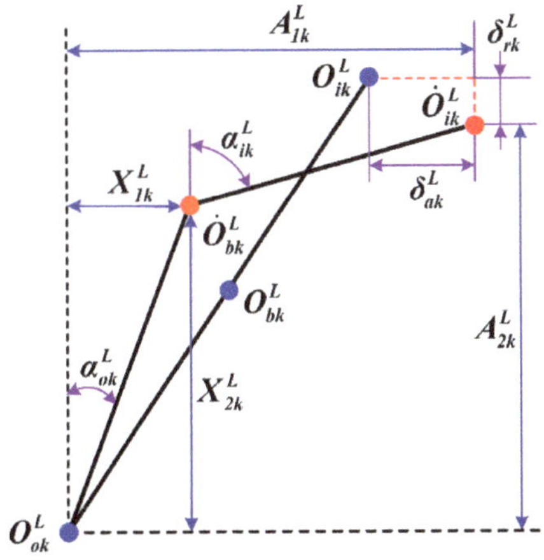

Figure 4. The geometric analysis of local ball of the left-side ball bearing for the double-row ball bearing in DB arrangement.

Furthermore, in order to ensure the stability and convergence of the subsequent iterative calculation, the ball raceway's actual contact states need to be judged to cover the possible ball raceway separation [35]:

$$\begin{cases} \delta^L_{ik} = 0 \\ \delta^L_{ik} = \text{eps} \end{cases} \; if \; \begin{matrix} \left[(A^L_{1k} - X^L_{1k})^2 + (A^L_{2k} - X^L_{2k})^2\right]^{1/2} \le (r_i - 0.5D) \\ \left[(X^L_{1k})^2 + (X^L_{2k})^2\right]^{1/2} \le (r_o - 0.5D) \end{matrix} \tag{22}$$

In the above expression, a small real number of eps is set to ensure the positive contact deformation between the ball and outer ring (i.e., the ball never separates from the outer ring due to the inertia force).

In order to shorten the article space, this paper will not repeat the derivation process of the ball inside the right-side ACBB.

2.2. Iterative Calculation of the Proposed Model

In order to calculate the load distribution and deformation of DR-ACBB, two kinds of nonlinear equations need to be solved simultaneously: (1) the equilibrium equations of balls (i.e., Equation (18), and the number of equations is 4*Z); (2) the ring equilibrium equations (i.e., Equation (7)/Equation (11)/Equation (14), and the number of equations is 5). Therefore, how to construct a reasonable iterative algorithm to ensure the efficient and stable solution of the above equations is also a key research direction in this paper.

As shown in Figure 5, the detailed calculation process based on the two-layer nested iterative algorithm used in this paper is presented. The inner iteration is used to solve the ball equilibrium equations and the outer iteration is used to solve the ring equilibrium equations. Through further observation, both the displacement vector **d** and stiffness matrix **K** of DR-ACBB play an important role in the iterative operation. The displacement vector **d** is continuously modified through the stiffness matrix **K** until the iteration errors are less than the set threshold value, and the stiffness matrix **K** needs to be updated continuously by the calculation results of the inner iteration. Furthermore, the parallel calculation scheme can be used in ball mechanical equations solutions to further improve the calculation efficiency.

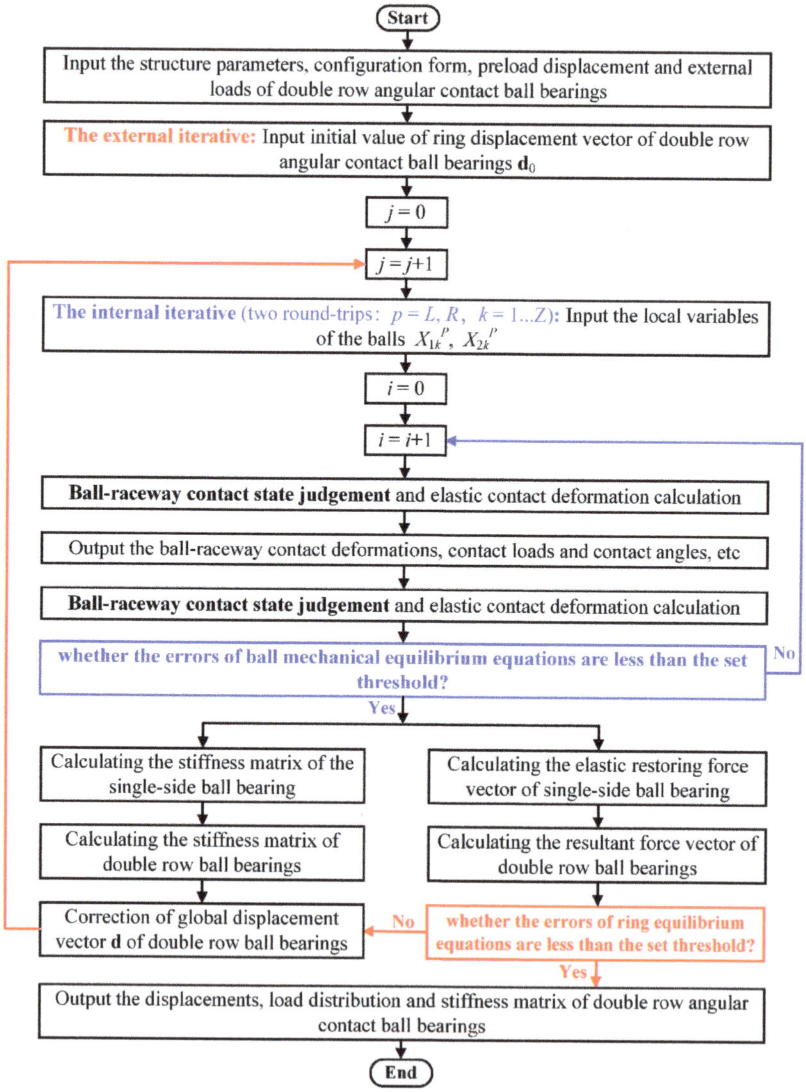

Figure 5. The detailed iterative calculation flow diagram for DR-ACBB.

2.3. Analytical Formulation of the Stiffness Matrix of DR-ACBB

To calculate the analytical expression of the stiffness matrix of DR-ACBB, the explicit relationship of the stiffness matrices between DR-ACBB and SR-ACBBs need to be determined first. As the typical multi-degree-of-freedom nonlinear system, the variation stiffness matrix **K** of DR-ACBB are given as the Jacobian matrix of the external load vector **F** to the relative displacement vector **d**:

$$\mathbf{K} = \frac{\partial \mathbf{F}}{\partial \mathbf{d}} = \frac{\partial \left(F_x, F_y, F_z, M_y, M_z\right)}{\partial \left(\delta_x, \delta_y, \delta_z, \theta_y, \theta_z\right)} \tag{23}$$

Taking DR-ACBB under a DB configuration as an example, according to Equations (7), (9), and (10), the above expression can be further extended as follows:

$$[\mathbf{K}]_{\mathrm{DB}} = \left[\frac{\partial \mathbf{F}}{\partial \mathbf{d}}\right]_{\mathrm{DB}} = \mathbf{N}_1 \frac{\partial \mathbf{F_L}}{\partial \mathbf{d}} + \mathbf{N}_2 \frac{\partial \mathbf{F_R}}{\partial \mathbf{d}} \tag{24}$$

with

$$\begin{cases} \frac{\partial \mathbf{F_L}}{\partial \mathbf{d}} = \frac{\partial \mathbf{F_L}}{\partial \mathbf{d_L}} \frac{\partial \mathbf{d_L}}{\partial \mathbf{d}} = \mathbf{K_L} \mathbf{N}_1^T \\ \frac{\partial \mathbf{F_R}}{\partial \mathbf{d}} = \frac{\partial \mathbf{F_R}}{\partial \mathbf{d_R}} \frac{\partial \mathbf{d_R}}{\partial \mathbf{d}} = \mathbf{K_R} \mathbf{N}_2^T \end{cases} \tag{25}$$

where $\mathbf{K_L}$ and $\mathbf{K_R}$ denote the stiffness matrix of two SR-ACBBs. Substituting Equation (25) into Equation (24), one can obtain:

$$[\mathbf{K}]_{\mathrm{DB}} = \left[\frac{\partial \mathbf{F}}{\partial \mathbf{d}}\right]_{\mathrm{DB}} = \mathbf{N}_1 \mathbf{K_L} \mathbf{N}_1^T + \mathbf{N}_2 \mathbf{K_R} \mathbf{N}_2^T \tag{26}$$

Similarly, the stiffness matrices of DR-ACBB in the DF and DT configurations are given as follows:

$$[\mathbf{K}]_{\mathrm{DF}} = \left[\frac{\partial \mathbf{F}}{\partial \mathbf{d}}\right]_{\mathrm{DF}} = \mathbf{N}_3 \mathbf{K_L} \mathbf{N}_3^T + \mathbf{N}_4 \mathbf{K_R} \mathbf{N}_4^T \tag{27}$$

$$[\mathbf{K}]_{\mathrm{DT}} = \left[\frac{\partial \mathbf{F}}{\partial \mathbf{d}}\right]_{\mathrm{DT}} = \mathbf{N}_1 \mathbf{K_L} \mathbf{N}_1^T + \mathbf{N}_4 \mathbf{K_R} \mathbf{N}_4^T \tag{28}$$

Then, taking the left-side ACBB of DR-ACBB in a DB arrangement, a brief review of the analytical formulation of SR-ACBB is presented.

At first, the stiffness matrix of SR-ACBB can also be calculated by the Jacobian matrix of the restoring load vector $\mathbf{F_L}$ and displacement vector $\mathbf{d_L}$ at the local coordinate center points O_L. Considering that the relationships of the external forces and deformations of SR-ACBB are determined by both the explicit and the implicit equations, the intermediate variables $\mathbf{x_k} = \{X_{1k}, X_{2k}, A_{1k}, A_{2k}\}$ are introduced to divide the stiffness matrix calculation into two steps:

$$\mathbf{K_L} = \sum_{k=1}^{Z} \left(\frac{\partial \mathbf{F_L}}{\partial \mathbf{x_k}} \frac{\partial \mathbf{x_k}}{\partial \mathbf{d_L}} \right) \tag{29}$$

The detailed derivation process for the differential operations of the explicit and the implicit equations can refer to [26,27].

Similarly, the stiffness matrix of the right-side ACBB can also be calculated, and then the complete analytical expression of the stiffness matrix of DR-ACBB under different configurations can be obtained by Equations (26)–(28).

3. Numerical Simulation and Discussions

In this section, the influence of the speeds, external loads, and configuration forms on the nonlinear stiffness variation of DR-ACBB 3210 (The detailed parameters is given in Table 1) is discussed. In addition, unless otherwise specified, the axial clearance of the split inner ring for the initial preload of DR-ACBB is δ_p = 12 μm (i.e., it corresponds approximately to the static preload force of 340 N), and the operating rotating speed is 5000 rpm.

Table 1. The key geometrical parameters of DR-ACBB 3210.

Parameters	320
Curvature radius (inner-raceway) r_i (mm)	4.54
Curvature radius (outer-raceway) r_o (mm)	4.54
Contact diameter (inner-raceway) d_i (mm)	61.22
Contact diameter (outer-raceway) d_o (mm)	78.78
Ball number Z	12
Ball diameter D (mm)	8.73
Pitch diameter dm (mm)	70
Preload displacement δ_p(μm)	12
Radial clearance (μm)	100
The ball center distance (mm)	15

3.1. Analysis of Nonlinear Stiffness Characteristic of the Axially Loaded DR-ACBB

At first, as shown in Figure 6, the variation curves of the axial ring displacements and axial load distributions with the axial external load for DR-ACBB in three different configurations are presented. One can find that, due to the symmetry of the structure for DR-ACBB in DB and DF configurations, they have the same change curve, and similar phenomena also occur for DR-ACBB under pure radial load conditions in the subsequent analysis, while DR-ACBB under a DT configuration have better unidirectional load carrying capacities (i.e., the deformation under the same load is smaller). furthermore, by further observing Figure 6b, with the increase in the axial load, two SR-ACBBs inside DR-ACBBs under DB and DF configurations show different force states; that is, the axial restoring force generated by the left-side ACBB gradually increases, and the axial restoring force generated by the right-side ACBB decreases until it completely disappears. While for DR-ACBB under a DT configuration, since the installation direction of two SR-ACBBs is the same, the axial external force is evenly shared by two SR-ACBBs, so the stiffness and stiffness variation of DR-ACBB under a DT configuration satisfies the linear superposition principle. Therefore, the follow-up research in this paper is mainly aimed at a nonlinear stiffness characteristics analysis of DR-ACBBs under DB and DF configurations. In fact, DR-ACBB under a DT configuration cannot be used alone because the axial restoring force generated by the initial preload cannot be balanced without the external load.

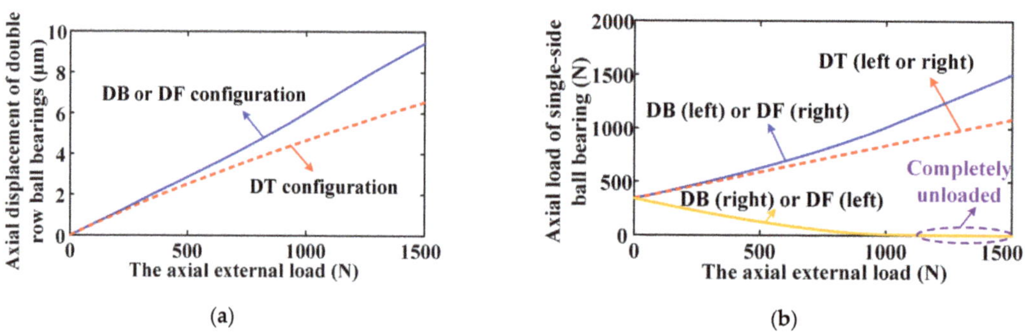

Figure 6. The result curves of the axial displacements and load distributions versus the axial load for DR-ACBB under different configurations (N = 1000 rpm): (**a**) the axial displacement; (**b**) the axial load distribution.

Then, the influence of rotating speeds on the axial and radial stiffness variation rules versus axial load for DR-ACBBs under DB and DF configurations are shown in Figure 7 (i.e., DR-ACBBs under DB and DF configurations show the same stiffness variation curves). It can be found that when DR-ACBB operated at a low-speed range (i.e., 1000 rpm), both the axial and radial stiffness of DR-ACBB show a similar two-stage variation characteristics; that is, the bearing stiffness decreases in the light load range and increases in the heavy load range, which indicates that DR-ACBBs under DB and DF configurations show the nonlinear spring characteristics of soft first and then hard with the increase in axial load. By further observing Figure 8a, the dividing point of the above two-stage stiffness change curves for DR-ACBB at 1000 rpm is approximately located at the complete unloaded point of the right-side ACBB. Furthermore, as the rotating speed increases to 5000 and 8000 rpm, the axial external load range corresponding to stiffness attenuation gradually decreases until it disappears; at the same time, the phenomenon of the completely unloaded single-side ACBB is also not found, as shown in Figure 8b,c. In summary, DR-ACBB under DB and DF configurations show nonlinear soft-spring stiffness characteristics at a low-speed range and light (axial) load condition, while they show a nonlinear hard spring stiffness characteristic at relatively high-speed ranges or heavy (axial) load condition.

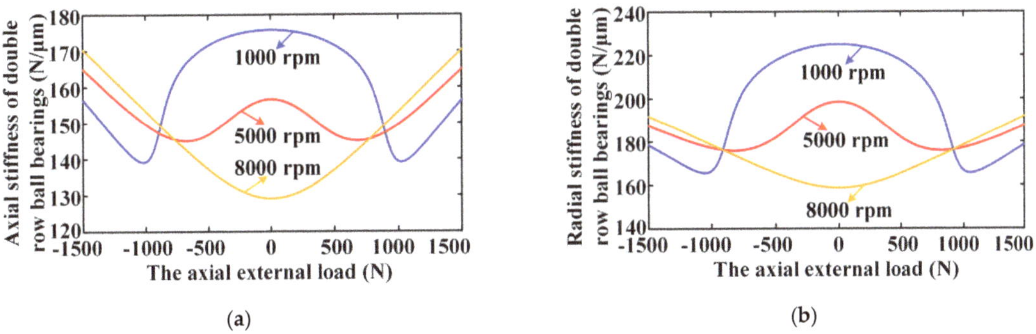

Figure 7. The axial and radial stiffness variation curves versus the axial external load for DR-ACBB at three different speeds: (**a**) the axial stiffness; (**b**) the radial stiffness.

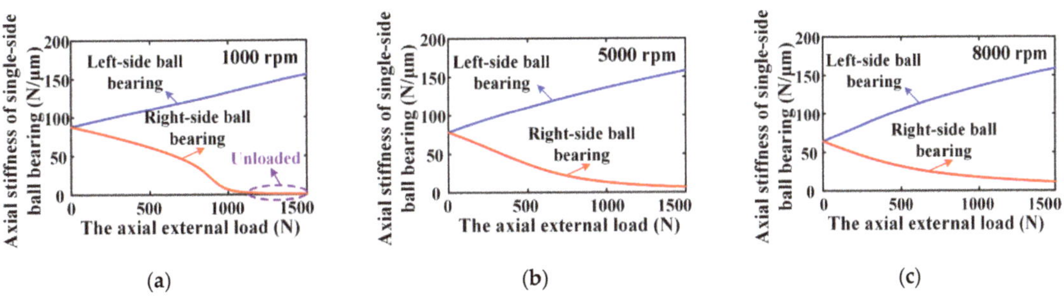

Figure 8. The change curves of the axial stiffness of two SR-ACBBs versus the axial external load for DR-ACBB under DB configuration: (**a**) rotating speed = 1000 rpm; (**b**) rotating speed = 5000 rpm; (**c**) rotating speed = 8000 rpm.

In addition, the comparison results of the angular stiffness versus axial external load of DR-ACBBs under DB and DF configurations are presented in Figure 9. One can find that the DR-ACBB under a DB configuration have a bigger angular stiffness and show a similar varying tendency to radial stiffness, while for DR-ACBB under a DF configuration, the bending moment resistance is poor due to the small angular stiffness, and the change trend of angular stiffness in its attenuation range is gentler.

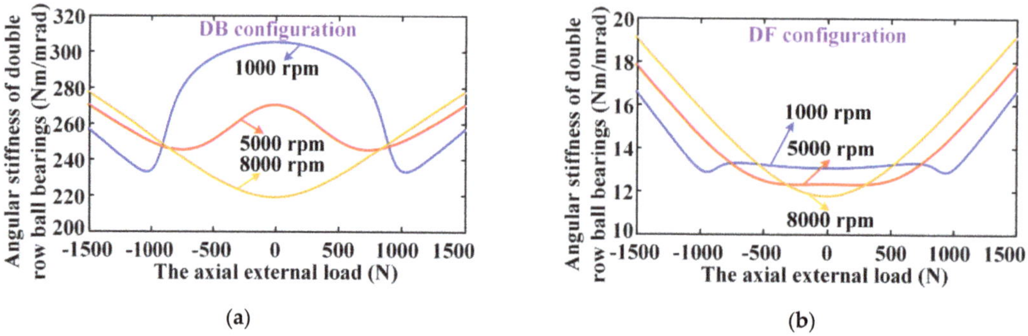

Figure 9. The comparison results of the angular stiffness versus axial external load of DR-ACBB under DB and DF configurations: (**a**) DB configuration; (**b**) DF configuration.

Lastly, the influences of rotating speed on axial stiffness and the actual preload of DR-ACBB without external load are presented in Figure 10a. One can find that although the actual preload increases with rotating speed due to the action of ball centrifugal forces and the fixed-position preload mechanism [33], the axial stiffness of DR-ACBB still shows a significant attenuation tendency with rotating speed. Furthermore, it can be seen from Figure 10b that the initial preload plays a key role on the nonlinear stiffness variations of DR-ACBB. One can find that, increasing the preload can not only improve the overall stiffness of DR-ACBB but also expand the axial load range corresponding to the stiffness attenuation. Furthermore, the influence strength of initial preload decreases obviously at the large axial load condition, and the stiffness of DR-ACBB is mainly determined by the external load at this time.

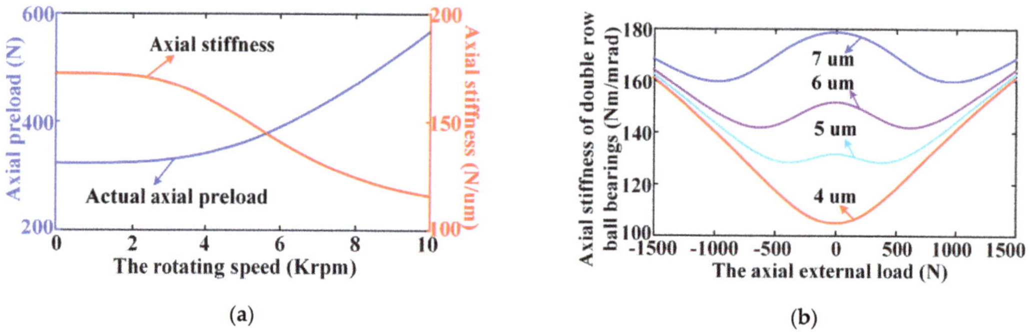

Figure 10. The effects of the speed and preload on the nonlinear stiffness variations of DR-ACBB: (**a**) the actual preload and axial stiffness varying with the speed; (**b**) the axial stiffness varying with axial load of DR-ACBB with different initial preloads ($\delta_p/2$).

3.2. Analysis of Nonlinear Stiffness Characteristic of the Radially Loaded DR-ACBB

In this section, the nonlinear stiffness characteristics of DR-ACBB subjected to the radial load will be detailed, discussed, and analyzed.

First, as shown in Figure 11, the variation curves of the radial stiffness along the Y-axis and Z-axis versus the external load along the Y-axis for DR-ACBB under DB and DF configurations at static conditions are given (i.e., DR-ACBBs under DB and DF configurations still have the same radial stiffness variation rules). Similar to the nonlinear stiffness variation of the axially loaded DR-ACBB, the stiffness varying with radial load shows a multi-stage variation characteristic, and DR-ACBB also show the nonlinear spring characteristics of soft first and then hard with the increase in radial load. Furthermore, the changing rules of the ball number in the loaded area for SR-ACBB are also given in Figure 11. One can find that some of the rolling balls are unloaded under the action of a large radial load, which has an important impact on the size and variation trend of the stiffness of DR-ACBB.

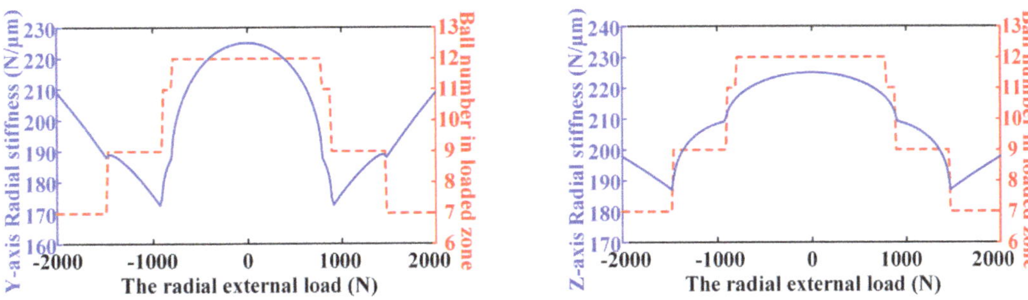

(**a**) (**b**)

Figure 11. The results of the radial stiffness and ball number in loaded area versus radial load of DR-ACBB at static: (**a**) Y-axis radial stiffness; (**b**) Z-axis radial stiffness.

On this basis, Figure 12 gives the influence rules of the rotating speed on the radial stiffness versus the radial load for DR-ACBB under DB or DF configurations. One can find that the stiffness variation trends of DR-ACBB in two different radial directions are slightly different; however, on the whole, they still show the following law: DR-ACBB under DB and DF configurations present a nonlinear soft-spring stiffness characteristic of a low-speed range and light (radial) load condition, while they show a nonlinear hard spring stiffness characteristic at a high-speed range or heavy (radial) load condition.

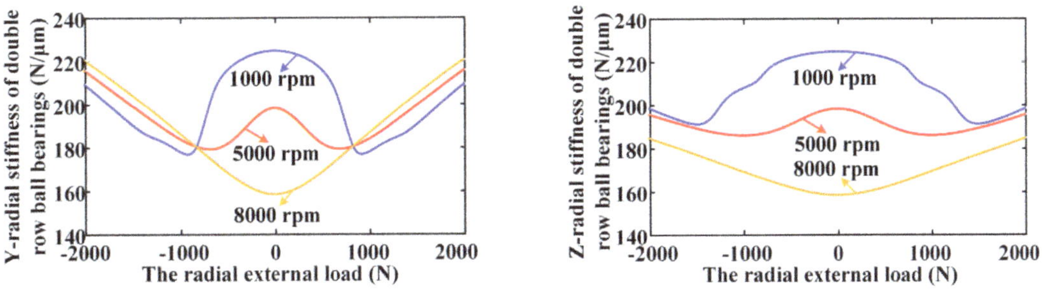

(**a**) (**b**)

Figure 12. The results of the radial stiffness versus radial load for DR-ACBB under three different rotating speeds: (**a**) Y-axis radial stiffness; (**b**) Z-axis radial stiffness.

In addition, the comparison results of the angular stiffness versus radial external load of DR-ACBBs under DB and DF configurations are presented in Figure 13. One can find that DR-ACBB under a DB configuration still have a bigger angular stiffness and show a similar varying tendency to radial stiffness, while the angular stiffness of DR-ACBB under a DF configuration is relatively small, showing a gentle change trend in its attenuation range.

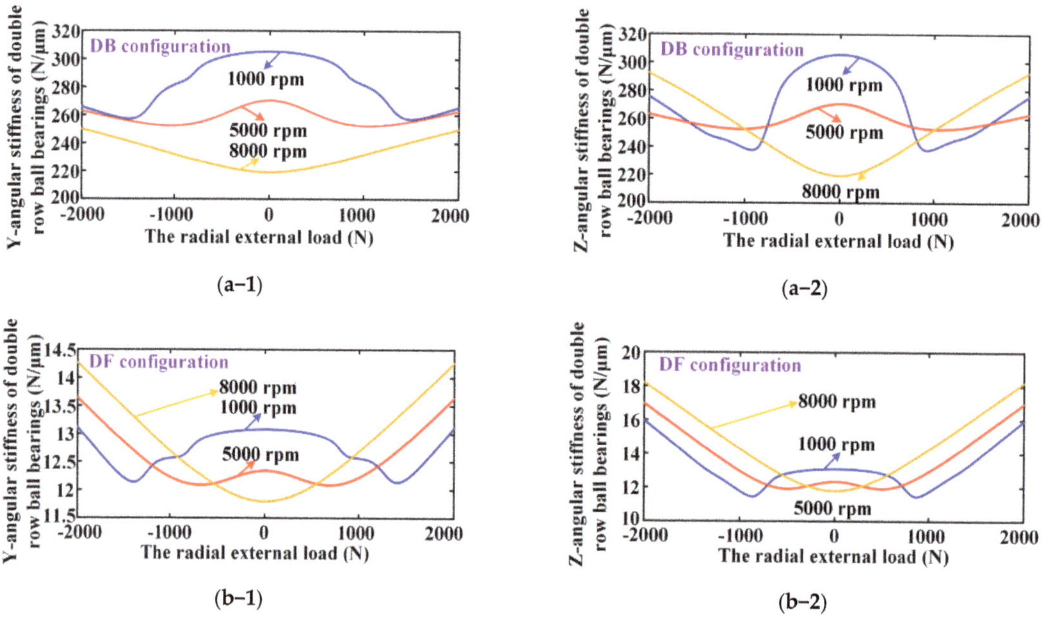

Figure 13. The comparison results of the angular stiffness versus radial load of DR-ACBB under DB and DF configurations: (**a–1**,**a–2**) DB configuration; (**b–1**,**b–2**) DF configuration.

Lastly, the influences of the initial preload on the radial stiffness variations versus radial load of DR-ACBB are given in Figure 14. One can find that increasing the preload can also improve the overall stiffness and expand the radial load range corresponding to the stiffness attenuation of DR-ACBB. Similar to the axial-loaded condition, the influence strength of the initial preload also decreases obviously at the large radial load condition, and the stiffness of DR-ACBB is mainly determined by the external load at this time.

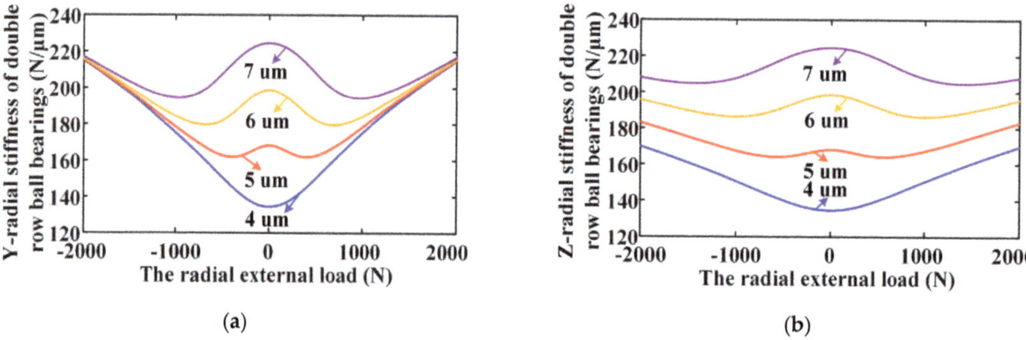

Figure 14. The effects of the initial preload on the radial stiffness variations of DR-ACBB: (**a**) Y-axis radial stiffness; (**b**) Z-axis radial stiffness.

3.3. Analysis of Nonlinear Stiffness Characteristic of the Combined-Loaded DR-ACBB

As shown in Figure 15, the nonlinear stiffness variation curves for DR-ACBB subjected to constant radial loads (i.e., 0 N, 100 N, 200 N, 300 N) and varying axial loads are presented. It can be found that both DR-ACBBs under DB and DF configurations show the nonlinear spring characteristics of soft first and then hard with axial load, and the axial load ranges corresponding to the stiffness softening decrease with constant radial load. Furthermore, compared to DR-ACBB under a DB configuration, nonlinear stiffness variation curves versus the axial loads of DR-ACBB under a DF configuration are more complex and show obvious multi-stage change trends under a large load range.

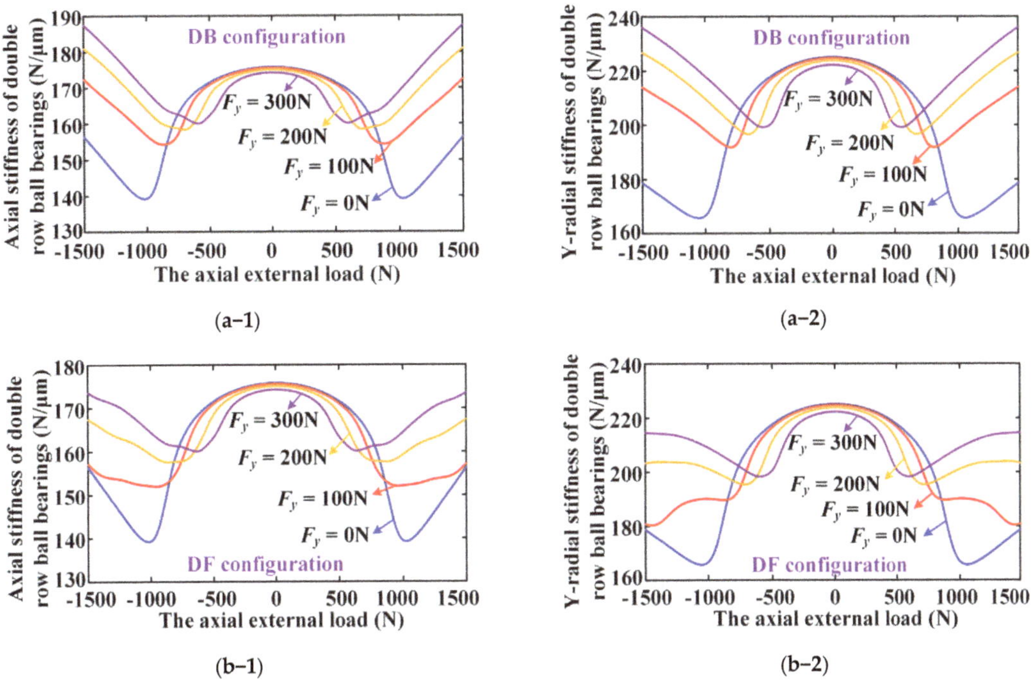

Figure 15. The results of the nonlinear stiffness variation for DR-ACBB subjected to constant radial loads and varying axial load (N = 1000 rpm): (**a**) axial stiffness; (**b**) radial stiffness.

Then, the nonlinear stiffness variation curves for DR-ACBB subjected to constant axial loads (i.e., 0 N, 100 N, 200 N, 300 N) and varying radial loads are presented in Figure 16. It can be found that both DR-ACBBs under DB and DF configurations show the nonlinear spring characteristics of soft first and then hard with axial load, and the axial load ranges corresponding to the stiffness softening decrease with constant axial load.

At last, we accurately simulate the actual bearing load conditions and illustrate the difference of the nonlinear stiffness characteristics of DR-ACBBs under DB and DF configuration. As shown in Figure 17, we assume that a rigid rotor system supported by DR-ACBB is subjected to radial load at the shaft end, and then the change curves of the radial and angular displacements and stiffness of DR-ACBB are calculated and presented in Figure 18. One can find that the radial and angular displacements of DR-ACBB under a DB configuration are less than those of DR-ACBB under a DF configuration, and the radial stiffness of DR-ACBBs under both DB and DF configurations shows a nonlinear soft spring stiffness characteristic under the light load range, while showing a nonlinear hard spring stiffness characteristic under the heavy load range, and the DB configuration shows a larger load range corresponding to the bearing stiffness attenuation. Furthermore, the angular stiffness of DR-ACBBs under two different configurations slightly fluctuates with radial external load, and the angular stiffness of DR-ACBB under DB configuration is far larger than that of DR-ACBB under a DF configuration. In summary, DR-ACBB under a DB configuration have better radial bearing capacity due to their large angular stiffness.

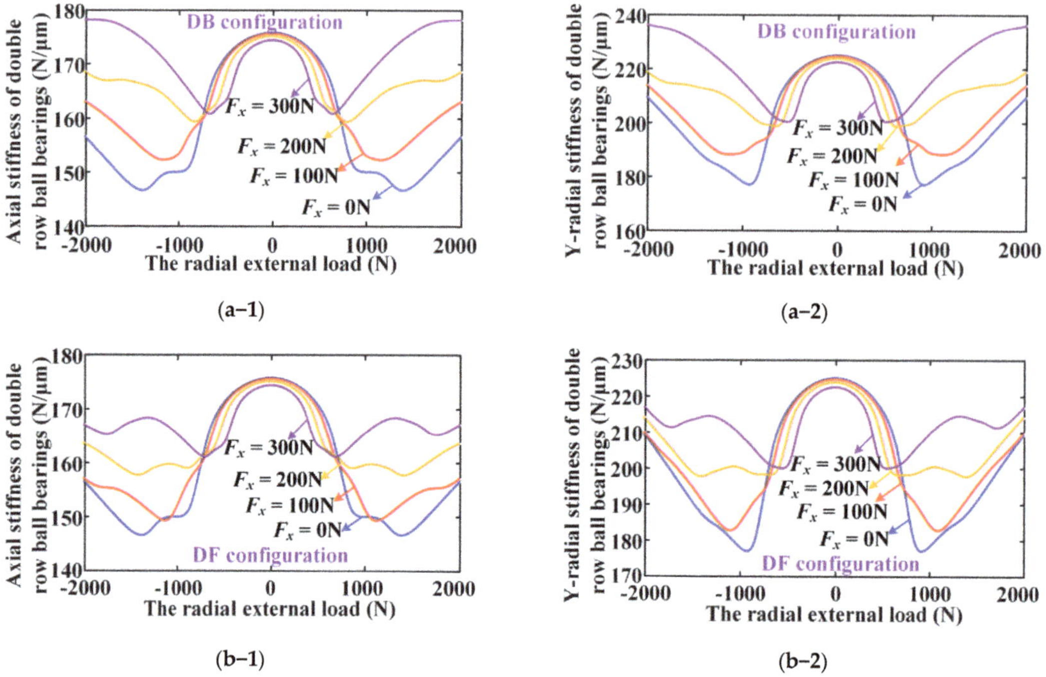

Figure 16. The results of the nonlinear stiffness variation for DR-ACBB subjected to constant axial loads and varying radial load (N = 1000 rpm): (**a**) axial stiffness; (**b**) radial stiffness.

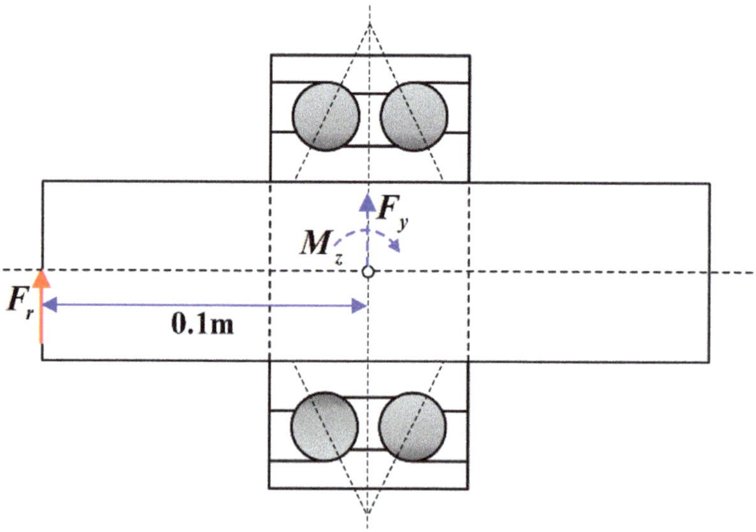

Figure 17. The schematic diagram of a rigid rotor system supported by the double-row ball bearing.

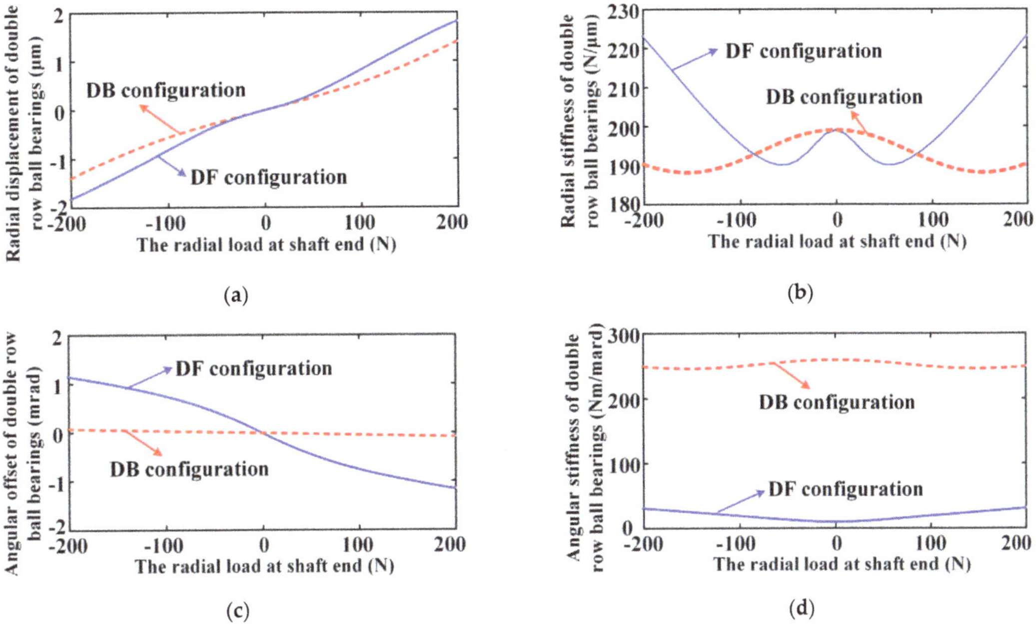

Figure 18. The radial and angular displacement and stiffness of double-row ball bearing in different configurations: (**a**) radial displacement; (**b**) radial stiffness; (**c**) angular offset; (**d**) angular stiffness.

4. Conclusions

In this paper, based on the improved quasi-static model of SR-ACBB, a general model is proposed to calculate and analyze the complex nonlinear stiffness characteristics of DR-ACBB under different arrangements, and the influences of the rotating speed, preload, and external force on the nonlinear stiffness characteristics of DR-ACBB under different arrangements are comparatively analyzed. This paper provides a theoretical foundation for the modeling and nonlinear dynamic analysis of the complex rotor-bearing system. Based on the above studies, the following conclusions are given:

(1) DR-ACBBs under DB and DF configurations have the same variation rule in axial and radial stiffness, and DR-ACBBs under DB and DF configurations show the nonlinear spring characteristics of soft first and then hard with the increase in external load;
(2) The SR-ACBB or part of the balls may be unloaded for DR-ACBB under the large load ranges, which further leads to the sudden change in the nonlinear stiffness characteristics of DR-ACBB;
(3) The initial preload has a great influence on the nonlinear stiffness characteristics of DR-ACBB, and it can effectively increase the external load range corresponding to the stiffness attenuation of DR-ACBB;
(4) DR-ACBB under a DB configuration have higher angular stiffness and bending moment resistance than those of DR-ACBB under a DF configuration.

Author Contributions: Conceptualization and methodology, B.F. and J.Z.; software, B.F.; formal analysis, K.Y.; investigation, J.H.; writing—original draft preparation, B.F.; writing—review and editing: J.Z., K.Y. and J.H. All authors have read and agreed to the published version of the manuscript.

Funding: This research was supported by the National Key R&D Program of China (Grant No. 2021YFB2011000), National Natural Science Foundation of China under Grant (No. 52205281), China Postdoctoral Science Foundation (Grant No. 2021M62551), and two-chain fusion high-end machine tool projects of Shaanxi Province under Grant (2021LLRh-01-02).

Data Availability Statement: Not applicable.

Conflicts of Interest: The authors declare no conflict of interest.

References

1. Bercea, I.; Nelias, D.; Cavallaro, G. A unified and simplified treatment of the non-linear equilibrium problem of double-row rolling bearings. Part 1: Rolling bearing model. *Proc. Inst. Mech. Eng. Part J.-J. Eng. Tribol.* **2003**, *217*, 205–212. [CrossRef]
2. Gunduz, A.; Singh, R. Stiffness matrix formulation for double row angular contact ball bearings: Analytical development and validation. *J. Sound Vib.* **2013**, *332*, 5898–5916. [CrossRef]
3. Gunduz, A.; Dreyer, J.T.; Singh, R. Effect of bearing preloads on the modal characteristics of a shaft-bearing assembly: Experiments on double row angular contact ball bearings. *Mech. Syst. Signal. Pr.* **2012**, *31*, 176–195. [CrossRef]
4. Petersen, D.; Howard, C.; Prime, Z. Varying stiffness and load distributions in defective ball bearings: Analytical formulation and application to defect size estimation. *J. Sound Vib.* **2015**, *337*, 284–300. [CrossRef]
5. Xu, T.; Yang, L.; Wu, W.; Wang, K. Effect of angular misalignment of inner ring on the contact characteristics and stiffness coefficients of duplex angular contact ball bearings. *Mech. Mach. Theory.* **2021**, *157*, 104178. [CrossRef]
6. Poplawski, J.V.; Maureillo, J.A. *Skidding in Lightly Loaded High Speed Ball Thrust Bearings*; ASME: New York, NY, USA, 1969.
7. Harris, T.A. An analytical method to predict skidding in thrust-loaded, angular-contact ball bearings. *J. Lubr. Technol.* **1971**, *93*, 17–23. [CrossRef]
8. Harris, T.A. Ball motion in thrust-loaded, angular contact bearings with Coulomb friction. *J. Lubr. Technol.* **1971**, *93*, 32–38. [CrossRef]
9. Harris, T.A. *Rolling Bearing Analysis*, 4th ed.; John Wiley and Sons, Inc.: New York, NY, USA, 2000.
10. Ding, C.A.; Zhou, F.Z.; Zhu, J.; Zhang, L. Raceway control assumption and the determination of rolling element attitude angle. *Chin. J. Mech. Eng.* **2001**, *37*, 58–61. [CrossRef]
11. Tomović, R. Calculation of the boundary values of rolling bearing deflection in relation to the number of active rolling elements. *Mech. Mach. Theory.* **2012**, *47*, 74–88. [CrossRef]
12. Tomović, R. Calculation of the necessary level of external radial load for inner ring support on q, rolling elements in a radial bearing with internal radial clearance. *Int. J. Mech. Sci.* **2012**, *60*, 23–33. [CrossRef]
13. Wang, W.Z.; Hu, L.; Zhang, S.G.; Zhao, Z.Q.; Ai, S.Y. Modeling angular contact ball bearing without raceway control hypothesis. *Mech. Mach. Theory.* **2014**, *82*, 154–172. [CrossRef]
14. Yan, K.; Wang, Y.; Zhu, Y.; Hong, J.; Zhai, Q. Investigation on Heat Dissipation Characteristic of Ball Bearing Cage and Inside Cavity at Ultra High Rotation Speed. *Tribol. Int.* **2016**, *93*, 470–481. [CrossRef]
15. Yang, Z.; Yu, T.; Zhang, Y.; Sun, Z. Influence of Cage Clearance on the Heating Characteristics of High-Speed Ball Bearings. *Tribol. Int.* **2017**, *105*, 125–134. [CrossRef]
16. Ren, X.; Zhai, J.; Ren, G. Calculation of radial load distribution on ball and roller bearings with positive, negative and zero clearance. *Int. J. Mech. Sci.* **2017**, *131*, 1–7.
17. Liu, J.; Shao, Y. Dynamic modeling for rigid rotor bearing systems with a localized defect considering additional deformations at the sharp edges. *J. Sound. Vib.* **2017**, *398*, 84–102. [CrossRef]
18. Liu, J. A dynamic modelling method of a rotor-roller bearing-housing system with a localized fault including the additional excitation zone. *J. Sound. Vib.* **2020**, *469*, 115144. [CrossRef]
19. Gargiulo, E.P. A Simple Way to Estimate Bearing Stiffness. *Mach. Des.* **1980**, *52*, 107–110.
20. Wardle, F.P.; Lacey, S.J.; Poon, S.Y. Dynamic and static characteristics of a wide speed range machine tool spindle. *Precis. Eng.* **1983**, *5*, 175–183. [CrossRef]
21. Lim, C.T.; Singh, R. Vibration transmission through rolling element bearings, part I: Bearing stiffness formulation. *J. Sound Vib.* **1990**, *139*, 179–199. [CrossRef]
22. Lim, C.T.; Singh, R. Vibration transmission through rolling element bearings, part II: System studies. *J. Sound Vib.* **1990**, *139*, 201–225. [CrossRef]
23. Lim, C.T.; Singh, R. Vibration transmission through rolling element bearings, Part III: Geared rotor system studies. *J. Sound Vib.* **1991**, *151*, 31–54. [CrossRef]
24. Houpert, L. A uniform analytical approach for ball and roller bearings calculations. *J. Tribol.* **1997**, *119*, 851–858. [CrossRef]
25. Hernot, X.; Sartor, M.; Guillot, J. Calculation of the stiffness matrix of angular contact ball bearings by using the analytical approach. *J. Mech. Des.* **2000**, *122*, 83–90. [CrossRef]
26. Sheng, X.; Li, B.; Wu, Z.; Li, H. Calculation of ball bearing speed-varying stiffness. *Mech. Mach. Theory.* **2014**, *81*, 166–180. [CrossRef]
27. Noel, D.; Ritou, M. Complete analytical expression of the stiffness matrix of angular contact ball bearings. *J. Tribol.* **2013**, *135*, 041101. [CrossRef]
28. Liu, J.; Tang, C.; Wu, H.; Xu, Z.; Wang, L. An analytical calculation method of the load distribution and stiffness of an angular contact ball bearing. *Mech. Mach. Theory.* **2019**, *142*, 103597. [CrossRef]
29. While, M.F. Rolling element bearing vibration transfer characteristics: Effect of stiffness. *J. Appl. Mech.* **1979**, *46*, 677–684. [CrossRef]

30. Yang, Z.H.; Li, B.T.; Yu, T.X. Influence of structural parameters and tolerance on stiffness of high-speed ball bearings. *Int. J. Precis. Eng. Man.* **2016**, *17*, 1493–1501. [CrossRef]
31. Yang, Z.; Chen, H.; Yu, T. Effects of rolling bearing configuration on stiffness of machine tool spindle, P. *I. Mech. Eng. C.-J. Mec.* **2018**, *232*, 775–785. [CrossRef]
32. Li, J.; Zhu, Y.; Yan, K.; Yan, X.; Liu, Y.; Hong, J. Research on the axial stiffness softening and hardening characteristics of machine tool spindle system. *Int. J Adv. Manuf. Tech.* **2018**, *99*, 951–963. [CrossRef]
33. Zhang, J.; Fang, B.; Zhu, Y.; Hong, J. A comparative study and stiffness analysis of angular contact ball bearings under different preload mechanisms. *Mech. Mach. Theory.* **2017**, *115*, 1–17. [CrossRef]
34. Zhang, J.; Fang, B.; Hong, J.; Wan, S.; Zhu, Y. A general model for preload calculation and stiffness analysis for combined angular contact ball bearings. *J. Sound. Vib.* **2017**, *411*, 435–449. [CrossRef]
35. Fang, B.; Zhang, J.; Yan, K.; Hong, J.; Wang, M. A comprehensive study on the speed-varying stiffness of ball bearing under different load conditions. *Mech. Mach. Theory* **2019**, *136*, 1–13. [CrossRef]

lubricants

MDPI

Article

A Generalized Bearing Dynamic with Adaptive Variation of Equation Numbers and Sliding Behavior Investigation

Shuaijun Ma [1], Yanjing Yin [1,2], Fei Chen [1,*], Bin Fang [1], Ke Yan [1,*] and Jun Hong [1]

[1] Key Laboratory of Education Ministry for Modern Design and Rotor-Bearing System, Xi'an Jiaotong University, Xi'an 710049, China

[2] Luoyang Bearing Science & Technology Co., Ltd., Luoyang 471003, China

[*] Correspondence: fei_chen2019@stu.xjtu.edu.cn (F.C.); yanke@mail.xjtu.edu.cn (K.Y.)

Abstract: The complex sliding behavior inside ball bearings seriously affects the mechanical system's performance. Current dynamic models for predicting this behavior suffer from poor generality and convergence. To address this issue, different interactions between the ball and raceway are proposed in this paper to simulate the dynamic behavior by analyzing the bearing assemblies' motion features under typical operating conditions. The number of variables and equations to be solved is determined adaptively according to the bearing load characteristics, thus improving the efficiency and convergence of the model solution. The good agreement between simulation results and experimental test results validates the reliability of the model. The sliding behavior at the ball/raceway interface under different conditions is further investigated. The results show that the heavy external loads can avoid severe sliding at the interface but shorten the bearing's fatigue. When the bearing is subjected to combined load conditions, the increased radial force inhibits bearing sliding while increasing the non-uniformity of the sliding velocity distribution.

Keywords: ball bearing; sliding behavior; dynamic model; geometric interaction

1. Introduction

Ball bearings are widely used in rotating machinery systems due to their ability to withstand various types of loads. Because of the complex motion of the bearing components, the sliding behavior at the ball/raceway interface is inevitable. Once this behavior becomes very severe, the surface wear of the balls or raceways will be accelerated, shortening bearing life and even causing catastrophic accidents in machine systems [1–4]. Furthermore, severe sliding inside the bearing generates a lot of heat, which drives the bearing into thermal instability and thus increases the wear of the interface [5–7]. As a result, the sliding behavior inside the ball bearing is required to be investigated to guide structural design and improvement of the systems.

Since current experimental methods cannot accurately monitor the dynamic behavior inside bearings, numerical simulation models have become a common approach for studying bearing sliding. The quasi-static model was first used by some scholars to conduct the sliding investigation. Based on this model, Xu [8] investigated the critical load of the bearing without integral sliding under pure axial force conditions. Following this, the sliding behavior at the ball/raceway interface under combined axial and radial loads was investigated by Liao [9,10] using the Hirano criterion [11]. In addition, Tong [12] improved the quasi-static model and further examined the effect of load, rotational speed, and mounting errors on the bearing sliding at the contact interface. Regrettably, the transient behavior of the bearing components cannot be simulated by the quasi-static model because many factors, including covering lubricants and the cage dynamic effect, are ignored. In particular, the action of the lubricant has a strong influence on the dynamic performance of the ball at the interface [13]. Therefore, the bearing dynamic model, being the most complete

Citation: Ma, S.; Yin, Y.; Chen, F.; Fang, B.; Yan, K.; Hong, J. A Generalized Bearing Dynamic with Adaptive Variation of Equation Numbers and Sliding Behavior Investigation. *Lubricants* **2023**, *11*, 96. https://doi.org/10.3390/lubricants11030096

Received: 27 December 2022
Revised: 16 February 2023
Accepted: 22 February 2023
Published: 23 February 2023

model available [14], has become one of the most effective means to investigate the sliding behavior inside bearings. Meeks [15,16] first established a new bearing dynamic model by extending Jones's quasi-static model [17,18]. Although this model can only investigate the cage's dynamic behavior, it does provide ideas for further work. Subsequently, Jain [19] used the load distribution obtained from a simplified quasi-static model, in conjunction with elastic fluid theory, to construct differential equations for the ball's rotational motion. The angular velocity features of the ball under a single operating condition were investigated by him. On this basis, Han [20,21] examined the sliding behavior at the ball/raceway interface under combined axial and radial load by considering the ball's gyroscopic moment. However, these above models used nonlinear equations rather than differential equations to constrain the ball's radial motion, suppressing the ball's high-frequency vibration and failing to accurately predict the dynamic behavior inside the bearing [22]. To address these deficiencies, Wang [23–26] employed six differential equations to completely determine the ball's complex motion and further developed a new bearing dynamic model. The sliding behavior at the contact interface under two typical operating conditions (pure axial force and combined axial and radial force) was investigated. Unfortunately, limited by the algorithm for solving the differential equations, the solution of this model depended heavily on the accuracy of the initial values [27]. To precisely simulate the dynamic behavior of the ball bearing components, these models must be solved in turn: the static model, the quasi-static model, and the dynamic model. The static model results are used as initial values for the quasi-static model interactions, and then the quasi-static model results are further used as initial values for the differential equations in the dynamic model. However, when these models are used to simulate the dynamic performance under various operating conditions, their solution efficiency and convergence can vary significantly [28]. The high-frequency dynamic behavior at the ball/raceway interface, in particular, can pose a significant challenge to model convergence for predicting the bearing's performance under certain harsh operating conditions. In addition, some scholars have noticed that the tribological behavior inside the contact interface has a non-negligible effect on the bearing's dynamic performance [29–31].

In summary, there is a lack of a general bearing dynamic model with efficient solutions and strong convergence under various operating conditions. Additionally, the sliding behavior under typical conditions has not been sufficiently investigated. To this end, different interactions between the ball and raceway are constructed for various operating conditions, and a general dynamic model of the ball bearing is further presented. By reducing the number of variables and equations, the solving efficiency and convergence of this model have been improved. On this basis, the sliding behaviors at the ball/raceway interface are investigated under three typical operating conditions: pure radial force, pure axial force, and combined axial and radial load.

2. Ball/Raceway Interaction Modeling

In this section, the corresponding numerical models are proposed for three different loading methods of ball bearing. Firstly, the spatial motion between the ball and ring under each loading method is thoroughly examined, and the relationship between motion and force, including the normal load and traction force at the ball/raceway interface, is established.

Four coordinate frames must be defined to describe the bearing components' motion in space, as shown in Figure 1. The first one (o-xyz) is called the inertial coordinate frame; the second one (o_i-$x_iy_iz_i$) is the inner-fixed coordinate frame; the third one (o_c-$x_cy_cz_c$) donates the cage-fixed coordinate frame; and the fourth one (o_{bj}-$x_{bj}y_{bj}z_{bj}$) represents the ball-fixed coordinate frame. To determine the relative motion between the components, the connection between these coordinate frames needs to be created. Furthermore, the azimuthal coordinate frame (o_{aj}-$x_{aj}y_{aj}z_{aj}$) is defined to simplify the geometric relationships based on the characteristics of the ball motion.

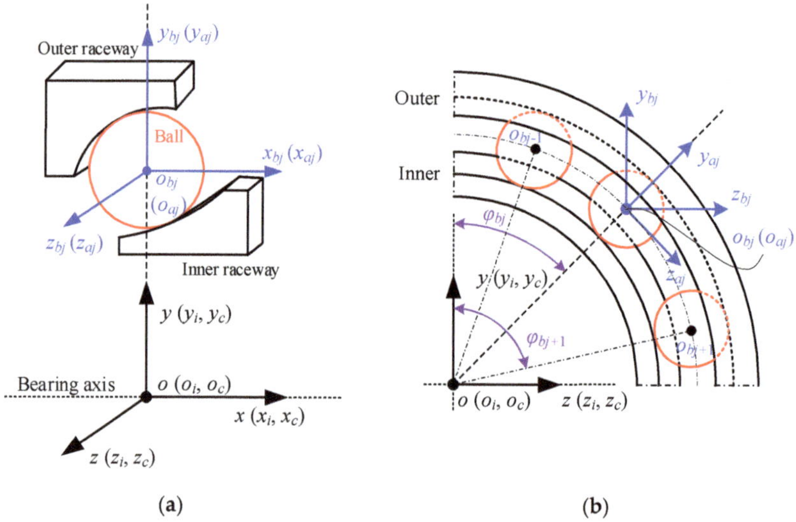

Figure 1. Bearing coordinate frames: (**a**) axial section; (**b**) radial section.

2.1. Pure Radial Force

(1) Normal load

The relative position between the ball and the raceway under pure radial force is presented in Figure 2. The outer ring is assumed to be fixed to the ground, while the inner ring moves in the radial plane under radial load and gradually contacts the ball. The inner ring does move axially because it is not subjected to axial force. The contact deformation between the j-th ball and raceways can be expressed as:

$$\delta_{ij} = \left(B_y + d_i/2 + r_i\right) - \left(X_y + d_m/2\right) - \left(r_i - D_w/2\right) \tag{1}$$

$$\delta_{oj} = \left(X_y + d_m/2\right) - \left(d_o/2 - r_o\right) - \left(r_o - D_w/2\right) \tag{2}$$

where $d_{i/o}$ represents the groove bottom circle diameter of the inner/outer ring. $r_{i/o}$ is the groove curvature radius of the inner/outer ring. D_w means the ball's diameter. d_m indicates the pitch diameter of the bearing, $d_m = (d_i + d_o)/2$. B_y is the displacement of the inner groove curvature center in the azimuthal plane along the y_{aj}-axis, defined by:

$$B_y = y_i \cos\varphi_{bj} + z_i \sin\varphi_{bj} \tag{3}$$

where φ_{bj} denotes the ball azimuth, $\varphi_{bj} = \arctan(z_{bj}/y_{bj})$.

X_y indicates the displacement of the ball's center in the azimuthal plane, determined by:

$$X_y = y_{bj}\cos\varphi_{bj} + z_{bj}\sin\varphi_{bj} - d_m/2 \tag{4}$$

The normal load between the ball and the raceway can be calculated by using the Hertzian point contact theory, formulated as [21,32]:

$$Q_{i/oj} = K_{i/o} \cdot \max\left(0,\ \delta_{i/oj}\right)^{1.5} \tag{5}$$

where $K_{i/o}$ denotes the contact deflection coefficient.

Furthermore, since the ball motion is confined in the radial plane, the working contact angle is always 0°.

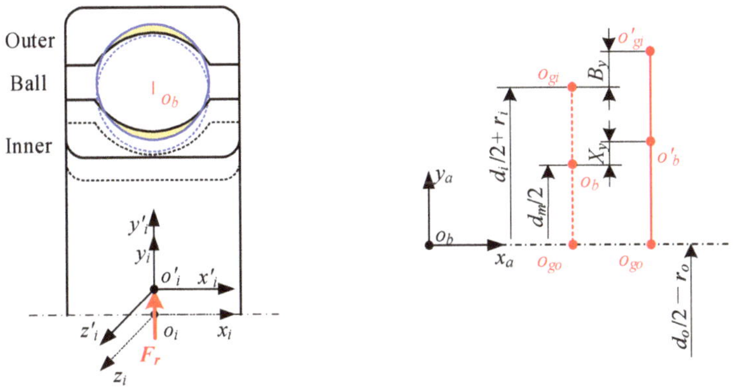

Figure 2. Relative position relationship between the ball and raceway under pure radial force: (**a**) schematic diagram of bearings assemblies; (**b**) positions of the ball's center and groove curvature centers.

(2) Traction force

The traction force inside the ball/raceway interface is critical to the bearing's dynamic performance. Unfortunately, the shear stress distribution in the contact zone is highly dispersed, making it difficult to calculate this force. The micro-element approach is generally used to first discretize the contact area, then solve for the shear stress at each point, and finally integrate to obtain the traction of the entire interface. Because of the presence of lubricant, the shear stress is a function of velocity and pressure. Therefore, the velocity and pressure distributions inside the contact zone must be specified before solving for the traction force. The distribution of these performance parameters in the contact zone under pure radial force is presented in Figure 3.

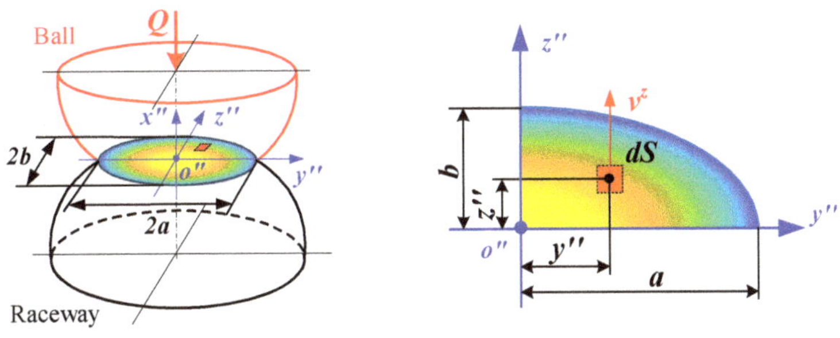

Figure 3. Distribution of stresses and velocities at ball/raceway interface under pure radial force: (**a**) schematic diagram of ball/raceway interface; (**b**) interface discretization.

The velocity distribution inside the contact interface is first investigated, presented in Figure 4. By observing Figure 4a, the velocity vector at any point (y_i'', z_i'') in the inner raceway can be obtained by the following formula:

$$\begin{bmatrix} v^{z''}_{(y_i'',z_i'')} \\ u^{z''}_{(y_i'',z_i'')} \end{bmatrix} = \begin{bmatrix} -(w_i - w_m) & -w_{ax} \\ (w_i - w_m)/2 & -w_{ax}/2 \end{bmatrix} \begin{bmatrix} (d_i/2 + r_i) + B_y - \sqrt{R_i^2 - y_i''^2} \\ \sqrt{0.25D_w^2 - y_i''^2} \end{bmatrix} \tag{6}$$

where w_i denotes the angular velocity of the inner ring. R_i indicates the curvature radius of the contact surface, $R_i = (2D_w \cdot r_i)/(D_w + 2r_i)$. w_m represents the ball orbital angular velocity,

$w_m = (-v_{by} \cdot \cos(\varphi_{bj}) + v_{bz} \cdot \sin(\varphi_{bj}))/(y_{bj}{}^2 + z_{bj}{}^2)^{0.5}$. w_{ax} is the ball's angular velocity about the x-axis in the azimuthal coordinate frame, $w_{ax} = w_{bx}$.

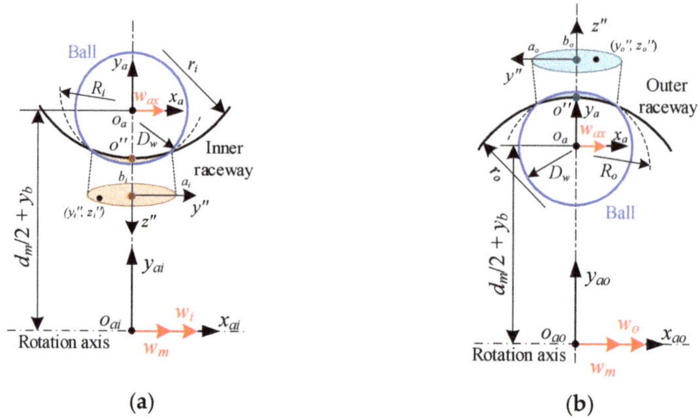

Figure 4. Velocities distribution at ball/raceway interface under pure radial force: (**a**) inner raceway; (**b**) outer raceway.

Similarly, the velocity vector at any point (y_o'', z_o'') in the outer raceway can be determined by:

$$\begin{bmatrix} v^{z''}_{(y_o'',z_o'')} \\ u^{z''}_{(y_o'',z_o'')} \end{bmatrix} = \begin{bmatrix} -(w_o - w_m) & w_{bx} \\ (w_o - w_m)/2 & w_{bx}/2 \end{bmatrix} \begin{bmatrix} (d_o/2 - r_o) - \sqrt{R_o^2 - y_o''^2} \\ \sqrt{0.25D_w^2 - y_o''^2} \end{bmatrix} \tag{7}$$

In addition, the stress at any point (y'', z'') inside the contact zone can be predicted using the Hertzian contact theory [23]:

$$p_{(y'',z'')} = p_{\max}\sqrt{1 - \left(\frac{y''}{a}\right)^2 - \left(\frac{z''}{b}\right)^2} \tag{8}$$

where $p_{\max}$ is the maximum contact stress. a and b represent the length of the semi-major axis and the length of the semi-minor axis of the contact ellipse, respectively.

As shown in Figure 3, the shear stress at this point can be further derived as follows [27]:

$$\tau_{(y'',z'')} = \frac{p_{(y'',z'')}\mu_{(y'',z'')}dy''dz''}{dS_{(y'',z'')}} \tag{9}$$

where S denotes the area of the micrometeoroid. μ represent the traction coefficient, which can be approximated by a five-parameter rheological model [26]:

$$\mu_{(y'',z'')} = \left(A + Bs_{(y'',z'')}\right)e^{s_{(y'',z'')}} + D \tag{10}$$

where A, B, C, and D are functions of lubricant physical parameters and dynamic performance parameters, etc. s indicates the ratio of the relative velocity to the rolling velocity, $s = v/u$.

The traction force between the ball and raceway can be estimated by integrating the shear stress, denoted as:

$$T^{z''}_{i/oj} = \int_{s_{i/o}} \tau^{z''}_{(y'',z'')}ds_{i/o} = \int_{-a_{i/o}}^{a_{i/o}}\int_{-b_{i/o}}^{b_{i/o}} p_{(y'',z'')}\mu_{(y'',z'')}dy''dz'' \tag{11}$$

(3) Differential equations

Since the bearing assemblies are assumed to be purely rigid bodies in this paper, the relationship between their motion and forces can be modeled by Newton's law and Euler's kinematical equations. When the bearing is subjected to radial force only, the inner ring has one degree of freedom (DOF) of translation, and the ball has two DOFs of translation and one DOF of rotation.

For the balls, the differential equations can be defined as follows:

$$m_b\ddot{y}_{bj} = (Q_{ij} + F_{cj} - Q_{oj})\cos\varphi_{bj} - \left(T_{ij} + T_{oj} - F_{dj} - Q_{cj}\right)\sin\varphi_{bj} \tag{12}$$

$$m_b\ddot{z}_{bj} = (Q_{ij} + F_{cj} - Q_{oj})\sin\varphi_{bj} + \left(T_{ij} + T_{oj} - F_{dj} - Q_{cj}\right)\cos\varphi_{bj} \tag{13}$$

$$I_b\dot{w}_{xbj} = \left(-T_{ij} + T_{oj}\right)\cdot D_w/2 \tag{14}$$

where m_b denotes the ball mass. I_b is the rotational inertia. F_{cj} indicates the ball centrifugal force, $F_{cj} = 0.5\cdot m_b\cdot d_m\cdot w_m{}^2$. F_{dj} means the oil and gas mixture mixing resistance. Q_{cj} represents the interaction between the ball and the cage pocket [33].

For the inner ring, the differential equations can be determined as follows:

$$m_i\ddot{y}_i = F_r - \sum_{j=1}^{Z}\left(-Q_{ij}\cos\varphi_{bj} + T_{ij}\sin\varphi_{bj}\right) \tag{15}$$

where m_i indicates the inner mass. Z is the number of the ball.

2.2. Pure Axial Force

(1) Normal load

Figure 5 presents the relative position between the ball and the raceway under pure axial force. The axial force moves the inner ring axially, first closing the radial clearance and forming the initial contact angle before entering the preload phase. The ball's motion changes from one-dimensional to two-dimensional due to the action of centrifugal load and axial force. Similar to pure radial force, the contact deformation between the j-th ball and raceway under pure axial force can be represented as:

$$\delta_{ij} = \sqrt{(B_x + 0.5P_a - X_x)^2 + (d_i/2 + r_i - d_m/2 - X_y)^2} - (r_i - D_w/2) \tag{16}$$

$$\delta_{oj} = \sqrt{X_x^2 + (X_y + d_m/2 - d_o/2 + r_o)^2} - (r_o - D_w/2) \tag{17}$$

where P_a denotes the axial clearance. B_x is the axial displacement of the inner ring and $B_x = x_i$. X_x and X_y are the displacements of the ball in the radial direction and axial direction, respectively, determined by:

$$X_x = x_{bj} \tag{18}$$

$$X_y = y_{bj}\cos\varphi_{bj} + z_{bj}\sin\varphi_{bj} - d_m/2 \tag{19}$$

Similarly, the Hertzian contact theory can be employed to establish the connection between the contact deflection and normal load.

The working contact angle under pure axial force can be further developed as:

$$\alpha_i = \arctan\frac{B_x + 0.5P_a - X_x}{d_i/2 + r_i - d_m/2 - X_y} \tag{20}$$

$$\alpha_o = \arctan\frac{X_x}{X_y + d_m/2 - d_o/2 + r_o} \tag{21}$$

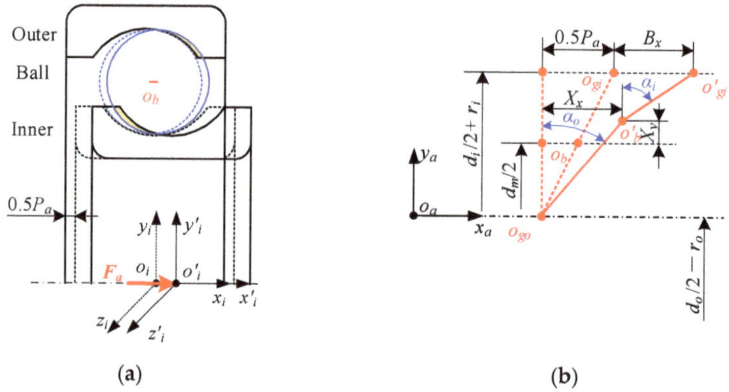

Figure 5. Relative position relationship between the ball and raceway under pure axial force: (**a**) schematic diagram of bearing assemblies; (**b**) positions of the ball's center and groove curvature centers.

(2) Traction force

Under pure axial force, a working contact angle is formed between the ball and the inner/outer raceway. Considering the ball's centrifugal force at high speed, the inner contact angle is larger than the outer contact angle, which can be seen in Figure 5. Figure 6 presents the velocity and stress distribution at the ball/raceway contact interface in the presence of the contact angle. As shown in Figure 6, the velocity distribution under pure axial force is significantly different from that under pure radial force, while the stress distribution is approximately the same for both. This means that the stresses at any point can continue to be predicted using Hertzian contact theory, while a new model needs to be proposed to elucidate the velocity distribution.

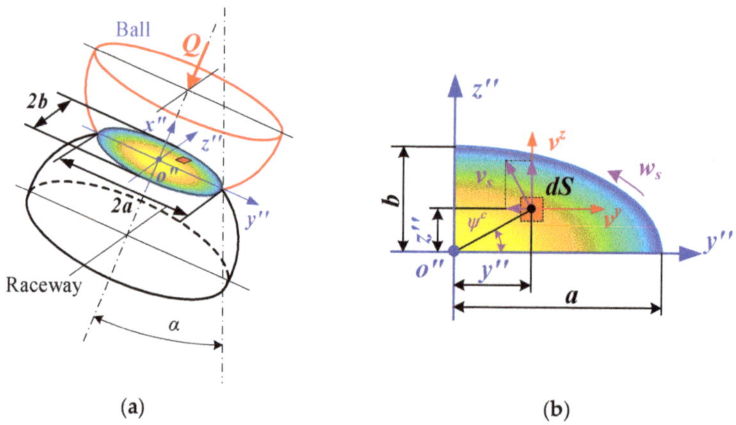

Figure 6. Distribution of stresses and velocities at ball/raceway interface under pure radial force: (**a**) schematic diagram of ball/raceway interface; (**b**) interface discretization.

The presence of the contact angle leads to a linear relative velocity component between the ball and raceway along the major axis of the contact ellipse. The velocity vector inside the contact interface also changes from one-dimensional to two-dimensional. Furthermore, the difference in the two contact angles on the ball generates a rotational velocity at this interface. Therefore, before integrating the traction force, both the linear velocity in each direction and the spin angular velocity inside the contact area must be clarified.

Similar to pure radial force, the velocity distribution inside the contact interface is displayed in Figure 7. As shown in Figure 7a, for any point (x_i'', y_i'') in the interface of the inner raceway, the translational velocity vector is given by:

$$\begin{bmatrix} v^{y''}_{(y_i'',z_i'')} \\ v^{z''}_{(y_i'',z_i'')} \\ u^{z''}_{(y_i'',z_i'')} \end{bmatrix} = \begin{bmatrix} \sqrt{0.25D_w^2 - y_i''^2} & 0 & 0 \\ 0 & -(d_i/2+r_i)/\cos\alpha_i + \sqrt{R_i^2 - y_i''^2} & \sqrt{0.25D_w^2 - y_i''^2} \\ 0 & \left[(d_i/2+r_i)/\cos\alpha_i - \sqrt{R_i^2 - y_i''^2}\right]/2 & \sqrt{0.25D_w^2 - y_i''^2}/2 \end{bmatrix} \begin{bmatrix} w_{az} \\ (w_i - w_m)\cos\alpha_i \\ w_{ax}\sin\alpha_i - w_{ay}\cos\alpha_i \end{bmatrix} \tag{22}$$

where w_{ay} and w_{az} are the angular velocities of the ball rotating about the y-axis and about the z-axis in the azimuthal coordinate frame, respectively, defined by:

$$\begin{bmatrix} w_{ay} \\ w_{az} \end{bmatrix} = \begin{bmatrix} \cos\varphi_{bj} & \sin\varphi_{bj} \\ -\sin\varphi_{bj} & \cos\varphi_{bj} \end{bmatrix} \begin{bmatrix} w_{by} \\ w_{bz} \end{bmatrix} \tag{23}$$

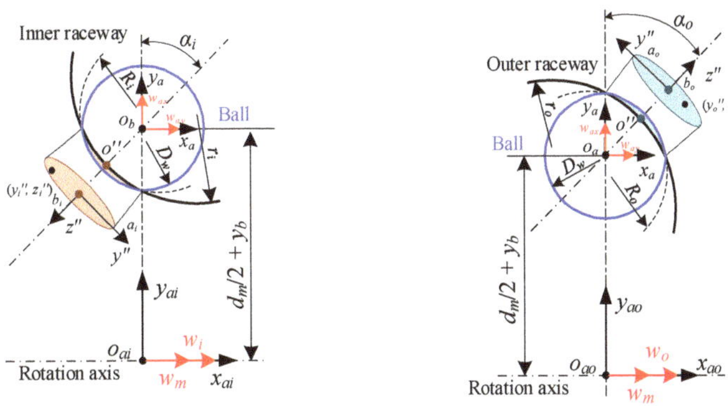

(a) (b)

Figure 7. Velocities distribution at ball/raceway interface under pure radial force: (**a**) inner raceway; (**b**) outer raceway.

In addition, the spin angular velocity inside the interface between the ball and the inner raceway can be deduced as follows:

$$w_s^i = (w_{ax}\cos\alpha_i + w_{ay}\sin\alpha_i) - (w_i - w_m)\sin\alpha_i \tag{24}$$

Similarly, the translational velocity vector and the spin angular velocity at any point (x_o'', y_o'') of the outer raceway can be obtained by:

$$\begin{bmatrix} v^{y''}_{(y_o'',z_o'')} \\ v^{z''}_{(y_o'',z_o'')} \\ u^{z''}_{(y_o'',z_o'')} \end{bmatrix} = \begin{bmatrix} \sqrt{0.25D_w^2 - y_o''^2} & 0 & 0 \\ 0 & (d_o/2-r_o)/\cos\alpha_o + \sqrt{R_o^2 - y_o''^2} & \sqrt{0.25D_w^2 - y_o''^2} \\ 0 & \left[(d_o/2-r_o)/\cos\alpha_o + \sqrt{R_o^2 - y_o''^2}\right]/2 & -\sqrt{0.25D_w^2 - y_o''^2}/2 \end{bmatrix} \begin{bmatrix} w_{az} \\ (w_o - w_m)\cos\alpha_o \\ w_{ax}\sin\alpha_o - w_{ay}\cos\alpha_o \end{bmatrix} \tag{25}$$

$$w_s^o = (w_{ax}\cos\alpha_i + w_{ay}\sin\alpha_i) - (w_o - w_m)\sin\alpha_i \tag{26}$$

As shown in Figure 6b, the relative sliding velocity vector is synthesized from the velocity vector and the spin velocity, given by [27]:

$$\begin{bmatrix} \Delta v^{y''}_{(y_{i/o}'',z_{i/o}'')} \\ \Delta v^{z''}_{(y_{i/o}'',z_{i/o}'')} \end{bmatrix} = \begin{bmatrix} v^{y''}_{(y_{i/o}'',z_{i/o}'')} \\ v^{z''}_{(y_{i/o}'',z_{i/o}'')} \end{bmatrix} + \begin{bmatrix} z_{i/o}'' \\ y_{i/o}'' \end{bmatrix} \cdot w_s^{i/o} \tag{27}$$

Since the ball velocity at the contact interface with respect to the raceway is a two-dimensional vector under pure axial force, its traction force needs to be obtained by integrating the shear stress in both directions [27,34].

$$\begin{bmatrix} T_{i/oj}^{y''} \\ T_{i/oj}^{z''} \end{bmatrix} = \begin{bmatrix} \int_{s_{i/o}} \tau^{y''}_{(y'',z'')} ds_{i/o} \\ \int_{s_{i/o}} \tau^{z''}_{(y'',z'')} ds_{i/o} \end{bmatrix} = \begin{bmatrix} \int_{-a_{i/o}}^{a_{i/o}} \int_{-b_{i/o}}^{b_{i/o}} p_{(y'',z'')} \mu^{y''}_{(y'',z'')} dy'' dz'' \\ \int_{-a_{i/o}}^{a_{i/o}} \int_{-b_{i/o}}^{b_{i/o}} p_{(y'',z'')} \mu^{z''}_{(y'',z'')} dy'' dz'' \end{bmatrix} \tag{28}$$

Furthermore, in addition to the sliding velocity inside the contact interface, there is a spin velocity around the contact normal, which produces the spin moment.

$$M_{i/oj}^{x''} = \int_{-a_{i/o}}^{a_{i/o}} \int_{-b_{i/o}}^{b_{i/o}} p_{(y'',z'')} \left(\mu^{z''}_{(y'',z'')} y'' - \mu^{z''}_{(y'',z'')} z'' \right) dy'' dz'' \tag{29}$$

(3) Differential equations

Once the bearing is subjected to axial force only, the inner ring has one degree of freedom (DOF) of translation, and the ball has three DOFs of translation and three DOFs of rotation.

For the balls, the differential equations can be defined as follows:

$$\begin{bmatrix} m_b \ddot{x}_{bj} \\ m_b \ddot{y}_{bj} \\ m_b \ddot{z}_{bj} \end{bmatrix} = \begin{bmatrix} 1 & 0 & 0 \\ 0 & \cos\varphi_{bj} & \sin\varphi_{bj} \\ 0 & -\sin\varphi_{bj} & \cos\varphi_{bj} \end{bmatrix} \begin{bmatrix} Q_{ij}\sin\alpha_{ij} - Q_{oj}\sin\alpha_{oj} - T_{ij}^{y''}\cos\alpha_{ij} + T_{oj}^{y''}\cos\alpha_{oj} \\ Q_{ij}\cos\alpha_{ij} - Q_{oj}\cos\alpha_{oj} + T_{ij}^{y''}\sin\alpha_{ij} + T_{oj}^{y''}\sin\alpha_{oj} - F_{cj} \\ T_{ij}^{z''} + T_{ij}^{z''} - F_{dj} - Q_{cj} \end{bmatrix} \tag{30}$$

$$\begin{bmatrix} I(\dot{w}_{bxj} - w_{byj}w_{bzj}) \\ I(\dot{w}_{byj} - w_{bzj}w_{bxj}) \\ I(\dot{w}_{bzj} - w_{bxj}w_{byj}) \end{bmatrix} = \begin{bmatrix} 1 & 0 & 0 \\ 0 & \cos\varphi_{bj} & \sin\varphi_{bj} \\ 0 & -\sin\varphi_{bj} & \cos\varphi_{bj} \end{bmatrix} \begin{bmatrix} \left(-T_{ij}^{y''}\cos\alpha_{ij} + T_{oj}^{y''}\cos\alpha_{oj}\right) \cdot D_w/2 + M_{ij}^{x''}\sin\alpha_{ij} - M_{oj}^{x''}\sin\alpha_{oj} \\ \left(+T_{ij}^{y''}\sin\alpha_{ij} - T_{oj}^{y''}\sin\alpha_{oj}\right) \cdot D_w/2 + M_{ij}^{x''}\cos\alpha_{ij} - M_{oj}^{x''}\sin\alpha_{oj} \\ -\left(T_{ij}^{y''} + T_{ij}^{y''}\right) \cdot D_w/2 \end{bmatrix} \tag{31}$$

For the inner ring, the differential equations can be determined as follows:

$$m_i \ddot{x}_i = F_a - \sum_{j=1}^{Z} \left(-Q_{ij}\sin\alpha_{ij} + T_{ij}\cos\alpha_{bj} \right) \tag{32}$$

where m_i indicates the inner mass.

2.3. Combined Axial and Radial Load

(1) Normal load

The position relationship of the bearing's assemblies with and without the combined axial and radial load is displayed in Figure 8. By observing Figure 8, it can be seen that when the bearing is in this operating condition, the inner displacement changes from the original one-dimensional vector to a two-dimensional vector, greatly increasing the complexity of the spatial position relationship.

Fortunately, the presence of radial force has no effect on the relative position between the ball and the outer raceway compared to the pure axial force condition, which means that the contact force and the contact angle between them can be obtained in this condition using the previous models. Therefore, the contact deformation and contact angle between the ball and inner raceway can be indicated as follows, respectively:

$$\delta_{ij} = \sqrt{(B_x + 0.5P_a - X_x)^2 + (d_i/2 + r_i + B_y - d_m/2 - X_y)^2} - (r_i - D_w/2) \tag{33}$$

As for traction force, it is perfectly possible to use the previous model of pure axial force to obtain the force under combined axial and radial load. The reason is that in both conditions, the contact interface between the ball and the raceway is a two-dimensional plane.

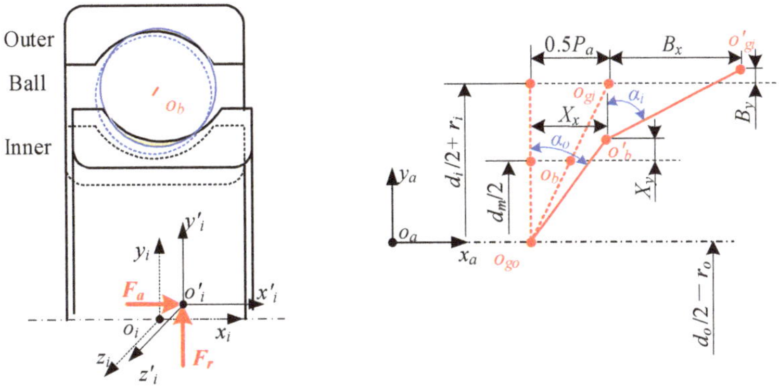

(a) (b)

Figure 8. Relative position relationship between the ball and raceway under pure axial force: (**a**) schematic diagram of bearings assemblies; (**b**) positions of the ball's center and groove curvature centers.

(2) Differential equations

When the bearing is subjected to combined axial and radial force, the inner ring has two degrees of freedom (DOFs) of translation, and the ball has three DOFs of translation and three DOFs of rotation. Furthermore, the differential equations of the ball are the same as those under pure axial force.

For the inner ring, the differential equations can be expressed as follows:

$$\begin{bmatrix} m_b \ddot{x}_{bj} \\ m_b \ddot{y}_{bj} \end{bmatrix} = \begin{bmatrix} F_a - \sum\limits_{j=1}^{Z} \left(-Q_{ij} \sin \alpha_{ij} + T_{ij} \cos \alpha_{bj} \right) \\ F_r - \sum\limits_{j=1}^{Z} \left(-Q_{ij} \cos \alpha_{ij} - T_{ij} \sin \alpha_{bj} \right) \sin \varphi_{bj} \end{bmatrix} \tag{34}$$

3. Theoretical Model Solving and Validation

In this section, the solution procedure of the previous numerical model is presented in detail. The validation of this model is further proven by comparing it with the experimental data of Han [35] and Pasdari [36]. Furthermore, the superiority of this model is highlighted by comparing it with the simulation results of Han [35] and Liu [26].

3.1. Model Solving

The flowchart for solving the general dynamic model of ball bearings is presented in Figure 9. This process can be divided into three steps, covering: parameters input and initialization, main calculations, and post-processing. Compared to the common ball bearing dynamics model, the improvement of this model is primarily reflected in the first two steps. After entering the parameters, the program quickly selects a model with a specific number of equations based on the load type entered. The issue of increasing the computational cost due to many unnecessary variables taking up memory is avoided, as is the problem of model divergence. Furthermore, because of the complex motion of the bearing components, it is necessary to ensure the accuracy of the initial values when solving the differential equations. This is commonly accomplished through the static model and the quasi-static model. The number of equations to be solved for the present model is listed in Table 1. The current model is capable of reducing iterative variables by reducing the dimensionality of the equation set, thus saving computational cost.

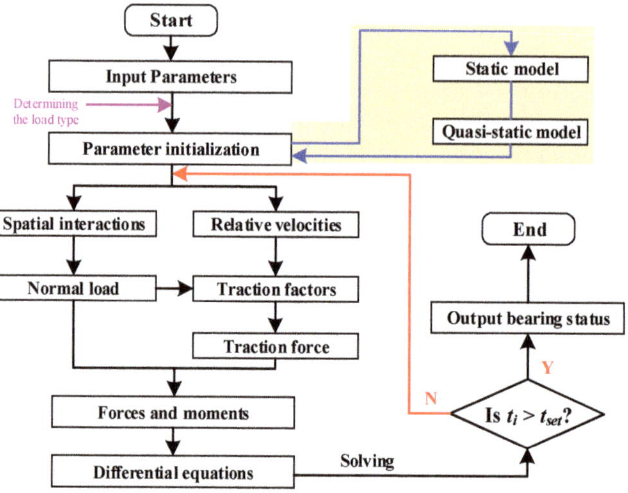

Figure 9. Flowchart for solving the general dynamic model of ball bearings.

Table 1. The number of equations to be solved for each operating condition.

	Static Model	Quasi-Static Model	Dynamic Model
	Nonlinear Equations	Nonlinear Equations	Differential Equations
Case: 1	1	1 + N	1 + 1 + 3 N
Case: 2	1	1 + 2	1 + 1 + 6 N
Case: 3	2	2 + 2 N	2 + 1 + 6 N
General model	2	2 + 2 N	2 + 1 + 6 N

3.2. Model Validation

To validate the reliability of the model proposed in this paper, its prediction results under two different operating conditions are compared with the experimental results [35,36], respectively. Since the bearing model for the pure axial force condition is approximately the same as the model for the combined axial and radial load, only the simulation results for the pure axial force are verified.

To verify the accuracy of the present model under pure radial force conditions, the driving speed of the bearing is kept at 1800 r/min and the radial force is incremented from 50 N to 300 N, and the model prediction results are compared with the experimental results in Ref. [35]. As for the pure axial force conditions, the radial force is increased from 10 N to 400 N at a driving speed of 4000 r/min, and the prediction is compared with the experimental results in Ref. [36]. The variations of cage speed with different loads for two operating conditions are described in Figure 10, respectively. The cage speed gradually increases and stabilizes with the growing load, indicating that the bearing sliding is suppressed. Furthermore, the good agreement between the model's predicted results and the experimental test results suggests its accuracy under three different operating conditions.

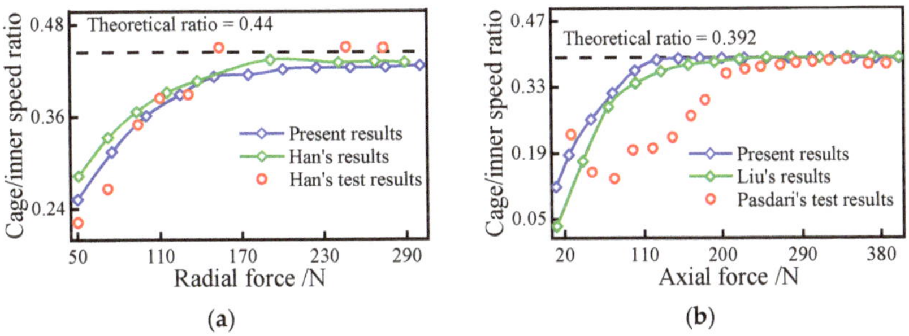

Figure 10. Comparison of simulation results with experimental data; (**a**) radial force; (**b**) axial force.

Moreover, the simulation results of Han's model [35] and Liu's model [26] are shown in Figure 10, and the number of equations to be solved for these models is listed in Table 1 as a general model. The results of the model in this paper do not differ much from those of other models, which proves that it reduces the number of variables and equations and improves efficiency without losing computational accuracy. Therefore, the superiority of the model in this paper over the general models is emphasized.

4. Results and Discussion

In this section, angular contact ball bearing 7008 is chosen to investigate the bearing skidding behavior under 3 operating conditions. Its structural parameters and material properties are listed in Table 2. The bearing speed is always kept at 5000 r/min.

Table 2. Structural parameters and material properties of 7008.

Name	Value
Pitch circle diameter d_m/mm	54
Inner curvature radius r_i/mm	3.990
Outer curvature radius r_o/mm	3.780
Ball diameter D_w/mm	7.000
Number of balls Z	18
Initial contact angle α_0/°	15
Cage pocket clearness C_b/mm	0.200
Ring material	Bearing Steel
Ball material	Bearing Steel
Cage material	Nylon

4.1. Case 1: Pure Radial Force

Applying radial forces of 50, 100, 150, and 200 to the bearing, the sliding behavior of ball bearings under pure radial force is first discussed. The variation of the sliding velocity at the ball/raceway under different loads is presented in Figure 11. Figure 11a,b displays the relationship between the sliding velocity and the normal load for a radial force of 50 N. As shown in Figure 11a, the ball's sliding velocity is strongly influenced by the load exerted on the ball. To further clearly illustrate the relationship between them, the two running cycles of the ball can be divided into four phases: AB, BC, CD, and DA stages. AB region and CD region denote the loaded zone, while the other two regions represent the unloaded zones. Interestingly, at the intersection of the two phases, the sliding velocity fluctuates considerably, indicating that the abrupt changes in the sliding velocity tend to occur near the boundary of the loaded zone. In addition, for the inner raceway, the steady value of the sliding velocity in the non-loaded zone is greater than in the loaded zone. The opposite phenomenon is observed in the outer raceway, as shown in Figure 11b. This is because the driving force provided by the inner raceway in the loaded zone is heavier than the resistance of the outer ring, whereas, in the unload zone, only the outer resistance exists.

Furthermore, the sliding velocity between the inner raceway and the ball is up to 7 m/s or more in the loaded zones, indicating severe sliding inside the bearing under a radial force of 50 N.

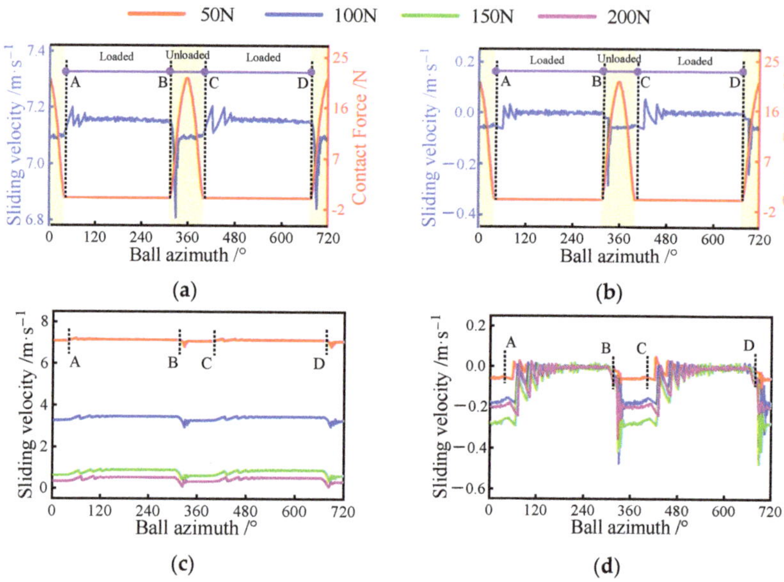

Figure 11. Sliding velocity variation at ball/raceway interface under pure radial force condition, A denotes the exit of the previous loaded area, B is the entrance of the current loaded area, C represents the exit of the current loaded area, D is the entrance of the next loaded area: (**a**) sliding velocity of the inner raceway for different azimuths; (**b**) sliding velocity of the outer raceway for different azimuths; (**c**) sliding velocity of inner raceway; (**d**) sliding velocity of the outer raceway.

To further discuss the effect of radial force on the bearing sliding, the trend of sliding velocity between the raceways and the ball under different radial loads is provided in Figure 11c,d. It can be seen that as the radial force increases, the sliding velocity between the inner raceway and the ball decreases significantly, while the amount of the change in sliding velocity at the loaded zone boundary increases. Interestingly, the radial load has little effect on the sliding velocity in the unloading zone of the outer raceway as compared to the inner raceway. However, in the load zone, the velocity increases with the gradual increase of the radial load and then decreases.

Furthermore, as an essential dynamic indicator, the traction factor has been also investigated. Figure 12 presents the variation of this factor with different loads under pure radial force conditions. By observing Figure 12a, it can be seen that the traction factor between the inner ring and the ball exists only in the load zone. This is because there is no contact load to form a lubricant film in the unloading zone. In addition, the factor first remains constant and then decreases with the gradual growth of the radial force in the loaded zone. A similar phenomenon can be observed in the outer raceway, as shown in Figure 12b. The reason for this is that when the radial load is too tiny to inhibit severe sliding inside the bearing, the traction factor is mainly determined by the sliding velocity. Once the bearing skidding is suppressed, the traction factor is a function of normal load and sliding velocity. Furthermore, the inhibition of sliding leads to an increase in the ball's centrifugal force, which helps to reduce the flotation of the traction factor between the ball and the outer raceway in the unloading zone.

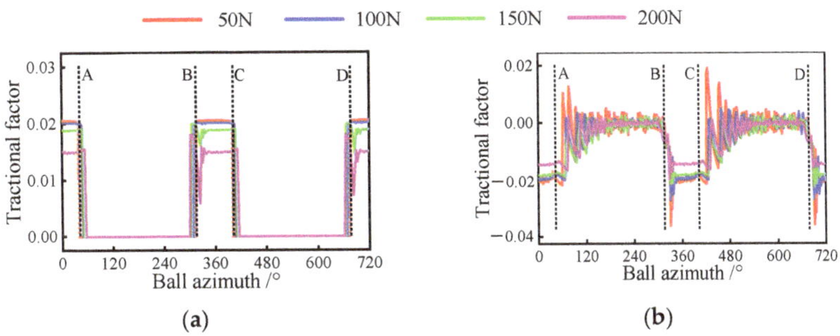

Figure 12. Tractional factor variation at ball/raceway interfaces under pure radial force condition, A denotes the exit of the previous loaded area, B is the entrance of the current loaded area, C represents the exit of the current loaded area, D is the entrance of the next loaded area: (**a**) tractional factor of inner raceway; (**b**) tractional factor of the outer raceway.

4.2. Case 2: Pure Axial Force

The sliding behavior at the ball/raceway interface under pure axial force is then considered. Figure 13a,b shows the sliding velocity on the ball at different azimuths when an axial force of 50 N is applied to the bearing. It can be seen that the velocity between the ball and the raceway remains constant at each revolution. In addition, the sliding velocity on the inner raceway is always greater than that of the outer raceway. This is due to the fact that the contact features of the ball at different azimuths are equal when the bearing is operating in pure axial condition. Therefore, the effect of axial force on the sliding behavior is further investigated with a certain ball.

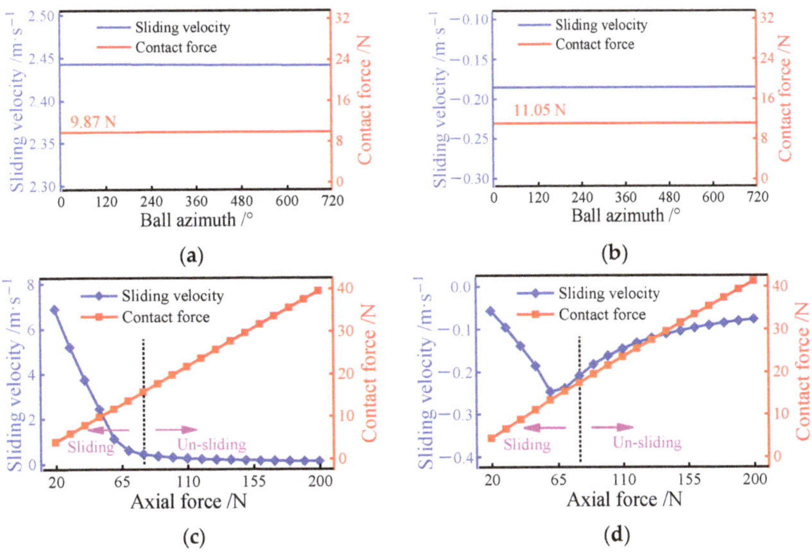

Figure 13. Sliding velocity variation at ball/raceway interface under pure axial force condition: (**a**) sliding velocity of the inner raceway for different azimuths; (**b**) sliding velocity of the outer raceway for different azimuths; (**c**) sliding velocity of inner raceway; (**d**) sliding velocity of the outer raceway.

Increasing the axial force from 20 N to 200 N, the change in sliding velocity on the ball is presented in Figure 13c,d. Similar to the pure radial force condition, the sliding inside the bearing is inhibited as the load is gradually increased. Once the axial force exceeds

80 N, the sliding velocity on the inner raceway stays in a small variation, while the outer velocity tends to become smaller. In addition, the sliding velocity on the inner raceway is gradually equal to the velocity of the outer, indicating that the skidding inside the bearing is completely suppressed. Interestingly, the inflection point of the sliding velocity on the outer occurs before the axial force reaches 80 N. This means that the normal load on the ball does not completely determine the degree of sliding at the ball/outer raceway interface.

The effect of the axial force on the traction factor is further discussed. As shown in Figure 14, when the axial load is less than 60 N, the traction factor of the inner and outer raceways keeps fluctuating around a large value. Meanwhile, the factor of the inner raceway is bigger than that of the outer raceway. This is because, for ball bearings subjected to axial loads only, the traction coefficient at the ball/raceway interface is greatly influenced by the slip-roll ratio. When the load is too tiny to suppress severe sliding inside the bearing, the slip-roll ratio at the interface will be larger, resulting in a larger value of the traction coefficient. Once the axial force exerted on the bearing is greater than 60 N, the traction factor of the inner and outer raceways tends to decrease. Although the traction factor tends to decrease when the axial force is between 60 N and 80 N, the sliding velocity is still high, which significantly reduces the bearing performance. This indicates that the load causing the shift in the traction factor is less than the critical load to prevent bearing sliding. As a result, it is better to choose an axial force of 80 N or greater as the bearing's normal working load to avoid serious sliding.

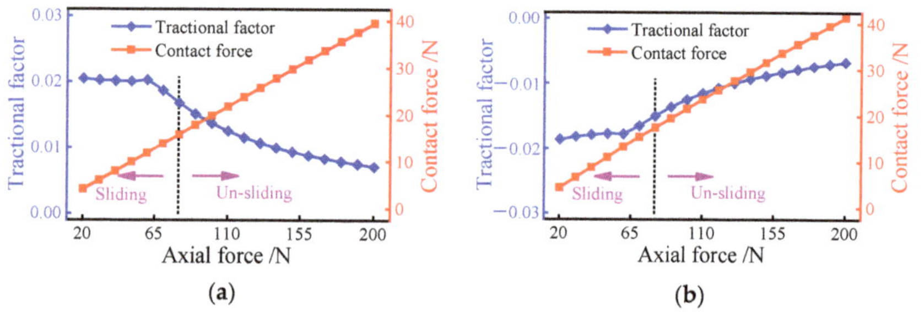

Figure 14. Tractional factor variation at ball/raceway interfaces under pure axial force condition: (**a**) tractional factor of inner raceway; (**b**) tractional factor of the outer raceway.

4.3. Case 3: Combined Axial and Radial Force

Under combined axial and radial force conditions, the load distribution on the ball is non-uniform due to the presence of radial load. Once the radial load is too larger, the ball experiences the loaded and unloaded zones in turn for each revolution. Therefore, the sliding behavior of the ball at different azimuths is significantly different. Figure 15a,b display the variation of the ball's sliding velocity for an axial force of 100 N and a radial force of 200 N. Similar to pure radial force condition, the ball goes through four stages in sequence for every two revolutions, covering: the AB phase, BC phase, CD phase, and DA phase. In the AB phase, the sliding velocity on the inner raceway tends to increase and the velocity of the outer remains constant; in the BC phase, the inner raceway's sliding velocity is smaller and stabilizes as the normal load increases, while the sliding velocity of the outer changes in the opposite trend. The sliding behavior of the ball changes in the other two phases in the same way as in the first two phases. In fact, the sliding variation at the ball/raceway interface is mainly influenced by its contact features.

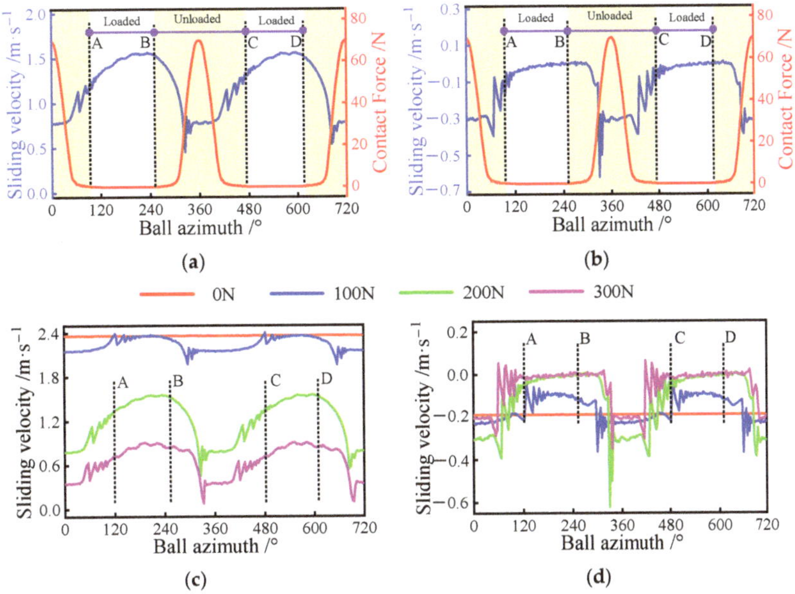

Figure 15. Sliding velocity variation at ball/raceway interface under combined axial and radial load conditions, A denotes the exit of the previous loaded area, B is the entrance of the current loaded area, C represents the exit of the current loaded area, D is the entrance of the next loaded area: (**a**) sliding velocity of the inner raceway for different azimuths; (**b**) sliding velocity of the outer raceway for different azimuths; (**c**) sliding velocity of inner raceway; (**d**) sliding velocity of the outer raceway.

The radial force is gradually increased from 0 N to 300 N to further investigate the sliding behavior under this condition. Figure 15c,d presents the changes in the sliding velocity under different radial forces. Although the floating range of the speed is small when the radial force is small, the overall sliding velocity inside the bearing is severe. With the gradual increase of the radial force, the variation of sliding velocity on the ball increases, but its maximum value decreases. In addition, a similar phenomenon can be observed at the ball/outer raceway interface.

As shown in Figure 16a,b, the traction factors of the outer and inner raceway are strongly influenced by the radial load in this case. Once the load exceeds 200 N, the traction factor in the inner raceway starts to appear as 0. This is due to excessive radial forces, which cause a portion of the loaded zone to become an unloaded zone. By observing Figure 16a, the unloaded zone gradually expands with the increase of radial force. It is noteworthy that the traction factor on the outer raceway is never zero regardless of the radial force, which is due to the non-zero contact force at the ball/outer raceway interface. Furthermore, when the bearing's sliding is inhibited, the traction factor gradually decreases as the load decreases, similar to the first two operating conditions.

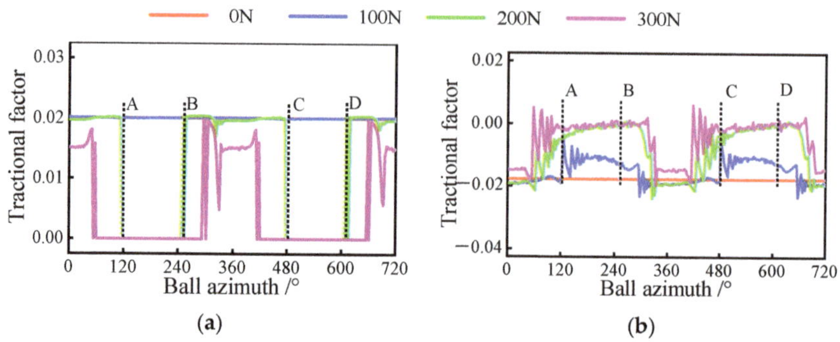

Figure 16. Tractional factor variation at ball/raceway interface under combined axial and radial load condition, A denotes the exit of the previous loaded area, B is the entrance of the current loaded area, C represents the exit of the current loaded area, D is the entrance of the next loaded area: (**a**) tractional factor of inner raceway; (**b**) tractional factor of the outer raceway.

5. Conclusions

In this paper, a generic ball-bearing dynamic model for predicting sliding behavior is presented. Specific geometric interactions between the ball and raceway are constructed based on the motion features between the components under different operating conditions. The number of iteration variables and equations are determined by the load vector exerted on the bearing to improve the convergence and efficiency of the model. The sliding behavior under three typical operating conditions is further discussed. The conclusions are as follows:

(1) When the bearing is subjected to pure axial force, the sliding velocity in the outer raceway first increases and then decreases as the load increases;
(2) Under combined load conditions, the heavy radial forces inhibit the bearing sliding while exacerbating the non-uniformity of the sliding distribution;
(3) Once the unloaded zone occurs, the sliding velocity between the ball and the inner raceway is less in the loaded zone than in the unloaded zone. While the velocity on the outer raceway has the opposite trend;
(4) The increased load helps to suppress severe sliding behavior inside the bearing but shortens the bearing fatigue life.

The presented investigations show that the increase in load helps to inhibit sliding inside the bearing, thus extending the bearing's fatigue life. However, excessive loads can lead to premature fatigue damage of the bearing and shorten its fatigue life. Therefore, during the bearing design process, dynamic models can be used to investigate bearing sliding to guide the optimization of its performance. In addition, the traction model for the ball/raceway interface in the proposed model is too simplified to ignore the viscous temperature and pressure effects within the interface. Hopefully, future research will remedy this deficiency.

Author Contributions: Conceptualization, S.M., F.C., B.F., K.Y. and J.H.; Data curation, S.M.; Formal analysis, S.M. and Y.Y.; Funding acquisition, J.H.; Investigation, S.M. and K.Y.; Methodology, S.M., and Y.Y.; Project administration, K.Y. and J.H.; Resources, J.H.; Software, S.M.; Supervision, K.Y.; Validation, Y.Y.; Visualization, Y.Y. and J.H.; Writing—original draft, S.M. and K.Y.; Writing—review & editing, K.Y., Y.Y. and J.H. All authors have read and agreed to the published version of the manuscript.

Funding: This research was founded by the National Key R&D Program of China (2020YFB2007901).

Data Availability Statement: Not applicable.

Conflicts of Interest: The authors declare no conflict of interest.

Abbreviations

Symbols	Meaning
δ	Contact deformation
$B_{x/y}$	Inner groove curvature center's displacement
$X_{x/y}$	Ball center's displacement
$Q_{i/o}$	Inner/outer normal load
Q_c	Ball/pocket interaction force
M^x	Spin moment
p	Contact stress
τ	Shear stress
μ	Traction coefficient
$T_{i/o}$	Inner/outer traction load
$\alpha_{i/o}$	Working contact angle
m	Mass
I	Rotational inertia
F_c	Ball centrifugal force
F_{dj}	Oil and gas mixture mixing resistance
$d_{i/o}$	Groove bottom circle diameter of inner/outer ring
$r_{i/o}$	Groove curvature radius of inner/outer ring
D_w	Ball diameter
d_m	Bearing pitch diameter
R_i	Curvature radius of the contact surface
Z	Ball number
P_a	Axial clearance
φ_{bj}	Ball azimuth
$K_{i/o}$	Contact deflection coefficient
v	Sliding velocity
u	Rolling velocity
w_m	Ball orbital angular velocity
w_{bx}	Ball rotation angular velocity
w_s	Spin velocity
p_{max}	Maximum contact stress
S	Micrometeoroid's area
A, B, C, D	Lubricant parameters
s	Slide-roll ratio
F_a	Axial load
F_r	Radial load
o-xyz	Inertial coordinate frame
o_i-$x_iy_iz_i$	Inner-fixed coordinate frame
o_c-$x_cy_cz_c$	Cage-fixed coordinate frame
o_b-$x_by_bz_b$	Ball-fixed coordinate frame
o_a-$x_ay_az_a$	Azimuthal coordinate frame

References

1. Cocks, M.; Tallian, T.E. Sliding Contacts in Rolling Bearings. *ASLE Trans.* **1971**, *14*, 32–40. [CrossRef]
2. Qian, D.; Xu, X.; Deng, S.; Jiang, S.; Hua, L. Sliding behavior of high speed ball bearings based on improved nonlinear dynamic model. *Proc. Inst. Mech. Eng. Part K J. Multi-Body Dyn.* **2021**, *235*, 627–640. [CrossRef]
3. Gupta, P.K. Minimum Energy Hypothesis in Quasi-Static Equilibrium Solutions for Angular Contact Ball Bearings. *Tribol. Trans.* **2020**, *63*, 1051–1066. [CrossRef]
4. Zhang, P.; Pan, A.; Yan, K.; Sun, J. High stability temperature sensors by CdTe quantum dots encapsulated in SiO2/PVA hybrids for bearing rotating elements. *Mater. Today Commun.* **2023**, *34*, 105456. [CrossRef]
5. Jorgensen, B.R.; Shin, Y. Dynamics of Machine Tool Spindle/Bearing Systems Under Thermal Growth. *J. Tribol.* **1997**, *119*, 875–882. [CrossRef]
6. Aramaki, H.; Shoda, Y.; Morishita, Y.; Sawamoto, T. The Performance of Ball Bearings With Silicon Nitride Ceramic Balls in High Speed Spindles for Machine Tools. *J. Tribol.* **1988**, *110*, 693–698. [CrossRef]
7. Harris, T. *Rolling Bearing Analysis*, 2nd ed.; John Wiley & Sons: New York, NY, USA, 1984.

8. Xu, T.; Xu, G.; Zhang, Q.; Hua, C.; Tan, H.; Zhang, S.; Luo, A. A preload analytical method for ball bearings utilising bearing skidding criterion. *Tribol. Int.* **2013**, *67*, 44–50. [CrossRef]
9. Liao, N.T.; Lin, J.F. Ball bearing skidding under radial and axial loads. *Mech. Mach. Theory* **2002**, *37*, 91–113. [CrossRef]
10. Liao, N.-T.; Lin, J.F. Rolling-Sliding Analysis in Ball Bearing Considering Thermal Effect. *Tribol. Trans.* **2006**, *49*, 1–16. [CrossRef]
11. Hirano, F. Motion of a Ball in Angular-Contact Ball Bearing. *ASLE Trans.* **1965**, *8*, 425–434. [CrossRef]
12. Oktaviana, L.; Tong, V.-C.; Hong, S.-W. Skidding analysis of angular contact ball bearing subjected to radial load and angular misalignment. *J. Mech. Sci. Technol.* **2019**, *33*, 837–845. [CrossRef]
13. Nelias, D.; Bercea, I.; Paleu, V. Prediction of Roller Skewing in Tapered Roller Bearings. *Tribol. Trans.* **2008**, *51*, 128–139. [CrossRef]
14. Cao, H.; Niu, L.; Xi, S.; Chen, X. Mechanical model development of rolling bearing-rotor systems: A review. *Mech. Syst. Signal Process.* **2018**, *102*, 37–58. [CrossRef]
15. Meeks, C.R.; Ng, K.O. The Dynamics of Ball Separators in Ball Bearings—Part I: Analysis. *ASLE Trans.* **1985**, *28*, 277–287. [CrossRef]
16. Meeks, C.R.; Tran, L. Ball Bearing Dynamic Analysis Using Computer Methods—Part I: Analysis. *J. Tribol.* **1996**, *118*, 52–58. [CrossRef]
17. Jones, A.B. Ball Motion and Sliding Friction in Ball Bearings. *Trans. ASME J. Basic Eng.* **1959**, *81*, 1–12. [CrossRef]
18. Jones, A.B. A General Theory for Elastically Constrained Ball and Radial Roller Bearings Under Arbitrary Load and Speed Conditions. *J. Basic Eng.* **1960**, *82*, 309–320. [CrossRef]
19. Jain, S.; Hunt, H. A dynamic model to predict the occurrence of skidding in wind-turbine bearings. *J. Phys. Conf. Ser.* **2011**, *305*, 012027. [CrossRef]
20. Han, Q.; Chu, F. Nonlinear dynamic model for skidding behavior of angular contact ball bearings. *J. Sound Vib.* **2015**, *354*, 219–235. [CrossRef]
21. Gao, S.; Chatterton, S.; Naldi, L.; Pennacchi, P. Ball bearing skidding and over-skidding in large-scale angular contact ball bearings: Nonlinear dynamic model with thermal effects and experimental results. *Mech. Syst. Signal Process.* **2020**, *147*, 107120. [CrossRef]
22. Gupta, P. *Advanced Dynamics of Rolling Bearings*; Springer: Boston, MA, USA, 2013. [CrossRef]
23. Wang, Y.; Wang, W.; Zhang, S.; Zhao, Z. Investigation of skidding in angular contact ball bearings under high speed. *Tribol. Int.* **2015**, *92*, 404–417. [CrossRef]
24. Wang, Y.; Wang, W.; Zhao, Z. Effect of race conformities in angular contact ball bearing. *Tribol. Int.* **2016**, *104*, 109–120. [CrossRef]
25. Yunlong, W.; Wenzhong, W.; Shengguang, Z.; Ziqiang, Z. Effects of raceway surface roughness in an angular contact ball bearing. *Mech. Mach. Theory* **2018**, *121*, 198–212. [CrossRef]
26. Liu, Y.; Wang, W.; Qing, T.; Zhang, Y.; Liang, H.; Zhang, S. The effect of lubricant temperature on dynamic behavior in angular contact ball bearings. *Mech. Mach. Theory* **2020**, *149*, 103832. [CrossRef]
27. Ma, S.; Li, W.; Yan, K.; Li, Y.; Zhu, Y.; Hong, J. A study on the dynamic contact feature of four-contact-point ball bearing. *Mech. Syst. Signal Process.* **2022**, *174*, 109111. [CrossRef]
28. Xu, H.; Wang, P.; Ma, H.; He, D.; Zhao, X.; Yang, Y. Analysis of axial and overturning ultimate load-bearing capacities of deep groove ball bearings under combined loads and arbitrary rotation speed. *Mech. Mach. Theory* **2021**, *169*, 104665. [CrossRef]
29. Li, B.L.; Zeng, L. Fractional Calculus Control of Road Vehicle Lateral Stability after a Tire Blowout. *Mechanika* **2021**, *27*, 475–482. [CrossRef]
30. Walczak, M.; Caban, J. Tribological characteristics of polymer materials used for slide bearings. *Open Eng.* **2021**, *11*, 624–629. [CrossRef]
31. Gil, L.; Przystupa, K.; Pieniak, D.; Kozłowski, E.; Antosz, K.; Gauda, K.; Izdebski, P. Influence of Contamination of Gear Oils in Relation to Time of Operation on Their Lubricity. *Appl. Sci.* **2021**, *11*, 11835. [CrossRef]
32. Fang, B.; Zhang, J.; Hong, J.; Yan, K. Research on the Nonlinear Stiffness Characteristics of Double-Row Angular Contact Ball Bearings under Different Working Conditions. *Lubricants* **2023**, *11*, 44. [CrossRef]
33. Wang, M.; Yan, K.; Zhang, X.; Zhu, Y.; Hong, J. A comprehensive study on dynamic performance of ball bearing considering bearing de-formations and ball-inner raceway separation. *Mech. Syst. Signal Process.* **2023**, *185*, 109826. [CrossRef]
34. Ma, S.; Yin, Y.; Chao, B.; Yan, K.; Fang, B.; Hong, J. A real-time coupling model of bearing-rotor system based on semi-flexible body element. *Int. J. Mech. Sci.* **2023**, *245*, 108098. [CrossRef]
35. Han, Q.; Li, X.; Chu, F. Skidding behavior of cylindrical roller bearings under time-variable load conditions. *Int. J. Mech. Sci.* **2018**, *135*, 203–214. [CrossRef]
36. Pasdari, M.; Gentle, C.R. Effect of Lubricant Starvation on the Minimum Load Condition in a Thrust-Loaded Ball Bearing. *ASLE Trans.* **1987**, *30*, 355–359. [CrossRef]

lubricants

MDPI

Article

Mixed Lubrication Analysis of Tapered Roller Bearings and Crowning Profile Optimization Based on Numerical Running-In Method

Renshui Cao, Hang Bai, Hui Cao, Yazhao Zhang and Yonggang Meng *

State Key Laboratory of Tribology, Tsinghua University, Beijing 100084, China
* Correspondence: mengyg@tsinghua.edu.cn; Tel.: +86-010-62773867

Abstract: Tapered roller bearings (TRBs) are widely used in heavy-load rotating machinery. One of the technical problems in TRBs is the existence of sharp spikes of the contact pressure in the vicinity of the two ends of the tapered rollers. To suppress the pressure spikes at the roller ends, a straight roller profile is crowned in cylindrical and tapered roller bearings. However, compared to cylindrical roller bearings, there are few studies on the profile modification of TRBs in the literature, and most of the publications on the EHD analysis of tapered rollers focused on a single roller, using traditional profiles such as logarithmic profiles, dub-off profiles and chamfer profiles. By using the numerical running-in method proposed and used in crowning profiles of cylindrical rollers by the authors, this paper provides the first ever asymmetric optimized profile solution for all TRBs rather than for just a single roller. The results show that the optimized profile has the best performance in smoothing contact pressure distribution in the axial direction compared with the conventional logarithmic profile and is a useful profile form with respect to the elimination of sharp pressure spikes. In addition, considering the effect of temperature and mixed lubrication, this paper analyzes the influences of different axial profiles under radial load (*Fr*), rotation speed (*N*) and standard deviation of roughness (*Rq*) conditions.

Keywords: tapered roller bearing; profile modification; running-in method; mixed lubrication

Citation: Cao, R.; Bai, H.; Cao, H.; Zhang, Y.; Meng, Y. Mixed Lubrication Analysis of Tapered Roller Bearings and Crowning Profile Optimization Based on Numerical Running-In Method. *Lubricants* **2023**, *11*, 97. https://doi.org/10.3390/lubricants11030097

Received: 2 January 2023
Revised: 18 February 2023
Accepted: 22 February 2023
Published: 24 February 2023

1. Introduction

Among many types of rolling element bearings, tapered roller bearings (TRB) are advantageous to operate under combined heavy radial and thrust loadings, and hence are widely used in rotating machinery, such as helicopter gearboxes, high-speed railway train axle boxes, wind turbines' main shafts and so on.

Over the decades, the numerical modeling of rolling element bearings has attracted significant attention [1–3]. Compared with other types of rolling element bearings, analyses of the kinematic and friction behavior of TRBs are more difficult due to their complex geometric structure. Therefore, in the literature, there are relatively few reports on TRB dynamics and elastohydrodynamic lubrication (EHL). Rahnejat and Gohar [4] reported the influence of misalignment to the radial contact pressure distributions on tapered rollers. They suggested careful axial profiling is important. Cretu et al. [5,6] developed a comprehensive dynamic model to analyze TRBs with six degrees of freedom. In addition to complex loads, the quasi-dynamic model considered centrifugal forces and roller gyroscopic moments. To obtain the film thickness and pressure distributions between tapered rollers and raceways, Yamashita et al. [7] presented an approximate fluid film lubrication model, which combined a quasi-static model and a raceway EHL model. In recent years, Zheng et al. [8] used a quasi-static model to investigate the influences of angular misalignment and frictional force on contact pressure distributions of the main shaft bearing installed in a modern wind turbine. Zhang et al. [9] explored the effects of roller skewing on the frictional torque of a dry-lubricated tapered roller bearing. Nguyen-Schaefer [10] presented a computational model consisting of many circular slices per rolling element of rollers in the TRB and applied the Levenberg–Marquardt's algorithm to solve the strongly nonlinear coupled equation systems. However, all of the studies mentioned above were carried out without considering the mixed lubrication of rollers in the TRB.

Considering the increasingly harsh working conditions of tapered roller bearings, investigating the mixed lubrication of finite line contacts has become more and more significant in recent years. The foundation of the line contact EHL theory has been paved by Petrusevich [11], Dowson and Higginson [12] and many others, including Gohar and Cameron [13] and Wymer and Cameron [14]. Bahadoran and Gohar [15] investigated the effects of different geometries of rollers on the EHL characteristics through experimental measurements, and they found that the film thickness is contractive near the roller ends. Kushwaha et al. [16,17] provided the EHL solutions of aligned and misaligned finite line contact. Then, they discussed the effect of transient conditions. Liu and Yang [18] developed the thermal EHL of finite line contacts under heavy loads. Based on the thermal EHL model, Yang and Yang [19] studied the lubrication performances of two tapered rollers located in the opposite orientations. Zhu et al. [20] presented a mixed EHL investigation considering realistic geometries (crowning, end corners and chamfers) and surface roughness effects on the behavior of finite line contact. Patir and Cheng [21,22] proposed the well-known average flow model to modify the Reynolds equation, which can consider the effect of surface roughness on lubrication performance. Then, Kogut and Etsion [23–25] presented the elastic–plastic model (KE model), which can calculate the asperity contact pressure in a statistic manner. Associated with this research into finite line contact and mixed lubrication, the modification of roller profiles has been studied and improved to reduce the end effect of stress concentration.

The famous logarithmic function to modify axial profile was introduced by Lundberg [26] in 1939. It can achieve the uniform pressure distribution of finite line contacts on the assumption of dry, static and elastic contact. Then, Johns and Gohar [27] improved Lundberg's function, and Fujiwara [28,29] put forward an optimized logarithmic profile. Cui and He [30] found a new logarithmic profile model of cylindrical roller bearings, which can avoid edge effects and increase the fatigue life of cylindrical roller bearings. Except for the logarithmic profile, the effects of other different types of profiles have also been investigated. Poplawski et al. [31] analyzed and compared four common roller profiles (flat roller profile, tapered crown roller profile, aerospace crown profile, and full crown roller profile) used in cylindrical roller bearing design and manufacturing. Najjari and Guilbault [32] studied the influence of seven common roller profile forms using the thermal EHL model. Recently, Zhang et al. [33] applied a numerical running-in method to modify the cylindrical roller profile, which enabled them to find an optimum profile leading to uniformly distributed asperity contact pressure in the roller axial direction without the need for any prior profile specifications.

It is worth noting that compared to cylindrical rollers, only a few studies considering the profile modification of tapered rollers [34,35] have been carried out, and these studies were only for single-roller contact, not for all tapered roller bearings. The present study attempts to employ the quasi-static model, the mixed lubrication model and the thermal effect equation to analyze the influences of different axial profiles on all tapered roller bearings. Based on the numerical running-in method recently proposed by Zhang et al. [33], a new asymmetric optimized profile of the tapered roller is found, which appears to be a useful profile form in respect to the elimination of the sharp contact pressure spikes near the ends of tapered rollers.

2. Mathematic Models

A quasi-static model and a mixed lubrication model are built for analyzing the mixed lubrication of a tapered roller bearing. The former is to find the load distribution inside a tapered roller bearing under given radial and axial bearing loads, while the latter is to analyze the mixed EHL of an individual tapered roller on a raceway. Moreover, the numerical running-in method for optimizing roller profiles is described in detail as well.

2.1. Quasi-Static Model of a Tapered Roller Bearing

Figure 1a shows the degrees of freedom (DOF) of a tapered roller bearing (TRB). The outer race (*OR*) is fixed in space with a center of O_B. The inner race (*IR*) has three DOFs; one of them is due to the bending moment M_b. Each rolling element (*RE*) has three DOFs as well. Thus, there are 3 + 3Z DOFs for computing the internal load distribution of the whole bearing, where Z is the number of rollers.

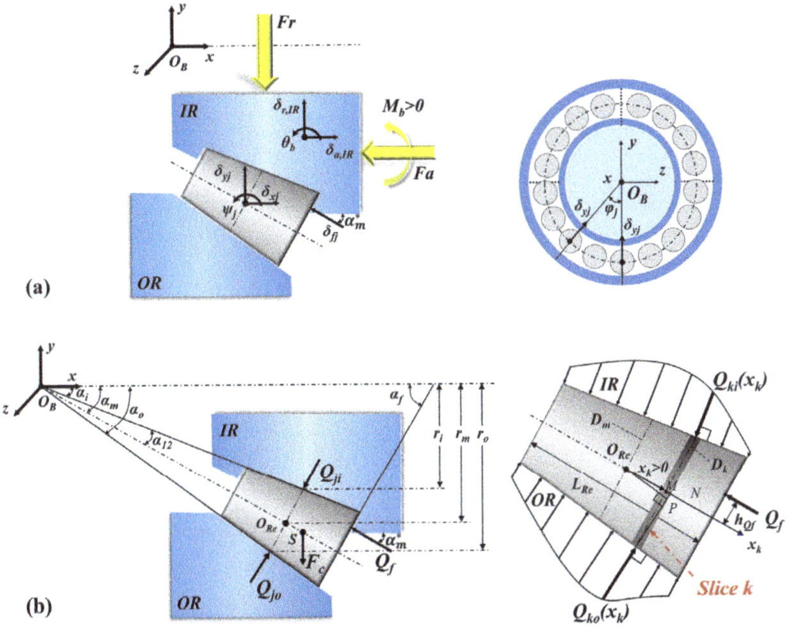

Figure 1. Description of the tapered roller bearing: (**a**) the degrees of freedom (DOF); and (**b**) the geometry features and the normal loads on a single element.

The geometry features and the normal loads on a single element are depicted in Figure 1b. Each roller is divided into n_s circular slices with the same thickness along the length L_{Re}. The resultant force Q_{ji} between the *IR* and the roller is the sum of normal loads acting on each slice.

$$Q_{ji}(j) = \sum_{k=1}^{n_s} Q_{ki}(k,j) = \frac{C_L}{n_s} \sum_{k=1}^{n_s} \widehat{\delta_{ki}}^{10/9} f_k(k) \quad (1)$$

in which C_L is the contact stiffness coefficient for two-side deformation of the roller on the *IR* and *OR*, $\widehat{\delta_{ki}}$ is the modified deformation on slice *k* of the *IR* of the roller #*j* and f_k is Reusner's correction factor of the load on slice *k* of the roller #*j*.

Similarly, the resultant force Q_{jo} between the *OR* and the roller can be written as

$$Q_{jo}(j) = \sum_{k=1}^{n_s} Q_{ko}(k,j) = \frac{C_L}{n_s} \sum_{k=1}^{n_s} \widehat{\delta_{ko}}^{10/9} f_k(k) \quad (2)$$

in which $\widehat{\delta_{ko}}$ is the modified deformation on slice *k* of the OR of the roller #*j*.

In the y-direction, the radial load F_r acting on the bearing equals to sum of the forces Q_{jo} acting on the OR and all rollers. Therefore, the nonlinear force balance equation relating to the unknown DOFs of $\delta_{r,IR}$ is expressed as

$$F_r - \frac{C_L}{n_s}\sum_{j=1}^{Z}\sum_{k=1}^{n_s}\widehat{\delta_{ko}}^{10/9} f_k(k)\cos\alpha_o\cos\varphi_j = 0 \tag{3}$$

where α_o is the outer race contact angle, and φ_j is the position angle of the roller #j.

Meanwhile, the axial load F_a acting on the bearing equals the sum of the forces Q_{ji} acting on the IR and all rollers in the x-direction. As a result, the nonlinear force balance equation relating to unknown DOFs $\delta_{a,IR}$ is given by

$$F_a - \frac{C_L}{n_s}\sum_{j=1}^{Z}\sum_{k=1}^{n_s}\widehat{\delta_{ko}}^{10/9} f_k(k)\sin\alpha_o = 0 \tag{4}$$

Each roller has a different bending moment M_{bj}. However, the sum of all moments M_{bj} equals the given bending moment M_b acting on the bearing in the direction z. Hence, the nonlinear moment balance equation relating to unknown DOFs θ_b, which is a function of bending deformation δ_{kM}, yields to

$$M_b - \frac{C_L{}'}{n_s}\sum_{j=1}^{Z}\sum_{k=1}^{n_s} l_{kM}\widehat{\delta_{kM}}^{10/9} f_k(k)\cos\varphi_j = 0 \tag{5}$$

where $C_L{}'$ is the contact stiffness coefficient for one-side deformation at the rib contact regime, l_{kM} is the moment arm, and $\widehat{\delta_{kM}}$ is the modified bending deformation on slice k.

Besides the above nonlinear equations relating to the three DOFs of IR, each roller also has three DOFs, δ_{yj}, δ_{xj} and ψ_j, in the x, y, and z directions, respectively. Using the force balance equations of all rollers, two sets of Z nonlinear equations in the directions of x and y can be written as

$$-Q_{ji}\cos\alpha_i + Q_{jo}\cos\alpha_o + Q_f\sin\alpha_m^* - F_c = 0 \tag{6}$$

$$-Q_{ji}\sin\alpha_i + Q_{jo}\sin\alpha_o - Q_f\cos\alpha_m^* = 0 \tag{7}$$

where α_i is the inner race contact angle, α_m^* is the rib contact angle and F_c is the centrifugal force.

Similarly, when the moments of all rollers are balanced in the direction z, the set of Z nonlinear equations for the DOF ψ_j are written as

$$\begin{gathered}(-\sum_{k=1}^{<n_{s,12}} l_{kL}Q_{ko} + \sum_{k\geq n_{s,12}}^{n_s} l_{kR}Q_{ko})\cos\varphi_j + (\sum_{k=1}^{<n_{s,12}} l_{kL}Q_{ki} - \sum_{k\geq n_{s,12}}^{n_s} l_{kR}Q_{ki})\cos\varphi_j - \\ F_c l_c\cos\varphi_j + Q_f h_{Qf}\cos\varphi_j + M_{bj}(j) = 0\end{gathered} \tag{8}$$

in which l_{kL} and l_{kR} are the moment arms $O_{Re}P$ shown in Figure 1b, $n_{s,12}$ equals $(n_s + 1)/2$, and l_c and h_{Qf} are the moment arm of the centrifugal force F_c and the rib force Q_f, respectively.

To sum up, a nonlinear equation system is derived by the 3 + 3Z equations written in Equations (3)–(8). The system describes a computational model that enables the calculation of the internal load distribution for a tapered roller bearing under given radial, axial and moment loads. The Levenberg and Marquardt method based on the Least Squares Method (LSM) was applied to solve the nonlinear equation system. More details of Equations (1)–(8) and the Levenberg and Marquardt method can be found in Reference [10].

2.2. Mixed Lubrication Model of Finite Line Contacts

After solving the normal forces acting on each roller in a TRB, a mixed lubrication model is needed to calculate the pressure, lubricant film thickness and temperature distributions within any roller/raceway contacts. Additionally, in the present study, TEHLs of finite line contacts are assumed under quasi-static and aligned conditions.

Taking a representative contact between a tapered roller and the IR for example, the governing equations for the mixed lubrication analysis are given as follows:

2.2.1. Velocity Relationship

Due to the contact angles of the TRB, the surface velocities in the contact area vary along the contact line. As shown in Figure 1b, the entrainment velocities of the roller–inner raceway can be expressed as

$$\vec{u_{ei}} = \frac{1}{2}\left[\omega_r \vec{r} + (\omega_i - \omega_c)\vec{r_i}\right] \tag{9}$$

where $\vec{r}$ and $\vec{r_i}$ are the radii of the roller and the inner race, ω_r is the angular velocity of the roller around its axis, ω_i is the angular velocity of the inner race, and ω_c is the angular velocity of the roller rotating with respect to the $O_B x$ axis. Assuming the outer race control and the slide–roll ratio s, ω_c and ω_r are expressed as

$$\begin{cases} \omega_c = \frac{1}{2}\frac{2-s}{2r_m+sr}\omega_i r_i \\ \omega_r = \frac{r_o}{r}\omega_c = \frac{1}{2}\frac{2-s}{2r_m+sr}\frac{r_o r_i}{r}\omega_i \end{cases} \tag{10}$$

where r equals $D_m/2$, r_i, r_m and r_o are shown in Figure 1b, and s is the slide–roll ratio between the roller and the IR.

2.2.2. Reynolds Equation

Based on the work of Patir and Cheng [21,22], the average flow Reynolds equation for rough surfaces is given as

$$\frac{\partial}{\partial x}(\Phi_x \frac{\rho h^3}{12\eta}\frac{\partial p_h}{\partial x}) + \frac{\partial}{\partial y}(\Phi_y \frac{\rho h^3}{12\eta}\frac{\partial p_h}{\partial y}) = u_{ei}\frac{\partial(\rho h)}{\partial x} + u_{ei} s R_q \frac{\partial \Phi_s}{\partial x} \tag{11}$$

in which Φ_x and Φ_y are the flow factors in the x-direction and y-direction, ρ and η are the density and the viscosity of the lubricant, respectively, h is the nominal film thickness, p_h is the hydrodynamic pressure, R_q is the composite standard deviation of roughness and Φ_s is the shear flow factor. In solving Equation (11), the pressure boundary conditions are written as

$$p_h(x_{in}, y) = p_h(x_{out}, y) = p_h(x, y_{in}) = p_h(x, y_{out}) = 0, \frac{\partial p_h(x_{out}, y)}{\partial x} = 0$$

2.2.3. Film Thickness Equation

For a finite line contact problem, the local lubricant film thickness h can be expressed as

$$h(x,y) = h_0 + g(x,y) + v(x,y) \tag{12}$$

where h_0 represents the approach between the two bodies, and $g(x,y)$ is due to the original geometry profile that can be calculated by

$$g(x,y) = R_x(y) - \sqrt{\delta(y) - x^2} \tag{13}$$

in which $R_x(y)$ is the equivalent radius along the contact line and $\delta(y)$ is the crown drop relating to the roller profile.

$v(x,y)$ is the sum of elastic deformations of contacting surfaces due to pressure, calculated by the well-known Boussinesq integration:

$$v(x,y) = \frac{2}{\pi E'} \iint_{\Omega} \frac{p_t(x',y')}{\sqrt{(x-x')^2 + (y-y')^2}} dx'dy' \tag{14}$$

where p_t is the sum of the hydrodynamic pressure p_h and the asperity contact pressure p_a.

However, Boussinesq integration is under half-space assumption, which leads to incorrect pressure increases near free edges. A correction factor, ψ, proposed by Guilbault [36], was applied to correct the shear and normal stress influence on displacements through the mirroring process and can be approximated by the following formulation:

$$\psi = 1.29 - \frac{1}{1-\nu}(0.08 - 0.5\nu) \tag{15}$$

in which ν is the Poisson's ratio.

2.2.4. Viscosity and Density Relationships

There are two types of lubricant in TRBs, either lubricating oils or greases. Grease consists of thickener and base oil, and its main advantages are ease in application and natural sealing ability. However, it is still difficult to model its rheological behavior under high shear rate, high pressure and non-constant temperature conditions precisely due to the complicated dynamics of its soap network in EHL contacts. Based on the experimental findings by Cen [37], the grease film thickness at EHL contacts can be calculated by using the base oil viscosity at higher entrainment velocities. Hence, the present study only considers base oil lubrication for the sake of simplicity.

In Equation (11), the oil lubrication viscosity is considered as a function of pressure and temperature, and one of the commonly used viscosity equations is the Roelands law [38]:

$$\eta(p_h,T) = \eta_0 \exp\left\{(\ln \eta_0 + 9.67)\left[-1 + (1 + 5.1 \times 10^{-9} p_h)^Z (\frac{T-138}{T_0-138})^{-S0}\right]\right\} \tag{16}$$

The density is also assumed to be dependent on pressure and temperature, usually expressed as [39,40]

$$\rho(p_h,T) = \rho_0 \left[1 + \frac{Ap_h}{1+Bp_h} + D(T-T_0)\right] \tag{17}$$

in which A, B and D are pressure–density coefficients. The parameters common to all results in this paper are: $A = 0.6 \times 10^{-9}\ \mathrm{Pa}^{-1}$, $B = 1.7 \times 10^{-9}\ \mathrm{Pa}^{-1}$ and $D = -0.00065\ \mathrm{K}^{-1}$.

In addition, oil lubrication usually behaves as a non-Newtonian fluid under high pressure and high shear rate. Hence, shear thinning should be considered in the TEHL calculation. Among several models of the shear thinning fluid behavior, the B-W model proposed by Bair and Winer [41] was used, written as

$$\dot{\gamma} = -\frac{\tau_{\lim}}{\eta(p_h,T)} \ln(1 - \frac{\tau}{\tau_{\lim}}) \tag{18}$$

where $\dot{\gamma}$ is the shear rate, τ is the shear stress and $\tau_{\lim}$ is the limiting shear stress, which can be illustrated as [42]

$$\tau_{\lim} = (\tau_{10} + \gamma_l p_h)\exp(\beta_l(\frac{1}{T} - \frac{1}{T_0})) \tag{19}$$

in which τ_{10} is the initial limiting shear stress, γ_l is the pressure coefficient corresponding to maximum friction coefficient and β_l is the temperature coefficient.

2.2.5. Asperity Contact Model

To calculate the asperity contact pressure under mixed lubrication, the stochastic rough surface contact model proposed by Kogut and Etsion (KE model) was used [23–25]. This model accounts for the elastic, first elastic–plastic, second elastic–plastic, and fully plastic deformation of asperities, and it can be written as

$$p_a = \frac{2}{3}\pi\beta_s K\omega_c^* H_d\left(\int_{d^*}^{d^*+\omega_c^*} I^{1.5} + 1.03\int_{d^*+\omega_c^*}^{d^*+6\omega_c^*} I^{1.425} + 1.4\int_{d^*+6\omega_c^*}^{d^*+110\omega_c^*} I^{1.263} + \frac{3}{K}\int_{d^*+110\omega_c^*}^{\infty} I^{1}\right)$$
$$\begin{cases} I^b = \left(\frac{z^*-d^*}{\omega_c^*}\right)^b \Phi^*(z^*)dz^*, K = 0.454 + 0.41\nu \\ \omega_c^* = \frac{\omega_c}{\sigma_0} = \frac{R_{as}}{\sigma_0}\left(\frac{\pi K H_d}{E'}\right)^2, d^* = h - \frac{1}{\sqrt{48\pi\beta_s}} \end{cases} \tag{20}$$

where β_s is the surface roughness parameter, which usually equals 0.05 [43], the hardness coefficient K is related to Poisson's ratio of the softer material (see CEB friction model [44]), ω_c^* is the critical interference value of the elastic and the elastoplastic deformation regime, R_{as} is the mean radius of asperity, H_d is the hardness of the softer material, z^* is the asperity height, $\Phi^*(z^*)$ is the asperity heights probability density function which is assumed to be Gaussian and d^* is the asperity separation. All the dimensionless values are normalized by the standard deviation of roughness σ_0 and denoted by *.

2.2.6. Thermal Effect

The calculation of 3D temperature distribution follows the energy equation within a mixed lubrication contact, which is given by

$$c\rho\left(u\frac{\partial T}{\partial x} + v\frac{\partial T}{\partial y} - w\frac{\partial T}{\partial z}\right) = k\frac{\partial^2 T}{\partial z^2} - \frac{T}{\rho}\frac{\partial \rho}{\partial T}\left(u\frac{\partial \rho}{\partial x} + v\frac{\partial \rho}{\partial y}\right) + \eta\left[\left(\frac{\partial u}{\partial z}\right)^2 + \left(\frac{\partial v}{\partial z}\right)^2\right] + \tau_a\sqrt{\left(\frac{\partial u}{\partial z}\right)^2 + \left(\frac{\partial v}{\partial z}\right)^2} \tag{21}$$

where c and k are the specific heat and conductivity of the lubricant, respectively, the shear stress of the asperity contact τ_a equals $\mu_a p_a$ and μ_a is the friction coefficient of the asperity contact.

The boundary conditions for surface temperature are written as

$$\begin{cases} T(x,y,0) = \frac{k_1}{\sqrt{\pi\rho_1 c_1 k_1 u_1}}\int_{-\infty}^{x} \left.\frac{\partial T}{\partial z}\right|_{x,y,0} \frac{ds}{\sqrt{x-s}} + T_0 \\ T(x,y,h) = \frac{k_1}{\sqrt{\pi\rho_2 c_2 k_2 u_2}}\int_{-\infty}^{x} \left.\frac{\partial T}{\partial z}\right|_{x,y,h} \frac{ds}{\sqrt{x-s}} + T_0 \end{cases} \tag{22}$$

2.2.7. Load Balance

In mixed lubrication, an external normal load acting on the roller/IR contact is balanced by the hydrodynamic pressure and asperity contact pressure together as follows:

$$Q_{ji} = \iint_{\Omega} p_h dxdy + \iint_{\Omega} p_a dxdy \tag{23}$$

where Q_{ji} is the external normal load, and Ω means the contact region.

2.3. Numerical Running-In Method

Running-in is a traditional and conventional technique widely used in industry for the improvement of the conformity of sliding and rolling contacts of machine components at the microscale under appropriate mild conditions in the initial operation phase. After a proper running-in process, the surface topographical, physical and chemical structures, including profiles, surface roughness, compositions and microstructures, transform from their initial as-finished forms due to wear, tribochemical reactions and phase transformations occurred during running-in too in some conditions. Differing from conventional running-in, which is carried out in a real machine system in the operation stage, numerical running-in aims to achieve a similar effect on modifications of surface profile as that of a physical running-in by the means of computer simulations in the design stage of machine elements. The approach was proposed by authors and first applied to the modification of cylindrical roller profile [33]. In the present paper, the numerical running-in method is used in tapered roller bearings to seek the optimal design of a tapered roller profile.

As presented in Reference [33], the wear depth in each step of iterations can be determined with the aid of Archard's wear law as follows:

$$\Delta w_d(x,y) = K_w \frac{p_a(x,y)\Delta S}{H_d} \tag{24}$$

where Δw_d is the wear depth increment, K_w is the wear coefficient, ΔS is the sliding distance increment and H_d is the material hardness.

When K_w, ΔS and H_d are constants, the wear depth increment is proportional to the asperity contact pressure. The optimized crown drop at position y is updated step-by-step, as shown below, until the calculated asperity contact pressure $p_a(0,y)$ meets a set convergence condition.

$$\delta''(y) = \delta'(y) + \Delta w_d(0,y) \tag{25}$$

in which $\delta'(y)$ is the crown drop at position y in the previous step and $\delta''(y)$ is the updated one.

The flowchart of the optimization process for the crowning profile is shown in Figure 2. At first, the internal load distribution in a TRB is calculated by using the quasi-static model described in Section 2.1 above. Then, the roller bearing the maximum load is focused on, and its pressure, lubricant film thickness and temperature distributions are analyzed with the mixed lubrication model in Section 2.2. The third step is to check the ratio of the maximum asperity contact pressure, $\max(p_a(0,y))$, to the mean asperity contact pressure, $\text{mean}(p_a(0,y))$. If the ratio is greater than 1.01, Equations (24) and (25) are used to update the crown drop at that point. Meanwhile, the minimum contact pressure, $\min(p_a(0,y))$, is also compared with the mean asperity contact pressure, $\text{mean}(p_a(0,y))$, and if it is less than the mean value, a negative crown drop is applied. Finally, a profile with uniform asperity contact pressure can be obtained, which is the optimal crowning profile.

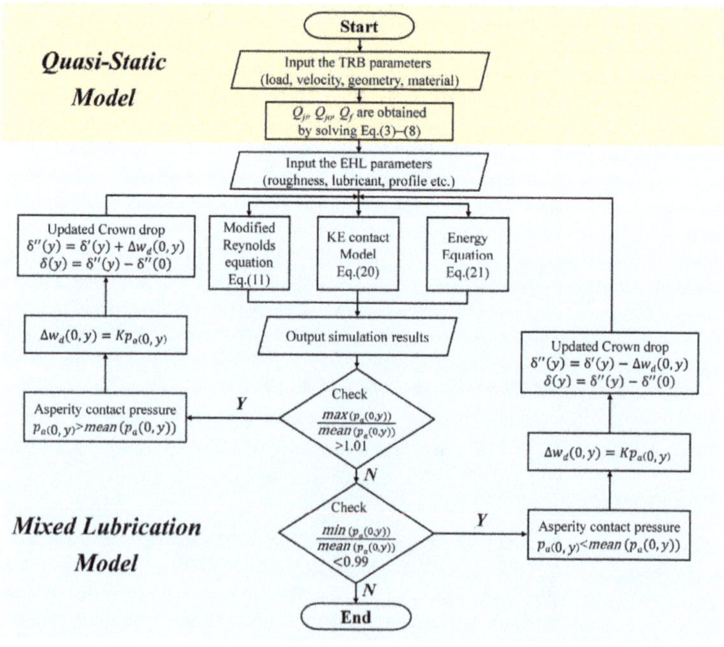

Figure 2. The flowchart of the running-in optimization of roller profile in TRB.

3. Numerical Model Validation

In Reference [34], Yang calculated the distributions of the EHL pressure and lubricant film thickness for three tapered rollers with different profiles rolling on an infinite plane based on the following dimensionless input parameters: $\overline{W} = 3 \times 10^{-5}$, $\overline{U} = 2 \times 10^{-11}$.

To validate the mixed lubrication model used in the study, numerical EHL simulations were performed with the same values of input parameters (shown in Table 1) as those in Reference [30]. As shown in Figure 3, the simulation results of the present study (denoted as current results in the figure) and Yang's results are generally in good agreement, except that the calculated lubricant film thickness is slightly smaller. The discrepancy is attributed to the thermal effect described in the present study (the main thermal parameters are listed in Table 2), while the EHL analysis in Reference [30] was isothermal. We can see that both pressure and film thickness distributions are asymmetric, owing to the difference in size between the bigger and smaller ends of the tapered rollers, and that the crown and dub-off profiles give rise to more uniform distributions of lubricant film thickness and pressure near the ends than the chamfer profile.

Table 1. Input parameters for the analyzed cases [34].

Parameter	Value	Parameter	Value
Half length of rollers, l, mm	20	Effective elastic modulus, E', GPa	228
Radius of rollers on the section $y = 0$, r, m	0.02	Material parameter, G, dimensionless	5000
Deflective angles of tapered rollers, β, (°)	10	Nominal maximum Hertzian pressure, P_H, GPa	0.5
Ambient viscosity of lubricant, η_0, Ns/m^2	0.08	Angular velocities of rollers, ω_a, rad/s	9.86
Ambient density of lubricant, ρ_0, kg/m^3	870	Angular velocities of plane, ω_b, rad/s	56.4

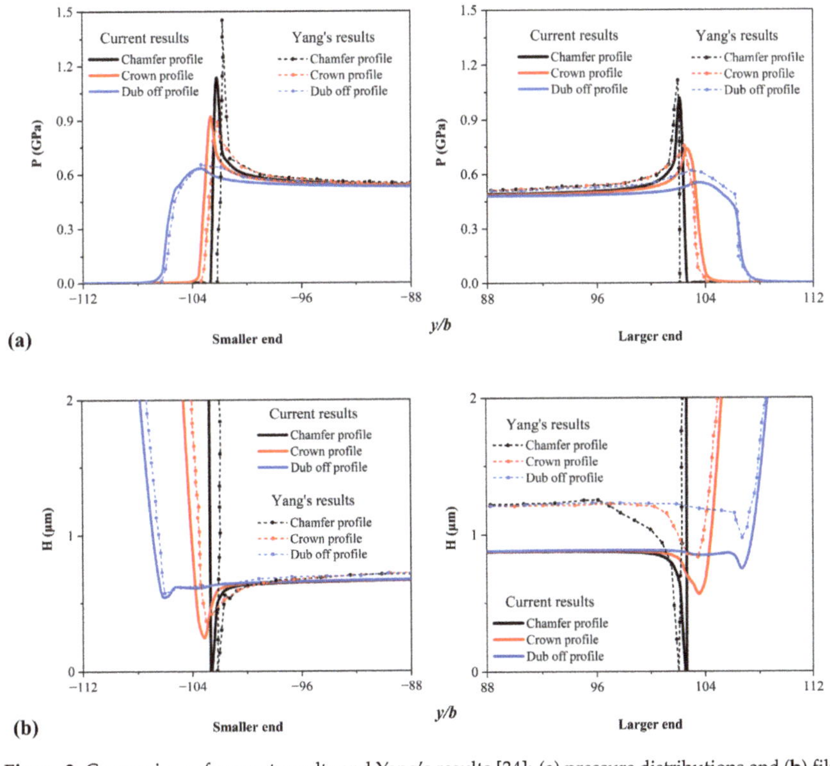

Figure 3. Comparison of current results and Yang's results [34]: (**a**) pressure distributions and (**b**) film thickness distributions.

Table 2. Parameters of the thermal effect of current model.

Parameter	Value
Ambient temperature, T_0, K	313
Specific heat of lubricant, c, J/kg K	1880
Specific heat of solids, c_1 and c_2, J/kg K	460
Thermal conductivity of lubricant, k, W/m K	0.145
Thermal conductivity of solids, k_1 and k_2, W/m K	46
Density of the solids, ρ_a and ρ_b, kg/m^3	7850
Thermos-viscosity index, β, K^{-1}	0.0585

4. Results and Discussion

4.1. Calculation Parameters of the Axial Box TRBs

TRBs are generally used in the axial boxes of trains on high-speed railways. Typical operation conditions of axial-box TRBs are listed in Table 3. The thermal parameters are the same as in Table 2, and the bearing rotation speeds corresponding to the steady running speeds of the high-speed railway are shown in Table 4.

Table 3. Main parameters of the TRB used in high-speed railway.

Parameter	Value	Parameter	Value
Constant axial load, F_a, KN	15	Length of rollers, L_{Re}, mm	50
Varying radial load, F_r, KN	10–50	Radius of rollers on the section $y = 0$, r, mm	13
Bearing rotation speed, N, 10^3 rpm	1–3	Mean pitch radius of bearing, r_m, mm	92.5
Bending moment, M_b, Nm	50	Outer race contact angle, α_o, (°)	12
Number of bearing rollers, Z	17	Inner race contact angle, α_i, (°)	9
Bearing diametric clearance, P_d, µm	30	Ambient temperature, T_0, K	353
Effective elastic modulus, E', GPa	226	Ambient viscosity of lubricant, η_0, Ns/m^2	0.014
Poisson's v ratio,	0.3	Composite standard deviation of roughness, R_q, µm	0.5
Slide–roll ratio, s	0.05	Hardness of the softer material, H_d, GPa	4.04
Material parameter, G, dimensionless	4241	Mean radius of asperity, R_{as}, µm	10

Table 4. The relationship between the bearing rotation speed and the high-speed railway speed.

The Bearing Rotation Speed	The High-Speed Railway Speed
1000 rpm	152 km/h
1500 rpm	228 km/h
2000 rpm	304 km/h
2500 rpm	380 km/h
3000 rpm	456 km/h

4.2. Internal Load Distributions

Figure 4 shows the calculated results of contact loads acting on each roller (half of rollers numbered as No.1 to No.9 are displayed for the sake of symmetry) under different radial loads and bearing rotation speeds. As the radial load increases, the rollers at the bottom bear higher loads no matter whether they contact with the inner or outer race. In contrast, the rollers at the top bear lighter loads. A minor difference in contact load between the inner and the outer raceways is caused by the centrifugal force F_c. As the bearing rotation speed increases, no significant change happens in internal load distributions, but huge differences in the TEHL results appear as well as the roller profile modifications, which are discussed later.

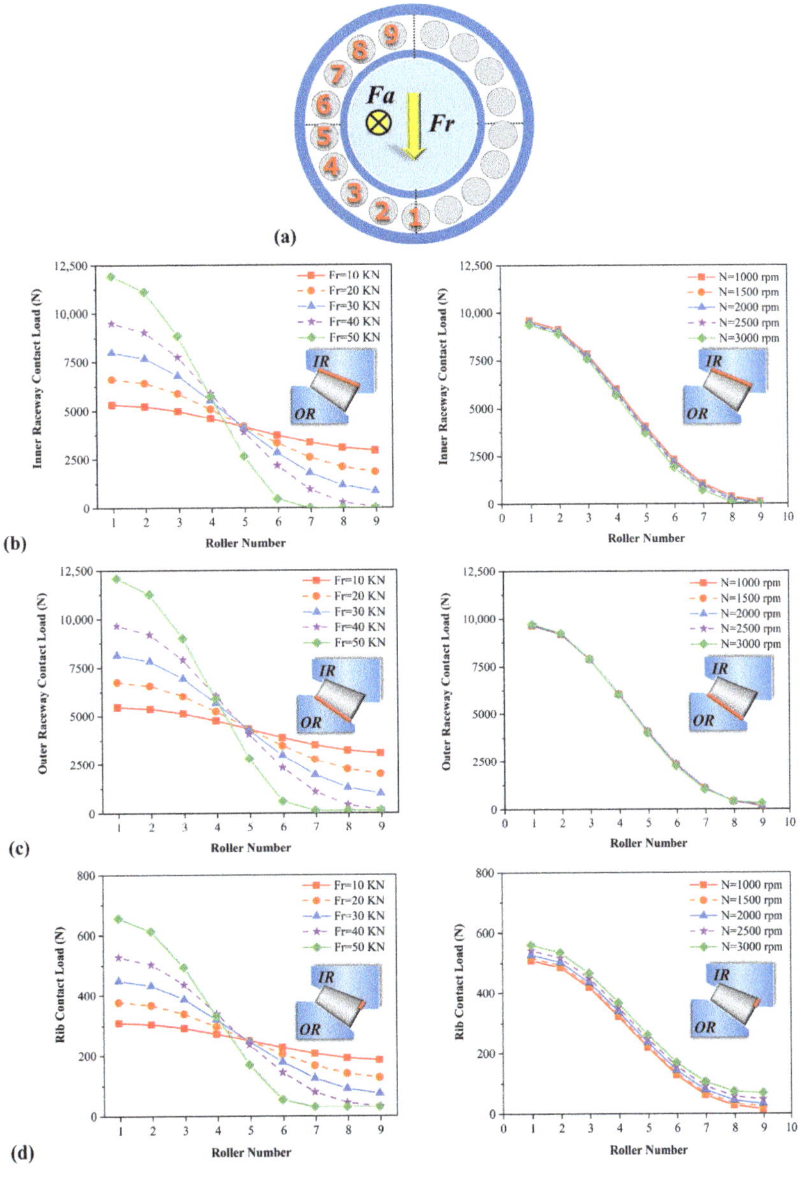

Figure 4. Contact loads between each component with varying radial loads and bearing rotation speeds: (**a**) description of roller number; (**b**) inner raceway contact load; (**c**) outer raceway contact load; and (**d**) rib contact load.

4.3. Effect of Different Axial Roller Profiles

In this section, the simulation results of the mixed lubrication of roller No.1 on the inner raceway—which is under the heaviest load among the all rollers—at typical working conditions ($F_a = 15$ KN, $F_r = 40$ KN, $N = 2000$ rpm) are presented, and the effect of roller profiles is demonstrated with three different forms: dub-off, logarithmic and running-in optimized profiles.

Figure 5a shows the diagrams of different roller profiles. For the roller with a dub-off profile, a rounding radius R is set as 30 mm and the rounding width L_d as 2.5 mm. The logarithmic profile is well-known for its uniformity of pressure distribution under static line contact conditions. It was proposed by Lundberg [26] and improved by Johns and Gohar [27], which can be expressed as a function

$$\delta(y) = \frac{W_r}{\pi L E'} \ln \frac{1}{1-(1-0.6066b/L)(2y/L)^2} \tag{26}$$

where W_r is the load acting on the roller and the inner raceway, L is the contact length, and b is the half width of the nominal Hertzian line contact.

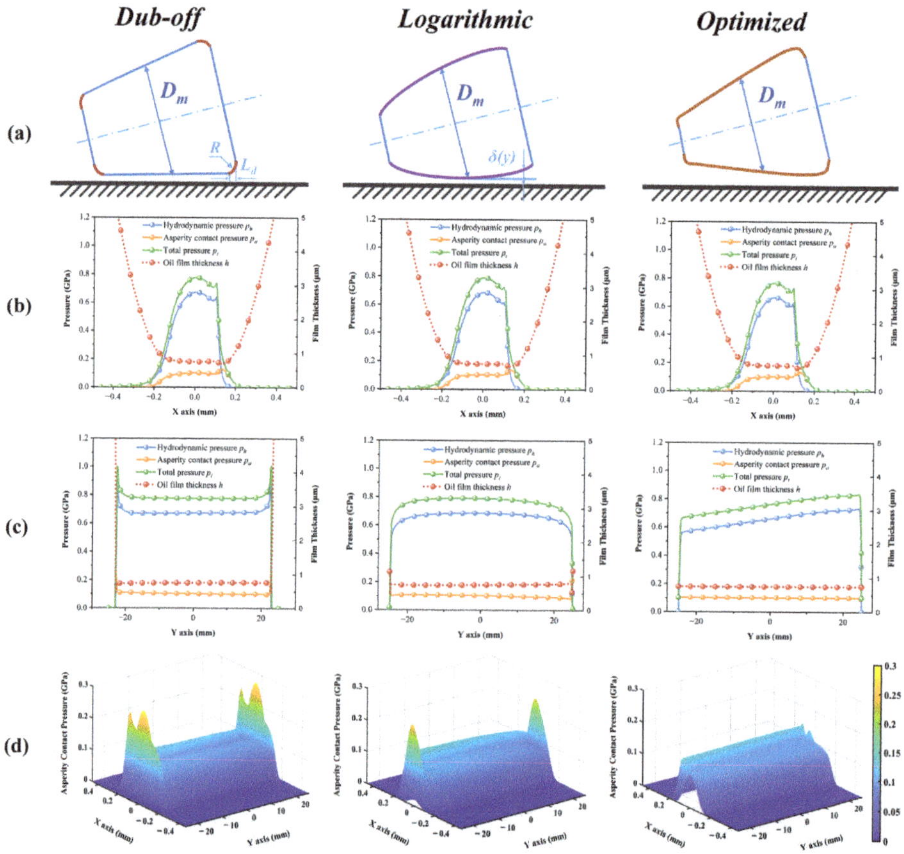

Figure 5. Pressure distributions and oil film thicknesses with different axial roller profiles under W = 7.513 × 10^{-5}, U = 5.367 × 10^{-11}, P_H= 0.78 GPa: (**a**) diagrams of different roller profiles; (**b**) rolling direction, y = 0; (**c**) axial direction, x = 0; and (**d**) 3D images of asperity contact pressure.

As shown in Figure 5b, there are no obvious differences between the three axial profiles along the rolling direction (y = 0). The maximum total pressure is about 0.7 GPa and close to the maximum Hertzian contact pressure. However, Figure 5c,d indicate that the differences in contact pressure are huge along the axial direction (x = 0). For the dub-off profile, no matter whether the pressure is hydrodynamic or asperity, it increases sharply at the edges. The maximum total pressure is about 0.9 Gpa, which is larger than the maximum Hertzian contact pressure. Though the hydrodynamic pressure does not rise sharply near the end with the logarithmic profile, the asperity contact pressure still has an obvious spike, which would result in severe local wear. Starting from the initial logarithmic profile, the axial profile was modified step by step by using the numerical running-in method described in

Section 2.3. The comparison is shown in Figure 6. Obviously, for the profile after running-in (denoted as the optimized profile in the figure), the spikes of asperity contact pressure in the vicinity of the roller ends are eliminated, resulting in a much more uniform asperity contact pressure distribution in the axial direction.

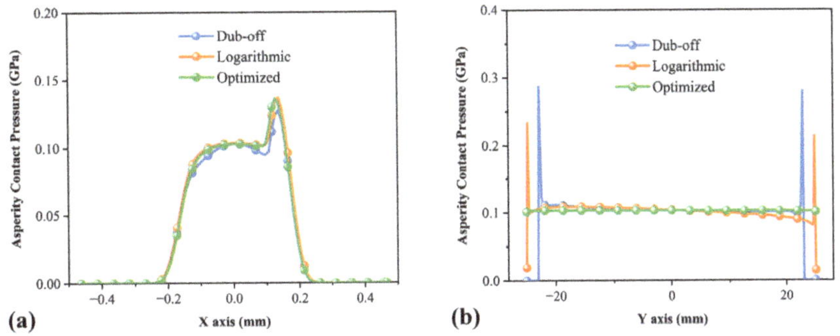

Figure 6. Asperity contact pressure distributions with different axial roller profiles: (**a**) rolling direction, $y = 0$; and (**b**) axial direction, $x = 0$.

4.4. Influences on Bearing Performance

To reveal the effect of profiles on bearing performance, all rollers in a TRB are assumed to be crowned in the same axial profile as roller No.1 with either dub-off, logarithmic or running-in optimized profiles, and the mixed lubrication of all rollers in a TRB was analyzed, respectively.

The asperity contact pressure distributions of each roller along the rolling direction are presented in Figure 7a–c. Like roller No.1 discussed in Section 4.3, the asperity contact pressure distributions of the other rollers have no obvious difference, although the profiles change. It should be noted that from roller No.1 to No.9, the contact load decreases, and hence, the contact half-width becomes smaller and asperity contact pressure becomes lower, which means that the lubrication state of the rollers changes from mixed lubrication to full-film lubrication.

Figure 7d–f display the distributions along the axial direction of the rollers. For the bearing of rollers with the dub-off profile, only the rollers with light loads have no contact pressure spikes near the ends, and the asperity contact pressures of the other rollers are non-uniform. Compared with the dub-off profile, the logarithmic profile gives rise to lower edge contact pressure, but most of the rollers still have non-uniform contact pressure. Because of the deflective angles of tapered rollers, the entrainment velocities along the axial direction are different, and the pressures near the smaller ends are slightly higher than those at the larger end. In contrast to the profiles mentioned above, it is demonstrated that the profile modified by the running-in method is beneficial to reducing the end effect of stress concentration for all of the rollers, not just for roller No.1.

In addition, we also compared the maximum asperity contact pressures and standard deviations of the contact pressure of all rollers among the three profiles. As shown in Figure 8, it is clear that the optimized profile is advantageous in both aspects over the other ones, especially in respect to the standard deviations, which means that the asperity contact pressure distributions are smoother. However, the effect of the optimized profile for roller Nos. 7 and 8 is worse than that of the logarithmic profile. The differences between the two types of profiles are small and ignorable because the loads acting on the two rollers are relatively light.

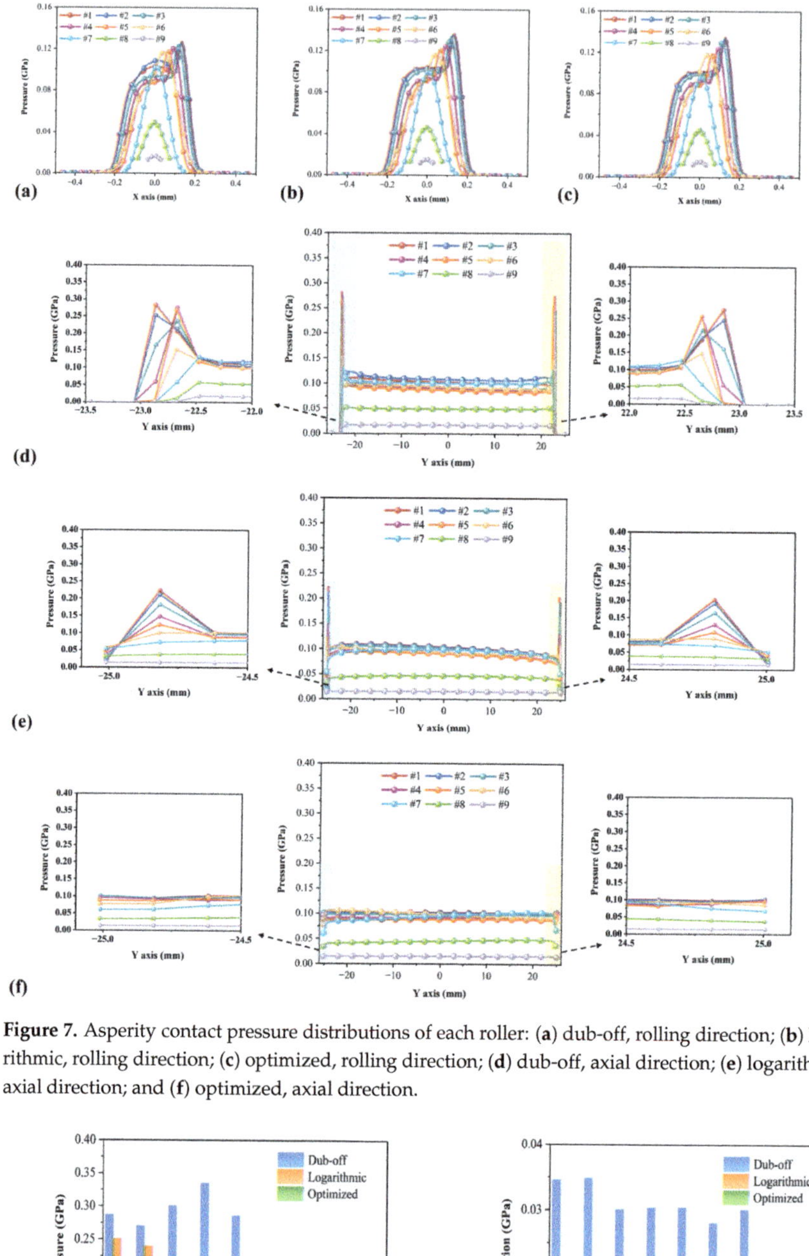

Figure 7. Asperity contact pressure distributions of each roller: (**a**) dub-off, rolling direction; (**b**) logarithmic, rolling direction; (**c**) optimized, rolling direction; (**d**) dub-off, axial direction; (**e**) logarithmic, axial direction; and (**f**) optimized, axial direction.

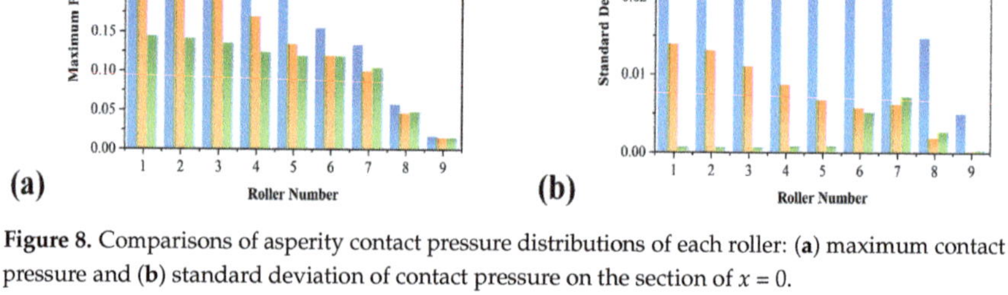

Figure 8. Comparisons of asperity contact pressure distributions of each roller: (**a**) maximum contact pressure and (**b**) standard deviation of contact pressure on the section of $x = 0$.

4.5. Effects of Other Factors on Asperity Contact Pressure Distributions

4.5.1. Radial Load (Fr) Effect

The radial load has a critical influence on mixed lubrication and asperity contact pressure distributions. Figure 9a indicates that the roller bearing under heavier loads needs larger crown drops at the smaller ends. Figure 9b–d compare the pressure distributions of the three profiles under different radial loads $Fr = 10$ KN, 30 KN, 50 KN. The contact pressures near the ends become higher as the radial load increases. In all cases, the optimized profile shows the best performance and effectively reduces the end effect.

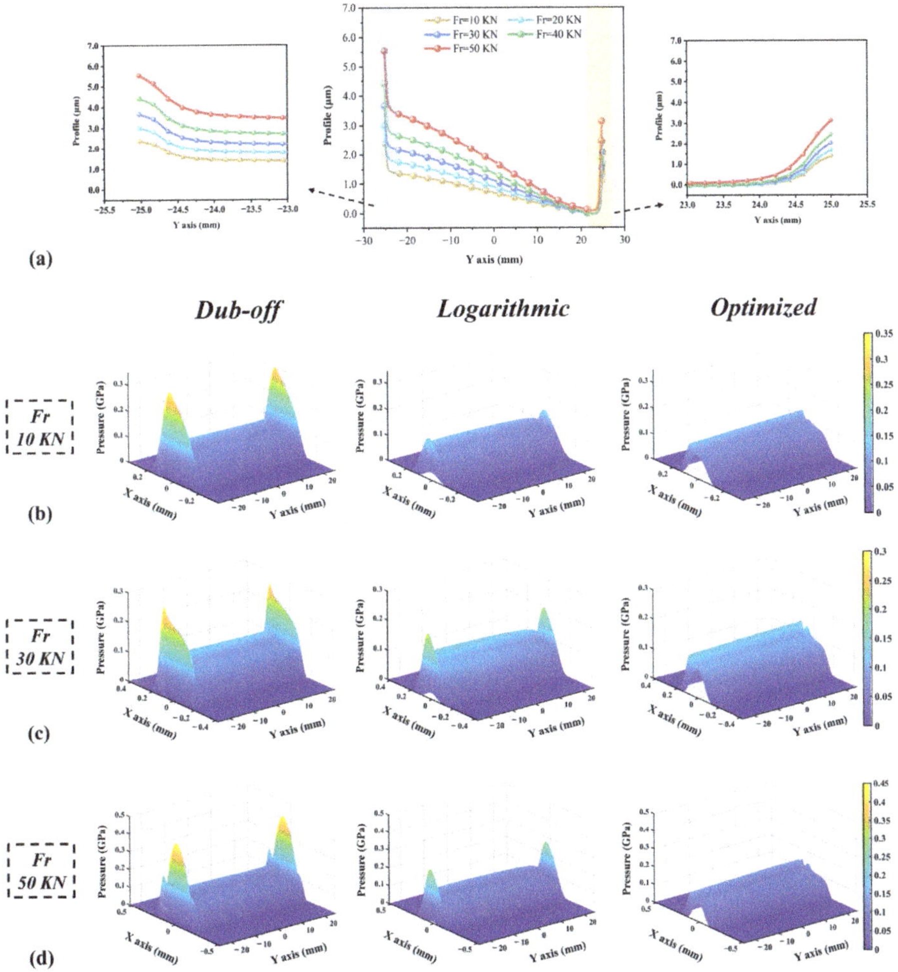

Figure 9. The effect of the radial load (Fr) under $N = 2000$ rpm, $Rq = 0.5$ μm: (**a**) The comparison of the profiles with different Fr; (**b**) contact pressures under $Fr = 10$ KN; (**c**) contact pressures under $Fr = 30$ KN; and (**d**) contact pressures under $Fr = 50$ KN.

4.5.2. Rotation Speed (N) Effect

In addition to the radial load, the rotation speed is also an important factor in determining the lubrication state. As discussed in Section 4.2, the increase in rotation speed has little influence on load distribution but has a great effect on the modified profile and TEHL performance, as shown in Figure 10. We can see that the higher the speed is, the larger the crown drop and the lower the contact pressure. Obviously, the optimized profile is better than the other profiles in terms of asperity contact pressure uniformity.

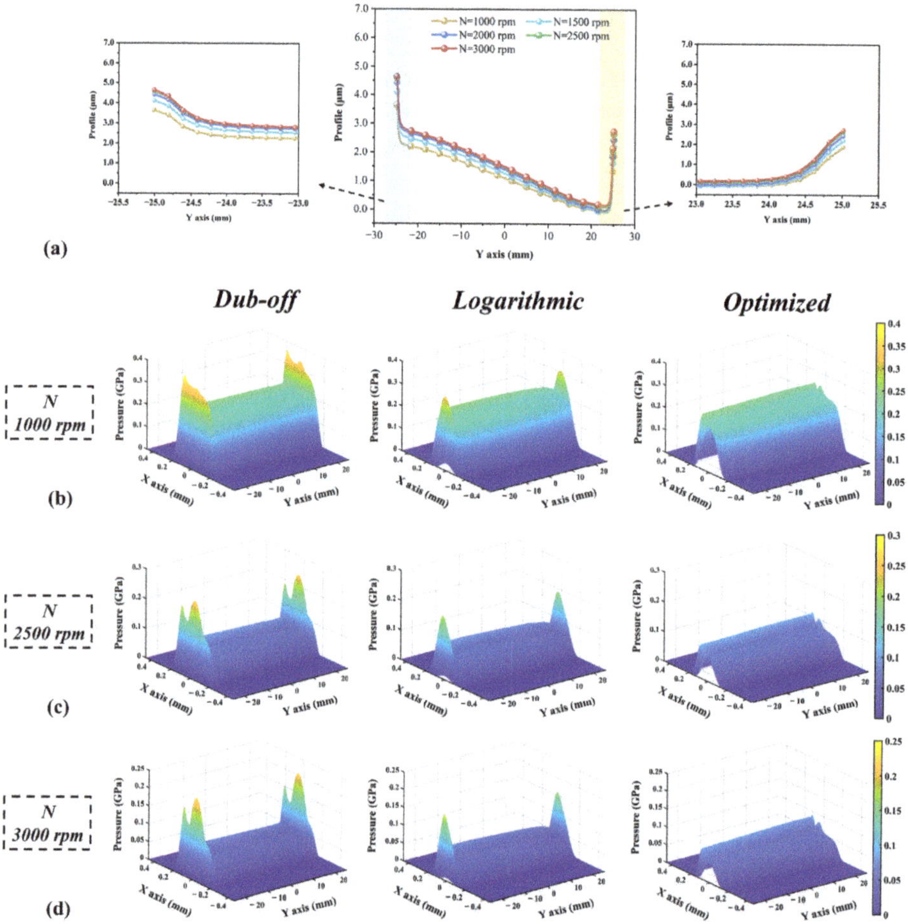

Figure 10. The effect of the rotation speed (N) under Fr = 40 KN, $Rq = 0.5$ μm: (**a**) The comparison of the profiles with different N; (**b**) contact pressures under $N = 1000$ rpm; (**c**) contact pressures under $N = 2500$ rpm; and (**d**) contact pressures under $N = 3000$ rpm.

4.5.3. Standard Deviation of Roughness (Rq) Effect

The optimized profiles are related to standard deviations of roughness, as shown in Figure 11a. When Rq equals 0.2 μm, non-uniform contact pressure only appears at the ends of the roller. It can be also found that with a smaller Rq, a larger crown drop is needed to achieve smoother pressure distributions. Moreover, due to the asymmetry of the tapered roller, the smaller end needs a larger amount of modification than the larger end. It is interesting to note that for the larger end (profile < 0), padding is needed instead of trimming to achieve a smooth asperity contact pressure distribution, as shown in Figure 11b–d.

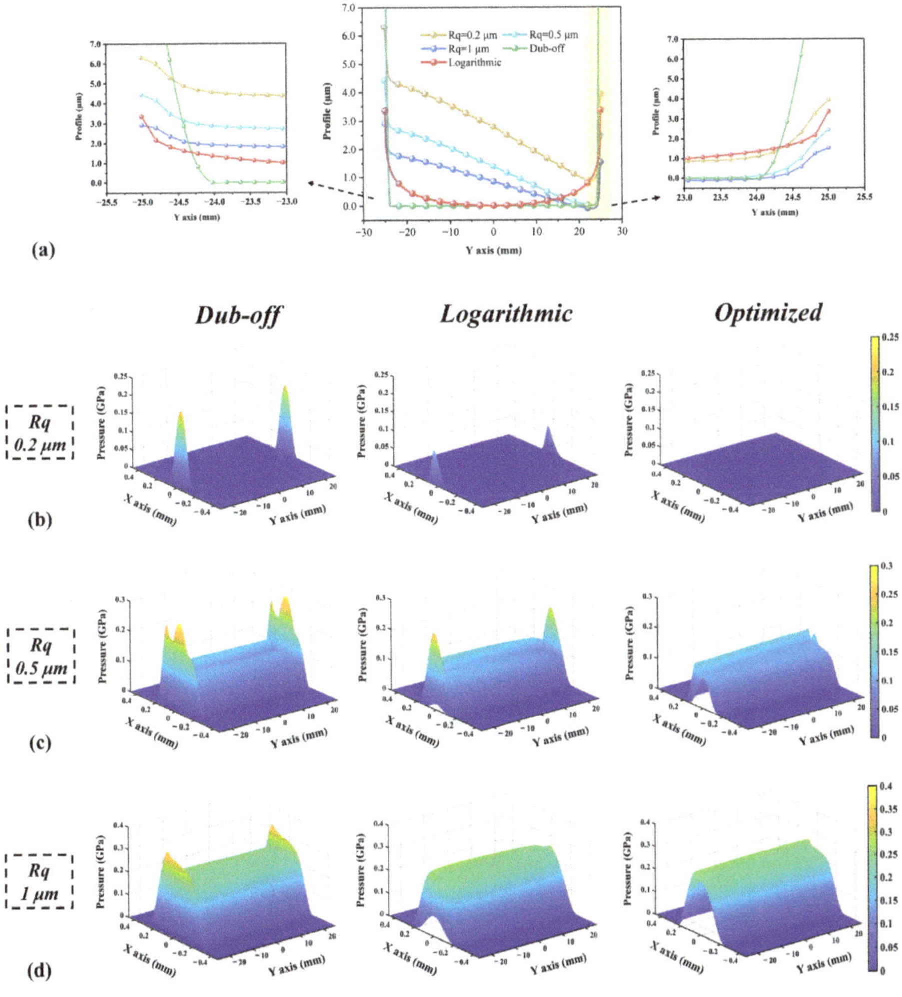

Figure 11. The effect of the standard deviation of roughness (Rq) under Fr = 40 KN, N = 2000 rpm: (**a**) The comparison of the profiles with different Rq; (**b**) contact pressures under $Rq = 0.2$ μm; (**c**) contact pressures under $Rq = 0.5$ μm; and (**d**) contact pressures under $Rq = 1$ μm.

5. Conclusions

For the purposes of making the asperity contact pressure of tapered rollers uniform and reducing the sharp spikes of the contact pressure in the vicinity of the roller ends, a numerical running-in method is applied, and a new asymmetric optimized profile is found.

Compared with the traditional profiles, it is clear that the optimized profile has immense advantages in terms of the asperity contact pressure uniformity and the elimination of the end effect of stress concentration not only for a single tapered roller but also for all tapered roller bearings. The numerical running-in method is proved to be suitable for various working conditions, and the optimized roller profile always shows better performance and effectively reduces the end effect. It is hoped that this method will be helpful in the improvement of tapered roller bearing design in the bearing industry.

Author Contributions: Conceptualization, Y.M.; methodology, R.C. and Y.Z.; software, R.C., H.B. and H.C.; validation, R.C.; formal analysis, R.C., H.B. and H.C.; writing—original draft preparation, R.C.; writing—review and editing, Y.M.; project administration, Y.M. All authors have read and agreed to the published version of the manuscript.

Funding: This research was partially supported by the National Natural Science Foundation of China, grant number 51635009.

Data Availability Statement: For more detailed data, please request data from the corresponding author or the first author.

Conflicts of Interest: The authors declare no conflict of interest.

Nomenclature

b	half width, m	u, u_1, u_2	rolling speed, m/s
c, c_1, c_2	specific heat of lubricant, upper solid and lower solid, J/(kg·K)	$\vec{u}_{ei}$	entrainment velocities of roller–inner raceway, m/s
$C_L, C_L{}'$	contact stiffness coefficient, N/m$^{10/9}$	$\overline{U}$	dimensionless velocity parameter
d^*	dimensionless asperity separation	$v(x,y)$	elastic deformation, m
D_m	pitch diameter, m	W_r	applied load, N
E'	effective elastic modulus, Pa	$\overline{W}$	dimensionless load parameter
f_k	Reusner's correction factor	z^*	dimensionless asperity height
F_a	axial load, N	Z	number of rollers
F_c	centrifugal force, N	α_i, α_o	race contact angle, (°)
F_r	radial load, N	α_m^*	rib contact angle, (°)
$g(x,y)$	original geometry profile, m	β	thermos–viscosity index
G	dimensionless material parameter	β_l	temperature coefficient
h	nominal film thickness, m	β_s	surface roughness parameter
h_0	approach between the two bodies, m	$\dot{\gamma}$	shear rate, s^{-1}
H_d	hardness of the softer material, Pa	γ_l	pressure coefficient
k, k_1, k_2	thermal conductivity of lubricant, upper solid and lower solid, W/(m·K)	ΔS	sliding distance interval, m
K	hardness coefficient, 0.454 + 0.41 v	Δw_d	wear depth interval, m
K_w	wear coefficient	$\delta(y)$	crown drop, m
l, h_{Qf}	moment arm, m	δ_{kM}	bending deformation, m
L	contact length, m	$\delta_{r,IR}, \delta_{a,IR}, \theta_b$	DOF of *IR*
L_d	rounding width, m	$\delta_{yj}, \delta_{xj}, \psi_j$	DOF of roller #*j*
L_{Re}	length of roller, m	$\widehat{\delta_{ki}}, \widehat{\delta_{ko}}$	modified deformation on the slice k, m
M_b	bending moment, N·m	η	viscosity of lubricant, Pa·s
n_s	circular slices	η_0	ambient viscosity of lubricant, Pa·s
N	bearing rotation speed, rpm	μ_a	coefficient of asperity contact
O_B	center of bearing	v	Poisson's ratio
p_a	asperity contact pressure, Pa	ρ	density of lubricant, kg/m^3
p_h	hydrodynamic pressure, Pa	τ	shear stress, Pa
p_t	total pressure, Pa	τ_a	shear stress of asperity contact, Pa
P_d	bearing diametric clearance, m	$\tau_{\lim}$	limiting shear stress, Pa
P_H	nominal maximum Hertzian pressure, Pa	τ_{10}	initial limiting shear stress, Pa
Q_{ji}, Q_{jo}	normal load between roller and race, N	φ_j	position angle of the roller, (°)
$\vec{r}, \vec{r_i}$	radius of roller and inner race, m	Φ_s	shear flow factor
R	rounding radius, m	Φ_x, Φ_y	flow factors
R_{as}	mean radius of asperity, m	$\omega_c, \omega_i, \omega_r$	angular velocities, rad/s

R_q, σ_0	composite standard deviation of roughness, m	ω_c^*	dimensionless critical interference
$R_x(y)$	equivalent radius, m	ψ	correction factor
s	slide–roll ratio	Ω	contact regime
T_0	ambient temperature, K		

References

1. Hong, S.-W.; Tong, V.-C. Rolling-element bearing modeling: A review. *Int. J. Precis. Eng. Manuf.* **2016**, *17*, 1729–1749. [CrossRef]
2. Cao, H.; Niu, L.; Xi, S.; Chen, X. Mechanical model development of rolling bearing-rotor systems: A review. *Mech. Syst. Signal Process.* **2018**, *102*, 37–58. [CrossRef]
3. Stacke, L.-E.; Fritzson, D.; Nordling, P. BEAST—A rolling bearing simulation tool. *Proc. Inst. Mech. Eng. Part K J. Multi-body Dyn.* **1999**, *213*, 63–71. [CrossRef]
4. Rahnejat, H.; Gohar, R. Design of profiled taper roller bearings. *Tribol. Int.* **1979**, *12*, 269–275. [CrossRef]
5. Creju, S.; Bercea, I.; Mitu, N. A dynamic analysis of tapered roller bearing under fully flooded conditions part 1: Theoretical formulation. *Wear* **1995**, *188*, 1–10. [CrossRef]
6. Creţu, S.; Mitu, N.; Bercea, I. A dynamic analysis of tapered roller bearings under fully flooded conditions part 2: Results. *Wear* **1995**, *188*, 11–18. [CrossRef]
7. Yamashita, R.; Dowson, D.; Taylor, C.M. An Analysis of Elastohydrodynamic Film Thickness in Tapered Roller Bearings. *Tribol. Ser.* **1997**, *32*, 617–637. [CrossRef]
8. Zheng, J.; Ji, J.; Yin, S.; Tong, V.-C. Internal loads and contact pressure distributions on the main shaft bearing in a modern gearless wind turbine. *Tribol. Int.* **2020**, *141*, 105960. [CrossRef]
9. Zhang, C.; Gu, L.; Mao, Y.; Wang, L. Modeling the frictional torque of a dry-lubricated tapered roller bearing considering the roller skewing. *Friction* **2019**, *7*, 551–563. [CrossRef]
10. Nguyen-Schäfer, H. *Computational Tapered and Cylinder Roller Bearings*; Springer: Cham, Switzerland, 2019. [CrossRef]
11. Petrusevich, A.I. Fundamental conclusions from the contact-hydrodynamic theory of lubrication. *Izv. Akad. Nauk SSSR* **1951**, *2*, 209–233.
12. Dowson, D.; Higginson, G.R. A Numerical Solution to the Elasto-Hydrodynamic Problem. *J. Mech. Eng. Sci.* **1959**, *1*, 6–15. [CrossRef]
13. Gohar, R.; Cameron, A. The Mapping of Elastohydrodynamic Contacts. *A S L E Trans.* **1967**, *10*, 215–225. [CrossRef]
14. Wymer, D.G.; Cameron, A. Elastohydrodynamic lubrication of a line contact. *Proc. Inst. Mech. Eng.* **1974**, *188*, 221–238. [CrossRef]
15. Bahadoran, H.; Gohar, R. Oil Film Thickness in Lightly-Loaded Roller Bearings. *J. Mech. Eng. Sci.* **1974**, *16*, 386–390. [CrossRef]
16. Kushwaha, M.; Rahnejat, H.; Gohar, R. Aligned and misaligned contacts of rollers to races in elastohydrodynamic finite line conjunctions. *Proc. Inst. Mech. Eng. Part C J. Mech. Eng. Sci.* **2002**, *216*, 1051–1070. [CrossRef]
17. Kushwaha, M.; Rahnejat, H. Transient concentrated finite line roller-to-race contact under combined entraining, tilting and squeeze film motions. *J. Phys. D: Appl. Phys.* **2004**, *37*, 2018–2034. [CrossRef]
18. Liu, X.; Yang, P. Analysis of the thermal elastohydrodynamic lubrication of a finite line contact. *Tribol. Int.* **2002**, *35*, 137–144. [CrossRef]
19. Yang, P.; Yang, P. Analysis on the thermal elastohydrodynamic lubrication of tapered rollers in opposite orientation. *Tribol. Int.* **2007**, *40*, 1627–1637. [CrossRef]
20. Zhu, N.; Wang, J.; Ren, N.; Wang, Q.J. Mixed Elastohydrodynamic Lubrication in Finite Roller Contacts Involving Realistic Geometry and Surface Roughness. *J. Tribol.* **2012**, *134*, 011504. [CrossRef]
21. Patir, N.; Cheng, H.S. An Average Flow Model for Determining Effects of Three-Dimensional Roughness on Partial Hydrodynamic Lubrication. *J. Lubr. Technol.* **1978**, *100*, 12–17. [CrossRef]
22. Patir, N.; Cheng, H.S. Application of Average Flow Model to Lubrication Between Rough Sliding Surfaces. *J. Lubr. Technol.* **1979**, *101*, 220–229. [CrossRef]
23. Kogut, L.; Etsion, I. Elastic-Plastic Contact Analysis of a Sphere and a Rigid Flat. *J. Appl. Mech.* **2002**, *69*, 657–662. [CrossRef]
24. Kogut, L.; Etsion, I. A Finite Element Based Elastic-Plastic Model for the Contact of Rough Surfaces. *Tribol. Trans.* **2003**, *46*, 383–390. [CrossRef]
25. Kogut, L.; Etsion, I. A Static Friction Model for Elastic-Plastic Contacting Rough Surfaces. *J. Tribol.* **2004**, *126*, 34–40. [CrossRef]
26. Lundberg, G. Elastische Berührung zweier Halbräume. *Forsch. Im Ingenieurwesen* **1939**, *10*, 201–211. [CrossRef]
27. Johns, P.; Gohar, R. Roller bearings under radial and eccentric loads. *Tribol. Int.* **1981**, *14*, 131–136. [CrossRef]
28. Fujiwara, H.; Kawase, T. Logarithmic Profile of Rollers in Roller Bearing and Optimization of the Profile. *Trans. Jpn. Soc. Mech. Eng. Ser. C* **2006**, *72*, 3022–3029. [CrossRef]
29. Fujiwara, H.; Kobayashi, T.; Kawase, T.; Yamauchi, K. Optimized Logarithmic Roller Crowning Design of Cylindrical Roller Bearings and Its Experimental Demonstration. *Tribol. Trans.* **2010**, *53*, 909–916. [CrossRef]
30. Cui, L.; He, Y. A new logarithmic profile model and optimization design of cylindrical roller bearing. *Ind. Lubr. Tribol.* **2015**, *67*, 498–508. [CrossRef]
31. Poplawski, J.V.; Peters, S.M.; Zaretsky, E.V. Effect of Roller Profile On Cylindrical Roller Bearing Life Prediction—Part II Comparison of Roller Profiles. *Tribol. Trans.* **2001**, *44*, 417–427. [CrossRef]

32. Najjari, M.; Guilbault, R. Edge contact effect on thermal elastohydrodynamic lubrication of finite contact lines. *Tribol. Int.* **2014**, *71*, 50–61. [CrossRef]
33. Zhang, Y.; Cao, H.; Kovalev, A.; Meng, Y. Numerical Running-In Method for Modifying Cylindrical Roller Profile Under Mixed Lubrication of Finite Line Contacts. *J. Tribol.* **2019**, *141*, 041401. [CrossRef]
34. Yang, P.; Yang, P.; Liu, X. Numerical Analysis of Isothermal EHL for Tapered Roller. *Tribology* **2005**, *25*, 456–460.
35. Chen, F.; Wang, J.; Zhang, G. Elastohydrodynamic lubrication of tapered roller with logarithmic profile. *Chin. J. Mech. Eng.* **2011**, *47*, 143–148. [CrossRef]
36. Guilbault, R. A Fast Correction for Elastic Quarter-Space Applied to 3D Modeling of Edge Contact Problems. *J. Tribol.* **2011**, *133*, 031402. [CrossRef]
37. Cen, H.; Lugt, P.; Morales-Espejel, G.E. On the Film Thickness of Grease-Lubricated Contacts at Low Speeds. *Tribol. Trans.* **2014**, *57*, 668–678. [CrossRef]
38. Roelands, C.J.A.; Winer, W.O.; Wright, W.A. Correlational Aspects of the Viscosity-Temperature-Pressure Relationship of Lubricating Oils (Dr In dissertation at Technical University of Delft, 1966). *J. Lubr. Technol.* **1971**, *93*, 209–210. [CrossRef]
39. Zhu, N.; Wen, S.-Z. A Full Numerical Solution for the Thermoelastohydrodynamic Problem in Elliptical Contacts. *J. Tribol.* **1984**, *106*, 246–254. [CrossRef]
40. Peiran, Y.; Shizhu, W. A Generalized Reynolds Equation for Non-Newtonian Thermal Elastohydrodynamic Lubrication. *J. Tribol.* **1990**, *112*, 631–636. [CrossRef]
41. Bair, S.; Winer, W.O. A rheological model for elastohydrodynamic contacts based on primary laboratory data. *J. Lubr. Technol.* **1979**, *101*, 258–265. [CrossRef]
42. Houpert, L.; Flamand, L.; Berthe, D. Rheological and Thermal Effects in Lubricated E.H.D. Contacts. *J. Lubr. Technol.* **1981**, *103*, 526–532. [CrossRef]
43. Masjedi, M.; Khonsari, M.M. Film Thickness and Asperity Load Formulas for Line-Contact Elastohydrodynamic Lubrication with Provision for Surface Roughness. *J. Tribol.* **2012**, *134*, 011503. [CrossRef]
44. Chang, W.-R.; Etsion, I.; Bogy, D.B. Static Friction Coefficient Model for Metallic Rough Surfaces. *J. Tribol.* **1988**, *110*, 57–63. [CrossRef]

lubricants

MDPI

Article

Analysis of Circulation Characteristics and Heat Balance of High-Speed Rolling Bearing under Oil-Air Lubrication

Xiqiang Ma [1,2,3], Mian Zhang [1], Fang Yang [1,2,3], Yujun Xue [1,2,3], Ruijie Gu [1,2,3] and Nan Guo [1,3,*]

1 School of Mechatronics Engineering, Henan University of Science and Technology, Luoyang 471003, China
2 Innovative Research Team of Livestock Intelligent Breeding and Equipment, Longmen Laboratory, Luoyang 471023, China
3 Henan Key Laboratory for Machinery Design and Transmission System, Henan University of Science and Technology, Luoyang 471003, China
* Correspondence: guonan@haust.edu.cn; Tel.: +188-3979-8351

Abstract: Aiming to solve the problem of oil-air lubrication failure caused by the high working temperature of high-speed rolling bearings, this study proposes a method, based on the theory of gas-solid two-phase flow and bearing tribology, of predicting the dynamic temperature rise of nonlinear high-speed rolling bearings under oil-air lubrication conditions. The accuracy of the fluid–structure coupling model is verified by comparing the temperature rise test results of angular contact ball bearing at different speeds. The characteristics of oil-air lubrication circulation and the relationship between the lubrication parameters and the heat balance of the high-speed rolling bearings have been studied. The results show that the gas supply pressure of the system has a significant influence on the continuity and fluctuation of the oil film in the oil pipe nozzle. The initial rise in temperature of the inner and outer rings of the bearing and the fluid domain has a speed threshold value, and the temperature increases linearly with the bearing speed. With the increase in the oil supply and lube oil viscosity of the system, the temperature rise of the outer ring of the bearing increases first, then decreases, and finally increases again. There is an optimal oil supply 5.5 mL and optimize viscosity 68 cSt for the bearing in the work condition.

Keywords: rolling bearing; liquid-solid coupling; thermal equilibrium; temperature rise

Citation: Ma, X.; Zhang, M.; Yang, F.; Xue, Y.; Gu, R.; Guo, N. Analysis of Circulation Characteristics and Heat Balance of High-Speed Rolling Bearing under Oil-Air Lubrication. *Lubricants* **2023**, *11*, 136. https://doi.org/10.3390/lubricants11030136

Received: 27 January 2023
Revised: 25 February 2023
Accepted: 3 March 2023
Published: 14 March 2023

1. Introduction

Under high-speed working conditions, rolling bearings suffer from problems such as increased friction heat generation, reduced fatigue life, cage and rolling element slippage, and rolling surface damage, which directly affect the working performance and life of the bearing system. An increase in the bearing temperature decreases the viscosity of the lubricating oil, the oil film thickness between the rolling element and the inner and outer ring raceways decreases, and the lubrication failure caused by the high temperature of the working surface leads to the failure of the spindle bearing. In addition, the temperature in the bearing system is the main factor affecting the bearing speed performance, which can be used as an important parameter to determine the bearing speed performance. Therefore, it is of great significance to study the fluid–solid coupling heat balance of high-speed rolling bearings under oil-air lubrication to investigate the bearing performance and its life.

At present, high-speed rolling bearings are widely used in railway bearings, motorized spindle bearings, engine main bearings, etc. Lubrication of a rolling bearing can efficiently reduce the heat generated by the internal friction in the bearing system. The type of lubrication method used in a rolling bearing has a significant influence on its performance under high-speed working conditions, which is the first problem to be solved in studying the heat balance of a high-speed rolling bearing.

Because of its good adhesion, sealing, lubricity and long life, grease is widely used in high-speed railway axle box bearings at home and abroad. Yang Fenglin [1] analyzed the friction power loss of axle box bearing, calculated the thermal conductivity of bearing, established the numerical model of temperature field of axle box bearing, studied the change of temperature field of axle box bearing under different working conditions, and studied the deterioration characteristics of grease on this basis. Allmaier H et al. [2] studied the potential ways to improve the service life of grease in rolling bearings, especially as the application of rail wheel bearings, and proposed that the basic lubrication process is the key to improve the service life of rail wheel bearings. Gao P [3] established a quasi-static mechanical model of the bearing according to the working conditions, and obtained the load distribution and kinematic parameters of the bearing. The temperature distribution of the railway double row tapered roller bearing under the test conditions was studied by finite element analysis, which was consistent with the test results.

Oil and gas cooling lubrication technology has the advantages of a high DN (the product of bearing diameter (D) and speed (N)) value, good lubrication performance, low pollution, and suitability for its use in high-load, high-speed, and high-temperature bearing work. Schubring [4] studied the two-phase flow characteristics of oil-air lubrication through experiments and obtained the variation of the oil film as a function of the lubrication parameters in oil pipelines under different working conditions. Jeng et al. and Gao et al. [5–8] developed stable oil supply test equipment for oil-air lubrication and an oil-air lubrication test bench for high-speed rolling bearings. The effect of the oil pipe length, gas supply pressure, oil supply interval, and viscosity of the lubricating oil on the oil supply stability of the oil-air lubrication system was studied, and the effects of radial load, oil supply, and rotational speed on the lubrication performance in the bearings were studied. Ramesh [9,10] et al. measured the thickness of the oil film between the rolling element and the raceway of the bearing under oil-air lubrication using the capacitance method on the oil-air lubrication test bench. They studied the influence of the heat transfer mode of the bearing on its temperature field distribution under oil-air lubrication and observed that the convective heat transfer is the main heat transfer mode of the bearing under oil-air lubrication. Wu [11] studied the oil-air lubrication and preload of machine tool spindle bearings, focusing on the influence of the design parameters on the oil-air lubrication effect, and obtained the optimal oil supply parameters of oil-air lubrication. Yuan and Julong [12] studied the lubrication performance of high-speed rolling bearings at different wear stages during service under oil-air lubrication. Their research showed that when the vibration value increased to 2–2.5 times of the initial wear value, the lubrication condition of the bearing can be improved by adjusting the oil supply interval reasonably. The oil-air parameters can be adjusted reasonably according to the changes in the working conditions at different wear stages in order to obtain the best lubrication strategy for rolling bearings under various working conditions. Yan and Bei [13] studied the migration and diffusion of the oil droplets in the bearing cavity using the coupling technology of the level-set function and the volume of fluid (VOF) method, analyzed the increase in the temperature and lateral oil-air lubrication mode of the outer ring bearing at different speeds, and predicted the lubricating oil flow and heat dissipation performance in the bearing cavity and key areas.

Compared with grease lubrication, oil-air lubrication has a wide application range and is not affected by temperature. It is suitable for poor conditions such as high temperature, heavy load, high speed, very low speed, and cooling water and dirt invading the lubrication point, and the lubrication effect is obvious.

Several research studies have been done on the temperature rise characteristics of the bearing system. Cui et al. [14] established the calculation model of the oil-gas two-phase flow and revealed the relationship between the temperature rise of the bearing and the speed, oil viscosity, radial clearance, preload, and air ratio. This study showed that the air ratio has the greatest influence on the increase in the temperature of the bearing, and the air ratio has an optimal value, at which the temperature rise of the bearing is the lowest. Li et al. [15–17] established an oil-air lubrication model of rolling bearings using the computational fluid dynamics (CFD) numerical simulation method and studied the influence of the inlet flow rate on the flow uniformity under oil-air lubrication.

They observed that the uneven distribution of oil in the bearing significantly affects the temperature rise and lubrication effect of the bearing. Li et al. [18–20] used the CFD method to analyze the flow velocity, temperature distribution, volume fraction, and heat transfer coefficient distribution of a two-phase medium in a bearing chamber and studied the flow and heat transfer law of the oil and gas two-phase medium in this chamber. Their results showed that the flow velocity of the two-phase medium increases first and then decreases with the increase in the radial height, and the lubricating oil in the bearing chamber is mainly distributed in the oil return pool and the outer wall of the chamber. Zheng [21] used the optimized thermal network model to estimate the thermal performance of the angular contact ball bearings under oil-air lubrication and obtained a numerical solution using the Newton-Raphson method. The change in the bearing temperature was tested to verify the model. Their results showed that the model exhibits high accuracy. Based on the artificial neural network and genetic algorithm, Wang [22] proposed a new prediction method for the temperature rise in the bearing by using the concept of universal coupling of the oil-air lubricated angular contact ball bearings. This prediction model has exhibited good accuracy, stability, and robustness.

Harris [23] applied heat transfer technology to study the temperature field of the bearings. Based on the research by Harris, Rumbarger et al. [24] studied the temperature field of high-speed cylindrical roller bearings. However, for the convective heat transfer coefficient of bearings, Rumbarger adopted a rough physical model of 'concentric rotating rings filled with a viscous medium' to simplify the heat transfer in the bearings. This method has some errors. Palmgren [25] developed a method for calculating the friction torque of high-speed cylindrical roller bearings and the viscous torque of lubricating fluid using a large number of tests, which laid the foundation for the study of heat generation in the bearing. Takeyama et al. [26] improved Palmgren's empirical formula to a certain extent and obtained the overall power loss of high-speed cylindrical roller bearings via experimental methods. They proposed an overall method of heat generation in the bearing and its heat transfer calculation method. However, this method can only be used in cylindrical roller bearings, which have certain limitations. In the working process of the bearing, the states of the contact area of each part are very different, and the overall method cannot reflect this difference. Fang B [27] used an experimental method to study the oil-air lubrication of high-speed rolling bearings. They studied the influence of various lubrication parameters on the temperature rise of the bearings using the single parameter method and obtained the optimal lubrication parameter values for a bearing with oil-air lubrication.

In this paper, based on the theory of elastohydrodynamic lubrication (EHL), thermodynamics and tribology, the VOF (volume of fluid) model of ANSYS 19.0 software has been used to study the oil-air distribution and oil-air lubrication characteristics in an oil-air lubrication pipeline of a high-speed rolling bearing. The simulation model of the fluid-solid coupling temperature field between the high-speed rolling bearing and the fluid domain under oil-air lubrication has been established. The relationship between the oil-air lubrication parameters and the thermal balance of the temperature field has been studied, which has important research significance and practical value for improving the speed performance of the high-speed bearing.

2. Analysis of the Circulation Characteristics of Oil-Air Lubrication in the Bearing

2.1. Establishment of the Nozzle Model for the Oil-Gas Lubricated Pipeline

The oil-air two-phase flow lubrication medium forms a lubricating oil film between the rolling element of the bearing and the raceway. The formed oil film can not only separate the two friction surfaces but also can bear a certain load. The continuity and stability of oil supply in oil-air lubrication are characterized by the change in the oil film continuity and the degree of oil fluctuation in the annular flow in the oil pipeline. In this study, based on the basic theory of gas-liquid two-phase flow, the VOF model has been used to study the characteristics of annular flow in a high-speed rolling bearing oil-air lubricated pipeline.

The two-phase flow distribution, heat transfer, and temperature of the oil-air two-phase flow in the bearing chamber were calculated using a finite element numerical simulation. The geometric model of the oil pipe and nozzle, based on their connection form in the oil and gas lubrication system, is shown in Figure 1.

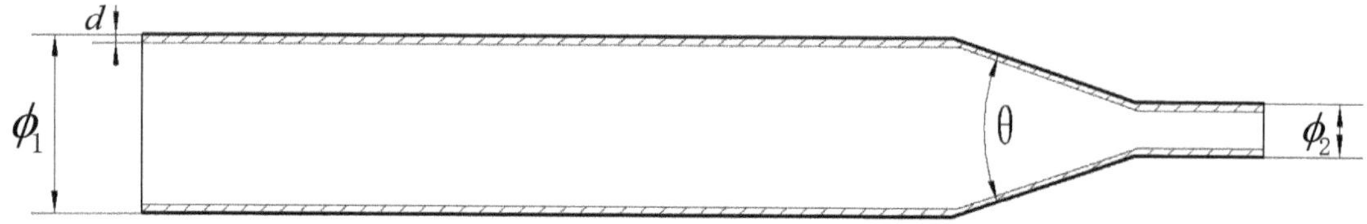

Figure 1. Geometric model of the oil pipeline and nozzle.

As shown in Figure 1, the pipe length was 1000 mm, the pipe diameter connected to the nozzle was 10 mm, the nozzle outlet diameter was 6 mm, and the initial oil film thickness was 0.5 mm, ignoring the effect of the nozzle length. The initial working environment in the pipeline was the standard atmospheric pressure, and considering the influence of gravity, VG68 turbine oil was selected as the lubricating oil. The density of the oil was 876 kg/m^3, and its dynamic viscosity was 0.058 Pa·s. The oil-air lubrication was set with air as the first phase and oil as the second phase. The gas and lubricating oil were incompressible, and there was no phase transition between the two phases.

2.2. *Analysis of the Oil-Gas Lubrication Characteristics*

In oil-air lubrication, the gas supply pressure of the oil-air system directly affects the oil-air velocity in the oil pipeline and has a huge influence on the uniformity and continuity of the oil supply at the lubrication point. The oil supply pressure of the system was set to be 3 MPa, the oil supply interval was 6 min, and the oil supplied within each interval was 5 mL. The state of the annular flow of oil inside the oil pipeline was analyzed when the gas supply pressure was 0.1, 0.3, 0.5, 0.7, and 0.9 MPa, respectively. The influence of the gas supply pressure in the system on the state of the annular flow oil film distribution in the oil pipeline and the nozzle and the change in the velocity field in the pipeline was studied.

Figure 2 shows the oil film distribution status of oil at the oil pipeline and nozzle when the gas supply pressure is 0.1 MPa. Red is the oil distribution status, and blue is the air distribution status. It can be seen from the figure that the oil can form a continuous and stable oil film on the pipe wall, and the oil at the nozzle has accumulated. On the one hand, the nozzle diameter is smaller, and on the other hand, the oil flow is relatively slow due to the smaller gas velocity under this pressure. This will not only cause oil blockage at the nozzle, but also reduce the gas volume fraction, thus reducing the speed of oil droplets when the oil leaves the nozzle. As the oil continues to block, the distribution volume of the oil leaving the nozzle is too large, which is likely to make it difficult for the oil droplets to pass through the wind curtain caused by the high-speed operation of the rolling bearing, thus reducing the lubrication efficiency.

It can be seen from Figure 3 that as the supply pressure increases, the degree of turbulence of the oil on the inner wall of the oil pipeline also increases. When the gas supply pressure is 0.3 MPa, the oil flow is stable, and a continuous oil film without any fractures is formed on the walls of the oil pipeline. When the gas supply pressure is 0.5 MPa, although the thickness of the oil film increases, the oil is continuous at the nozzle and is not atomized. Therefore, this oil will eventually enter the lubrication point in the form of separate dispersed oil droplets. When the gas supply pressure is 0.7 MPa, the oil film close to the wall fluctuates, the oil film breaks, and the oil film at the nozzle inlet appears to be mixed with oil and gas. This is mainly because the gas velocity is too large to produce atomization, and thus the lubricating oil is dispersed. When the supply pressure is 0.9 MPa, the degree of fluctuation is more prominent, the degree of turbulence of the oil increases, the oil film fluctuates strongly, and a stable and continuous annular flow cannot be formed. Consequently, the phenomenon of oil droplet entrainment occurs in the pipeline. The oil at the nozzle position is completely atomized by air, and it is impossible to form continuous and uniform lubrication at the lubrication point. This not only reduces the utilization rate of the lubricating oil but also reduces the lubrication efficiency.

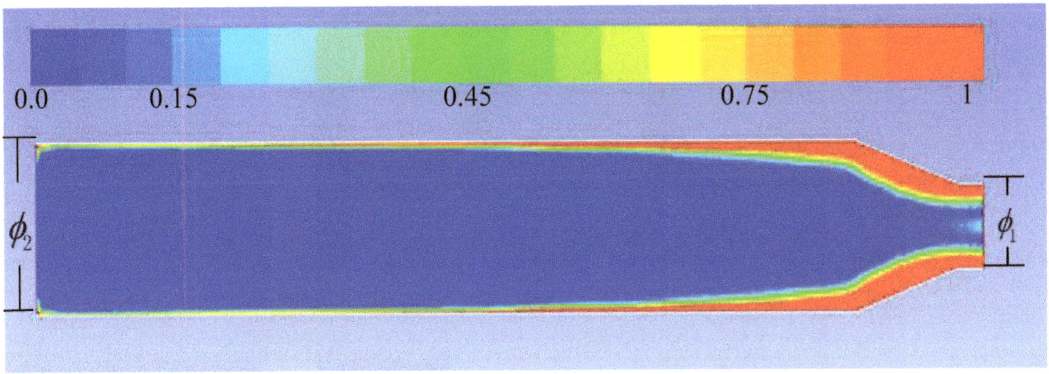

Figure 2. Oil-gas lubrication nozzle model and oil film distribution.

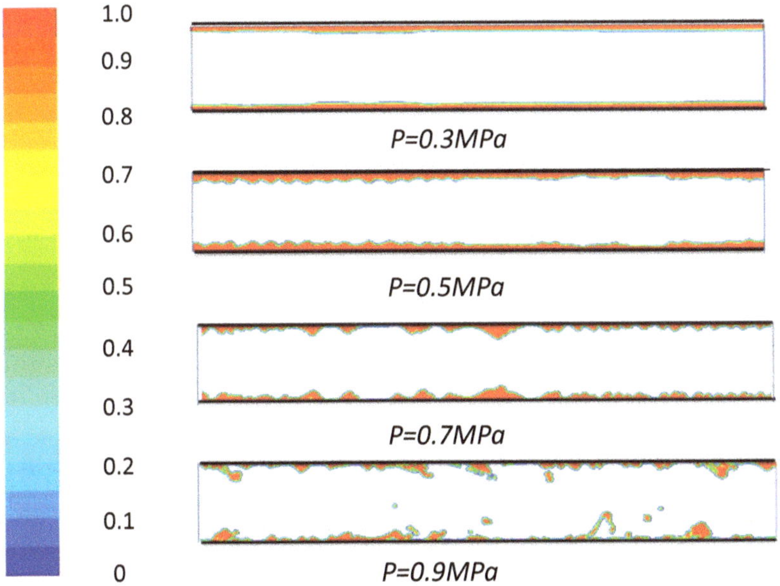

Figure 3. Distribution and fluctuation of the oil film in the oil pipeline and nozzle at different gas supply pressures.

3. Establishment of the Temperature Rise Model of the Rolling Bearing under Oil-Gas Lubrication

3.1. *Establishment of the Fluid-Solid Coupling Model for the bearing*

Taking a particular type of angular contact ball bearing as the research object, the main parameters were Φ130 mm × Φ200 mm × 34 mm. The fluid analysis software Ansys Fluent was used for studying the heat generation and heat transfer state of the bearing. The geometric model of the bearing is shown in Figure 4, in which Figure 4a shows the three-dimensional model of the bearing. The bearing cavity was extracted to obtain Figure 4b. The flow field model of the bearing cavity was obtained, and the inlet of the oil and gas nozzle was added to establish the fluid–solid mixed grid model of the bearing, as shown in Figure 4c.

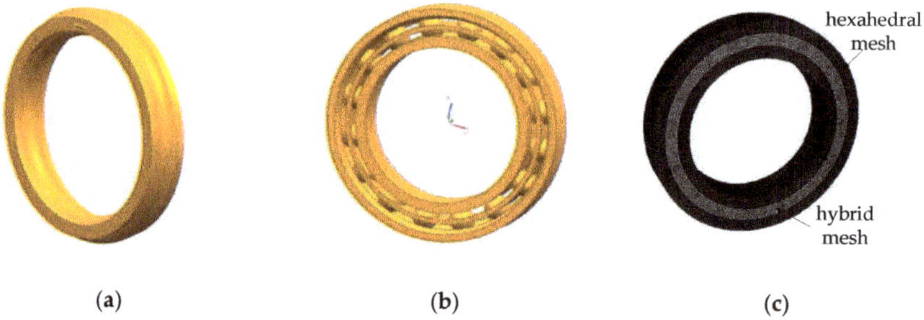

Figure 4. Geometric model of the angular contact ball bearing.

In the simulation, the working environment pressure was set to the standard atmospheric pressure. Considering the influence of the gravity of the bearing, the initial temperature of the bearing was 25 °C. A solver based on the pressure method, the κ-ε renormalization group turbulence model, the VOF two-phase flow model, and the heat transfer model was used. High-speed compressed gas was set as the first phase and the lubricating medium as the second phase. The oil and gas inlet was set as the pressure inlet, and the outlet pressure was set to the standard atmospheric pressure. The rotating grid model was used for the entire bearing cavity. The fluid domain and the inner ring of the bearing rotated at a particular speed. The outer ring of the bearing was relatively fixed, and its speed was zero. The iterative convergence residual of the momentum equation and the continuity equation was set to 10^{-3} and the iterative convergence residual of the energy equation was 10^{-7}. In the process of high-speed rotation of the bearing, the rotation speed of the rolling element was faster. It could be approximately considered that the calorific value of each rolling element was the same. The rolling element was approximately regarded as a solid ball with a radius of $D_m/2$. The heat generation rate, q, of the bearing was calculated using the following formula:

$$q = \frac{N_f}{\pi^2 D_m (D_b/2)^2} \quad (1)$$

where q is the heat generation rate of the bearing (in w/m^3), N_f is the internal heat of the bearing, the rolling element, D_b is the diameter of the toroidal section (in mm), and D_m is the diameter of the rolling element (in mm).

A heat source was applied to the rolling element and the inner and outer raceways as a set heat generation rate. Considering the convective heat transfer between the bearing and the fluid, the convective heat transfer coefficient between the bearing and the lubricating medium was calculated using the following formula:

$$\alpha_1 = 0.0986\left[\mathrm{n}\left(1 \pm \frac{D_b \cos\alpha}{D_m}\right)/\mathrm{v}\right]^{1/2} k{P_r}^{1/3} \quad (2)$$

where α_1 is the forced convection heat transfer coefficient inside the bearing, n is the bearing speed, D_b is the diameter of the rolling element (in mm), D_m is the average diameter of the bearing (in mm), α is the bearing contact angle (in rad), v is the average flow rate of the lubricant (in mm/s), k is the thermal conductivity of the lubricant, and P_r is the Prandtl number.

Because the thickness of the inner and outer rings of the angular contact ball bearing is much smaller than the width of the bearing, the inner and outer rings of the bearing can be approximately regarded as the cylinder wall. Considering the heat conduction between the inner ring of the bearing and the shaft, and the outer ring and the bearing seat, the conduction thermal resistance between the inner and outer rings of the bearing, and the bearing seat and the mandrel can be expressed as

$$R_1 = \frac{\ln(d_i/d)}{2\pi\lambda_q B} \tag{3}$$

$$R_2 = \frac{\ln(D/d_0)}{2\pi\lambda_q B} \tag{4}$$

where R_1 is the heat conduction resistance of the inner ring, R_2 is the heat conduction resistance of the outer ring, d is the inner diameter of the bearing (in mm), d_i is the diameter of the contact point between the inner ring raceway and the rolling element (in mm), $d_i = D_m(1 - \cos\alpha)$, d_0 is the diameter of the inner wall of the outer ring (in mm), D is the diameter of the outer wall of the outer ring (in mm), $d_0 = D_m + D_b \cos\alpha)$, and λ_q is the thermal conductivity of the inner and outer ring material of the bearing (in w/(m·°C)).

3.2. Experimental Verification

In order to verify the accuracy of the temperature rise simulation results of the rolling bearing, a particular type of angular contact ball bearing was used in the test bearing and the size as same as the simulation model. The lubrication method is oil and gas lubrication. In the test, the bearing was run in for one hour. When the bearing temperature dropped to room temperature, the temperature rise of the bearing at different speeds was tested.

Figure 5 shows an oil-air lubrication bearing test bench. The system was composed of three main parts: an oil-air lubrication system, a high-speed drive system, and a test piece system. The oil and gas lubrication system included an air compressor and filter, a cooling and drying device, an air storage tank, and an oil and gas generator. The high-speed drive system included a high-speed spindle, an inverter, and circulating water cooling equipment. The test piece system mainly included a test piece, an infrared thermometer, and a hydraulic loading device. In the entire test system, the air compressor continuously provided compressed air. The compressed air passed through the air filter to remove dust and particulate matter and other debris, and reached the gas storage tank. After the cooling and drying device removed the water vapor, it flowed through the oil and gas generator at a constant pressure. The oil-gas generator periodically injected lubricating oil into the oil-gas mixer in the form of oil droplets, which were transported to the lubrication point through the tubing. The frequency converter controlled the rotation speed of the high-speed spindle, realized stepless speed change, and drove the test piece through the coupling. Circulating water cooling equipment was used for cooling the motorized spindle built-in motor. A infrared thermometer was used to measure the temperature of the outer ring of the two bearings in the test piece. In the test, the main bearing was under-ring lubrication, and the test bearing was an angular contact ball bearing with oil-air lubrication.

In the experiment, the gas source was set to be 0.4–0.6 MPa, 500 L/min, the pressure of the oil and gas mixer was 1.0 MPa, and the reset pressure was 0.15 MPa. The experimental setup consisted of four distribution ports, each with a discharge of 0.015 mL/s, and the number of injections per hour was approximately 12 times (the number of injections could be changed according to the actual situation). The import tube had an inner diameter φ10 mm, a length of 2000 mm, and a horizontal spiral of 5 laps. The gas supply pressure was set to be 0.3 MPa, the oil supply interval was 6 min, the preload was 300 N, the lubricating oil viscosity was 68 cSt, and the oil supplied within each interval was 4–8 mL, respectively. When the outer ring of the test bearing was fixed and the inner ring speed changed from 3000 r/min to 10,560 r/min, the temperature rise of the outer ring of the bearing at different speeds was tested. The test results are presented below.

It can be seen from Table 1 that when the bearing speed is 3000 r/min, 4980 r/min, 8240 r/min and 10,560 r/min, respectively, compared with the test results, the simulation temperature rise error is less than 10%, indicating that the simulation model has high accuracy.

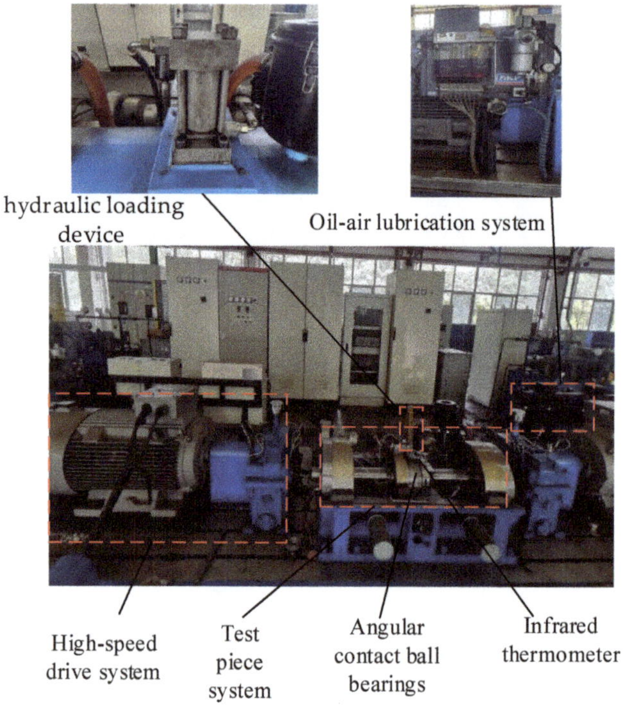

Figure 5. Oil-air lubrication test bench.

Table 1. Comparison of temperature rise test and simulation of bearing outer ring.

Revolution Speed r/min	Test Temperature Rise °C	Simulation Temperature Rise °C	Error
3000	21.7 °C	20.5 °C	5.5%
4980	30.2 °C	27.9 °C	7.6%
8240	40.7 °C	37.1 °C	8.8%
10,560	46.6 °C	41.9 °C	9.7%

4. Discussion of the Temperature Rise Characteristics of Rolling Bearing under Oil-Gas Lubrication

4.1. Influence of the Rotational Speed on the Temperature Rise Performance of the Bearing

The increase in the bearing speed aggravates the friction heat between the rolling element and the inner and outer rings, and also the change in the viscosity of the oil. Thus, the bearing speed has a significant influence on the temperature rise of the bearing. In the simulation system, the gas supply pressure was set to 0.3 MPa, the oil supply pressure was 3 MPa, the oil supply interval was 6 min, the preload was 300 N, and the viscosity of the lubricating oil was 68 cSt, as shown in Table 1. The simulation was carried out at the revolution speeds of 5000, 8000, 10,000, 12,000, 15,000, 18,000, and 20,000 *r*/min. The internal heat flow parameters of the bearing at different speeds are listed in Table 2.

Table 2. Internal heat flow parameters of the bearing at different rotational speeds.

Revolution Speed r/min	Heat Generation w	Heat Production Rate w/m^3	Coefficient of Convective Heat Transfer w/(m^2·K)	Maximum Temperature °C
5000	53.4	125,973	38.62	35.7
8000	78.6	185,421	41.31	42.3
10,000	106.2	250,530	46.25	50.4
12,000	138.1	325,784	52.25	63.5
15,000	176.3	415,899	58.32	81.4
18,000	216.9	511,677	63.24	96.6
20,000	263.6	621,844	71.58	110.2

Figure 6 shows the relationship between the speed and the temperature rise of the outer ring, the inner ring, and the fluid domain of the bearing. It can be seen from the figure that the temperature rise performance of the bearing at high speed is significantly affected by the speed. The temperature rise of the inner and outer rings of the bearing and the fluid domain increases with the increase in the bearing speed. In addition, the degree of increase of these parameters also gradually increases. The speed of the bearing is proportional to the temperature rise of the inner and outer rings of the bearing and the fluid domain. The discrete points in Figure 5 were linearly fitted and empirical formulae for the temperature of the bearing at different speeds were obtained as follows:

$$Y_{inner} = -10.86878 + 0.00397x \tag{5}$$

$$Y_{outer} = -9.65789 + 0.00339x \tag{6}$$

$$Y_{fluid} = -5026667 + 0.00227x \tag{7}$$

When y = 25 °C, the speed threshold of the outer ring, the inner ring, and the fluid domain of the bearing is 10,224, 9035, and 13,245 r/min, respectively. When the speed exceeds the threshold value, the temperature begins to rise linearly.

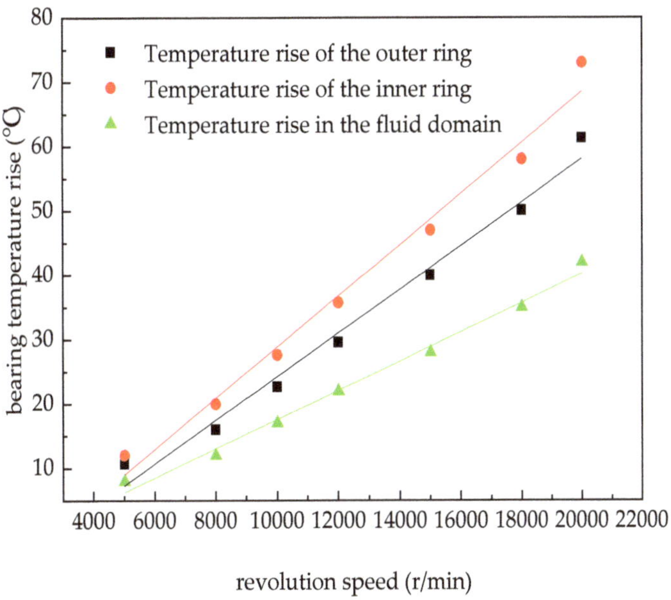

Figure 6. Effect of the rotational speed on the temperature rise of the bearing.

It can be seen from Figure 6 that the slope of the empirical formula for the temperature of the inner ring of the bearing is the largest and the trend of the temperature rise is the most obvious. This is because the faster the speed of the bearing, the more severe is the differential sliding between the rolling element and the inner and outer rings and the cage. The more severe the sliding friction and the rolling element spin friction, the more is the friction torque between the various parts, which leads to the increased heating of the bearing as a whole. At the same time, the internal vibration of the bearing is obvious when it rotates at high speed. Due to the oscillation of the oil film and the temperature rise of the lubricating oil, the lubrication of the bearing worsens, and some contact areas between the rolling element and the raceway are in mixed friction, resulting in a large amount of heat. The generated heat and the lubrication performance of the lubricant influence each other, resulting in increasingly poor bearing lubrication conditions and a sharp rise in the temperature of the bearing.

Under the high-speed operation of the bearing, due to the inertia force and heating, the ring will expand, resulting in a certain change in the diameter of the bearing groove bottom, which will lead to a change in the contact angle, change in the working performance of the bearing, and cause the temperature rise of the bearing. With the increase of bearing speed, the rolling component at the contact between the steel ball and the raceway increases, and the spin component of the steel ball also increases, which increases the rolling and sliding friction of the steel ball, resulting in increased heat generation and temperature of the bearing. During the operation of the bearing at high speed, the temperature of the bearing will change with the increase of the number of working cycles of the bearing. Because the bearing will generate heat when it works, the heat will also generate the temperature rise of the bearing, which will cause thermal expansion. The thermal deformation will increase, and the actual axial force of the bearing will increase due to the effect of thermal expansion. The increase of axial force further intensifies the friction heat generation. This is because, with the increase of the bearing speed, the rolling component of the contact point between the steel ball and the raceway increases, and the spin component of the contact point between the steel ball and the raceway increases. The rolling friction and sliding friction of the steel ball are increased, resulting in increased heat generation and temperature rise.

4.2. *Influence of the Preload on the Temperature Rise Performance of the Bearing*

If the bearing preload is too large, the contact stress between the bearing roller and the inner and outer tracks will increase, which will increase the friction between the two, and eventually lead to the high working temperature rise of the bearing. However, the preload is too small, and the bearing cannot run smoothly when rotating at high speed. Therefore, the preload of the bearing should be reasonably selected according to the load and service requirements of the bearing. When the operating parameters are an air supply pressure of 0.3 Mpa, oil supply pressure of 3 Mpa, oil supply interval of 6 min, rotational speed of 5000 r/min, lubricating oil viscosity of 68 cSt, and the preload is 100 N, 200 N, 300 N, 400 N, 500 N, 600 N, 700 N, respectively, the temperature field of the bearing is simulated.

It can be seen from Figure 7 that the bearing temperature rise increases linearly with the increase of preload. This is because when the bearing rotates at high speed, the friction caused by the elastic hysteresis of the bearing, the friction caused by the local differential sliding and the friction torque caused by the spin sliding of the ball along the ring groove will increase with the increase of the bearing preload, which will lead to the increase of the overall heat generation of the bearing and eventually the increase of the temperature rise of the bearing.

The bearing preload is in direct proportion to the temperature rise of the bearing outer ring. The empirical formula of the temperature rise of the bearing outer ring under different preloads can be obtained by linear fitting of discrete points:

$$Y_{temp\ rise} = 17.3 + 0.00736x$$

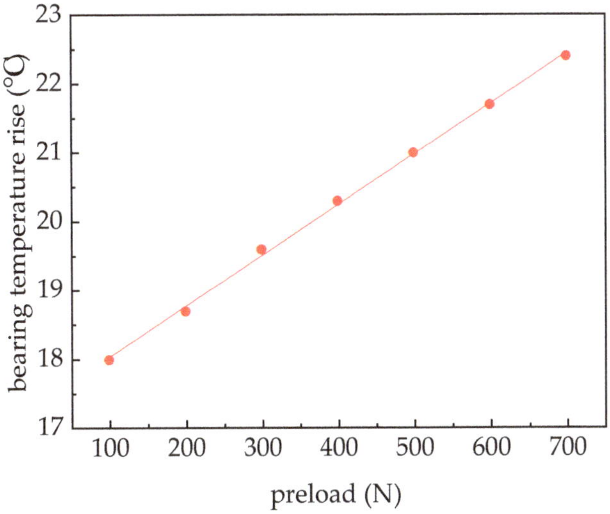

Figure 7. Effect of rotational preload on temperature rise of bearing.

4.3. Analysis of the Oil Supply on the Temperature Rise Performance of the Bearing

The lubrication oil supply plays a vital role in high-speed rolling bearings. The amount of oil supplied directly affects the thickness of the lubricating oil film within the bearing. It can not only reduce the direct contact friction heat between the roller and the inner and outer rings but can also take away a certain amount of heat in the process of circulating the oil supply. In this study, the effect of oil supply on the temperature rise of the bearing was studied. For performing this investigation, the following parameters were set: the gas pressure was 0.3 MPa, the oil supply interval was 6 min, the preload was 300 N, the viscosity of lubricating oil was 68 cSt, and the speed of the bearing was 10,000 r/min. The oil supplied within each interval of 4–8 mL was used.

Figure 8 shows the change in the temperature rise of the bearing as a function of the oil supply of the lubrication system. It can be seen from the figure that the temperature rise of the bearing increases first and then decreases with increasing oil supply, and finally increases once again. This is because at the beginning (oil supply is 4–4.5 mL), it may be in a lean state, and the friction between the roller and the inner and outer rings generates more heat. As the viscous torque of the lubricating oil increases, the heat generated by the stirring of the lubricating oil also increases; thus, increasing the temperature of the outer ring of the bearing. As the oil supply continues to increase (4.5–5.5 mL), the oil film between the bearing rolling element and the raceway gradually becomes complete, the friction heat of the bearing decreases, the heat taken away by the lubricating oil increases, and the temperature rise of the bearing decreases. In addition, the stirring friction gradually increases and reaches an equilibrium state at a certain value of oil supply. As a result, the temperature rise of the bearing is the lowest, and the oil supply is also optimum for the bearing. However, as the oil supply continues to increase (5.5–8 mL), the bearing is in an oil-rich state, and thus an excessive amount of oil is present for forming the oil film. This large amount of lubricating oil increases the heat generated due to the stirring of the bearing. When it is greater than the heat taken away after improving the lubrication conditions, the temperature rise of the bearing increases with increasing oil supply. Therefore, the change in the bearing temperature at high speed is sensitive to the change in the oil supply. When the lowest temperature rise point of the outer ring of the bearing is at approximately 5.5 mL of the oil supply, a good oil film can be formed in the bearing.

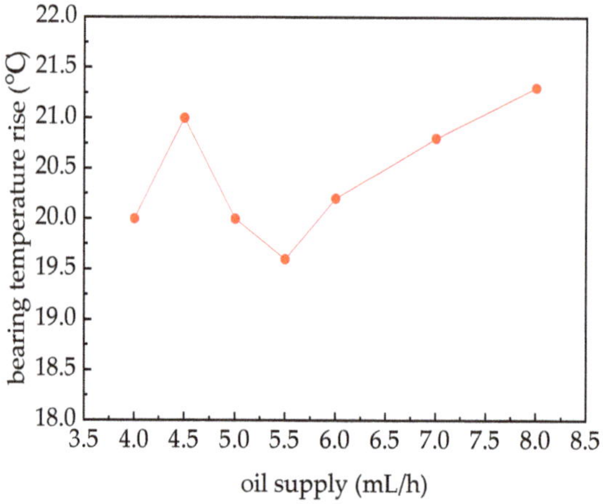

Figure 8. Effect of the oil supply on the temperature rise of the bearing.

4.4. Analysis of the Lubricant Viscosity on the Temperature Rise Performance of the Bearing

When the bearing rotates at high speed, the viscosity of the lubricating oil has a direct relationship with the oil film thickness and the heat generation of the bearing. When the bearing speed is 10,000 r/min, respectively, it maintains the air supply pressure of the lubrication system at 0.3 MPa, the preload at 300 N, and the single nozzle inlet, conduct simulation analysis on the bearing temperature rise characteristics, and obtain the influence curve of the lubricating oil viscosity on the bearing temperature rise.

It can be seen from Figure 9 that, within a certain viscosity range, the influence of lubricating oil viscosity on bearing temperature rise is small. However, when the viscosity exceeds a certain value, the bearing temperature rise gradually increases. The change trend of lubricating oil viscosity on bearing temperature rise characteristics is consistent with the theory of elastohydrodynamic lubrication. However, when the viscosity of the lubricating oil is too large, the friction torque related to the viscosity of the lubricant increases, the heat generation increases, and the temperature rise of the bearing increases. Lubricating oil with higher viscosity is easy to form oil film, but at the same time, the heat generated by viscous torque is also increasing, so the viscosity of lubricating oil should be reasonably selected according to the actual working conditions such as bearing speed and load. At the same higher speed, the lubricating oil has the best viscosity value.

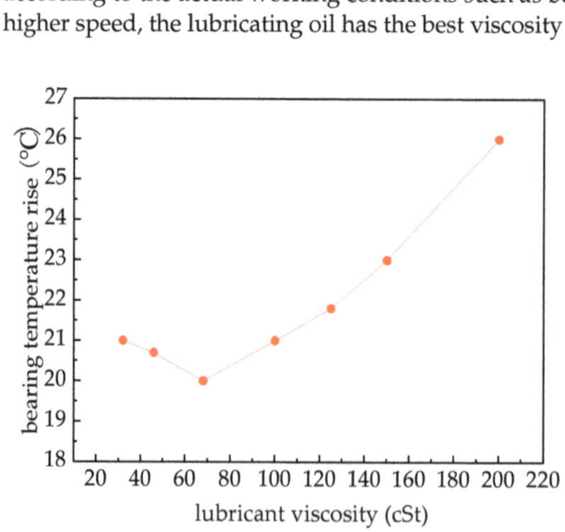

Figure 9. Effect of the lubricating oil viscosity on the temperature rise of the bearing.

5. Conclusions

In this study, based on the basic theory of the gas-liquid two-phase flow, the VOF model has been used to study the distribution of oil and gas in the oil-air lubrication pipeline of a high-speed rolling bearing. The fluid-solid coupling temperature field simulation model of a bearing under the oil-air lubrication condition has been established, and the relationship between the fluid-solid coupling heat balance of the high-speed rolling bearing under this lubrication condition has been studied. The following are the conclusions of this study:

(1) The air supply pressure in the oil-air lubrication system has a huge influence on the continuity and fluctuation of the oil film in the nozzle. If the gas pressure is too low, oil accumulation occurs in the nozzle. If the supply pressure is too large, the greater the speed of the gas in the pipeline, the more the probability of fluctuation and fracture of the oil film.

(2) The speed of rotation has a great influence on the temperature rise of high-speed rolling bearing. With the increase of the speed shaft, the temperature rise of the bearing inner ring is faster than that of the bearing outer ring and the bearing inner fluid domain. The temperature rise of the three increases linearly with the bearing speed, and the initial temperature rise has a threshold value of rotation speed. In addition, with the increase of preload, the temperature rise of bearing outer ring also shows a linear increase trend.

(3) In high-speed rolling bearings, with the increase of oil supply and lubricating oil viscosity, the bearing temperature rise decreases first and then increases. The system oil supply and lubricating oil viscosity have a certain impact on the temperature rise of high-speed rolling bearings, and there is an optimal oil supply and viscosity value.

Author Contributions: Author Contributions: Conceptualization, M.Z.; methodology, M.Z. and N.G.; software, M.Z.; validation, X.M. and M.Z.; investigation, F.Y.; resources, F.Y.; data curation, N.G.; writing—original draft preparation, M.Z.; writing—review and editing, R.G.; supervision, Y.X.; project administration, X.M.; funding acquisition, X.M. All authors have read and agreed to the published version of the manuscript.

Funding: This research was funded by National Key R & D Program of China (Grant No.2021YFB2011000); Henan Science and Technology Project (212102210365); Research Program in University of Henan Province (Grant No. 21A460014) and Major Science and Technology Project of Henan Province (221100220100).

Data Availability Statement: Data is contained within the article.

Conflicts of Interest: The authors declare no conflict of interest.

References

1. Yang, F.L. Research on Temperature Field and Grease Degradation Mechanism of High-Speed Locomotive Axle Box Bearing. *Henan Univ. Sci. Technol.* **2020**, *32*, 1345.
2. Allmaier, H. Increase Service Life for Rail Wheel Bearings—A Review of Grease Lubrication for This Application. *Lubricants* **2022**, *10*, 36. [CrossRef]
3. Gao, P.; Tang, W.; Cui, Y.; Wang, Y.; Mo, G.; Yin, J. Theoretical and Experimental Investigation on Thermal Characteristics of Railway Double-Row Tapered Roller Bearing. *Energies* **2022**, *15*, 4217. [CrossRef]
4. Schubring, D.; Shed, T. Two-phase wavy-annular flow in small tubes. *Int. J. Heat Mass Transf.* **2009**, *52*, 1619–1622. [CrossRef]
5. Jeng, Y.; Gao, C. Investigation of the ball-bearing temperature rise under an oil-air lubrication system. *Proc. Inst. Mech. Eng.* **2001**, *215*, 139–148. [CrossRef]
6. Jiang, L.; Lyu, Y.; Gao, W.; Zhu, P.; Liu, Z. Numerical investigation of the oil–air distribution inside ball bearings with under-race lubrication. *Proc. Inst. Mech. Eng. Part J J. Eng. Tribol.* **2022**, *236*, 499–513. [CrossRef]
7. Deng, S.; Zhao, G.; Qian, D.; Jiang, S.; Hua, L. Investigation of Oil-Air Flow and Temperature for High-Speed Ball Bearings by Combining Nonlinear Dynamic and Computational Fluid Dynamics Models. *J. Tribol.* **2022**, *144*, 071204. [CrossRef]
8. Zhang, C.; Zhou, J.; Meng, X. Temperature rise of magnetorheological fluid sealing film in a spiral grooved mechanical. *Ind. Lubr. Tribol.* **2022**, *74*, 683–691. [CrossRef]
9. Ramesh, K.; Yeo, S.H.; Zhong, Z.; Yui, A. Ultra-high-speed thermal behavior of a rolling element upon using oil-air mist lubrication. *J. Mater. Process. Technol.* **2002**, *127*, 191–198. [CrossRef]
10. Yeo, S.H.; Ramesh, K.; Zhong, Z.W. Ultra-high-speed grinding spindle characteristics upon using oil/air mist lu brication. *Int. J. Mach. Tools Manuf.* **2002**, *42*, 815–823. [CrossRef]

11. Wu, C.-H.; Kung, Y.-T. A parametric study on oil/air lubrication of a high-speed spindle. *Precis. Eng.* **2005**, *29*, 162–167. [CrossRef]
12. Yuan, J.; Ma, L.; Wang, J.; Lv, B.; Wang, D. Oil air lubrication strategy for different wear stages of high speed rolling bearings. *Harbin Gongye Daxue Xuebao/J. Harbin Inst. Technol.* **2022**, *3*, 80–87.
13. Yan, B.; Dong, L.; Yan, K.; Chen, F.; Zhu, Y.; Wang, D. Effects of oil-air lubrication methods on the internal fluid flow and heat dissipation of high-speed ball bearings. *Mech. Syst. Signal Process.* **2021**, *151*, 1884–2022. [CrossRef]
14. Cui, L.; Wan, J. Study on temperature control method of high speed ball bearing with oil-air lubrication. In Proceedings of the International Conference on Information Science and Control Engineering, Changsha, China, 21–23 July 2017; pp. 1263–1267.
15. Li, Y.; Yang, Z.; Chen, F.; Zhao, J. Effect of air inlet flow rate on flow uniformity under oil-air lubrication. *Lubr. Tribol.* **2018**, *70*, 282–289. [CrossRef]
16. Zhang, H.; Huang, J.; He, Z.; Guo, S.; Bai, C. Coupling effect of unbalanced magnetic pull and ball bearing on nonlinear vibration of motor rotor system. *J. Vib. Control* **2022**, *28*, 665–676.
17. Fei, J. Transient Thermal Analysis Model of Damaged Bearing Considering Thermo-Solid Coupling Effect. *Sensors* **2022**, *22*, 8171.
18. Li, Y. Flow and thermal analysis of oil air two-phase medium in bearing chambe. *J. Aerosp. Power* **2021**, *2*, 606–615.
19. Ma, S.J.; Yin, Y.J.; Chao, B.; Yan, K.; Fang, B.; Hong, J. A Real-time Coupling Model of Bearing-Rotor System Based on Semi-flexible Body Element. *Int. J. Mech. Sci.* **2023**, *245*, 108098. [CrossRef]
20. Zhao, J.; Sheng, W.; Li, Z.; Zhang, H.; Zhu, R. Effect of lubricant selection on the wear characteristics of spur gear under oil-air mixed lubrication. *Tribol. Int.* **2022**, *167*, 107382. [CrossRef]
21. Zheng, D.; Chen, W.; Li, M. An optimized thermal network model to estimate thermal per formances on a pair of angular contact ball bearings under oil-air lubrication. *Appl. Therm. Eng.* **2018**, *131*, 328–339.
22. Wang, B.; Chang, X. Temperature rise prediction of oil-air lubricated angular contact ball bearings using artificial neural network. *Recent Pat. Mech. Eng.* **2019**, *12*, 248–261. [CrossRef]
23. Harris, T.A. *Rolling Bearing Analysis*; John Wiieyand Sons, Inc.: New York, NY, USA, 1990.
24. Sawicki, J.T.; Johansson, S.A.; Rumbarger, J.H.; Sharpless, R.B. The effect of structural distortion on the fatigue life of large-diameter ball bearing. *Ball Bear. Fatigue Life Finite Elem. Method* **2007**. Available online: https://engagedscholarship.csuohio.edu/enme_facpub/47/ (accessed on 25 January 2023).
25. Palmgren, A. *Ball and Roller Bearing Engineering*, 3rd ed.; SKF Industries Inc.: Philadelphia, PA, USA, 1959; Volume 2, pp. 231–321.
26. Okada, Y.; Takeyama, H.; Sato, M. Experimental implication of celiac ganglionotropic invasion of pancreatic-cancer cells bearing c-ret proto-oncogene with reference to glial-cell-line-derived neurotrophic factor. *Math. Probl. Eng.* **2020**, *81*, 67–73. [CrossRef]
27. Fang, B.; Zhang, J.; Hong, J.; Yan, K. Research on the Nonlinear Stiffness Characteristics of Double−Row Angular Contact Ball Bearings under Different Working Conditions. *Lubricants* **2023**, *11*, 44. [CrossRef]

lubricants

Article

Study on the Effect of Oil Supply on the Sound Field Characteristics of Full Ceramic Ball Bearings under Oil Lubrication

Jian Sun, Xin Fang, Jinmei Yao *, Renyun Guan, Zhe Zhang and Guangxiang Zhang

School of Mechanical Engineering, Shenyang Jianzhu University, Shenyang 110168, China
* Correspondence: yaojinmei06@126.com

Abstract: To study the full sound field distribution characteristics of full ceramic ball bearings, reduce the radiation noise of the bearings, and improve their service performance. In this paper, the sound field distribution characteristics of 6206 silicon nitride ceramic deep groove ball bearings are studied under different oil supplies. A mathematical model of the sound field distribution of full ceramic ball bearings under oil lubrication is established, and the validity of the model is verified by experimental data. The bearing cavity simulation model of the full ceramic ball bearing is established, and the influence of oil supply on the operation characteristics of the full ceramic ball bearing is studied. Through theoretical and experimental research, the circular distribution law of the noise signal of ceramic ball bearings under different oil supplies is revealed. It is found that there is an optimal fuel supply when the speed and load are constant. Under optimal oil supply lubrication conditions, the full ceramic ball bearing has the minimum radiation noise, and the bearing exhibits optimal lubrication state, vibration and temperature rise characteristics. The new contribution of this paper: with the increase in oil supply, the sound pressure level of radiation noise of full ceramic ball bearings decreases and then increases. The research results reveal the radiation noise mechanism of full ceramic ball bearings, which is of great significance for enriching its theory and method.

Keywords: full ceramic ball bearing; oil supply; radiated noise; sound field distribution; mathematical model

Citation: Sun, J.; Fang, X.; Yao, J.; Guan, R.; Zhang, Z.; Zhang, G. Study on the Effect of Oil Supply on the Sound Field Characteristics of Full Ceramic Ball Bearings under Oil Lubrication. *Lubricants* **2023**, *11*, 146. https://doi.org/10.3390/lubricants11030146

Received: 6 February 2023
Revised: 10 March 2023
Accepted: 15 March 2023
Published: 19 March 2023

1. Introduction

To study the influence of oil supply on the sound field distribution characteristics of full ceramic ball bearings under different working conditions. The oil supply required for the full ceramic ball bearing to generate the minimum radiation noise is obtained, which reduces the radiation noise and improves its service performance. With the progress of science and technology, the use environment and conditions of ball bearings are more and more demanding, such as high speed, high temperature, corrosion resistance, strong magnetism, oil-free lubrication, etc. [1,2]. As a new high-end product, full ceramic ball bearings have the advantages of large stiffness, good thermal stability, corrosion and wear resistance, high operation accuracy and long service life [3–5]. With the increasingly harsh operating conditions of bearings, the friction and impact between internal components of full ceramic ball bearings intensify, and the problem of radiated noise is gradually highlighted. Radiated noise has a large impact on the sound quality of bearings, limiting the development of equipment in the high speed and quiet direction [6]. Bearing lubrication effect has an important impact on the running performance and service life of bearings. Good lubrication can reduce friction and improve bearing performance. Different lubrication states will also produce different sound field characteristics of bearings [7,8]. Among them, the oil supply for bearings directly determines the lubrication state of the bearing, thus affecting the bearing acoustic characteristics. Therefore, it is of great significance to study the influence of oil supply on the sound field characteristics of full ceramic ball bearings and to explore the oil supply required to produce minimum radiated noise [9,10].

In recent years, experts and scholars at home and abroad have conducted a lot of research on the sound field and radiated noise characteristics of ball bearings. Yan Haipeng studied the radiated noise characteristics of a high-speed ceramic ball bearing motorized spindle. Through grinding tests on the spindle, it was found that speed is one of the main factors affecting the radiated noise of full ceramic ball bearings [11,12]. From the perspective of fault analysis of double-row tapered roller bearings, Li Defa conducted a mechanism study between the bearing state and acoustic emission signal through theoretical analysis and test comparison, which provided theoretical support for the application of acoustic emission perception in bearing state detection [13]. Xiong Shi analyzed the relationship between the sound pressure and acoustic power of the shaft-hull coupling structure under different bearing stiffness [14]. Zhang Qitao established a mathematical model for calculating the center motion trajectory of the rolling body of deep groove ball bearings and analyzed the influence of rotation speed and radial load on the noise size at a fixed point, as well as the change law of noise along the axis direction of the rolling bearing, combined with the acoustic theory and specific examples [15]. Under the condition of low speed and heavy load, Wang Jiaxu studied the influence of different friction coefficients, loads and rotation speeds on the transient characteristics of water-lubricated bearings using a finite element model, determined the vibration frequency and effective value of vibration acceleration of the bearings, and carried out experimental verification. This provides a theoretical basis for studying the vibration and noise mechanism of water-lubricated bearings [16]. Peng Liqiang considered the oil supply at the lubricating oil inlet as a variable parameter and solved the mathematical model with the finite difference method to analyze the influence of the oil supply on the static lubrication characteristics of the inner and outer oil film of floating ring bearings [17]. Nam Jaehyeon measured the vibration and sound pressure of the contact surfaces under two lubrication conditions, the change in friction coefficients, and analyzed the frictional noise mechanism of the lubricant under clean or polluted surfaces [18]. Zhang Qitao established a noise calculation model for deep groove ball bearings. The effect of the number of ripples, the ripple amplitude, the bearing speed, the bearing load, and the ball size error on the noise of the fixed measuring point bearings was studied through numerical calculations [19]. Jahagirdar studied the effect of noise on the statistical moment of a bearing vibration signal. The distribution function of the inner, outer, and sphere defects was tested using K-S tests. The change in noise level and its effect on the statistical moment were verified [20]. Botha established a sound radiation model for bearing related research objects, analyzed the characteristics of noise sources, and determined the location and mechanism of the main noise sources for fault detection [21].

The mechanism of bearing vibration is very complex, and the vibration characteristics of ball bearings directly affect the accuracy and radiation noise. P. K. Gupta [22] studied the vibration characteristics in rolling bearings, defining two characteristic intrinsic frequencies. They are the elastic boundary contact frequency and the bearing motion frequency, and their existence is verified by computer analysis simulations and available experimental data. Lynagh N. [23] presented a detailed model of bearing vibration, including the effect of contact spring non-linearity in balls-to-raceways contacts. This model was successfully employed for the recognition of complex real-time vibration spectra of a precision routing spindle obtained by accurate non-contact sensors. Based on the hydrodynamic lubrication model, contact force model, and wear prediction model, R.B. Randall [24] simulated the vibration signal generated by the relative motion between the journal and the bearing. D. Abbound [25] compared the application of the the minimum entropy deconvolution (MED)and the spectral kurtosis (SK) method and the cyclostationary method in actual rolling bearing vibration signal fault detection. It is found that, in most cases, both methods can detect bearing faults.

As can be seen from the research results in recent years, most studies on the radiated noise of ball bearings take the noise as the judgment basis for bearing fault analysis [26–28] or consider it as a whole sound source and calculate it based on experimental and finite element methods. The sound field model of ball bearings and the mechanism of sound field distribution are not established. However, there is little research on the influence of full ceramic ball bearing lubrication on sound field characteristics.

Therefore, this paper takes the silicon nitride full ceramic ball bearing as the research object, aiming at the problems of radiated noise generated by bearings when running at high speed. Based on the radiation noise mechanism, the full sound field distribution characteristics of full ceramic ball bearings are discussed. The influence of oil supply on the sound field distribution characteristics of full ceramic ball bearings under different working conditions is investigated. By establishing the simulation model of the bearing cavity of the full ceramic ball bearing, the influence of different oil supplies on the oil phase distribution in the bearing cavity, temperature rise, and vibration acceleration of the bearing is obtained. Based on the mathematical model of sound field characteristics, the influence of oil supply on the sound field characteristics of full ceramic ball bearings is analyzed by combining the simulation results and test data. The research results can provide theoretical support and experimental reference for the analysis of the sound field characteristics and noise signal transmission mechanism of full ceramic ball bearings.

2. Mathematical Model of Sound Field Characteristics of Full Ceramic Ball Bearings

2.1. Contact Analysis between Ceramic Balls and Bearing Rings

In general, the bearing inner ring rotates with the shaft. The outer ring is assembled on the bearing housing and does not rotate with the rotation of the inner ring, but it can still produce vibration. The noise mainly comes from the friction and impact vibration caused by interaction between components during its operation. In order to study the contact vibration characteristics of the ceramic ball and ring, the contact between the ball and the inner ring and the outer ring of the ball bearing can be regarded as a point-to-surface contact. In addition, the lubrication problem of the ball/outer ring tribology system can be regarded as a point contact elastohydrodynamic lubrication problem plane when the ball bearing's contact with the micro area is taken as the research object.

The Hertz point contact elasticity theory studies the local stress and strain distribution of two smooth ellipsoids after contact under pressure [29].

According to the Hertz point contact elasticity theory, the contact force Q is:

$$Q = K\delta^{3/2} \tag{1}$$

where δ is the contact deformation, K is the contact stiffness, and the calculation formula is as follows:

$$\begin{aligned} K &= \tfrac{1}{3}E'(\Sigma\rho)^{\frac{1}{2}}\left(\tfrac{2}{\delta^*}\right)^{\frac{3}{2}} \\ \delta^* &= \tfrac{2F}{\pi}\left(\tfrac{\pi}{2\kappa^2 E}\right)^{\frac{1}{3}} \end{aligned} \tag{2}$$

where E' represents the elastic modulus parameter, κ represents the eccentricity parameter of the ellipse, and E and F are the first and second types of complete ellipse integral, respectively. Therefore, the contact force and the length of the long radius and short radius of the ellipse in the contact area can be found by calculating the contact deformation.

When analyzing the contact deformation between the ceramic balls and rings, the surface elastic deformation should be considered. According to the elastic theory, the elastic displacement of each point in the vertical direction can be deduced as [30]:

$$\begin{aligned} v(x) &= -\tfrac{2}{\pi E}\int_{s_1}^{s_2} p(s)\ln{(s-x)}^2 ds + c \\ \tfrac{1}{E} &= \tfrac{1}{2}\left(\tfrac{1-{v_e}^2}{E_e} + \tfrac{1-{v_r}^2}{E_r}\right) \end{aligned} \tag{3}$$

where $v(x)$ is the elastic displacement in the vertical direction. For the elastohydrodynamic lubrication (EHL) problem, $p(s)$ is the fluid pressure distribution; s is the additional coordinate on the x-axis, representing the distance between $p(s)$ and the origin of the coordinate; s_1

and s_2 are the starting point and ending point of the load, respectively; c is an undetermined constant, which can usually be incorporated into h_0; E is the equivalent elastic modulus; E_e is the elastic modulus of the outer raceway; E_r is the elastic modulus of ball bearings; v_e is the Poisson's ratio of the outer raceway; and v_r is the Poisson's ratio of ball bearings.

2.2. Vibration Differential Equations of Ceramic Balls

As the key component of a full ceramic ball bearing, the ceramic ball has contact force with the cage, inner ring and outer ring, so the stress of ceramic balls is complicated. Especially when considering the influence of the ceramic ball's diameter error on the operating state of the bearings, the interaction between ceramic balls and rings with larger ball diameters and smaller ball diameters is more complicated [31].

In this paper, it is assumed that the cage pocket holes are uniformly distributed along the circumference of the cage, and there is no other dimensional error.

Figure 1 shows the acting force on the ceramic ball when the full ceramic ball bearing runs at high speed. α_{ij} and α_{oj} represent the contact angle between the ceramic ball and the inner and outer orbits, respectively. Q_{ij} and Q_{oj} represent the normal contact force between the ceramic ball and the inner and outer orbit, respectively. $T_{\eta ij}$, $T_{\eta oj}$, $T_{\xi ij}$ and $T_{\xi oj}$ represent the traction force between the ceramic ball and the contact surface of the inner and outer orbit. Q_{cxj} and Q_{cyj} represent the components of the impact force occurring between the ceramic ball and the cage along the coordinate direction. G_{byj} and G_{bzj} represent the gravity component of the ceramic ball along the coordinate direction. $P_{\eta j}$ and $P_{\xi j}$ indicate the friction force acting on the surface of the ceramic ball, including rolling friction force and sliding friction force. F_{bxj}, F_{byj} and F_{bzj} represent the hydrodynamic viscous resistance component acting on the ceramic ball. $F_{\eta ij}$, $F_{\eta oj}$, $F_{\xi ij}$ and $F_{\xi oj}$ represent the rolling friction force between the ceramic ball and the inner and outer raceway. J_x, J_y and J_z represent the component of inertia of the ceramic ball rotating about its own center. ω_{xj}, ω_{yj} and ω_{zj} represent the angular velocity of the ceramic ball in its coordinate system along the respective coordinate direction and represent the angular acceleration of the ceramic ball in its coordinate system along their respective coordinate directions [32].

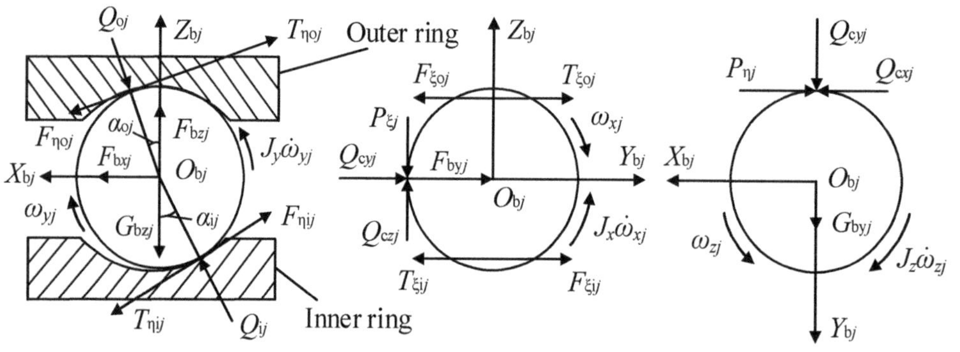

Figure 1. Stress diagram of ceramic ball bearing.

Where D_W is the diameter of the ceramic ball; m_b is the mass of the ceramic ball; $\ddot{x}_{bj}$, $\ddot{y}_{bj}$ and $\ddot{z}_{bj}$ represent the Jth ceramic ball in the coordinate system {O, X, Y, Z}; ω_{bxj}, ω_{byj} and ω_{bzj} represent the angular velocity of the ceramic ball in the coordinate system {O, X, Y, Z}; $\dot{\omega}_{bxj}$, $\dot{\omega}_{byj}$ and $\dot{\omega}_{bzj}$ represent the angular acceleration of the ceramic ball in the coordinate system {O, X, Y, Z}; $\dot{\theta}_{bj}$ represents the orbital velocity of the ceramic ball in the coordinate system {O, X, Y, Z}; and I_b is the moment of inertia of the ball in the coordinate system {O, X, Y, Z}.

The vibration differential equation of the ceramic ball can be described as:

$$\begin{array}{l} F_{byj} - F_{\xi oj} + T_{\xi oj} + G_{byj} + Q_{cyj} - F_{Dj} = m_b \ddot{y}_{bj} \\ F_{bzj} - F_{\eta oj} \sin \alpha_{oj} + T_{\eta oj} \sin \alpha_{oj} - Q_{oj} \cos \alpha_{oj} - G_{bzj} + Q \\ (T_{\eta oj} - F_{\eta oj}) \frac{D_W}{2} - M_{gyj} - J_y \dot{\omega}_{yj} = I_b \dot{\omega}_{byj} + I_b \omega_{bzj} \dot{\theta}_{bj} \\ [(T_{\xi oj} - F_{\xi oj}) \sin \alpha_{oj} - P_{\eta j}] \frac{D_W}{2} + M_{gzj} - J_z \dot{\omega}_{zj} = I_b \dot{\omega}_{bzj} + I_b \omega_{byj} \dot{\theta}_{bj} \end{array} \quad (4)$$

2.3. Properties of Lubricating Oil Properties

Good lubrication plays a crucial role in the operating condition of the bearing. If there is not enough lubrication, dry friction or boundary friction will occur between the ceramic balls and the raceway, causing friction and wear and producing more noise and even various failures. The total amount of oil supplied for full ceramic ball bearings during operation is:

$$q_m = \sum_{j=1}^{N} q_{ij} + \sum_{j=1}^{N} q_{oj} \quad (5)$$

where N represents the number of ceramic balls, and q_{ij} and q_{oj} represent the oil supply required by the ceramic ball in the contact area with the inner and outer ring, respectively. If q_j represents the oil supply required for the ceramic ball in the contact area with the inner or outer ring, then:

$$q_j = \rho_m h_c \mu_m \quad (6)$$

where ρ_m represents the lubricating oil density in the contact area; μ_m represents the average speed of the contact surface, that is, the equivalent speed; and h_c represents the thickness of the oil film in the contact area. Based on the results of Hamrock–Dowson [33], the lubricating oil film thickness can be calculated by the following formula:

$$h_c = 2.69 R_x U^{0.67} G^{0.53} W^{-0.067} \left(1 - 0.61 e^{-0.73}\right) \quad (7)$$

where $U = \eta_0 \mu_m / E' R_x$ represents the dimensionless velocity; η_0 represents the lubricant viscosity at 20 °C under the standard atmospheric pressure; $R_x = D_W(1 \mp \gamma_b)/2$ represents the equivalent radius of curvature, where the negative sign is used to calculate the equivalent radius of curvature of the inner ring and the positive sign is used to calculate the equivalent radius of curvature of the outer ring; $G = E' c_\eta$ represents the dimensionless elastic modulus; c_η represents the viscosity pressure coefficient; $W = Q/E' R_x^2$ represents the dimensionless load; and Q represents the contact load.

The most important factors affecting the viscosity of the lubricating oil are the pressure and temperature of the lubricating oil. The relationship between the lubricating oil viscosity and the lubricating oil temperature and pressure is [34]:

$$\eta = \eta_0 \exp\left\{ (\ln \eta_0 + 9.67) \times \left[(1 + 5.1 \times 10^{-9} p)^{0.68} \left(\frac{T_1 - 138}{T_0 - 138} \right)^{-1.1} - 1 \right] \right\} \quad (8)$$

where η_0 is the initial viscosity of the lubricating oil, T_0 is the ambient temperature, and T_1 is the actual temperature of the oil film.

2.4. Establishment of Sound Field Points of the Full Ceramic Ball Bearing

In order to facilitate the analysis of the sound field distribution characteristics of the full ceramic ball bearing, 24 field points are selected in the circumferential direction to calculate the radiated noise, as shown in Figure 2. All field points are evenly distributed in the same plane, the distance from the plane to the bearing plane is 50 mm, and the distance from each field point to the bearing axis is 210 mm. Therefore, the radial distance of each field point is 210 mm, the axial distance is 50 mm, and the angle between two adjacent field points is 15°. Here, the first field point is placed in the 12 o'clock direction and defined as 0° position angle or field point 1. The remaining field points are numbered counterclockwise.

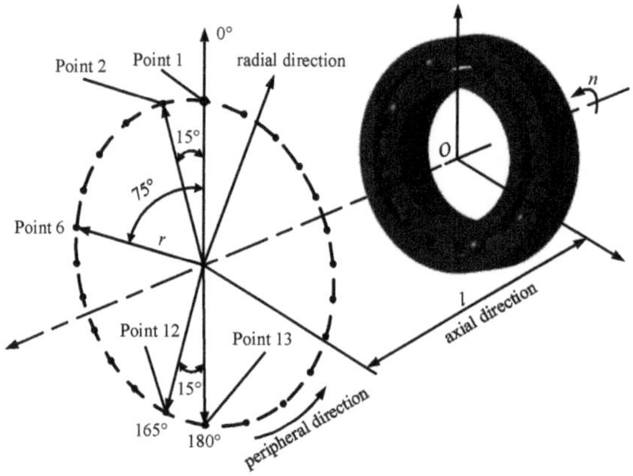

Figure 2. The distribution of acoustic field points in the circumferential direction of all ceramic ball bearing.

2.5. Characterization of the Sound Field

2.5.1. Sound Pressure Level of the Sound Field

In general, the sound pressure level is the main index used to evaluate and analyze sound radiation. When analyzing the radiated noise of a certain point in the sound field of full ceramic ball bearings, the effective value of sound pressure is generally used for analysis. The effective sound pressure level can be calculated by:

$$p_e = \sqrt{\frac{1}{T}\int_0^T p^2 dt} \tag{9}$$

where p_e represents the effective sound pressure level value, T represents sampling time, and p represents instantaneous sound pressure.

Therefore, the effective SPL at the analyzed field point can be obtained by Equation (2):

$$L = 20\lg\frac{p_e}{p_{ref}} \tag{10}$$

where p_{ref} refers to the reference sound pressure. In this paper, the reference sound pressure value is 2×10^{-5} Pa.

2.5.2. Directivity of the Sound Field

The sound field directivity of radiated noise is also an important index for evaluating the sound field performance and noise environment. In order to further analyze the sound field distribution characteristics of full ceramic ball bearings, the directivity of the sound field is used to characterize the distribution law of the sound field. At different positions of the sound field, there will be different SPLs (sound pressure levels). Even if the field point is the same distance from the sound source, SPLs of different field points will still be different, which indicates that the sound source has directivity and can also be expressed by the directivity of the sound field. Directional angle and directional level can be used to describe the directivity of the sound field [35]. Directional angle and directional level are called directivity parameters of the sound field, and they can be defined as:

$$DA = \theta_{\mathrm{SPL}}(n, r, l) \tag{11}$$

$$DL = \frac{SPL_{max}(n, r, l)}{SPL_{ave}(n, r, l)} \tag{12}$$

where SPL_{max} represents the maximum pressure level in the whole circumferential direction of a given radial distance r or represents the maximum pressure level in the angle range of a certain position at a given bearing speed n, radial distance r, and axial distance l. SPL_{ave} represents the average sound pressure level over the entire circumference of a given radial distance r. θ_{SPL} represents the angle of position where the maximum sound pressure level occurs. In the circumferential direction, the field point with the greatest radiated noise is also called the sensitive field point. Therefore, the direction of the position angle where the sensitive field point is located is the directional direction of the sound field.

2.6. Calculation Process

The numerical calculation process of this paper is shown in Figure 3.

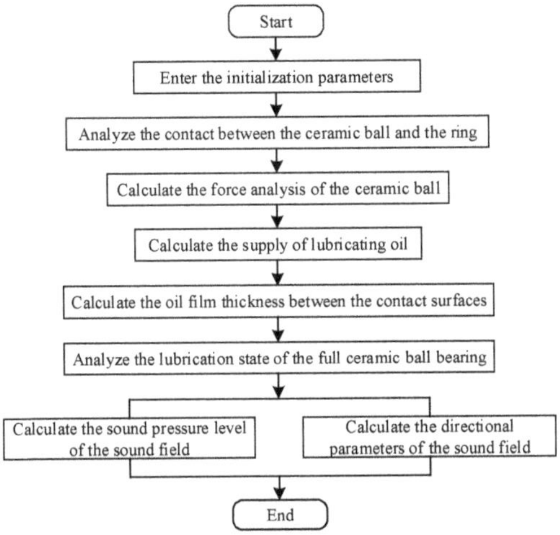

Figure 3. Flow chart of simulation calculation.

3. Simulation Analysis of the Influence of Oil Supply on Radiated Noise of the Full Ceramic Ball Bearing

3.1. Simulation Analysis of the Influence of Oil Supply on the Circumferential Sound Field Distribution of the Full Ceramic Ball Bearing

In order to analyze the influence of the lubrication state on the radiated noise of full ceramic ball bearings, taking the oil supply as the lubrication index, the influence of different oil supplies on the radiated noise distribution of full ceramic ball bearings is discussed. In the circumferential direction with a radius of 210 mm, 24 uniformly distributed field points are selected for analysis. The included angle between two adjacent field points is 15°, and the distance between each field point and the bearing plane is 50 mm. The bearing speed is set at 18,000 r/min and 24,000 r/min for calculation. The preload is set as 350 N, and the radiated noise of the full ceramic ball bearing is calculated when the oil supply is 0.010 mL/min, 0.015 mL/min, 0.020 mL/min, 0.025 mL/min, 0.030 mL/min and 0.035 mL/min respectively [36,37]. After calculation, the radial noise distribution in the circumferential direction of the full ceramic ball bearing with different oil supplies is shown in Figure 4.

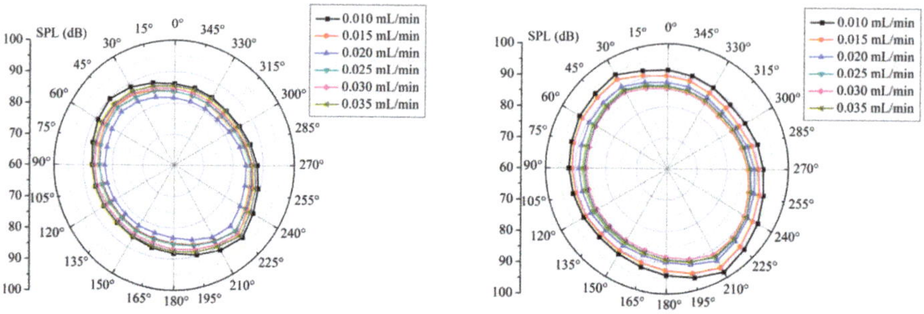

(a) (b)

Figure 4. Sound field distribution in the circumferential direction under different oil supplies. (**a**) n = 18,000 r/min; (**b**) n = 24,000 r/min.

It can be found that under different oil supplies, the radiated noise of the full ceramic ball bearing presents uneven distribution in the circumferential direction and shows an obvious direction of the sound field. With the increase in oil supply, the radiated noise decreases first and then increases. It can also be seen from the figure that at different speeds, the amount of oil supply required for small radiated noise is different. When the speed is 18,000 r/min, there is a relatively small SPL at the oil supply of 0.020 mL/min, while when the speed is 24,000 r/min, there is a relatively small SPL at the oil supply of 0.030 mL/min.

3.2. Simulation Analysis of the Influence of Oil Supply on the Characterization of Different Regions of the Sound Field of Full Ceramic Ball Bearings

In the range of 15~75° of the upper left semicircle, because of the small contact force between the ceramic balls and the ring, the impact effect of the bearing ceramic balls, the ring and the cage will produce a large impact noise, so this area will produce a large radiated noise and is called the impact load zone. In the range of 195~255° of the lower right semicircle, due to the large contact force between the ceramic balls and the ring, the violent interaction between the ceramic balls and the bearing inner and outer ring produces a large friction noise [38,39], so this area will also produce a relatively large radiation noise and is called the friction load zone. Other regions are called stationary load zone because of the relatively small friction and impact and relatively stable radiation noise changes.

It can be seen from Figure 5 that the radiated noise in the friction load zone is higher than that in the impact load zone, and their variation trends are similar to the increase in oil supply. When the rotational speed is 18,000 r/min, the variation trend first decreases and then increases. When the rotational speed is 24,000 r/min, the SPL decreases with the increase in oil supply. Furthermore, when the oil supply is large, the amplitude of SPL increase tends to decrease.

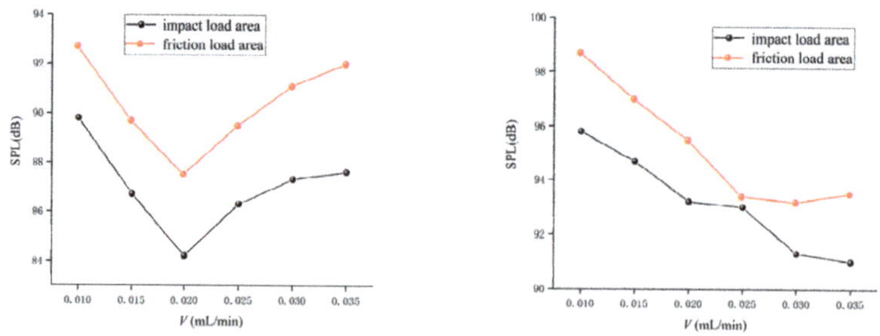

(a) (b)

Figure 5. Variation of maximum sound pressure level in impact load zone and friction load zone with oil supply. (**a**) n = 18,000 r/min; (**b**) n = 24,000 r/min.

In order to improve the analysis of how oil supply affects the directivity of the sound field, the variation trend of the directional level of the sound field in the impact load zone and friction load zone with the oil supply at different rotational speeds is simulated and calculated.

As can be seen from Figure 6, when the rotational speed is 18,000 r/min, the directional level of the impact load zone gradually decreases with the increase in the oil supply, while the directional level of the friction load zone firstly decreases and then increases. When the rotational speed is 24,000 r/min, it is the opposite. With the increase in oil supply, it can be found that the radiated noise of the full ceramic ball bearing has different directional levels in different zones.

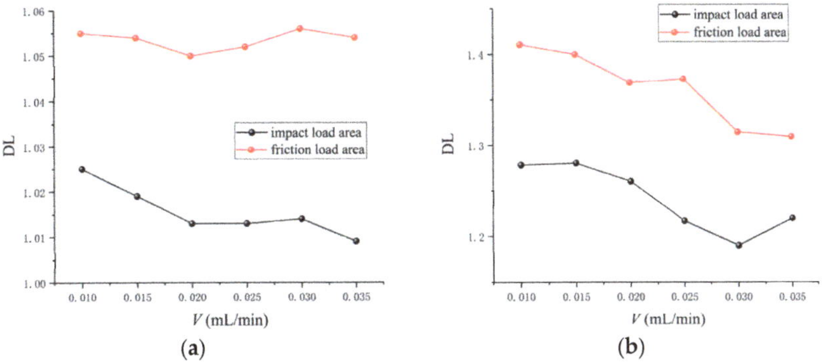

Figure 6. Variation of the directional level(DL) in impact load zone and friction load zone with oil supply. (**a**) n = 18,000 r/min; (**b**) n = 24,000 r/min.

Based on the above analysis, under the preload of 350 N applied to the bearing, when the rotational speed is 18,000 r/min, the oil supply required for the bearing to produce the minimum radiation noise is 0.020 mL/min. When the rotational speed is 24,000 r/min, the oil supply required for the bearing to produce the minimum radiation noise is 0.030 mL/min. With the increase in rotational speed, the oil supply of the full ceramic ball bearing when the minimum radiation noise is generated also increases.

4. Simulation Analysis of the Influence of Oil Supply on the Operation Characteristics of the Full Ceramic Ball Bearing

4.1. Geometric Modeling and Meshing

In this paper, the 6206 silicon nitride ceramic ball bearing is taken as the research object. The solid model of the 6206 deep groove ball bearing is drawn using three-dimensional modeling software, including the rolling element, bearing inner and outer ring, and cage, and then the patrs are assembled. Figure 7 shows the bearing model.

The bearing model is imported into the pre-processing software, and the fluid domain of the bearing is extracted. The research focus of this paper is the fluid domain part of the bearing, so the solid domain part of the bearing is suppressed, only the fluid domain part of the bearing is retained, and two inlets are added. The inlets are located above and below the end face of the model, the diameter is 2 mm, and the outlet is located at the other end face of the model. The bearing cavity model is meshed. Figure 8 is the bearing cavity grid. The number of elements in the overall calculation domain is 1,300,381, the number of nodes is 260,735, and the minimum mass of the grid is greater than 0.2, which meets the calculation requirements.

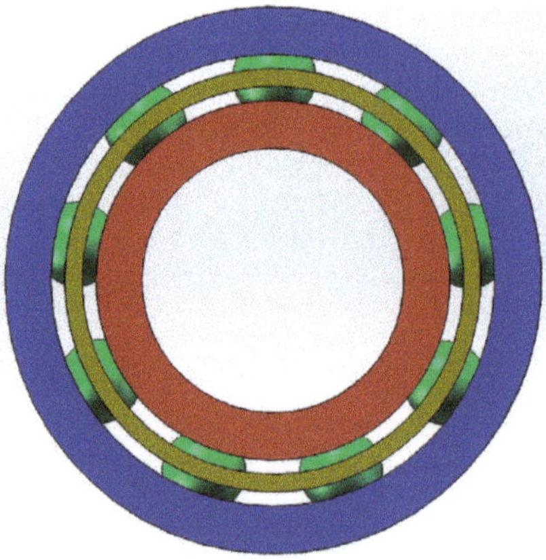

Figure 7. Bearing model.

Figure 8. Bearing cavity grid.

4.2. Boundary Condition Setting

The bearing cavity grid model is imported into Fluent for simulation analysis. The simulation parameters are set as follows:

(1) The lubricating inlet is set to speed inlet, and the inlet speed is set. The outlet is set to the pressure outlet, and the value is standard atmospheric pressure.

(2) The outer ring of the bearing is set as a stationary wall surface, and the surfaces in contact with the inlet and outlet areas are also set as stationary wall surfaces. Considering the revolution motion of the rolling element and the cage, the inner ring, the rolling element and the cage are set as the rotating wall surface, ignoring the spin motion of the rolling element, and the cage speed is the same as the rolling element.

(3) The heat flux at the corresponding speed is loaded on the inner and outer rings and balls.

(4) The VOF model [40] is used to solve the flow state of the oil and gas phases. Air is the main phase, and lubricating oil is the secondary phase. In the VOF method, the volume fraction α is used to mark the volume fraction of the oil phase. Then, $\alpha = 0$ represents an

oil-free unit, and $\alpha = 1$ represents an oil-filled unit. If $0 < \alpha < 1$, it represents the interface between the oil phase and the gas phase.

4.3. Simulation Analysis of the Influence of Oil Supply on the Lubrication Characteristics of Full Ceramic Ball Bearings

When the bearing runs at high speed, the lubricant distribution and content of the key lubrication areas near the contact area, cage pocket hole, and rolling element surface are vital to the overall lubrication of the bearing and its operational reliability.

Figure 9 shows the distribution of oil volume fraction on the circumference under different oil supplies when the rotational speed is 24,000 r/min. In the bearing cavity, the lubricating oil is subjected to centrifugal force for radial movement and also circumferential movement by the influence of the cage and inner ring. The high-speed spin balls make the lubricating oil distribute in the contact area near the ball. The distributing oil is not uniformly distributed in the circumferential direction, and the content varies with the azimuth angle.

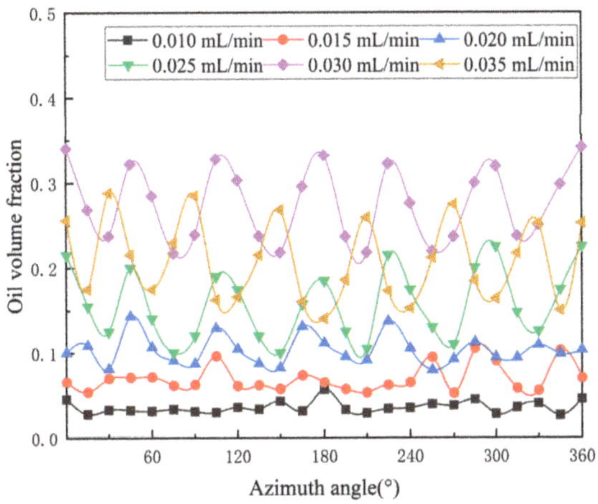

Figure 9. The distribution of oil volume fraction on the circumference under different oil supplies.

The average oil volume fraction inside the bearing under different oil supplies is shown in Figure 10. Figure 11a,b, respectively, show the oil volume fraction distribution inside the full ceramic ball bearing under the conditions of a rotation speed of 18,000 r/min and oil supply of 0.020 mL/min, and a rotation speed of 24,000 r/min and oil supply of 0.030 mL/min. The results show that with the increase in oil supply, the average oil volume fraction increases first and then decreases. When the bearing oil supply is small, the bearing cavity is in a starved state, and the oil film thickness of the two-phase flow is thin and incomplete, which is not conducive to bearing lubrication and heat transfer. When the oil supply increases, the oil film thickness of the two-phase flow increases and becomes more complete, enhancing the lubrication performance and decreasing the bearing temperature. When the oil supply increases to a certain amount, the oil film thickness of the two-phase flow reaches the maximum, and the bearing is in the best lubrication state. As the oil supply continues to increase, the oil film thickness decreases and gradually becomes incomplete. This is because the excess lubricating oil inside the bearing will cause churning and heat generation, resulting in a sharp rise in the bearing temperature and a decrease in the lubricating oil viscosity, thus thinning the oil film.

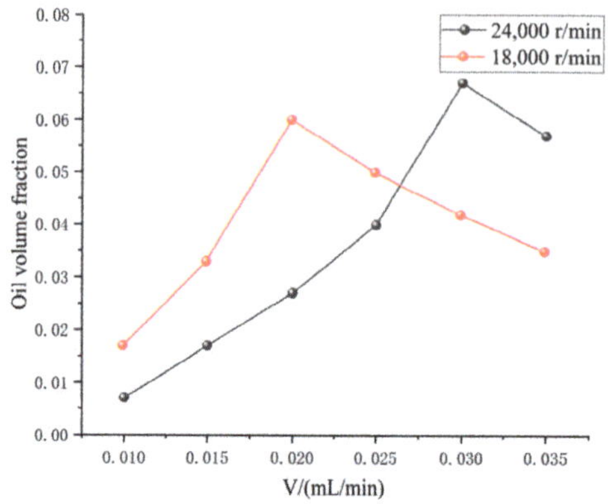

Figure 10. The average oil volume fraction inside the bearing under different oil supplies.

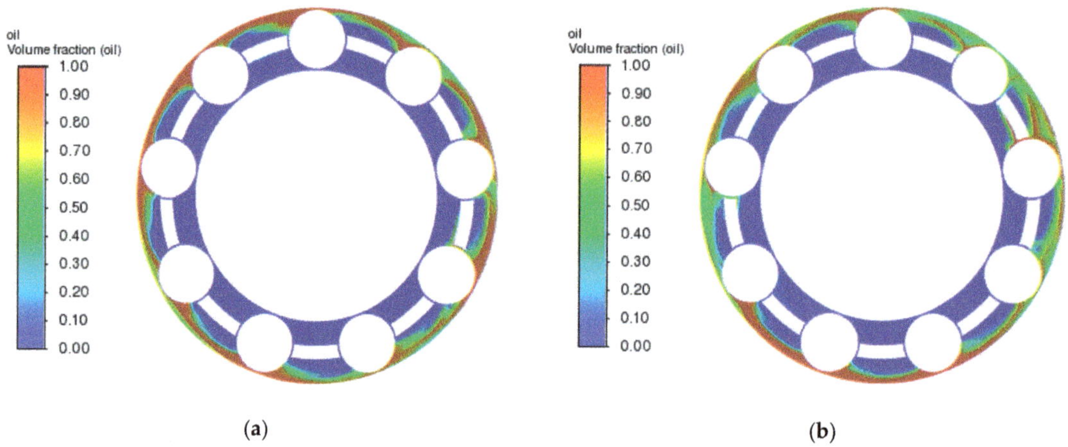

Figure 11. Oil phase distribution in bearing cavity under different working conditions. (**a**) n = 18,000 r/min, v = 0.020 mL/min; (**b**) n = 24,000 r/min, v = 0.030 mL/min.

Figure 12 shows the influence of different oil supplies on the temperature rise of the outer ring of the full ceramic ball bearing at the rotational speed (18,000 r/min and 24,000 r/min). Figure 13a,b, respectively, show the oil phase distribution in the full ceramic ball bearing under the conditions of a rotation speed of 18,000 r/min and oil supply of 0.020 mL/min, and a rotation speed of 24,000 r/min and oil supply of 0.030 mL/min. With the increase in oil supply, the temperature rise decreases first and then increases. Each speed has a minimum oil supply, and the temperature rise increases with the increase in speed.

Figure 14 shows the influence of different oil supplies on the vibration acceleration of the outer ring of the full ceramic ball bearing at the rotational speed of 18,000 r/min and 24,000 r/min. The vibration acceleration decreases as the oil supply increases, and there is an optimum oil supply that minimizes the vibration acceleration. In addition, the vibration acceleration increases with the increase in speed. When the oil supply is low and the oil film thickness is low, little oil is introduced into the contact area of the rollers and raceways, which leads to oil pressure fluctuations and high bearing vibration. As the oil supply increases, the amount of oil in the bearing cavity increases, creating a thick oil film that helps reduce vibration.

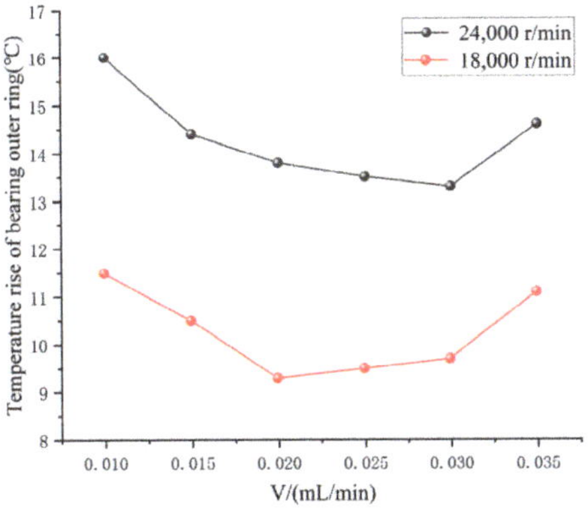

Figure 12. Temperature change in bearing outer ring under different oil supply.

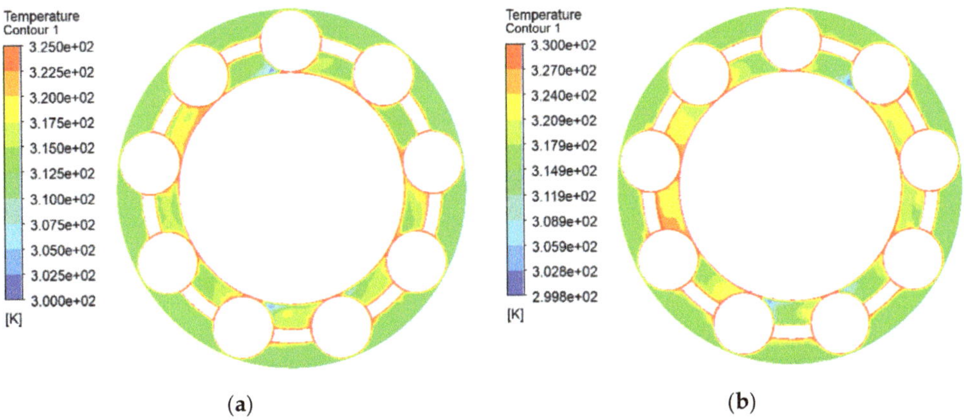

Figure 13. The temperature distribution in the bearing chamber under different working conditions. (**a**) n = 18000 r/min, v = 0.020 mL/min; (**b**) n = 24000 r/min, v = 0.030 mL/min.

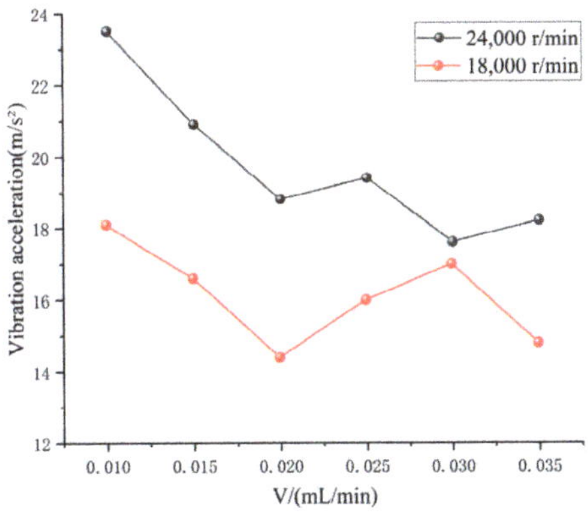

Figure 14. Vibration acceleration of bearing outer ring under different oil supplies.

5. Experimental Analysis of the Influence of Oil Supply on the Sound Field Distribution of the Full Ceramic Ball Bearing

5.1. Experimental Scheme and Method

According to the analysis of the simulation results, we set the working conditions in the experiment as the preload adjusted to 350 N and two set rotational speeds. Six groups of radiated noise tests with different oil supplies are carried out. During the test, the settings of other parameters are the same as during the simulation calculation.

The influence of the oil supply on the vibration and noise characteristics of the full ceramic ball bearing under oil lubrication conditions is tested by using the upgraded rolling bearing lubrication vibration test platform. The test device and the test process are shown in Figure 15.

Figure 15. Experimental equipment and signal acquisition equipment.

The full ceramic ball bearing in this test is a P4 ultra-precision ball bearing. The type is a 6206 deep groove ball bearing. The material of the bearing inner and outer ring and balls is silicon nitride ceramic. The material of the cage is PVX. Its structural parameters are shown in Table 1.

Table 1. Structural parameters of the test bearing.

Parameter Name/Unit	Bearing Data
Bearing bore diameter/mm	30
Bearing outside diameter/mm	62
Bearing width/mm	16
Amount of balls	9
Elastic modulus of silicon nitride material/GPa	300
Poisson's ratio of silicon nitride material	0.27

5.2. Analysis of Test Results of the Influence of Oil Supply on the Circumferential Sound Field Distribution of Full Ceramic Ball Bearings

Figure 16 shows the circumferential distribution of the radiated noise of the full ceramic ball bearing with different oil supplies. With the increase in oil supply, the change in bearing radiation noise is more complicated, and the radiation noise varies greatly in the impact load area and the friction load area. When the oil supply increases from 0.010 mL/min to 0.035 mL/min, the change in radiated noise is first large, then small, and finally gentle. In addition, it can be seen from the figure that the overall variation trend of radiated noise with oil supply is consistent with the simulation results.

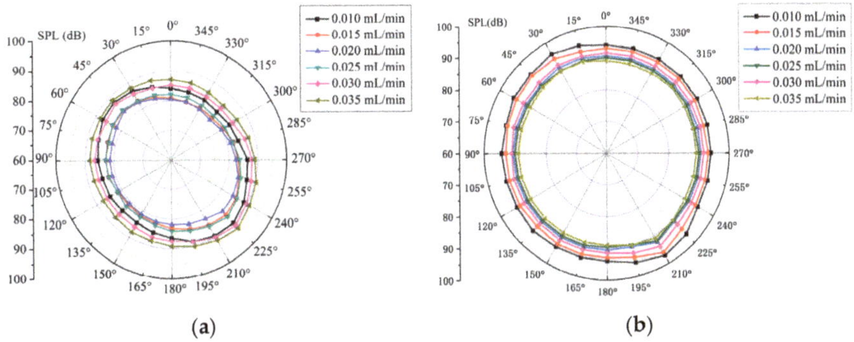

Figure 16. Test results of circumferential sound field distribution with different oil supplies. (**a**) n = 18,000 r/min; (**b**) n = 24,000 r/min.

5.3. Influence of Oil Supply on the Characterization of Different Sound Field Regions of the Full Ceramic Ball Bearing

Figure 17 shows the curves of the sound pressure level of bearing radiated noise in the direction of the impact load and friction load with the change in oil supply. It can be seen from the figure that the radiation noise in the impact load area is weaker than that in the friction load area. When the rotational speed is 18,000 r/min, the maximum sound pressure level decreases first and then increases with the increase in oil supply in the impact load zone and friction load zone. The minimum sound pressure level is 83.15 dB and 84.83 dB when the oil supply is 0.020 mL/min. When the speed is 24,000 r/min, the change curves of the impact load and friction load area are consistent, and both reach the minimum sound pressure level when the oil supply is 0.030 mL/min.

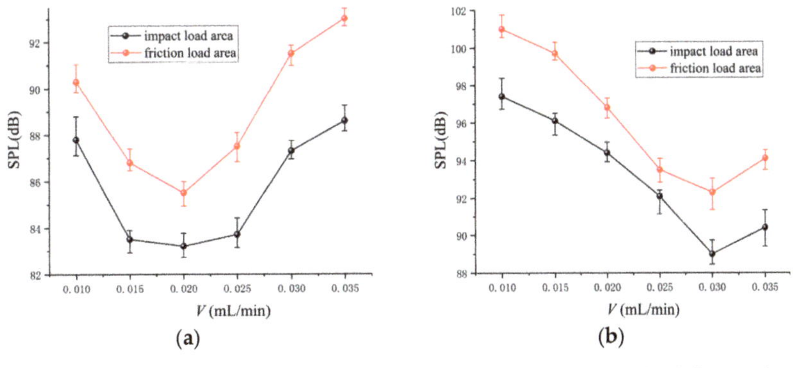

Figure 17. Test results of SPL in impact load zone and friction load zone under different oil supplies. (**a**) n = 18,000 r/min (**b**) n = 24,000 r/min.

Figure 18 is the curve of the directional level of the sound field of bearing radiated noise in the impact load zone and friction load zone with the change in oil supply. It can be seen from the figure that, compared with the friction load zone, the sound field directivity of the impact load area is relatively weak. With the increase in oil supply, the directional level of the sound field in the impact load zone tends to weaken at different rotational speeds. In the friction load zone, more complex changes in the directivity of the sound field When the oil supply is small (0.010 mL/min), the directivity of the sound field tends to weaken with the increase in the oil supply, but when the oil supply reaches a certain value, the directivity of the sound field is enhanced.

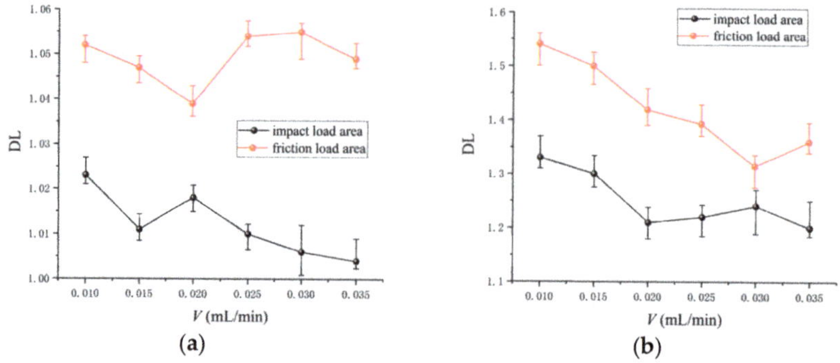

Figure 18. Test results of DL in impact load zone and friction load zone under different oil supplies. (**a**) n = 18,000 r/min (**b**) n = 24,000 r/min.

The increase in lubricating oil reduces the impact and has a certain absorption effect on the radiated noise. Due to the increase in lubricating oil, the contact friction between the ceramic balls and the ring decreases, which leads to the above changes in the sound field directivity of the friction load area. Moreover, the oil film absorbs the noise to some extent, which results in a decrease in the radiated noise in this area, and the radiated noise in the circumferential direction also becomes more uniform, so the directivity of the sound field is weakened. However, when there is more lubricating oil, it will not only increase the friction resistance between bearing assemblies and the lubricating oil film but also make the temperature rise greatly, reducing the gap between the ceramic balls and the ring and increasing the friction radiation noise. Too much lubricating oil will also destroy the lubrication state so that the radiated noise in the whole circumferential direction increases and tends to be uniform, resulting in a slight weakening of the sound field directivity.

In conclusion, although the ceramic material has a small thermal expansion coefficient, it still has a certain amount of deformation at a higher temperature. When the lubricating oil is too much, the temperature of the full ceramic ball bearing will rise faster, and the temperature of the inner ring is higher than that of the outer ring so that the clearance between ceramic balls and the ring decreases slightly, which leads to the increase in the radiated noise of the full ceramic ball bearing. In addition, the cage has a large thermal expansion coefficient, which is prone to produce large deformation at high temperatures, leading to the increase in the clearance and the gap between the pocket hole and the ceramic ball, which reduces the stability of the cage and intensifies the impact between the pocket hole and the ceramic ball.

5.4. Analysis of the Influence of Oil Supply on the Sound Field Characteristics of the Full Ceramic Ball Bearing

Figure 19 shows the variation curve of the SPL of the sound field directivity with the increase in oil supply for the full ceramic ball bearing. It can be seen from the figure that at different speeds, the maximum sound pressure level of the radiated noise of the full ceramic ball bearing decreases first and then increases with the increase in the oil supply, and there is a relatively large radiated noise at higher speeds. In the process of increasing the oil supply from 0.020 mL/min to 0.035 mL/min, the radiation noise at the speed of 18,000 r/min showed a gradually increasing trend. This indicates that when the fuel supply is greater than a certain amount, increasing the oil supply still leads to an increase in radiation noise, but the degree of increase is weaker and weaker.

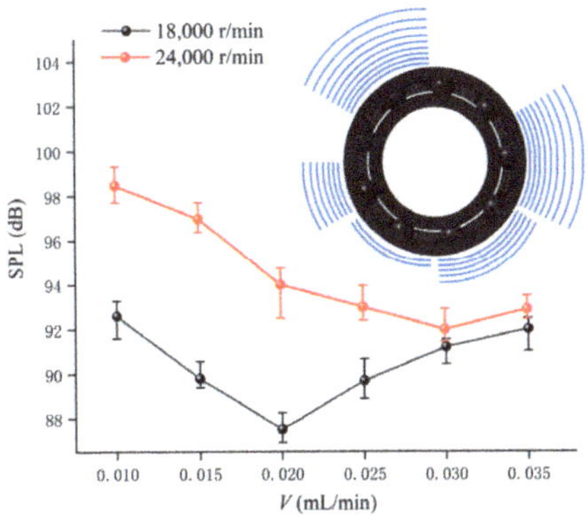

Figure 19. The change in SPL in sound field directivity with different oil supplies.

Figure 20 shows the variation of radiated noise sound pressure levels in the circumferential direction with different oil supplies. Combined with Figures 19 and 20, when the rotational speed is 24,000 r/min, with the increase in oil supply, the changing trend of the sound pressure level in the circumferential direction of radiated noise is consistent with the changing trend of the maximum sound pressure level. When the oil supply is 0.030 mL/min, the radiation noise of the full ceramic ball bearing is evenly distributed in the circumferential direction. When the speed is 18,000 r/min, and the oil supply is 0.020 mL/min, the change in sound pressure level in the circumferential direction is small. This shows that the distribution of the radiated noise of the full ceramic ball bearing is uniform in the circumferential direction.

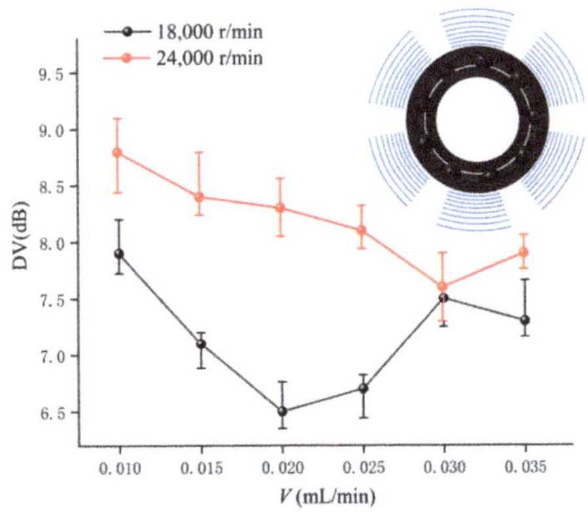

Figure 20. The variation of sound pressure level in the circumferential direction with different oil supplies.

Figure 21 shows the variation curve of the DL of the sound field directivity with the oil supply. Combined with Figures 19 and 21, it can be seen that in the range of oil supply from 0.010 mL/min to 0.025 mL/min, the directional level of the sound field at 24,000 r/min is larger relative to the sound pressure level than that at the speed of 18,000 r/min. However, when the oil supply is greater than 0.025 mL/min, the directivity of the sound field at 18,000 r/min is increasing, while that at 24,000 r/min is still decreasing, resulting in the directivity of the sound field at 24,000 r/min being gradually lower than that at 18,000 r/min, but the sound pressure level at high speed is still relatively high.

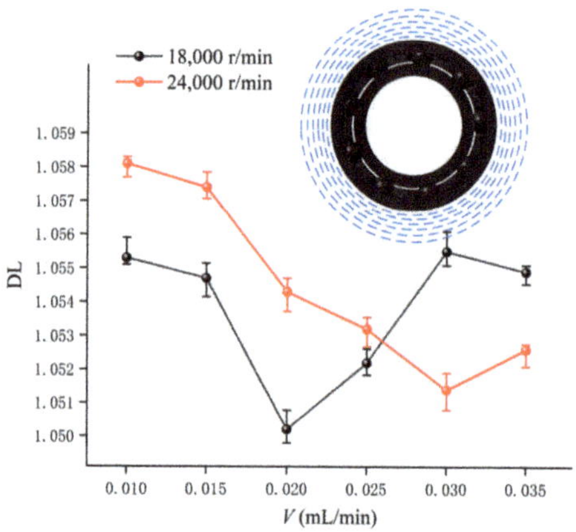

Figure 21. The change in the DL in sound field directivity with different oil supplies.

5.5. Comparative Analysis of the Test Results and Simulation Results

Figure 22 shows the simulation results and experimental results at different oil supplies. It is calculated that the relative error of radiation noise in the circumferential direction is less than 4.0% under different fuel supplies. When the oil supply is 0.030 mL/min and 0.035 mL/min, the test results are higher than the experimental results. According to the absolute value of the calculated relative error, when the oil supply is 0.020 mL/min, the relative error in the direction of 210° is the largest, which is 3.98%. When the oil supply is 0.030 mL/min, the relative error of the position angle of 45° is the smallest, which is 0.16%. It can be seen that the error at this time is very small.

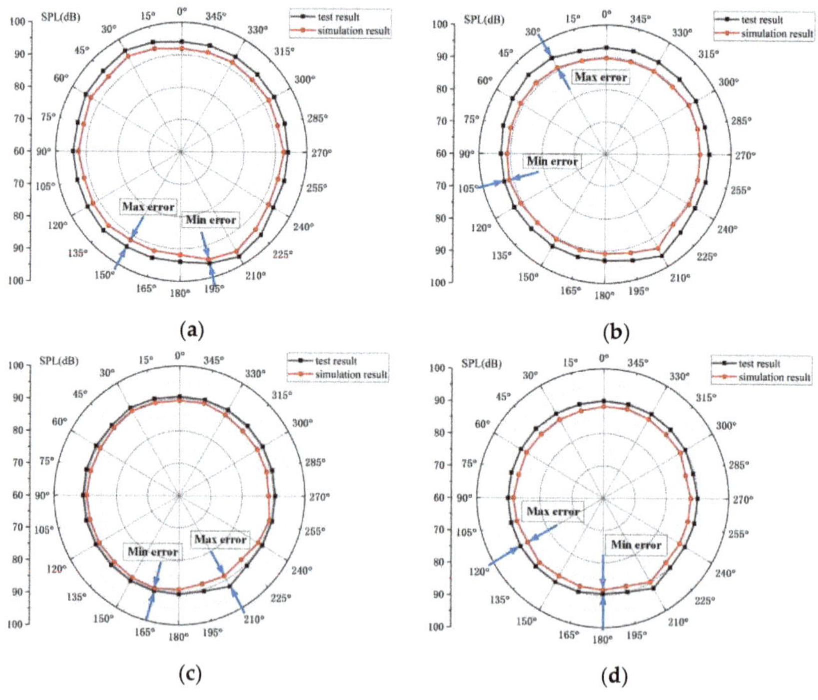

Figure 22. *Cont.*

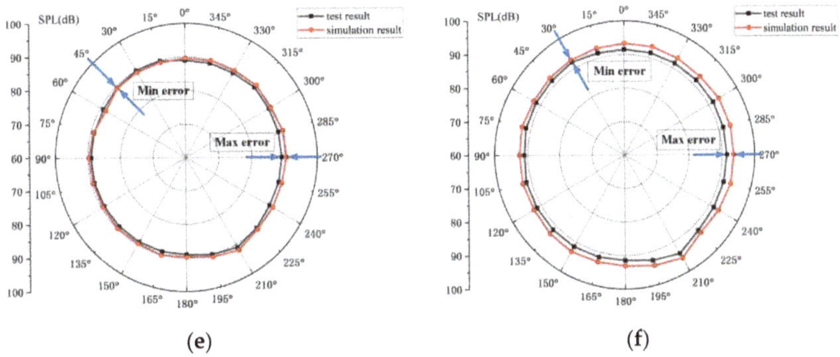

Figure 22. Test results and simulation results at different oil supplies. (**a**) v = 0.010 mL/min; (**b**) v = 0.015 mL/min; (**c**) v = 0.020 mL/min; (**d**) v = 0.025 mL/min; (**e**) v = 0.030 mL/min; (**f**) v = 0.035 mL/min.

Because the working equipment of bearings (such as the spindle and bearing housing) has a certain barrier and absorption effect on the transmission of the bearing sound signal, the general simulation results are greater than the test results, but in the case of large oil supply, the test results are greater than the simulation results. This is because in the test process, the full ceramic angular contact ball bearing is installed in the spindle, and the heat generated in its operation is not easy to dissipate, leading to a higher temperature rise in the bearing so that deviation appears in the results. In addition, the outer ring is in contact with the bearing housing, and the bearing housing is in contact with the outside air. The temperature is generally lower than the inner ring, the expansion of the inner ring is larger, the thermal expansion coefficient of the silicon nitride ceramic ball is smaller than the ring, and the deformation of the ball is very small so that the bearing clearance is reduced leading to a large friction noise. Additionally, the increase in temperature also makes the gap between the outer ring and the bearing housing larger, resulting in larger vibration noise, so that when the oil supply is larger, the test results are greater than the calculation results. Comparing the simulation calculation results with the test results, the error is within the acceptable range, and with the increase in the oil supply, the simulation results are consistent with the changing trend of the test results, so the accuracy and validity of the simulation results of the radiated noise of the full ceramic ball bearing with the oil supply are verified.

6. Conclusions

In this paper, the influence of lubricating oil supply on the sound field distribution characteristics of full ceramic ball bearings is studied. Some of the conclusions are as follows:

(1) Under different oil supplies, the sound pressure signal of the full ceramic ball bearing shows an uneven distribution in the circumferential direction. With the increase in oil supply, it shows a trend of decreasing and then increasing. With the increase in rotational speed, the amount of oil supply when the bearing produces the minimum radiation noise also increases. At a given speed and preload, there is an optimal amount of oil supply to minimize the radiation noise of full ceramic ball bearings.

(2) Under different oil supplies, the oil phase distribution in the circumferential direction is not uniform in the full ceramic ball bearing cavity. With the increase in oil supply, the oil volume fraction increases first and then decreases, while the temperature rise and vibration acceleration of the bearing outer ring decreases first and then increases. There is an optimal oil supply to ensure the lubrication performance of the bearing reaches the optimal state.

(3) At different rotational speeds, the maximum sound pressure level and sound pressure level variation of the full ceramic ball bearing shows a trend of decreasing and

then increasing with the increase in oil supply and are relatively larger at higher rotational speeds. The relative error between the simulation results and the test results of the radiated noise in the circumferential direction at different oil supply amounts does not exceed 4%.

Author Contributions: Conceptualization, R.G.; methodology, Z.Z.; software, G.Z.; writing—original draft preparation, J.S. and X.F.; writing—review and editing, J.Y. All authors have read and agreed to the published version of the manuscript.

Funding: The authors acknowledge the collective support granted by the National Natural Science Foundation of China (Grant No. 52105196), the Education Department of Liaoning Province (Grant No. LJKMZ20220936), Young and Middle-aged Innovation Team of Shenyang (Grant No. RC210343). National defense science and technology innovation zone plan: ultra-high precision ceramic bearing (No. 20-163-00-TS-006-002-11).

Data Availability Statement: Not applicable.

Conflicts of Interest: The authors declare no conflict of interest.

Nomenclature

c	An undetermined constant
c_η	The viscosity pressure coefficient
D_W	The diameter of the ceramic ball
E	Equivalent elastic modulus
E_e	The elastic modulus of the outer raceway
E_r	The elastic modulus of ball bearings
E'	The elastic modulus parameter
E, F	The first and second types of complete ellipse integral
$F_{bxj}, F_{byj}, F_{bzj}$	The hydrodynamic viscous resistance component acting on the ceramic ball
$F_{\eta ij}, F_{\eta oj}, F_{\xi ij}, F_{\xi oj}$	The rolling friction force between the ceramic ball and raceway
G	The dimensionless elastic modulus
G_{byj}, G_{bzj}	The gravity component of the ceramic ball along the coordinate direction
h_c	The thickness of the oil film in the contact area
I_b	Moment of inertia of the ball
j	The jth ball
J_x, J_y, J_z	The component of inertia of the ceramic ball rotating about its own center
K	The contact stiffness
m_b	The mass of the ceramic ball
N	The number of ceramic balls
$\{O; X, Y, Z\}$	The inertial coordinate system of the bearing
p	Instantaneous sound pressure
p_e	Effective sound pressure level value
p_{ref}	The reference sound pressure
$p(s)$	The fluid pressure distribution
$P_{\eta j}, P_{\xi j}$	The friction force acting on the surface of the ceramic ball
q_{ij}, q_{oj}	The oil supply required by the ceramic ball in the contact area with raceway
Q	Normal contact force between the ball and raceway
Q_{cxj}, Q_{cyj}	The components of the impact force between the ceramic ball and the cage along the coordinate direction
Q_{ij}, Q_{oj}	Normal contact force between the ceramic ball and orbit
Rx	The equivalent radius of curvature
s	The distance between $p(s)$ and the origin of the coordinate
s_1	The starting point of load
s_2	The ending point of load
T	Sampling time
T_0	The ambient temperature
T_1	The actual temperature of oil film
$T_{\eta i}, T_{\eta oj}, T_{\xi ij}, T_{\xi o}$	Traction force of the contact surface between the ceramic ball and orbit

U	The dimensionless velocity
$v(x)$	The elastic displacement in the vertical direction
v_e	The Poisson's ratio of the outer raceway
v_r	The Poisson's ratio of ball bearings
W	The dimensionless load
α	Volume fraction of oil phase
α_{ij}, α_{oj}	Contact angle between the ceramic ball and orbit
δ	The contact deformation
η_0	The lubricant viscosity at 20 °C under the standard atmospheric pressure
$\dot{\theta}_{bj}$	Orbital velocity of the ceramic ball
θ_{SPL}	The angle of position where the maximum sound pressure level occurs
κ	The eccentricity parameter of the ellipse
μ_m	The average speed of the contact surface
ρ_m	The lubricating oil density in the contact area
$\omega_{bxj}, \omega_{byj}, \omega_{bzj}$	The angular velocity of the ceramic ball
$\dot{\omega}_{bxj}, \dot{\omega}_{byj}, \dot{\omega}_{bzj}$	The angular acceleration of the ceramic ball
$\omega_{xj}, \omega_{yj}, \omega_{zj}$	The angular velocity of the ceramic ball in its coordinate system
$\ddot{x}_{bj}, \ddot{x}_{bj}, \ddot{x}_{bj}$	The displacement acceleration of the ceramic ball
Subscript i	Represents the inner ring
Subscript o	Represents the outer ring

Abbreviations

EHL	Elastohydrodynamic lubrication
SPL	Sound pressure level
DL	Directional level
DV	Directional variation
VOF	Volume of fluid

References

1. Yao, J.; Wu, Y.; Sun, J.; Xu, Y.; Wang, H.; Zhou, P. Research on the metamorphic layer of silicon nitride ceramic under high temperature based on molecular dynamics. *Int. J. Adv. Manuf. Technol.* **2020**, *109*, 1249–1260. [CrossRef]
2. Yao, J.; Wu, Y.; Sun, J.; Tian, J.; Zhou, P.; Bao, Z.; Xia, Z.; Gao, L. Friction and wear characteristics of silicon nitride ceramics under dry friction condition. *Mater. Res. Express* **2021**, *8*, 035701. [CrossRef]
3. Sun, J.; Wu, Y.; Zhou, P.; Li, S.; Zhang, L.; Zhang, K. Simulation and experimental research on Si_3N_4 ceramic grinding based on different diamond grains. *Adv. Mech. Eng.* **2017**, *9*, 9–14. [CrossRef]
4. Han, X.X.; Xu, C.H.; Jin, H.; Xie, W.H.; Meng, S.H. An experimental study of ultra-high temperature ceramics under tension subject to an environment with elevated temperature, mechanical stress and oxygen. *Sci. China Technol. Sci.* **2019**, *62*, 1349–1356. [CrossRef]
5. Sun, J.; Zhou, P.; Wu, Y.H.; Zhang, K.; Zhang, L.X. Analysis on the factors of surface morphologies on Si_3N_4 ceramic internal grinding. *Int. J. Eng. Res. Afr.* **2017**, *31*, 44–52. [CrossRef]
6. Kevin, L.; Muhammad, K. Interdependence of friction, wear, and noise: A review. *Friction* **2021**, *9*, 1319–1345.
7. Wang, J.; Zhuang, W.; Liang, W.; Yan, T.; Li, T.; Zhang, L.; Li, S. Inorganic nanomaterial lubricant additives for base fluids, to improve tribological performance: Recent developments. *Friction* **2022**, *10*, 645–676. [CrossRef]
8. Zhu, W.; Zhang, X.; Lu, Y.; Wang, D. Numerical study on influence of lubrication mode on lubrication performance of ball bearing. *J. Propuls. Technol.* **2019**, *40*, 892–901.
9. Chen, C.; Li, J.; Yu, Y.; Xue, Y. Research on Temperature Rise Characteristics of High Speed Angular Contact Ball Bearings in Oil-Gas Lubrication. *Mach. Des. Manuf.* **2021**, 216–221,227. [CrossRef]
10. Nichols, B.R.; Fittro, R.L.; Goyne, C.P. Steady-state tilting-pad bearing performance under reduced oil supply flow rates. *J. Tribol.* **2018**, *140*, 051701. [CrossRef]
11. Yan, H.; Hu, B.; Niu, H.; Zhu, J.; Wu, Y. Radiation noise characteristics in grinding workpieces by the motorized spindle with full ceramic ball bearings. *Manuf. Technol. Mach. Tool* **2021**, *4*, 111–114,125.
12. Yan, H.; Wu, Y.; Wang, H. Analysis of Radiation Noise of High Speed Angular Contact Ceramic Ball Bearing Electric Spindle. *Manuf. Technol. Mach. Tool* **2019**, *4*, 145–148.
13. Li, D.; Qi, H.; Hou, D.; Kong, F.; Hong, M.; Han, D. Research on Acoustic Emission Detection Mechanism of Axle Box Bearing State of EMU. *J. Mech. Eng.* **2021**, *57*, 153–160. [CrossRef]
14. Xiong, S.; Zhou, R. Influence of bearing stiffness on radiated noise of ship hull. *Ship Ocean Eng.* **2017**, *46*, 86–89+93.

15. Zhang, Q.; An, Q. Calculation method of motion noise of inner ring and rolling element of deep groove ball bearings. *J. East China Univ. Sci. Technol. (Nat. Sci. Ed.)* **2018**, *44*, 935–944.
16. Wang, J.; Qiu, Q.; Zhou, G.; Li, J.; Xu, T. Vibration and noise analysis and experimental study of water-lubricated bearings. *J. Hunan Univ. (Nat. Sci. Ed.)* **2015**, *42*, 53–58.
17. Peng, L.; Zheng, H.; Shi, Z. Influence of oil supply on static and dynamic lubrication characteristics of floating ring bearing. *Mach. Tool Hydraul.* **2019**, *47*, 6–9.
18. Nam, J.; Baek, J.; Do, H.; Kang, J. Experimental investigation of friction noise on lubricated contact. *J. Mech. Sci. Technol.* **2017**, *31*, 5751–5760. [CrossRef]
19. Zhang, Q.; Yang, J.; An, Q. Noise Calculation Method for Deep Groove Ball Bearing With Considering Raceway Surface Waviness and Roller Size Error. *Front. Mech. Eng.* **2018**, *4*, 13. [CrossRef]
20. Jahagirdar, A.; Mohanty, S.; Gupta, K.K. Study of noise effect on bearing vibration signal based on statistical parameters. *Vibroeng. Proc.* **2018**, *21*, 26–31. [CrossRef]
21. Botha, J.D.M.; Shahroki, A.; Rice, H. An implementation of an aeroacoustic prediction model for broadband noise from a vertical axis wind turbine using a CFD informed methodology. *J. Sound Vib.* **2017**, *410*, 389–415. [CrossRef]
22. Gupta, P.K.; Winn, L.W.; Wilcock, D.F. Vibrational Characteristics of Ball Bearings. *J. Lubr. Tech.* **1977**, *99*, 284–287. [CrossRef]
23. Lynagh, N.; Rahnejat, H.; Ebrahimi, M.; Aini, R. Bearing induced vibration in precision high speed routing spindles. *Int. J. Mach. Tools Manuf.* **2000**, *40*, 561–577. [CrossRef]
24. Haneef, M.D.; Randall, R.B.; Smith, W.A.; Peng, Z. Vibration and wear prediction analysis of IC engine bearings by numerical simulation. *Wear* **2017**, *384*, 15–27. [CrossRef]
25. Abboud, D.; Elbadaoui, M.; Smith, W.A.; Randall, R.B. Advanced bearing diagnostics: A comparative study of two powerful approaches. *Mech. Syst. Signal Process.* **2019**, *114*, 604–627. [CrossRef]
26. Yang, J.; Wu, C.; Shan, Z.; Liu, H.; Yang, C. Extraction and enhancement of unknown bearing fault feature in the strong noise under variable speed condition. *Meas. Sci. Technol.* **2021**, *32*, 105021. [CrossRef]
27. Xiao, M.; Zhang, C.; Wen, K.; Xiong, L.; Geng, G.; Wu, D. Bearing fault feature extraction method based on complete ensemble empirical mode decomposition with adaptive noise. *J. Vibroeng.* **2018**, *20*, 2622–2631. [CrossRef]
28. Mishra, C.; Samantaray, A.K.; Chakraborty, G. Rolling element bearing fault diagnosis under slow speed operation using wavelet de-noising. *Measurement* **2017**, *103*, 77–86. [CrossRef]
29. Fang, B.; Zhang, J.; Hong, J.; Yan, K. Research on the Nonlinear Stiffness Characteristics of Double− Row Angular Contact Ball Bearings under Different Working Conditions. *Lubricants* **2023**, *11*, 44. [CrossRef]
30. Wang, M.; Yan, K.; Tang, Q.; Guo, J.; Zhu, Y.; Hong, J. Dynamic modeling and properties analysis for ball bearing driven by structure flexible deformations. *Tribol. Int.* **2023**, *179*, 108163. [CrossRef]
31. Ma, S.; Zhang, X.; Yan, K. A Study on Bearing Dynamic Features under the Condition of Multiball–Cage Collision. *Lubricants* **2022**, *10*, 9. [CrossRef]
32. Ma, S.; Yin, Y.; Chao, B.; Yan, K.; Fang, B.; Hong, J. A Real-time Coupling Model of Bearing-Rotor System Based on Semi-flexible Body Element. *Int. J. Mech. Sci.* **2023**, 108098. [CrossRef]
33. Wang, H.; Ha'n, Q.; Zhou, D. Nonlinear dynamic modeling of rotor system supported by angular contact ball bearings. *Mech. Syst. Signal Process.* **2017**, *85*, 16–40. [CrossRef]
34. Peiran, Y.; Shizhu, W. A Generalized Reynolds Equation for Non-Newtonian Thermal Elastohydrodynamic Lubrication. *J. Tribol.* **1990**, *112*, 631–636. [CrossRef]
35. Zhang, Y.; Zhang, J.; Li, T.; Zhang, L. Investigation of the aeroacoustic behavior and aerodynamic noise of a high-speed train pantograph. *Sci. China (Technol. Sci.)* **2017**, *60*, 561–575. [CrossRef]
36. Zhang, J.; Fang, B.; Hong, J.; Zhu, Y. Effect of preload on ball-raceway contact state and fatigue life of angular contact ball bearing. *Tribol. Int.* **2017**, *114*, 365–372. [CrossRef]
37. He, P.; Gao, F.; Li, Y.; Wu, W.; Zhang, D. Research on optimization of spindle bearing preload based on the efficiency coefficient method. *Ind. Lubr. Tribol.* **2020**, *73*, 335–341. [CrossRef]
38. Rehman, W.U.; Jiang, G.; Luo, Y.; Wang, Y.; Khan, W.; Rehman, S.U.; Iqbal, N. Control of active lubrication for hydrostatic journal bearing by monitoring bearing clearance. *Adv. Mech. Eng.* **2018**, *10*, 1687814018768142. [CrossRef]
39. Ding, C.; Zhu, H.; Sun, G.D.; Zhou, Y.K.; Zuo, X. Chaotic characteristics and attractor evolution of friction noise during friction process. *Friction* **2018**, *6*, 47–61. [CrossRef]
40. Hirt, C.W.; Nichols, B.D. Volume of fluid (VOF) method for the dynamics of free boundaries. *J. Comput. Phys.* **1981**, *39*, 201–225. [CrossRef]

lubricants

Article

Digital Twin-Driven Thermal Error Prediction for CNC Machine Tool Spindle

Quanbo Lu [1], Dong Zhu [2], Meng Wang [1,3] and Mei Li [1,*]

1 School of Information Engineering, China University of Geosciences, Beijing 100083, China; luquanbo111@163.com (Q.L.); wangmeng2003@126.com (M.W.)
2 Sevnce Robotics Co., Ltd., Chongqing 401123, China; wangchengcheng@sevnce.com
3 School of Mechanical Engineering, Tangshan Polytechnic College, Tangshan 063299, China
* Correspondence: maggieli@cugb.edu.cn

Abstract: Traditional methods for predicting thermal error ignore the correlation between physical world data and virtual world data, leading to the low prediction accuracy of thermal errors and affecting the normal processing of the CNC machine tool (CNCMT) spindle. To solve the above problem, we propose a thermal error prediction approach based on digital twins and long short-term memory (DT-LSTM). DT-LSTM combines the high simulation capabilities of DT and the strong data processing capabilities of LSTM. Firstly, we develop a DT system for the thermal characteristics analysis of a spindle. When the DT system is implemented, we can obtain the theoretical value of thermal error. Then, the experimental data is used to train LSTM. The output of LSTM is the actual value of thermal error. Finally, the particle swarm optimization (PSO) algorithm fuses the theoretical values of DT with the actual values of LSTM. The case study demonstrates that DT-LSTM has a higher accuracy than the single method by nearly 11%, which improves the prediction performance and robustness of thermal error.

Keywords: digital twin; thermal error; CNCMT; spindle; LSTM

Citation: Lu, Q.; Zhu, D.; Wang, M.; Li, M. Digital Twin-Driven Thermal Error Prediction for CNC Machine Tool Spindle. *Lubricants* **2023**, *11*, 219. https://doi.org/10.3390/lubricants11050219

Received: 12 April 2023
Revised: 27 April 2023
Accepted: 9 May 2023
Published: 14 May 2023

1. Introduction

The performance of CNCMT is largely measured by its machining precision. Error is an important factor that influences machining precision. Generally, the error of CNCMT has a geometric error, thermal error, and force-induced error. Thermal error accounts for the highest proportion [1]. According to the existing literature, the thermal error is the major manufacturing error of CNCMT. With the improvement of the manufacturing level, the geometric accuracy of CNCMT has improved, and geometric errors have gradually decreased. The impact of thermal error on processing and manufacturing has become increasingly prominent. Therefore, exploring the thermal error of the CNCMT spindle system is a hot topic.

The spindle system is the main subsystem of CNCMT. The quick prediction and accurate compensation of thermal error can enhance the machining precision of CNCMT. Currently, there are two methods for constructing a thermal error model for CNCMT. One is the LSTM-based thermal error model. The other is the DT-based thermal error model [2]. However, the large heat sources and rapid speed changes of the CNCMT spindle system lead to significant time-varying thermal errors [3]. Based on a single thermal error modeling method, it is difficult to achieve a real-time and accurate reflection of the physical spindle system, which seriously hinders the efficient operation of CNCMT.

We propose a hybrid thermal error prediction method based on DT and LSTM. DT-LSTM improves the real-time reflection, accuracy, and robustness of thermal error prediction for the CNCMT spindle system. The rest of the paper is organized as follows. Section 2 introduces the related works of thermal error prediction. Section 3 presents the implementation of DT-LSTM. Section 4 shows a case study on how to predict thermal error based on DT-LSTM. Finally, conclusions are drawn in Section 5.

2. Related Works

2.1. LSTM-based Thermal Error Prediction

Zimmermann et al. designed a new self-adaptive approach for thermal error prediction [4]. Liang et al. developed an LSTM-based thermal error prediction model for heavy-duty CNCMT [5]. Li et al. reviewed LSTM thermal error modeling methods for MTs [6]. Li et al. proposed a thermal error prediction approach for electrical spindles using an optimized extreme learning machine algorithm [7]. Liao et al. established a robust thermal error model for spindles based on the improved fruit fly optimization algorithm [8]. Kumar et al. designed a real-time LSTM-based predictive model for thermal perception [9]. Li et al. established a multiple regression approach, which used nut temperature and ambient temperature as independent variables of the model [10]. Abdulshahed et al. considered the impact of nut temperature and bearing temperature on a thermal error and established a thermal error approach using gray neural networks [11]. Zhu et al. added additional temperature key points to prediction models (e.g., multiple regression models and neural network models), which enhanced the prediction accuracy of the approach [12]. Yang et al. established a finite element approach to the thermal error and the nut temperature rise, which predicted the spindle thermal error [13].

An LSTM-based model with high prediction performance is established when it obtains sufficient input and output data. If the data is not comprehensive enough, the established approach will be difficult to adapt to various situations. The poor robustness is characteristic of LSTM-based modeling. To obtain excellent robustness, the LSTM-based error prediction approach should combine with the DT system.

2.2. DT-based Thermal Error Modeling

Some scholars designed a DT system to predict thermal error. Liu et al. presented a DT system of thermal error prediction for gear profile grinders [14]. Ma et al. presented a self-learning-empowered thermal error prediction approach for machine tools based on DT [15]. Xiao et al. developed a DT system to analyze the thermal characteristics of the MT spindle [16]. Liu et al. designed an error prediction model for the MT spindle using DT [17]. To enhance the prediction performance of thermal error, Yi et al. designed a DT-based system [18]. Liu et al. designed a comprehensive machining thermal error approach based on DT [19]. Lunev et al. assessed the thermal performance of metal foams based on DT [20]. Kuprat et al. presented a thermal characterization analysis approach using DT for a power semiconductor [21].

The DT model based on hypothetical operating conditions is inconsistent with the actual operating conditions of the equipment, which leads to inconsistent models and low prediction accuracy. Therefore, we use data-based models to correct DT simulation data. It enhances the prediction performance of thermal errors and the processing precision of CNCMT.

3. DT-LSTM

3.1. DT

DT creates virtual models of real objects. DT combines models, data, and integration technologies. It achieves the coverage of the entire product lifecycle process and the connectivity and interaction between physical space and information space [22]. Grieves first proposed the concept of DT and defined the 3D model of DT (e.g., physical product, virtual product, and connection) [23]. NASA has successfully applied DT to aircraft health management. Tao et al. introduced DT into the field of intelligent manufacturing and presented the concept of DT workshop [24], which promoted the research and development of DT. The evolution characteristics of the spindle DT system are complex, dynamic, and stochastic. Therefore, we detect and correct the thermal boundary of physical equipment and map it to virtual entities.

The actual thermal characteristic of physical equipment is obtained via finite element simulation, which improves the accuracy of the thermal characteristic.

3.2. LSTM

LSTM is a time cycle network. When there is a time series relationship between the processed task and time, LSTM has excellent processing and prediction performance. The spindle thermal deformation has a time series characteristic. Therefore, LSTM can predict spindle thermal error.

LSTM has the forget gate, input gate, and output gate [25]. At the previous time, the preservation degree of the unit state is determined by the forget gate. At the current time, the preservation degree of the unit state is determined by the input gate. The output gate determines the output degree of the unit state to the current output value. Figure 1 shows the network structure of LSTM.

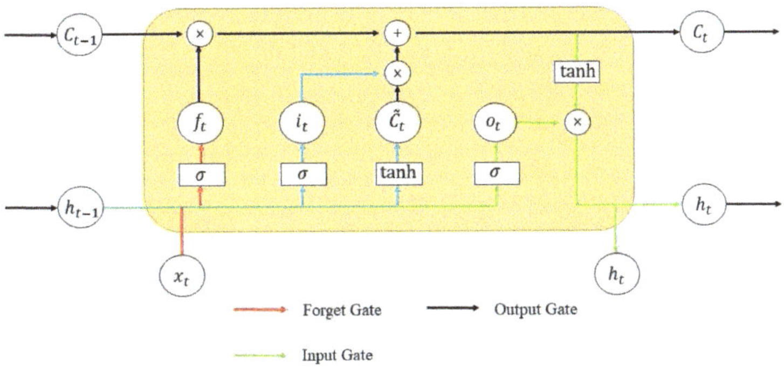

Figure 1. The architecture of LSTM.

The forget gate is given by [26] as follows:

$$f_t = \sigma\left(W_f \cdot [h_{t-1}, x_t] + b_f\right), \quad (1)$$

where f_t denotes the forget information; σ denotes the sigmoid function; W_f represents the weight coefficient; h_{t-1} represents the last moment output; x_t represents the input at the time t; b_f represents the offset.

The input gate formulate is given by [26] as follows:

$$i_t = \sigma(W_i \cdot [h_{t-1}, x_t] + b_i), \quad (2)$$

$$\widetilde{C} = \tanh(W_c \cdot [h_{t-1}, x_t] + b_C), \quad (3)$$

where i_t and $\widetilde{C}_t$ denote the input information; tanh denotes the tanh function; W_i and W_C represent the weight coefficient; b_i and b_C represent the offset.

The updated cell information is given by [26] as follows:

$$C_t = f_t \cdot C_{t-1} + i_t \cdot \widetilde{C}_t, \quad (4)$$

where C_{t-1} is the old cell information.

The output formula of LSTM is given by [26] as follows:

$$o_t = \sigma(W_o \cdot [h_{t-1}, x_t] + b_o), \quad (5)$$

$$h_t = o_t \cdot \tanh(C_t), \quad (6)$$

where W_o denotes the weight coefficient; b_o denotes the offset; h_t denotes the current moment output.

3.3. DT-LSTM

3.3.1. Framework

Figure 2 shows the framework of DT-LSTM. DT-LSTM combines LSTM and DT to obtain high prediction accuracy. Based on material characteristics and operating conditions, a multi-domain DT model for the CNCMT spindle is established. The temperature field is simulated using the working condition mapping of the CNCMT spindle. The internal temperature state of the CNCMT spindle is calculated as a virtual sensing signal. Meanwhile, the thermal error prediction approach using LSTM is performed on the actual signal. Finally, the PSO algorithm is applied to combine theoretical value and actual value. The LSTM observation result modifies the DT simulation result.

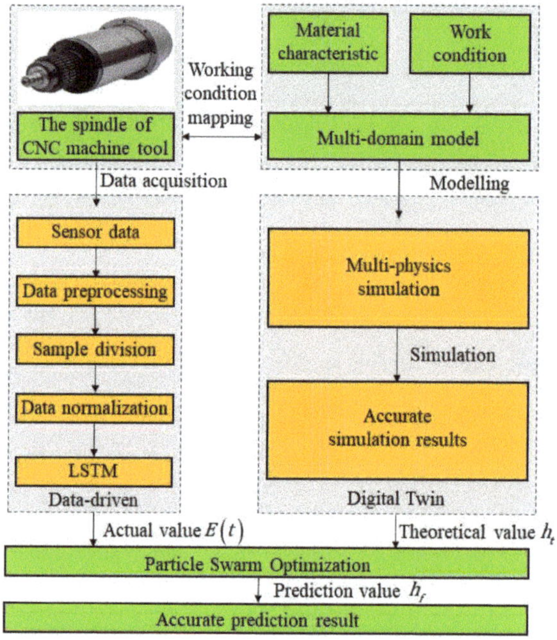

Figure 2. The framework of DT-LSTM.

3.3.2. Implementation

(1) The implementation of the DT model

DT model implementation is shown in Figure 3. During DT model building, multi-domain knowledge (e.g., thermal boundary condition, thermal deformation field, and thermal characteristic theory) must be considered simultaneously. Multi-domain modeling software contains ANSYS and UG. Therefore, object models from the spindle system can be constructed and embedded into a unified multi-domain model.

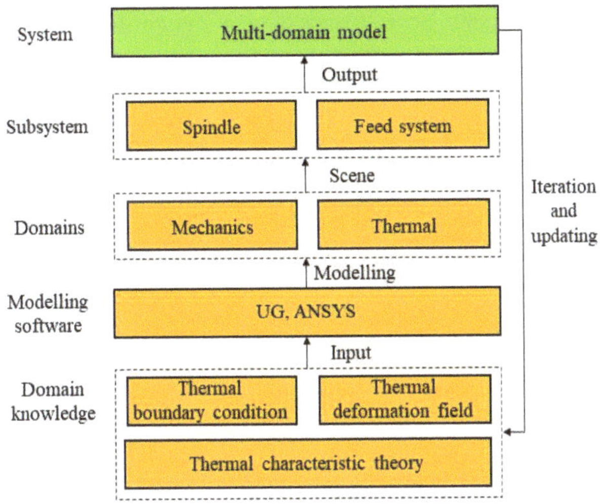

Figure 3. The implementation of the DT model.

1. Thermal boundary condition

a. Bearing heat calculation

The main heat source of the CNCMT spindle system is the bearing, and the friction torque of the bearing generates the heat. The generated heat is given by [27] as follows:

$$Q_1 = 1.047 \times 10^{-4} nM, \quad (7)$$

where Q_1 represents the generated heat of bearing; n represents the speed of bearing; M represents the friction torque of bearing.

M is equal to the sum of the load-unrelated friction torque M_0 and the load-related friction torque M_1.

$$M = M_0 + M_1, \quad (8)$$

$$M_0 = 10^{-7} f_0 (v_0 n)^{2/3} D_m^3 M_0 \cdot v_0 n \geq 2000, \quad (9)$$

$$M_0 = 160 f_0 D_m^3 \cdot v_0 n < 2000, \quad (10)$$

where f_0 represents the friction constant; v_0 represents the kinematic viscosity; D_m represents the average diameter of bearing [27].

$$M_1 = 0.0013 f_1 (X_0 F_r + Y_0 F_a / C_0)^{10/3} D_m, \quad (11)$$

where f_1 represents the load constant; C_0 represents the basic load rating; X_0 represents the distribution coefficient of the radial load; Y_0 represents the distribution coefficient of the axial load; F_r represents the radial load; F_a represents the axial load [27].

b. Screw-nut heat calculation

The heating principle of the lead screw-nut is identical to the bearing. The calculation formula is given by [27] as follows:

$$Q_2 = 0.12 \pi n' M', \quad (12)$$

where Q_2 represents the generated heat of the lead screw-nut; n' represents the speed of the lead screw-nut; M' represents the friction torque of the lead screw-nut.

The calculation formula of $M\prime$ is given by [27] as follows:

$$M\prime = 2z(M_e + M_g)\cos\beta, \tag{13}$$

where z is the ball number; β is the helical angle of the lead screw spiral raceway; M_e is the friction resistance torque; M_g is the geometric sliding friction torque.

The calculation formula of z is given by [27] as follows:

$$z = i\left(\frac{\pi d}{d_0} - 3\right), \tag{14}$$

where i represents the coefficient; d_0 represents the ball diameter; d represents the nominal diameter of the lead screw.

The calculation formula of M_e and M_g are given by [27] as follows:

$$M_e = m_\beta \sqrt[3]{\frac{4Q^4}{\vartheta \sum \rho}}, \tag{15}$$

$$\begin{array}{l} M_g = 0.08\frac{f \cdot m_\alpha^2}{R}\sqrt[3]{\frac{16Q^5}{(\vartheta \sum \rho)^2}} \\ \vartheta = \frac{8}{3\left(\frac{1-u_1^2}{E_1} + \frac{1-u_2^2}{E_2}\right)} \\ \sum \rho = \rho_{11} + \rho_{12} + \rho_{21} + \rho_{22} \\ R = R_1 R_2/(R_1 + R_2) \end{array}, \tag{16}$$

where f is the sliding friction coefficient; m_α and m_β are the eccentricity coefficient; Q is the radial pressure borne by a single sphere; E_1 and E_2 are the elastic modulus; u_1 and u_2 are the poison's ratio of the material; R_1 and R_2 are the radius of curvature of the ball and raceway; ρ_{11}, ρ_{12}, ρ_{21}, and ρ_{22} are the principal curvature of the ball and raceway.

c. Convective heat transfer coefficient

When the external surface of a spindle system directly contacts the air, we can find that convective heat transfer occurs. In the CNCMT spindle system, it is influenced by many factors and is difficult to obtain accurately. The most effective method is to calculate the convective heat transfer coefficient h_c according to the Nusselt criterion. The formula of is given by [27] as follows:

$$\begin{array}{l} h_c = \lambda Nu/l_c \\ Nu = 0.133\mathrm{Re}^{2/3}\mathrm{Pr}^{1/3} \end{array}, \tag{17}$$

where λ represents the air thermal conductivity; Nu represents the Nusselt number; l_c represents the feature size; Re represents the Reynolds number; Pr represents the fluid Prandtl number.

d. Thermal resistance calculation

In the CNCMT spindle system, the bearing is used as the heat source. Based on the heat transfer theory, the heat will undergo convective heat transfer from high temperature to low temperature. There is thermal resistance at the locations where different components contact. According to the thermal boundary condition of the bearing, the contact thermal resistance R_1 can be given by [27] as follows:

$$R_1 = \frac{\Psi}{4\lambda_1 a} + \frac{\Psi}{4\lambda_2 a}, \tag{18}$$

where Ψ is a geometric factor, which is related to the contact area; a is the long half shaft the bearing contact ellipse; λ_1 is the thermal conductivity of the rolling ball; λ_2 is the thermal conductivity of the loop.

The contact thermal resistance R_2 between the bearing inner race and rotor is given by [27] as follows:

$$R_2 = \frac{L(\lambda_3 + \lambda_4)}{AA_r 2\lambda_3\lambda_4}, \tag{19}$$

where λ_3 determines the thermal conductivity of the bearing inner race; λ_4 determines the rotor thermal conductivity; A_r represents the dimensionalized actual contact area; A represents the nominal contact area; L represents the spatial thickness of the joint surface gap.

2. Thermal characteristic analysis and modeling theory

The temperature field is the temperature distribution of all points on the CNCMT spindle. The expression of the temperature field is $T = f(x, y, z, t)$, which indicates that the temperature of a certain point on the CNCMT spindle is a function of the spatial location (x, y, z) and time t of the point. The temperature field that changes over time is called a transient temperature field. After reaching a certain time, the temperature field of the CNCMT spindle does not change, and then it forms a steady temperature field. The expression for the steady temperature field is $T\prime = f(x, y, z)$.

The differential equation of heat conduction is defined as a calculation formula for the temperature field of an object. We randomly select a microelement hexahedron from a thermally conductive object, as shown in Figure 4. ϕ represents the heat energy consumed per unit volume per unit time.

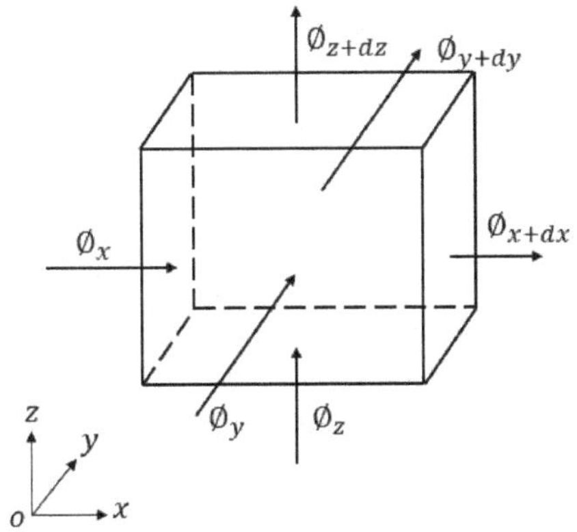

Figure 4. The microelement hexahedron.

Assuming that there is a heat source inside the object, the differential equation of the heat conduction for a unit body is given by [27] as follows:

$$\rho c \frac{\partial t}{\partial \tau} = \lambda \left(\frac{\partial^2 t}{\partial x^2} + \frac{\partial^2 t}{\partial y^2} + \frac{\partial^2 t}{\partial z^2} \right) + q_v, \tag{20}$$

where q_v is the internal heat source density.

The simplest initial condition is an initial uniform distribution of initial temperature, which is given by [27] as follows:

$$t = f(x, y, z, \tau) = t_0 \ (Constant), \tag{21}$$

The boundary conditions are used to guide the temperature distribution of a hot object, which can generally be divided into three categories.

a. The temperature distribution at any time on the boundary of a given heat conductor [27].

$$t = t(x, y, z, \tau)(On\ the\ boundary\ of\ \Gamma_1) \tag{22}$$

The temperature is uniformly distributed at the boundary [27].

$$t = t(\tau)(On\ the\ boundary\ of\ \Gamma_1) \tag{23}$$

b. The heat flux density at any time on the boundary of a given heat conductor [27].

$$\lambda_x \frac{\partial t}{\partial x} n_x + \lambda_y \frac{\partial t}{\partial y} n_{xy} + \lambda_z \frac{\partial t}{\partial z} n_z = q(x, y, z, \tau)\ (On\ the\ boundary\ of\ \Gamma_2) \tag{24}$$

The heat flux density is uniformly distributed at the boundary [27].

$$\lambda_x \frac{\partial t}{\partial x} n_x + \lambda_y \frac{\partial t}{\partial y} n_{xy} + \lambda_z \frac{\partial t}{\partial z} n_z = q(\tau)\ (On\ the\ boundary\ of\ \Gamma_2) \tag{25}$$

c. The convective heat transfer coefficient h between the boundary of the thermal conductor and the surface fluid, and the temperature of the surface fluid t_B are given [27].

$$\lambda_x \frac{\partial t}{\partial x} n_x + \lambda_y \frac{\partial t}{\partial y} n_{xy} + \lambda_z \frac{\partial t}{\partial z} n_z = h(t_B - t)\ (On\ the\ boundary\ of\ \Gamma_3) \tag{26}$$

The fluid temperature t_B and the convective heat transfer coefficient h at the boundary interface are known, while the temperature t and temperature change rate $\frac{\partial t}{\partial n}$ at the boundary interface are unknown.

Assuming the temperature field solution domain is Ω, and $\Gamma = \Gamma_1 + \Gamma_2 + \Gamma_3$, Γ is the entire boundary of the domain Ω.

The transient temperature field is given by [27] as follows:

$$t(x, y, z, 0) = t(x, y, z). \tag{27}$$

When $\frac{\partial t}{\partial \tau} = 0$, the steady heat conduction equation is given by [27].

$$\frac{\partial}{\partial x}(\lambda_x \frac{\partial t}{\partial x}) + \frac{\partial}{\partial y}\left(\lambda_y \frac{\partial t}{\partial y}\right) + \frac{\partial}{\partial z}\left(\lambda_z \frac{\partial t}{\partial z}\right) + q_v = 0\ (In\ \Omega\ ). \tag{28}$$

When the thermal boundary condition and the heat generation rate are known, the temperature field distribution of the hexahedral microelement can be calculated and used for finite element modeling.

During the movement of the CNCMT spindle, heat conduction exists among the parts of the CNCMT spindle system. It is most evident in the nut and bearing seat. Therefore, the temperature fields under the heat source of the nut and bearing seat are calculated, respectively. The temperature field under a nut heat source is obtained iteratively by Equation (29) [27].

$$\begin{cases} \Delta Q(t) = Q_n(t) - Q_c(t) - Q_t(t) \\ \Delta Q(t) = c\rho L_i S \Delta T_{L_i}(t) \\ Q_n = 0.12\pi f_w v_0 n M_w \\ Q_c(t) = h \times S\prime \times \left(T_{L_i}(t) - T_f(t)\right)\Delta t \\ Q_t(t) = \lambda \times S \times \frac{\left(T_{L_i}(t) - T_{L_{i+1}}(t)\right) + \left(T_{L_i}(t) - T_{L_{i-1}}(t)\right)}{L} \times \Delta t \end{cases} \tag{29}$$

where ΔQ represents the $L_i{}'$ D-value between the generated heat and the dissipation heat; Q_n represents the frictional heat generation of L_i; Q_c represents the heat transfer rate between L_i and the environment; Q_t represents the heat conduction between L_i and two side elements t; c represents the lead screw specific heat capacity; ρ represents the lead screw density; S represents the lead screw equivalent cross-sectional area; $T_{L_i}(t)$ represents the $L_i{}'$s temperature rise; f_w represents the coefficient associated with the screw-nut class and lubrication method; v_0 represents the kinematic viscosity; n represents the lead screw rotational speed; h represents the convective heat transfer coefficient; $S\prime$ represents L_i's convective heat transfer area; $T_f(t)$ represents the air temperature around the lead screw, which approximately replaces with the temperature T_b of the bed; λ represents the lead screw thermal conductivity.

The temperature field of the bearing block under the heat source is given by [27] as follows:

$$\Gamma_{br}(P_y, t) = \left(\chi e^{-\sqrt{\frac{2h}{kR}}P_y} + \zeta e^{-\sqrt{\frac{2h}{kR}}P_y}\right) \times \left(1 - e_f\left(\frac{P_y}{2\sqrt{\alpha t}}\right)\right), \tag{30}$$

where T_{br} represents the bearing block temperature; P_y represents the distance from the heat source on the lead screw; χ and ζ are the unidentified coefficient; k represents the thermal conductivity; R represents the lead screw radius; α represents the lead screw thermal conductivity; $e_f(P_y)$ is given by [27] as follows:

$$e_f(P_y) = \frac{2}{\sqrt{\pi}} \int_0^{P_y} e^{-\lambda^2} d\lambda. \tag{31}$$

Under the joint action of the lead screw nut and the heat source of the bearing seat, the thermal error at any time t is given by [27] as follows:

$$E(t) = \sum_{i=1}^{M} \lambda_s \times (\Delta T_{L_i}(t) + \Delta \Gamma_{br}(t)) \times L_i, \tag{32}$$

where λ_s represents the thermal expansion coefficient; M represents the discrete segment number of the lead screw.

3. Mathematical model of the thermal deformation field

Figures 5 and 6 show the structural diagram and the thermal elongation of the CNCMT spindle, respectively. B_1B_2 and L_1 are the distance of two bearings of the CNCMT spindle. WB_1 and L_2 are the distance of a bearing and a mandrel. Because of the uneven temperature distribution of the CNCMT spindle system, the thermal deformation satisfies the following relationship [28].

$$\Delta L = \alpha \int_0^l T(x) dx, \tag{33}$$

where Δx represents the elongation of the spindle; α represents the linear expansion coefficient; l represents the length of the main shaft; $T(x)$ represents the CNCMT spindle temperature distribution function. $T(x) = T_{max}(x/L_2)$. T_{max} is the temperature at B_1.

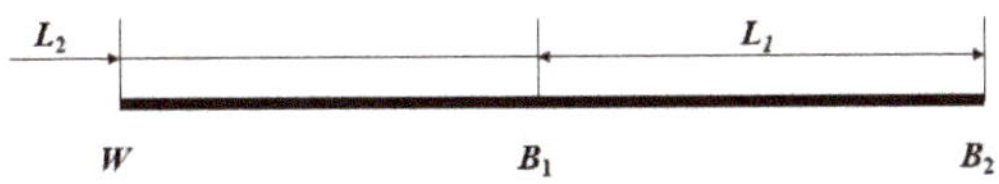

Figure 5. The structural diagram of the CNCMT spindle.

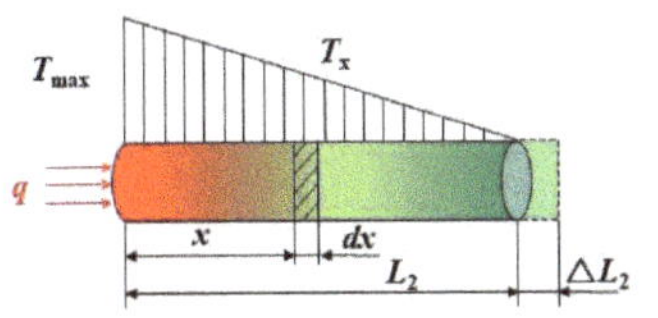

Figure 6. The thermal elongation of the CNCMT spindle.

(2) The implementation of the LSTM model

Figure 7 shows the specific implementation process of LSTM. The inputs of the thermal error model are the screw-nut temperature, bearing temperature, ambient temperature, and motor temperature. First, it is necessary to preprocess the data. Then, the preprocessed data is separated into the training group and the test group. Next, we normalize the feature data. Finally, an LSTM network is constructed and trained to predict thermal error.

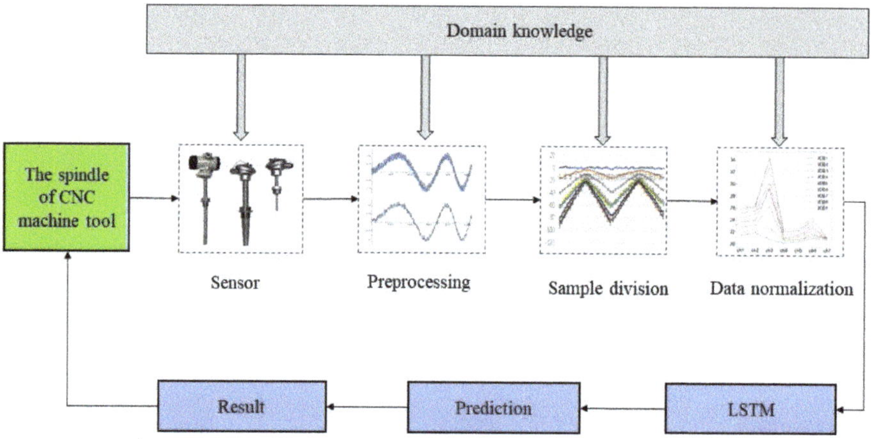

Figure 7. The implementation process of LSTM model.

(3) The implementation of DT-LSTM

Using the fusion method, the output of LSTM is taken as the systematic observation value to correct the theoretical and empirical derivation results driven by the DT model. The PSO used in this paper is a fusion algorithm. The steps of DT-LSTM are shown in Figure 8.

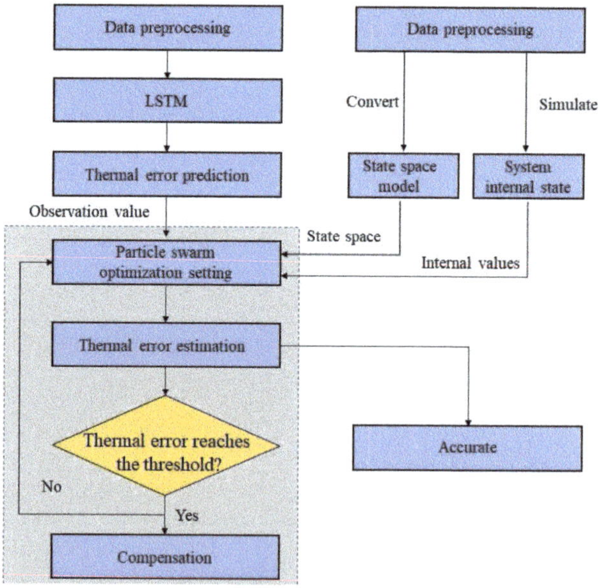

Figure 8. The steps of DT-LSTM.

i. An LSTM model for the CNCMT spindle system is established, and the predicted thermal error obtained from the model is used as an observation value.

ii. According to the temperature variation rules in the DT model, it is converted into a temperature space model for initialization based on the fusion algorithm, and the internal state of the system is calculated using model simulation.
iii. The fusion algorithm is initialized based on the temperature space model, and the observed values are used to modify the theoretical values obtained from the system model simulation and reasoning. We can obtain more accurate thermal error prediction values.
iv. We judge whether the thermal error reaches the threshold value based on the analysis results of the fusion algorithm. If the thermal error reaches the threshold value, we should make appropriate compensation. Otherwise, return to ii to repeat the iteration.

4. Case Study

4.1. Design of Experiment Platform

Figure 9 shows that the CNCMT spindle system is composed of the motor, coupling, bearing, lead screw, and screw-nut. The motor provides the power for the feed system. The coupling links the motor and the lead screw to transmit torque. The bearings at both ends support the rotation of the lead screw. The lead screw and screw-nut comprise a lead–screw-nut pair. It enables the conversion from rotary motion to linear motion. The guide rail drives the workbench to perform the linear motion.

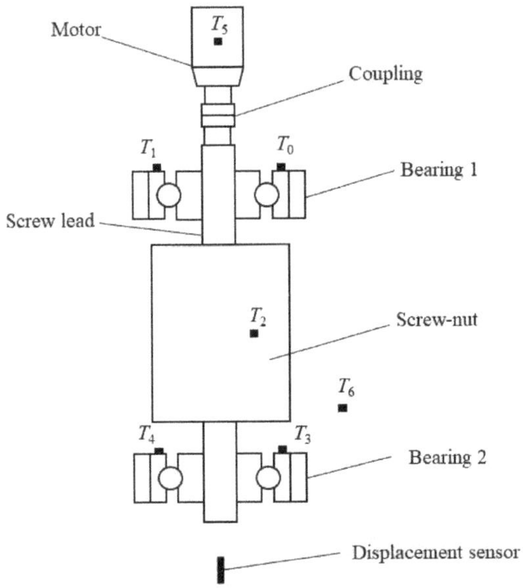

Figure 9. The structure of the CNCMT spindle system.

Seven STT-Pt100 temperature sensors are arranged on the CNCMT spindle system to measure the temperature changes. To obtain the measurement value of the axial thermal deformation, we arrange a displacement sensor on the CNCMT spindle's Z-end. The CNCMT speed is 2000 r/min. The temperature point locations are shown in Table 1.

Table 1. The location of temperature measuring points.

T_0	T_1	T_2	T_3	T_4	T_5	T_6
Bearing 1	Bearing 1	Screw-nut	Bearing 2	Bearing 2	Motor	Surrounding

4.2. The Optimization of the Temperature Measurement Points

The optimized temperature measurement points can reduce the temperature variables. Meanwhile, it can reduce the model computing resources and improve the model computing accuracy. We use the fuzzy clustering method to optimize the temperature measurement points. The specific method is seen in the literature [29]. We set the number of the cluster center as five. The iteration times are set to 100. The convergence precision is set to 1×10^{-6}. The blur coefficient is set to two. Therefore, the grouping results of the temperature data can be obtained. Seven temperature measurement points are divided into five groups: T_0, T_1; T_2; T_3, T_4; T_5; T_6.

Based on the above grouping results, we use statistical theory to optimize the temperature measurement points. We use the similarity coefficient method to calculate the correlation coefficient between each temperature measurement point and the thermal error [29].

$$r_{x_iE} = \frac{\sum\limits_{j=1}^{n}\left(x_{ij} - \overline{x_i}\right)\left(E_j - \overline{E}\right)}{\sqrt{\sum\limits_{J=1}^{n}\left(x_{ij} - \overline{x_i}\right)^2}\sqrt{\sum\limits_{J=1}^{n}\left(E_j - \overline{E}\right)^2}}, \quad (34)$$

where r_{x_iE} represents the correlation coefficient; x_{ij} represents the measuring point temperature; i represents the i-th measuring point, totaling m measuring points; j represents the measurement number at the same measuring point; n represents the measurement number at the same measuring point; x_i represents the average temperature of all n measurements at the measurement point i; $\overline{E}$ represents the average thermal error of all measured values at the same measuring point; E_j represents the j-th thermal error measurement value.

Through the calculation of Equation (34), we can obtain the correlation coefficient of each temperature measurement point. The correlation coefficients are shown in Table 2. Figure 10 shows the correlation curve of "E—x".

Table 2. The correlation coefficient of temperature measurement points.

Temperature Measurement Point	Correlation Coefficient	Temperature Measurement Point	Correlation Coefficient
T_0	0.9498	T_4	0.9565
T_1	0.9489	T_5	0.8866
T_2	0.9503	T_6	0.8815
T_3	0.9555		

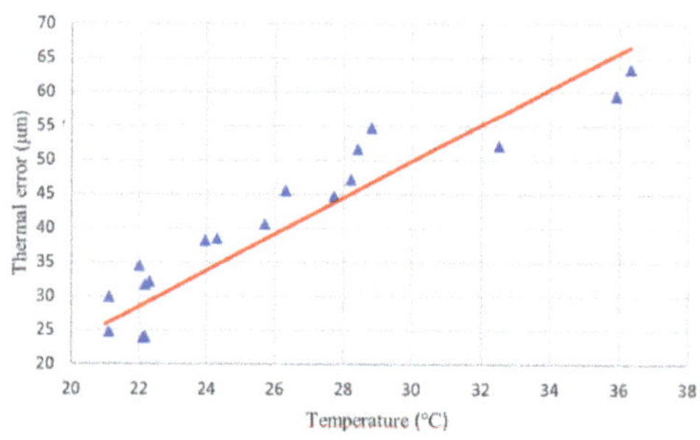

Figure 10. The correlation curve of "E—x".

Comparing the coefficients of each group, we select the optimal temperature measurement point of each group. From the values of the correlation coefficients in Table 2, we select T_0, T_2, T_4, T_5, and T_6 as the temperature input data for the thermal error modelling.

4.3. *DT-LSTM-based Thermal Error Prediction Approach for CNCMT Spindle*

4.3.1. The Realization of the DT Model

In order to construct a DT model of the CNCMT spindle system, Table 3 shows the material parameters of the CNCMT spindle system.

Table 3. The material parameters of the CNCMT spindle system.

Part	Spindle	Bearing
Material	GCr15SiMn	GCr15
Density/(kg/m^3)	7810	7830
Modulus of elasticity E/Pa	2.06×10^{11}	2.19×10^{11}
Poisson's ratio μ	0.3	0.3
Specific heat capacity C/(J·(kg·K)$^{-1}$)	460	160
Thermal conductivity/(W·(m·K)$^{-1}$)	60.5	81
Coefficient of thermal expansion	1.2×10^{-5}	1.25×10^{-5}

To facilitate the establishment of the DT model of the CNCMT spindle system, we use UG software to appropriately simplify the three-dimensional model of the spindle (e.g., smaller holes, non-critical grooves, smaller fillets, and chamfers). The simplified model is imported into Workbench software. Then, in order to simplify the calculation process, a two-dimensional axisymmetric model is drawn based on the cross-sectional dimensions. Finally, to ensure the convergence of the entire model and the accuracy of the temperature distribution results, the grid division of the heat transfer concentration area near the bearing is relatively dense, and the grid division of the inner cavity and outer surface edge areas of the spindle is relatively sparse. After adding temperature measuring points, the DT system starts to measure the spindle temperature and correct the thermal boundary. Figure 11 shows the temperature field of the CNCMT spindle system at 60 min. Figure 12 shows the thermal deformation field of the CNCMT spindle system at 60 min.

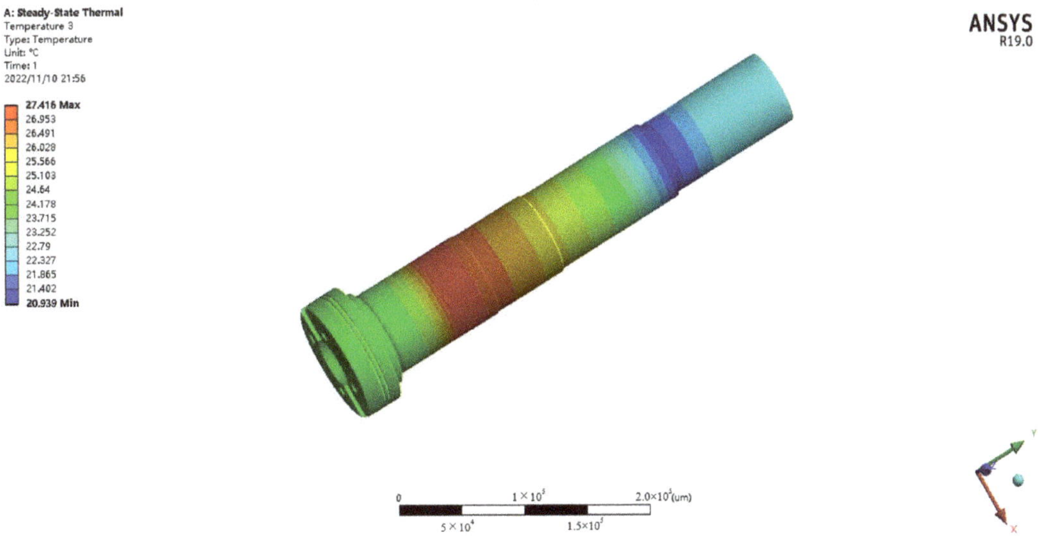

Figure 11. The temperature field of the CNCMT spindle.

Figure 12. The thermal deformation field of the CNCMT spindle's Z-end.

Figures 13–17 show the temperature rise at the key temperature measurement points. Figure 18 shows a comparison between the simulated thermal deformation and the actual thermal deformation of the spindle's Z-end. We can see that the simulated temperature accuracy of DT is above 98%, and the thermal deformation simulation accuracy is up to 95%. It proves that the spindle DT model can reflect the actual thermal characteristics.

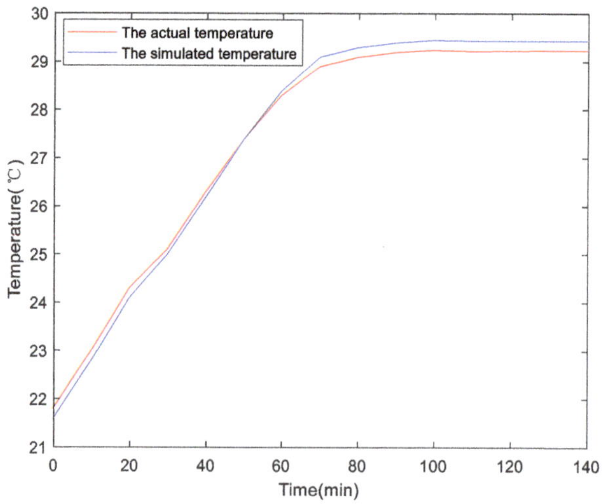

Figure 13. The temperature rise of T_0.

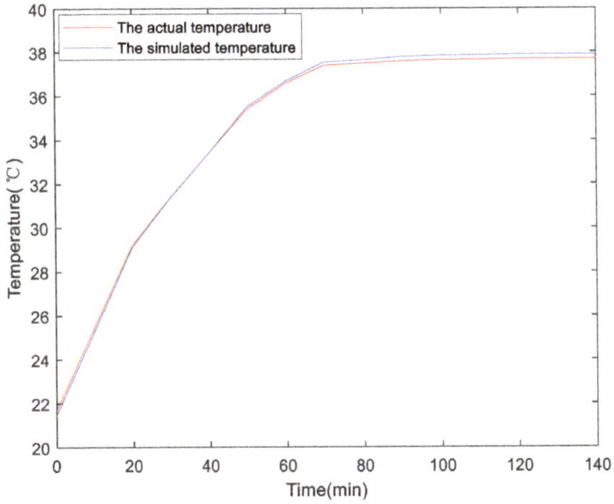

Figure 14. The temperature rise of T_2.

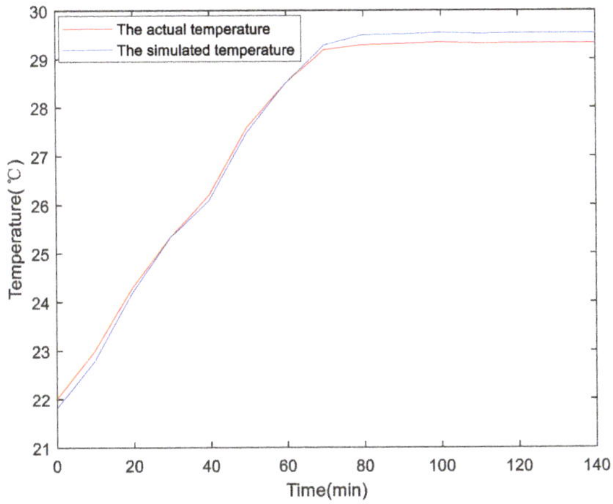

Figure 15. The temperature rise of T_4.

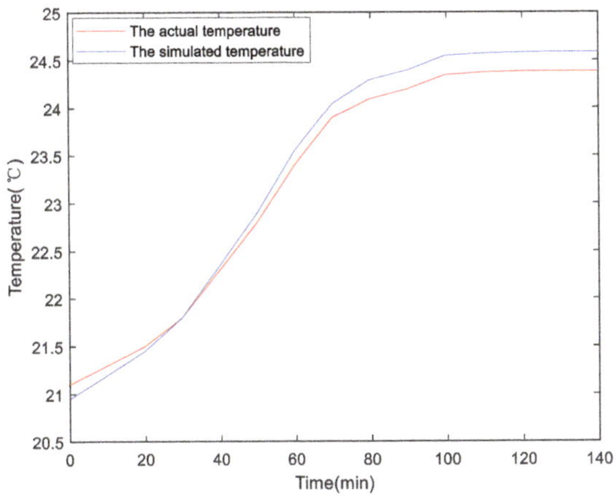

Figure 16. The temperature rise of T_5.

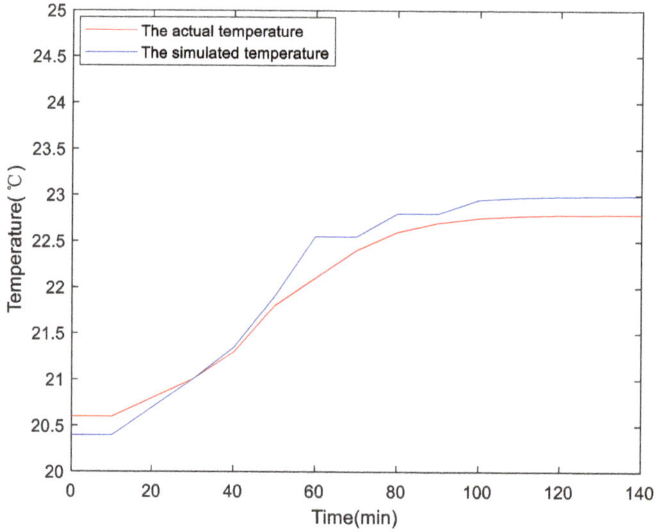

Figure 17. The temperature rise of T_6.

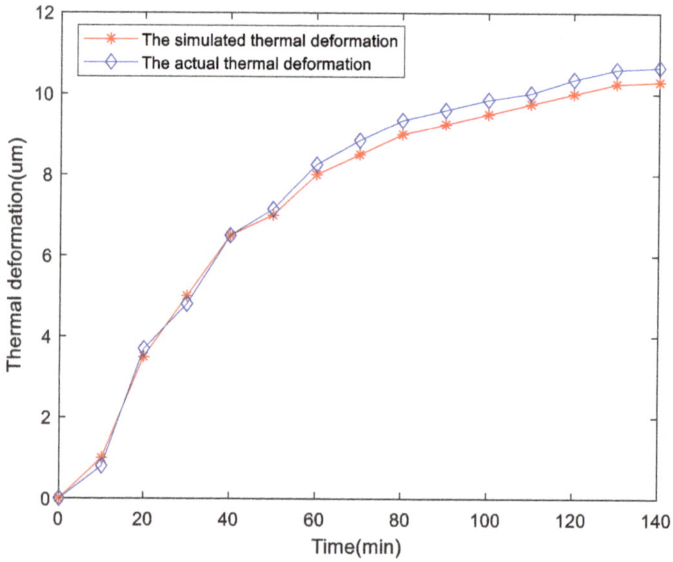

Figure 18. The thermal deformation of CNCMT spindle's Z-end.

4.3.2. The Realization of the LSTM Model

First, we smooth the collected data. Taking temperature data as an example, Figure 19 depicts one of the collected temperature data. The curve has a lot of noise. Therefore, it needs to be smoothed. To reduce computational complexity, we adopt the moving average filtering method. The result of smoothing is shown in Figure 20. After noise elimination, we chose 70% for the training group and 30% for the test group. The normalization formula is givenby [30,31].

$$x_i' = \frac{x_i - x_{\min}}{x_{\max} - x_{miin}} \quad (35)$$

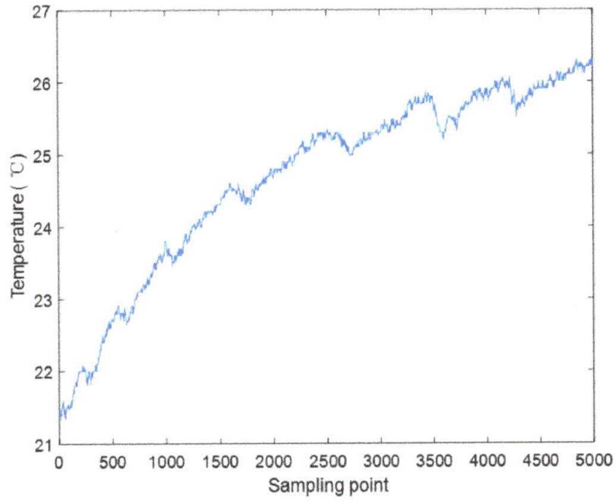

Figure 19. The collected temperature data.

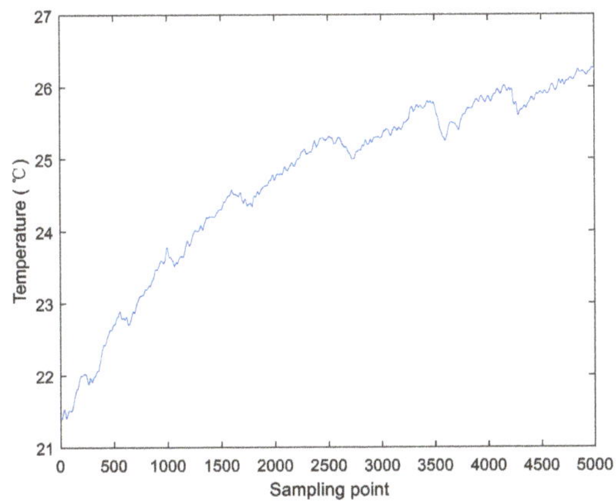

Figure 20. The collected temperature data after smoothing.

This paper builds an LSTM neural network based on the Python framework. Due to the complexity and depth of the model structure, it is necessary to simplify the model structure. In order to select a suitable model structure, we conduct the relevant experiments. As shown in Table 4, we can obtain the maximum residual error by setting different LSTM layers and hidden node numbers. The maximum residual error of the LSTM with two layers and twelve hidden nodes is the smallest, which has the highest accuracy.

Table 4. The maximum residual error corresponding to different LSTM layers and hidden node numbers.

Model Structure	LSTM Two-Layer Maximum Residual Error (μm)	LSTM Three-Layer Maximum Residual Error (μm)	LSTM Four-Layer Maximum Residual Error (μm)
eight hidden nodes	10.5	7	16
twelve hidden nodes	6	9.5	19
sixteen hidden nodes	8	11.5	24.2
twenty hidden nodes	9.6	10	16.7

After repeated experiments, we choose the LSTM structure with four layers, which includes one input layer, two hidden layers, one output layer, six input layer nodes, twelve hidden layer nodes, and two output layer nodes. We use the gradient descent method to find the optimal solution, and the training results display the maximum residuals of the predicted value. We set the number of iterations as 1000 and the learning rate as 0.1. The model parameters are randomly initialized. The root mean square error (RMSE) convergence curve during training is shown in Figure 21.

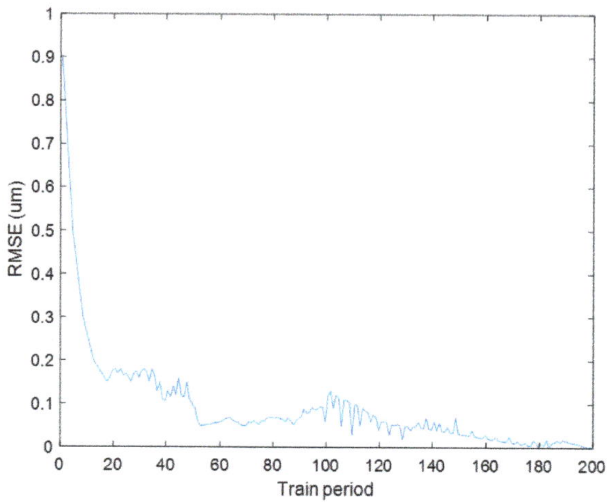

Figure 21. The RMSE convergence curve of LSTM.

4.3.3. The Realization of DT-LSTM

Equation (32) shows the theoretical value of the CNCMT spindle thermal error, which is used as the system state equation in the PSO to initialize the algorithm. Meanwhile, the actual value of the thermal error obtained by LSTM is shown in Equation (6), which is used as the system observation value in the PSO.

The number of particles is set to 150. The DT-LSTM Algorithm 1 is shown below.

Algorithm 1. DT-LSTM for the CNCMT spindle's thermal error prediction

Input: The theoretical prediction value of DT and the actual prediction value of LSTM
Output: The particles prediction value
(1) Initialize the parameters and particles
(2) $E(t) = \sum_{i=1}^{M} \lambda_s \times (\Delta T_{L_i}(t) + \Delta \Gamma_{br}(t)) \times L_i$
(3) $h_t = o_t \cdot \tanh(C_t)$
for 1 = 1:150
(4) Sample from (2)
(5) Calculate the thermal error prediction value h_{ti} of particles by (3)
(6) Calculate the weight $E(i)$ of each particleend
(7) Normalize the weight
(8) Resample according to the normalized weight
(9) Output the CNCMT spindle's thermal error prediction value h_f

4.4. The Analysis of Experiment Result

The thermal prediction value of different methods is shown in Figure 22. Figure 23 shows the comparison of different hybrid algorithms, which proves the superior performance of PSO. When we use single prediction methods (e.g., DT and LSTM), there is a significant error between the predicted curve and the actual curve. When we use DT-LSTM, the predicted curve is close to the actual curve, and the prediction accuracy is improved. DT-LSTM overcomes the model inconsistency of DT and the poor adaptability of LSTM. DT-LSTM has higher accuracy than the single method by nearly 11%, which improves the prediction performance and robustness of thermal error. Table 5 shows the quantitative evaluation of different methods. Table 6 shows the accuracy comparison of different methods. The result indicates that DT-LSTM has a higher accuracy at all stages compared with DT and LSTM. The average accuracy of LSTM is 98%. Meanwhile, DT-LSTM shows better prediction performance compared with the other method.

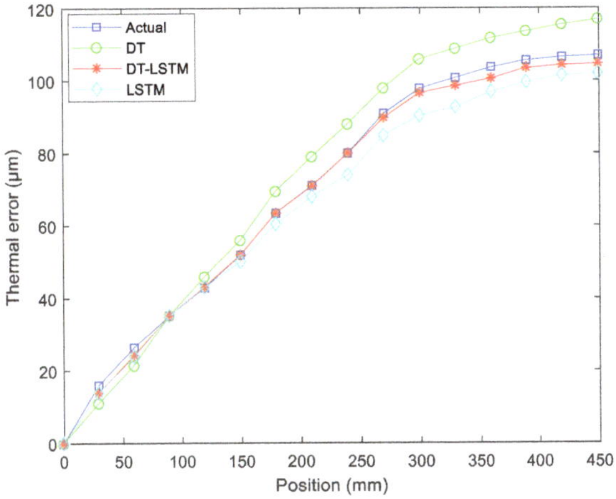

Figure 22. The thermal prediction value of different methods.

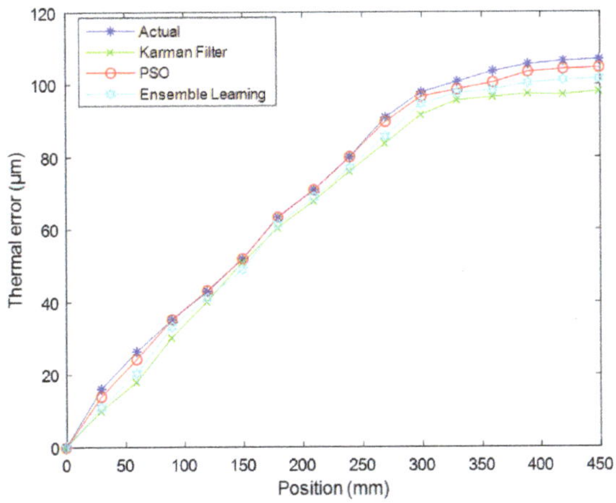

Figure 23. The comparison of different hybrid algorithms.

Table 5. The quantitative evaluation of different methods.

	Start Stage Accuracy	Middle Stage Accuracy	End Stage Accuracy	Average Accuracy
DT	91%	88%	82%	87%
LSTM	98%	90%	85%	91%
DT-LSTM	100%	99%	95%	98%

Table 6. The accuracy comparison of different methods.

Method	Accuracy
Co-Simulation-Based DT [18]	96%
SVM [14]	97.16%
DT-LSTM	98%

5. Conclusions

To enhance the prediction performance of CNCMT spindle thermal error, we propose a novel thermal error prediction method using DT-LSTM. Firstly, we establish a DT system to simulate the CNCMT spindle thermal characteristics. Based on the simulated result, we can obtain the theoretical value of thermal error. Then, LSTM is constructed to analyze the experimental data. The output of LSTM is the actual value of thermal error. Finally, we use the particle swarm optimization algorithm to fuse the theoretical values of DT with the actual values of LSTM. DT-LSTM is compared with the single method, and the accuracy of DT-LSTM is greater than 98%. Therefore, DT-LSTM can predict the CNCMT spindle thermal error effectively and provide a theoretical basis for the CNCMT spindle thermal error compensation. This paper only verifies the method for the spindle. In the future, DT-LSTM will be applied to the other components of CNCMT. Real time is a crucial aspect of thermal error prediction. The simulation of the DT physical performance model consumes many computing resources and time. In the future, we will improve the real-time performance and computational efficiency of DT simulation.

Author Contributions: Methodology, D.Z.; Resources, M.W.; Writing—original draft, Q.L.; Writing—review & editing, M.L. All authors have read and agreed to the published version of the manuscript.

Funding: This research was supported by Hebei University Science and technology research project: QN2022201. This work was also funded by 2021 Graduate Innovation Fund Project of China University of Geosciences, Beijing: ZD2023YC041.

Data Availability Statement: Not applicable.

Conflicts of Interest: The authors declare no conflict of interest.

References

1. Li, Y.; Tian, H.; Liu, D.; Lu, Q.B. Thermal error analysis of five-axis machine tools based on five-point test method. *Lubricants* **2022**, *10*, 122. [CrossRef]
2. Peng, J.; Yin, M.; Cao, L.; Liao, Q.; Wang, L.; Yin, G. Study on the spindle axial thermal error of a five-axis machining center considering the thermal bending effect. *Precis. Eng.* **2022**, *75*, 210–226. [CrossRef]
3. Ouerhani, N.; Loehr, B.; Rizzotti-Kaddouri, A.; Santo De Pinho, D.; Limat, A.; Schinderholz, P. Data-driven thermal deviation prediction in turning machine-tool-a comparative analysis of machine learning algorithms. *Procedia Comput. Sci.* **2022**, *200*, 185–193. [CrossRef]
4. Zimmermann, N.; Lang, S.; Blaser, P.; Mayr, J. Adaptive input selection for thermal error compensation models. *CIRP Ann.* **2020**, *69*, 485–488. [CrossRef]
5. Liang, Y.C.; Li, W.D.; Lou, P.; Hu, J.M. Thermal error prediction for heavy-duty CNC machines enabled by long short-term memory networks and fog-cloud architecture. *J. Manuf. Syst.* **2022**, *62*, 950–963. [CrossRef]
6. Li, Y.; Yu, M.; Bai, Y.; Hou, Z.; Wu, W. A review of thermal error modeling methods for machine tools. *Appl. Sci.* **2021**, *11*, 5216. [CrossRef]
7. Li, Z.; Wang, B.; Zhu, B.; Wang, Q.; Zhu, W. Thermal error modeling of electrical spindle based on optimized ELM with marine predator algorithm. *Case Stud. Therm. Eng.* **2022**, *38*, 102326. [CrossRef]
8. Liao, Q.; Yin, Q.; Xie, L.; Yin, G. Improved exponential model for thermal error modeling of machine-tool spindle based on fruit fly optimization algorithm. *Proc. Inst. Mech. Eng. C J. Mech.* **2022**, *236*, 6912–6922. [CrossRef]
9. Kumar, T.S.; Kurian, C.P. Real-time data based thermal comfort prediction leading to temperature setpoint control. *J. Ambient Intell. Hum. Comput.* **2022**, 1–12. [CrossRef]
10. Li, Z.; Fan, K.; Yang, J.; Zhang, Y. Time-varying positioning error modeling and compensation for ball screw systems based on simulation and experimental analysis. *Int. J. Adv. Manuf. Technol.* **2014**, *73*, 773–782. [CrossRef]
11. Abdulshahed, A.M.; Longstaff, A.P.; Fletcher, S.; Potdar, A. Thermal error modelling of a gantry-type 5-axis machine tool using a grey neural network model. *J. Manuf. Syst.* **2016**, *41*, 130–142. [CrossRef]

12. Zhu, X.; Liu, Q.; Zhang, X.; Jiang, X.; Lou, P. Robustness analysis of the thermal error model for a CNC machine tool. In Proceedings of the 2016 8th International Conference on Intelligent Human-Machine Systems and Cybernetics (IHMSC), Cairo, Egypt, 27–28 August 2016; pp. 510–513.
13. Yang, J.; Zhang, D.; Mei, X.; Zhao, L.; Ma, C.; Shi, H. Thermal error simulation and compensation in a jig-boring machine equipped with a dual-drive servo feed system. *Proc. Inst. Mech. Eng. B J. Eng.* **2015**, *229*, 43–63. [CrossRef]
14. Liu, J.; Gui, H.; Ma, C. Digital twin system of thermal error control for a large-size gear profile grinder enabled by gated recurrent unit. *J. Ambient Intell. Hum. Comput.* **2023**, *14*, 1269–1295. [CrossRef]
15. Ma, C.; Gui, H.; Liu, J. Self learning-empowered thermal error control method of precision machine tools based on digital twin. *J. Intell. Manuf.* **2023**, *34*, 695–717. [CrossRef]
16. Xiao, J.; Fan, K. Research on the digital twin for thermal characteristics of motorized spindle. *Int. J. Adv. Manuf. Technol.* **2022**, *119*, 5107–5118. [CrossRef]
17. Liu, K.; Song, L.; Han, W.; Cui, Y.; Wang, Y. Time-varying error prediction and compensation for movement axis of CNC machine tool based on digital twin. *IEEE Trans. Ind. Inform.* **2021**, *18*, 109–118. [CrossRef]
18. Yi, H.; Fan, K. Co-simulation-based digital twin for thermal characteristics of motorized spindle. *Int. J. Adv. Manuf. Technol.* **2023**, *125*, 4725–4737. [CrossRef]
19. Liu, J.; Ma, C.; Gui, H.; Wang, S. A four-terminal-architecture cloud-edge-based digital twin system for thermal error control of key machining equipment in production lines. *Mech. Syst. Signal Process.* **2022**, *166*, 108488. [CrossRef]
20. Lunev, A.; Lauerer, A.; Zborovskii, V.; Léonard, F. Digital twin of a laser flash experiment helps to assess the thermal performance of metal foams. *Int. J. Therm. Sci.* **2022**, *181*, 107743. [CrossRef]
21. Kuprat, J.; Pascal, Y.; Liserre, M. Real-Time thermal characterization of power semiconductors using a PSO-based digital twin approach. In Proceedings of the 2022 24th European Conference on Power Electronics and Applications (EPE'22 ECCE Europe), Hanover, Germany, 5–9 September 2022; pp. 1–8.
22. Liu, R.J.; Li, H.S.; Lv, Z.H. Modeling methods of 3D model in digital twins. *CMES Comp. Model. Eng.* **2023**, *136*, 985–1022. [CrossRef]
23. Grieves, M.W. Product lifecycle management: The new paradigm for enterprises. *Int. J. Prod. Dev.* **2005**, *2*, 71–84. [CrossRef]
24. Tao, F.; Zhang, H.; Liu, A.; Nee, A.Y. Digital twin in industry: State-of-the-art. *IEEE Trans. Ind. Inform.* **2018**, *15*, 2405–2415. [CrossRef]
25. Smagulova, K.; James, A.P. A survey on LSTM memristive neural network architectures and applications. *Eur. Phys. J. Spec. Top.* **2019**, *228*, 2313–2324. [CrossRef]
26. Korstanje, J. LSTM RNNs. In *Advanced Forecasting with Python: With State-of-the-Art-Models Including LSTMs, Facebook's Prophet, and Smazon's DeepAR Apress*; Apress: Berkeley, CA, USA, 2021; pp. 243–251.
27. Pope, J.E.; Pope, E. *Rule of Thumb for Mechanical Engineers-A Manual of Quick, Accurate Solutions to Everyday Mechanical Engineering Problems*; Gulf Professional Publishing: Houston, TX, USA, 1997; pp. 18–50.
28. Chen, Z.C.; Chen, Z.N. *Termal Characteristics Foundation of Machine Tools*; Machinery Industry Press: Beijing, China, 1989.
29. Ruspini, E.H.; Bezdek, J.C.; Keller, J.M. Fuzzy clustering: A historical perspective. *IEEE Comput. Intell. Mag.* **2019**, *14*, 45–55. [CrossRef]
30. Fang, B.; Zhang, J.; Hong, J.; Yan, K. Research on the nonlinear stiffness characteristics of double-row angular contact ball bearings under different working conditions. *Lubricants* **2023**, *11*, 44. [CrossRef]
31. Ma, S.; Yin, Y.; Chao, B.; Yan, K.; Fang, B.; Hong, J. A real-time coupling model of bearing-rotor system based on semi-flexible body element. *Int. J. Mech. Sci.* **2023**, *245*, 108098. [CrossRef]

lubricants

MDPI

Article

Eccentric Rotor Drop Dynamics Study of Vertical Maglev Bearing System

Xiaoxu Pang [1,2,*], Dingkang Zhu [1,*], Ming Qiu [1,2,3], Dongfeng Wang [4] and Xinlong Wang [1]

1 School of Mechanical and Electrical Engineering, Henan University of Science and Technology, Luoyang 471003, China
2 High-Level Bearing Henan Synergy Innovation Center, Luoyang 471003, China
3 Henan Collaborative Innovation Center for Advanced Manufacturing of Machinery and Equipment, Luoyang 471003, China
4 Luoyang Bearing Science & Technology Co., Ltd., Luoyang 471003, China
* Correspondence: pxx8308@163.com (X.P.); zdkkong@163.com (D.Z.)

Abstract: When considering the problem of a vertical magnetic levitation bearing system, the rotor eccentric fall is more likely to cause the failure of the protective bearing. In this paper, a rotor drop collision model and a protective bearing dynamics model are constructed. It compares and analyzes the evolution of collision force values of the rotor eccentric drop as well as the non-eccentric drop. Further, this paper discusses the law of influence of three factors, rotor quality, rotational speed, and axial protection clearance, on the collision characteristics of the protected bearing in eccentric and non-eccentric cases. It has also experimentally verified this characteristic of rotor speed. The results show that compared with the non-eccentric condition, the axial impact force and radial impact forces of the rotor in the eccentric condition increase by 14% and 114%, respectively. Compared with the non-eccentric condition, with the increase in rotor quality, the axial and radial impact force increase by 68% on average, and the axial depth amplitude of the rotor increases by 350%. With the increase of rotor speed, the axial impact force without an eccentric drop is basically unchanged; the axial impact force of an eccentric drop increases slightly, and the radial impact force increases by 110%. With the increase of axial protection clearance, the radial displacement vibration of the rotor axis increases; the average increase of the maximum axial force is 120 N, and the average increase of the maximum radial force is 100%.

Keywords: vertical magnetic levitation system; protective bearing; collision characteristics; eccentric drop

Citation: Pang, X.; Zhu, D.; Qiu, M.; Wang, D.; Wang, X. Eccentric Rotor Drop Dynamics Study of Vertical Maglev Bearing System. *Lubricants* **2023**, *11*, 246. https://doi.org/10.3390/lubricants11060246

Received: 26 April 2023
Revised: 26 May 2023
Accepted: 30 May 2023
Published: 3 June 2023

1. Introduction

The magnetic suspension bearing can suspend the rotor with the magnetic field force without lubrication and friction and can use the unbalanced compensation method to control and optimize the rotor position using a high-precision system [1]. It is a mechatronics support device [2] that can be applied and can realize active control.

After the failure of the magnetic levitation system, the rotor and the bearing inner ring will cause a huge vibration impact, and the bearing roller will slip and lead to rubbing wear. In 1997, when the EC conducted the experiment on the aero-engine with the magnetic suspension bearing system at 18,000 r/min, the 150 kg rotor fell to the protective bearing due to the failure of the magnetic suspension system, resulting in a huge impact. This led to the breakage of the protective bearing and severe damage to the magnetic levitation bearing system [3].

2. Literature Review

As an important line of defense to protect the whole machine after the magnetic suspension system fails, the protective bearing can make a passive response when the system fails. In order to improve the drop resistance of the protective bearing and the reliability during service, it is necessary to study the dynamics of the collision of the rotor falling into the protective bearing. Heshmatallah Mohammad Khanlo et al. [4] studied the contact force between the inner axis and the auxiliary bearing and revealed various nonlinear dynamical behaviors, such as periodic, quasi-periodic, periodic, and chaotic vibrations, and jumping phenomena. The results show that the speed parameters, axial speed ratio, and gravity parameters have a significant impact on the dynamic response. Wu Guoqing et al. [5] studied the support structure of the maglev wind turbine and constructed a drop simulation model to explore the influence of different structural parameters of protective bearing on the rotor track. Kong Yanan et al. [6] simulated the rotor drop process of the magnetic bearing system to explore the influence trend of the eccentric rotor speed and dynamic balance accuracy level on the drop collision force. Zhu Yili et al. [7] proposed using the elastic ring to buffer the shock and vibration caused by the rotor drop and to simulate the drop model under different installation positions of the rotor. The results show that increasing the elastic ring on the rotor can effectively reduce the collision force and amplitude during the drop. Wei Peng et al. [8] simulated the drop collision force of the rotor at high speed, explored the test machine, and gave a method to measure the collision force. Zhao Jingxiong et al. [9] used finite element software to simulate the rotor drop process, explored the collision characteristics of the auxiliary bearing in the helium fan experimental bench, and analyzed the intrusion depth and wear influence between the bearing inner ring and the rolling body. Patrick S. Keogh et al. [10] studied the transient thermal response of protective bearings, and the results showed that this method had some reference value for predicting the life span of protective bearings. Ma Zilin [11] selected full-loaded ceramic ball angular contact bearings without cages as protective bearings for horizontal magnetic bearing systems. This paper analyzed the rotation speed of the inner ring of the bearing, the axial collision force of the inner ring, the contact force between adjacent spheres, and the sliding speed of the raceway between the ball and the inner ring during the rotor dropping process and concluded that the fully-loaded ceramic ball has the margin to bear further load at high speed. Lin Ma et al. [12] established a dynamic model of the rotor and the protective bearing. They analyzed the dynamic behavior of the axis trajectory diagram through the three-dimensional bifurcation diagram and two-parameter bifurcation diagram, optimized the structural parameter design of the rotor system, and determined its reasonable and stable working interval. Yang Guojun et al. [13] discussed the design characteristics of the magnetic bearing control system of HTR-10GT and then studied the influence of AMB stiffness on the critical speed of the rotor. Zhu Changsheng [14] established the active electromagnetic bearing internal rotor in the finite element method as the basis of kinematics formulas, mainly analyzing the maglev bearing after the failure of the rotor drop dynamics, generally for the instantaneous nonlinear characteristics, and discussed the application of protective bearing support damping and support stiffness. Li [15] designed the vertical magnetic bearing system by using two kinds of protective bearings with a full ball and a cage without a cage to conduct a rotor drop simulation, analyzed the internal temperature rise of bearings, rotor drop space trajectory, and the influence of changes in axial and radial contact forces, and finally concluded that the ceramic ball hybrid bearing with a cage was a more suitable anti-impact scheme.

Many achievements have been made in the research of rotor falling under non-eccentric conditions after the failure of the maglev bearing system, but research on rotor falling dynamics with an eccentric angle is scarce. However, due to the inherent disorder of the magnetic levitation bearing system during high-speed operation and the problem of low control accuracy, the rotor is in a state of inclined rotation and revolution motion superposition. Therefore, in the actual manufacturing or motor-driven rotor high-speed rotation, the rotor drop under the eccentric condition is more practical. In this paper, we investigate the eccentric rotor drop dynamics of the vertical magnetic levitation system that is more in line with the actual working conditions; it constructs a rotor drop collision model and a protective bearing dynamics model. This paper makes a comparison of the magnitude of the collision force between the rotor drop and the protective bearing under eccentric and non-eccentric conditions, and the effect of different operating parameters of the rotor on the collision force and the radial displacement of the rotor axis is also investigated.

3. Research Methodology

3.1. Working Principle of Vertical Maglev Bearing System

Figure 1 is the vertical maglev bearing system structure diagram. The model is composed of a rotor, motor, two radial magnetic bearings, a group of symmetrical configurations of axial magnetic bearings, upper and lower protective bearings, and an upper collision adjustment pad. In this paper, considering the rotor eccentricity, high-speed heavy load, and large impact load conditions, the 71913C angular contact ball bearing (without cage) is installed face to face as the upper protective bearing, bearing the main axial and radial impact forces. The lower protective bearing is a 61809 deep groove ball bearing, which only bears part of the radial impact force to reduce the rotor swing amplitude when the rotor swings inertia, and only plays a radial protection role. The specific parameters of the two bearings are shown in Table 1:

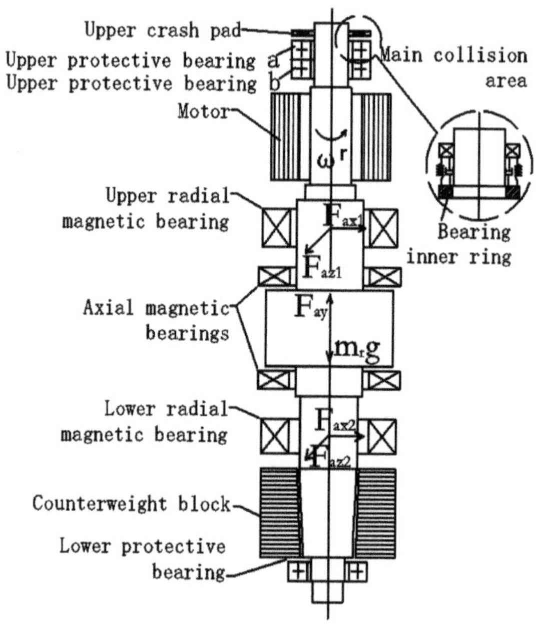

Figure 1. Schematic diagram of a shaft system structure of vertical magnetic levitation bearing system.

Table 1. Input boundary conditions.

Type	Model Number	Bore Diameter (mm)	Outside Diameter (mm)	Width (mm)	Cr (kN)	Cor (kN)
Deep groove ball bearing	61809	45	58	7	6.40	5.60
Angular contact ball bearing	71913C	65	90	13	20.8	21.2

The outer ring of the protective bearing of the upper and lower parts of the rotor of the maglev bearing system is placed on a special tooling, and the inner ring is in a stable state under gravity (in the simplified model, it is considered that the outer ring is in a fixed state and the inner ring is in a floating state). In normal operation, the rotor is suspended by the magnetic force of axial and radial magnetic bearings, and the motor drives the rotor to rotate at high speed. The weight of the whole rotor can be changed by increasing or decreasing the weight of the counterweight block so as to analyze different working conditions. However, due to the performance and accuracy problems of the magnetic bearing control system, the magnetic bearing disconnection caused by the control failure will cause the rotor to drop off in high-speed rotation, so the protective bearing is set as an efficient defense line to avoid the rotor collision and damage the whole machine. There is a gap between the upper and lower protective bearings and the rotor without force action. At this time, the force of the rotor is gravity $m_r g$, the upper and lower radial magnetic bearing support forces F_{ax} and F_{az}, and the axial magnetic bearing supports force F_{ay}. The falling rotor colliding with the inner ring of the protective bearing will produce complex nonlinear behavior on the axial displacement and collision force of the rotor, so the analysis of nonlinear multi-factor coupling dynamics becomes complex [16–18]. The following mathematical model is established for this.

3.2. The Collision Model between the Rotor and the Protective Bearing

In the collision process, the rotor is assumed to be a rigid body. Only elastic deformation occurs after the collision between the rotor and the inner ring of the protective bearing, and the collision process can be converted into a spring-damping model.

The mechanical calculation formula of the linear spring is

$$F_n = kx + c\dot{x} \tag{1}$$

where k is the contact stiffness, c is the contact damping, and x is the deformation degree. The linear mathematical model presented in this paper is sufficient to evaluate the amplitude-frequency response of a magnetic bearing under multi-frequency excitation of a rotor drop collision by self-programming a multi-degree-of-freedom magnetic levitation nonlinear energy acquisition system formed by adding linear spring vibrators [19,20].

Figure 2a is the schematic diagram of the position of the rotor and the rolling bearing, where o is the coordinate origin, namely, the equilibrium position of the stable suspension of the rotor. Furthermore, o_1 is the rotor axis; set the initial unilateral gap between the rotor and the bearing inner ring to be s, and the intrusion depth of the rotor and the inner ring to be τ_1:

$$\tau_1 = \sqrt{(x_r - x_{b1})^2 + (z_r - z_{b1})^2} - s \tag{2}$$

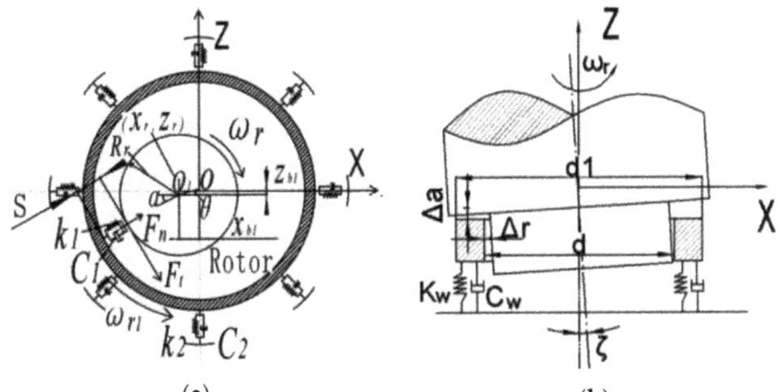

Figure 2. Simplified collision model of rotor and protective bearing. (**a**) Radial collision; (**b**) axial impact.

The relative compression deformation between the bearing inner ring and the steel ball is τ_2:

$$\tau_2 = \sqrt{{x_{b1}}^2 + {z_{b1}}^2} \tag{3}$$

where the axis-coordinate value of the rotor is (x_r, z_r). x_{b1} is the offset of the bearing hub axis X direction. z_{b1} is the offset of the bearing inner ring axis in the Z direction.

When $\tau_1 > 0$, the rotor is in contact with the inner ring of the bearing, and then the rotor will receive radial contact force and tangential friction force from the inner ring on the contact surface. A relevant equation of state can be obtained using force analysis on the rotor:

$$m_r\ddot{x}_r = -F_{n1}\cos\alpha' - F_{t1}\sin\alpha' \tag{4}$$

$$m_r\ddot{z}_r = -m_r g + F_{t1}\cos\alpha' - F_{n1}\sin\alpha' \tag{5}$$

$$\alpha' = -\arctan\frac{z_r - z_{b1}}{x_r - x_{b1}} \tag{6}$$

where F_{n1} is the radial contact force between the rotor and the bearing inner ring, F_{t1} is the tangential friction between the rotor and the bearing inner ring, and α' is the contact phase angle between the rotor and the bearing inner ring.

We analyzed the torque balance of the rotor. In the contact stage, relative to the axis, the balanced rotor is subjected to friction torque, so there is

$$J_r\ddot{\theta}_r = -F_{t1}R_r \tag{7}$$

According to the spring damping system model, the contact force between the rotor and the inner ring can be expressed as

$$F_{n1} = k_1\tau_1^{e_1} + c_1V_{n1} \tag{8}$$

$$V_{n1} = \frac{(\dot{x}_r - \dot{x}_{b1})(x_r - x_{b1}) + (\dot{z}_r - \dot{z}_{b1})(z_r - z_{b1})}{\sqrt{(x_r - x_{b1})^2 + (z_r - z_{b1})^2}} \tag{9}$$

where k_1 is the radial contact stiffness coefficient between the rotor and the bearing inner ring; e_1 is the contact coefficient; c_1 is the contact damping between the rotor and the bearing inner ring; and V_{n1} is the normal velocity of the rotor opposite to the bearing inner ring. The contact stiffness coefficient can be solved according to the calculation method of line contact in Hertz's theory [21]. According to the simplified model, the collision contact between the rotor and the inner ring can be considered the contact between the rotor and the fixed ring. When the material between the rotor and the fixed ring is steel, the contact width between them is the width of the inner ring end face l = 3.5 mm: k_1 = 3.378 × 10^8 (N/mm), e_1 = 10/9.

In addition, research on damping has not yet formed a relatively mature calculation method, and damping parameters are generally obtained by testing. Here, contact damping is commonly used [10]: c_1 = 1000 (N·s/mm)

Figure 2b is the axial collision model. K_w is supporting stiffness; C_w is the contact damping; Δa is the axial protection gap; Δr is the radial protection gap; ζ is the rotor eccentric angle; k_t is the eccentric angle fixed constant; d is the bearing bore diameter; and d_1 is the diameter of the inner ring retaining edge.

The axial collision force is

$$F_a = K_{ca}(|l_a - l_{ia}| - \Delta_a)\cos(k_t - \zeta) \tag{10}$$

$$K_{ca} = 1.2 \times 10^{11}\sqrt{\pi(d_1^2 - d^2)} \tag{11}$$

In formula: l_a is the axial displacement of the rotor, and l_{ia} is protecting the axial displacement of the bearing inner ring.

3.3. Dynamics Model of Protective Bearings

The relative displacements of the inner and outer rings in radial and axial directions are τ_r and τ_a, respectively. We specify that the position of the scroll is $\theta = 0$. Figure 3 shows the center of the ball and the curvature of the inner and outer grooves. After any steel ball is loaded, the axis and radial coordinates of the curvature center of the inner groove Q'_{da} at position θ_a are [22]:

$$A_{ya} = BD_W \sin\alpha + \tau_a \quad (12)$$

$$B_{ya} = BD_W \cos\alpha + \tau_r \cos\theta_q \quad (13)$$

where: D_W is the bearing steel ball diameter; $B = g_q - 1 + g_w$, g_q, g_w, respectively, is the inner and outer ring of the groove radius of curvature coefficient; general ball bearing g value between 0.52~0.53; α is the contact angle before loading.

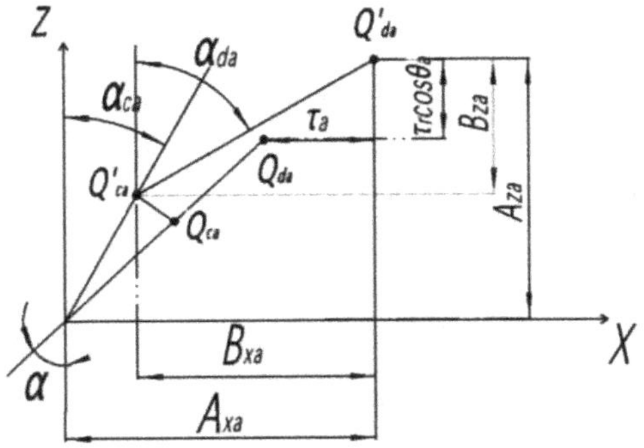

Figure 3. The position of the center of a steel ball relative to the center of curvature of the groove.

The deformation of the steel balls is, respectively,

$$\tau_{ca} = Q'_{ca}Q'_{da} - Q_{ca}Q_{da} = \sqrt{B_{xa}^2 + B_{za}^2} - (g_q - 0.5)D_W \quad (14)$$

$$\tau_{da} = Q'_{ca}Q'_{da} - Q_{ca}Q_{da} = \sqrt{(A_{xa} - B_{xa})^2 - (A_{za} - B_{za})^2} - (g_w - 0.5)D_W \quad (15)$$

In this formula, B_{xa} is the force, and B_{za} is the speed; Q_{ca} and Q_{da} are the center of the ath steel ball before the load and the corresponding curvature center. Because the outer ring is fixed, the center of curvature Q'_{ca} does not change, and Q'_{ca} and Q'_{da} are the center of the ath steel ball and the corresponding center of curvature after being loaded, respectively.

The contact load of the steel ball and the inner and outer loops is

$$Q_{da} = K_{da}\tau_{da}^{\frac{3}{2}} \quad (16)$$

$$Q_{ca} = K_{ca}\tau_{ca}^{\frac{3}{2}} \quad (17)$$

where: K_{da} and K_{ca} can be calculated from formula (5-5) in literature [23], and their values are related to the contact angle.

The contact angle of the ath steel ball and the inner and outer ring rolling path is

$$\cos\alpha_{da} = \frac{B_{za}}{(g_q - 0.5)D_W + \tau_{da}} \quad (18)$$

$$\cos \alpha_{da} = \frac{B_{xa}}{(g_q - 0.5)D_W + \tau_{da}} \tag{19}$$

$$\cos \alpha_{ca} = \frac{A_{za} - B_{za}}{(g_w - 0.5)D_W + \tau_{ca}} \tag{20}$$

$$\sin \alpha_{da} = \frac{A_{xa} - B_{za}}{(g_q - 0.5)D_W + \tau_{da}} \tag{21}$$

4. Results

4.1. Comparative Analysis of Collision Characteristics under Non-Eccentric and Eccentric Working Conditions

The simplified rotor drop collision model of the vertical magnetic suspension bearing system is simulated and analyzed by using self-programming multi-body dynamics software. That is, the multivariate nonlinear second-order differential equations obtained from the above simultaneous equations were solved by using the Runge-Kutta method of 4–5 orders. The calculation accuracy was set as 1×10^6, and the solution was solved from the rotor to 0.5 s after the drop. Due to the large axial impact force caused by the vertical fall of the vertical rotor, the bearing outer ring is set with a fixed pair to be pre-tightened and fixed to prevent the axial displacement of the protective bearing outer ring. Moreover, the upper collision pad is fixed on the rotor in the form of a fixed pair. The simulated radial protection gap is 0.2 mm, and the contact stiffness is 1×10^5; the contact damping is 70; the coefficient of dynamic friction and static friction are 0.1 and 0.3, respectively, and the time is 0.4 s or 0.5 s. In the subsequent analysis, the above initial input parameters are used for dynamic simulation analysis.

Figure 4 shows a simplified schematic diagram of the rotor movement under eccentric working conditions. The rotor coupled the rotational and rotational motion states, which were superimposed on each other and acted on the axis locus together [24–26]. Most of the axial impact force in the upper angle contact ball bearing group is borne by protective bearing *a*. Rotor axis displacement, protective bearing an inner ring, shaft between rotors, and radial impact force are the main research parts of this paper.

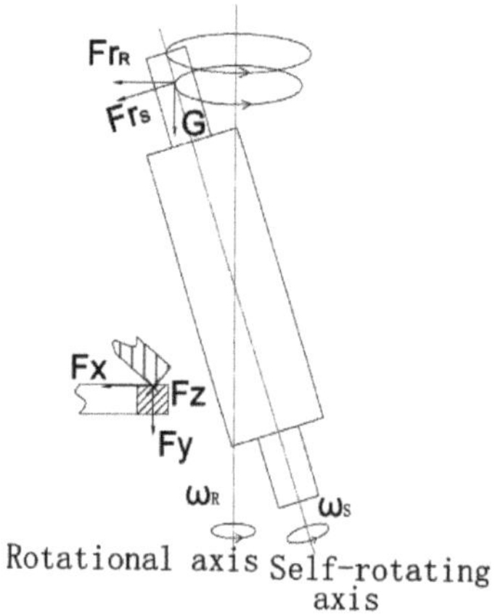

Figure 4. Schematic diagram of the rotor simplified model under eccentric working conditions.

Due to the large distance between the radial plane of the rotor axis and the radial plane of the protective bearing *a*, there is a motion phase difference between the displacement at the rotor axis and that at the protective bearing *a*. Therefore, this paper does not study the displacement at the rotor axis, and the axis studied in this paper is the axis that protects a bearing a with the radial plane. The rotor drop axis displacement and impact force under eccentric conditions are more in line with the actual conditions. The following is a comparative analysis of the impact characteristics of the rotor that falls under non-eccentric conditions.

Figures 5–7 are comparative analyses of non-eccentric and eccentric conditions under the rotating speed of 20,000 r/min, axial and radial protection clearance of 0.2 mm, 45 kg rotor, and eccentric angle of 0.05°.

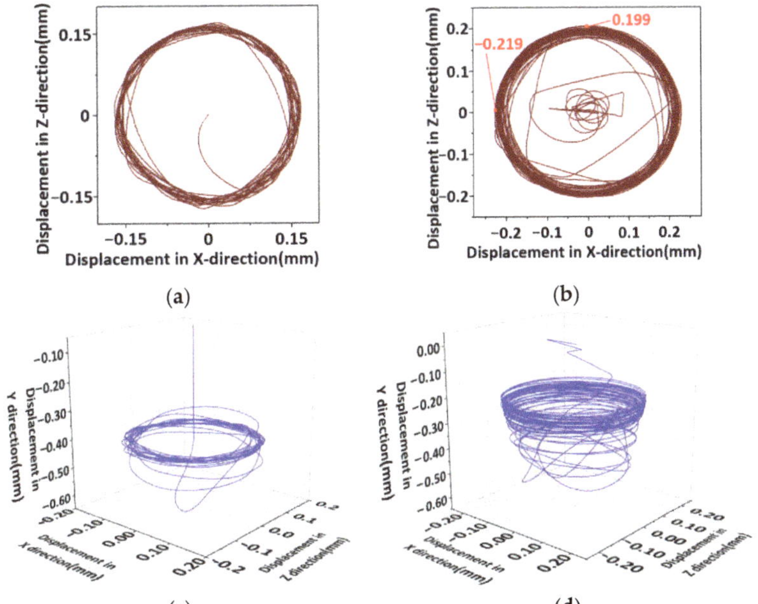

Figure 5. Comparison of eccentric-free and eccentric 0.05° rotor axis. (**a**) Non-eccentric rotor radial axial displacement; (**b**) Radial axial displacement of the eccentric rotor; (**c**) Space-wise displacement of the non-eccentric rotor axis; (**d**) The spatial displacement of the eccentric rotor axis.

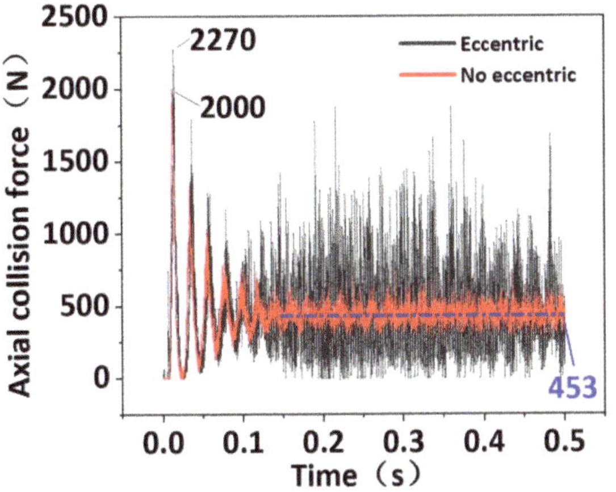

Figure 6. Comparison of the trend of the axial collision force at 0.05° with and without eccentricity.

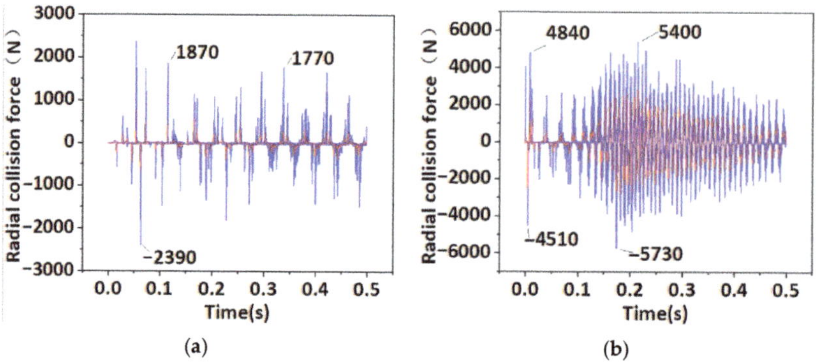

Figure 7. Comparison of non-eccentric and eccentric 0.05° rotor radial collision force. (**a**) Non-eccentric rotor radial collision force; (**b**) Radial collision force of the eccentric rotor.

As can be seen from Figure 5a,c, the axis locus after rotor fall and collision under a non-eccentric working condition is relatively stable, and there is almost no chaotic trajectory in the circle. The motion trajectory of the rotor in the radial direction is approximately fitted to a circle with a diameter of 0.158 mm, which has a small motion amplitude and fast stabilization speed.

According to Figure 5b,d, in the radial plane of the trajectory, the ellipse fitted by the rotor trajectory has a long axis of 0.225 mm and a short axis of 0.2 mm. Because the radial protection clearance of the magnetic levitation bearing system is 0.2 mm, it can be seen that the rotor shows a polygonal trajectory due to rebound under eccentric conditions, which indicates that the rotor will have friction collision with the inner ring of the protective bearing in the radial direction. The rotor falls behind, in the axial upward, after the collision bounce height fluctuation. It can be seen that the track-intensive area is concentrated in the upper part, and the overall lower track clutter is very significant. This is caused by the oblique elliptic surface of the drop space trajectory of the rotor axis under eccentric conditions.

Figure 6 shows the trend comparison of the maximum axial impact force on the rotor under the two working conditions over time. The maximum impact force appears in the first impact. It is observed that the falling collision force (red line) in unbiased conditions will soon stabilize at around rotor gravity of 450 N, as shown in the blue line in the figure. The maximum impact force of the rotor falling 2270 N under eccentric conditions is 14% higher than that under non-eccentric conditions, and its variation trend before 0.1 s is similar to that under non-eccentric conditions. In 0.1 s~0.5 s, the collision force with an average value of 967 N, which is 100% higher than that of the former, continues to appear and does not stabilize to 450 N at 0.5 s.

Corresponding to Figure 5d, it can be seen that the increase of axial impact force under eccentric conditions will increase the reaction force on the rotor from the bearing inner ring, resulting in the rise of the rotor rebound height. The rotor axis locus is no longer a circle in a horizontal radial direction but appears as an oblique elliptic plane in space. As the value of the late collision force continues to be high, the rotor rebound height is still higher than that in the non-eccentric condition and finally presents the trajectory as shown in Figure 5d.

It can be seen from Figure 7a that the maximum radial impact force does not occur in the first impact. Compared with the axial impact force, the maximum radial impact force does not gradually stabilize with time. According to Figure 5a, the maximum displacement of the rotor axis locus is 0.158 mm, less than the radial protection clearance of 0.2 mm. Furthermore, because the inner ring of the protective bearing also has a radial floating instability phenomenon, the rotor and the inner ring have periodic collisions, resulting in radial collision force. According to Figure 7b, the high numerical radial impact force of the rotor falling under eccentric conditions appears before 0.1 s, while it is very intensive after 0.1 s. The peak radial force of 5400 N is 130% higher than that of 2390 N in non-eccentric conditions, and the total radial force value is 140% higher than that in non-eccentric conditions. This is the elliptical trajectory caused by the superimposed rotation and revolution of the rotor. When the linear velocity of the rotor axis increases, the centrifugal force increases, and the radial impact component is generated after the collision between the rotor and the inner ring, which makes the eccentric radial impact force multiply compared with the non-eccentric conditions.

It can also be seen from Figure 5b that the radial impact force generated by the collision between the front rotor and the inner ring of the protective bearing at 0.1 s makes the trajectory of the axis center have a polygon trajectory. As shown in Figure 7b, the radial collision force of the latter part is dense, and the collision between the rotor and the inner ring leads to a higher frequency of polygon trajectories, which will make the effect of the quasi-synthetic circle of the center of mass worse. As shown in Figure 5b, the moving distance of the centroid on the radial plane increases from 0.15 mm to 0.219 mm; that is, the long axis of the trajectory shape is 0.219 mm, and the short axis is 0.199 mm, which approximates to form an ellipse.

In conclusion, under certain initial conditions of the external boundary, the spatial form of rotor eccentricity makes the drop impact force of the rotor colliding with the inner ring of the protective bearing produce axial and radial impact components. In the late 0.1 s, the values of both remain high, and the relative slip between the inner ring and the rotor also causes changes in radial force and abnormal vibration, accompanied by a certain reverse vorticity [27–29]. This will lead to a larger range of longitudinal rebound displacement or penetration displacement and a larger unidirectional radial displacement than the trajectory of the non-eccentric drop conditions. Finally, the elliptic trajectory is synthesized, and a collision occurs, which is more unstable.

It can be seen that the rotor drop under the actual eccentric condition has greater impact-wear on the protective bearing. Under the conditions of heavy load, high speed, and large gap between tooling, it is easy to make the magnetic levitation system scrap and fail. The following will be analyzed and discussed from the following three aspects: the influence of rotor quality, rotational speed, and the gradient rise of axial protection clearance on the fall collision under the condition of five eccentric angles.

4.2. *Effect of the Rotor Quality on the Collision*

Table 2 shows the control parameters of working conditions with different rotor quality conditions, and five groups of control simulation are set.

Table 2. Input boundary conditions.

Name	1	2	3	4	5
Rotor quality (kg)	25	35	45	55	65
Axial protection clearance (mm)			0.25		
Rotor rotation speed (r/min)			20,000		

It can be seen from Figure 8a that the distance between the position of the final stop and the zero of the rotor in the non-eccentric condition increases with the increase of the rotor quality. The trend of the axial displacement curve is similar to that of axial impact force under non-eccentric conditions in Figure 6. It can be seen from Figure 8b that the displacement trend of the rotor falling before 0.1 s in the eccentric condition is similar to that in the non-eccentric condition. After 0.1 s, the longitudinal rebound amplitude increases due to rotor collision. This corresponds to the phenomenon that the axial impact force remains high after 0.1 s under the eccentric working condition in Figure 6.

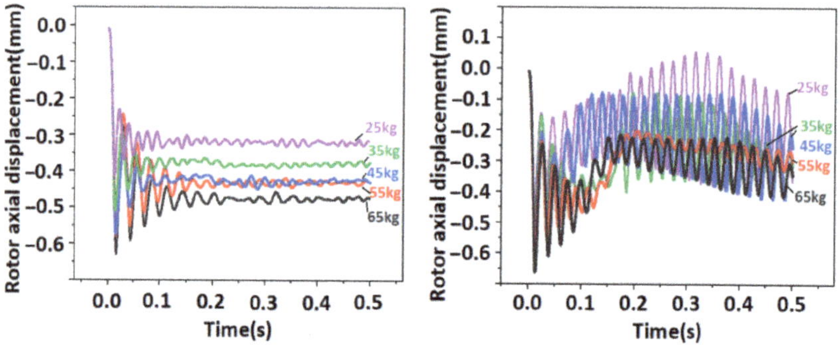

(a) (b)

Figure 8. Comparison of non-eccentric and eccentric 0.05° rotor radial collision force. (**a**) Axial displacement of the rotor drops of different qualities under non-eccentric working conditions. (**b**) Axial displacement of rotor drops of different qualities at 0.05° eccentric conditions.

As can be seen from Figure 9a, the maximum axial displacement of rotor drop in both non-eccentric and eccentric conditions increases with the increase of rotor quality, and the maximum axial displacement in the eccentric condition is greater than that in the non-eccentric condition. Compared with the non-eccentric condition, the quality increases by 160%, and the displacement increases by 50%, indicating that this trend is nonlinear. In Figure 9b, it can be seen that the harmonic motion generated with rotor collision and rebound in the non-eccentric working condition within 0.2~0.5 s enters the stable stop state. With the increase of rotor quality, the depth amplitude is basically stable at 0.053 mm, while the average depth amplitude in the eccentric condition is increased by 350% compared with that in the non-eccentric condition, and the amplitude decreases with the increase of rotor quality [30].

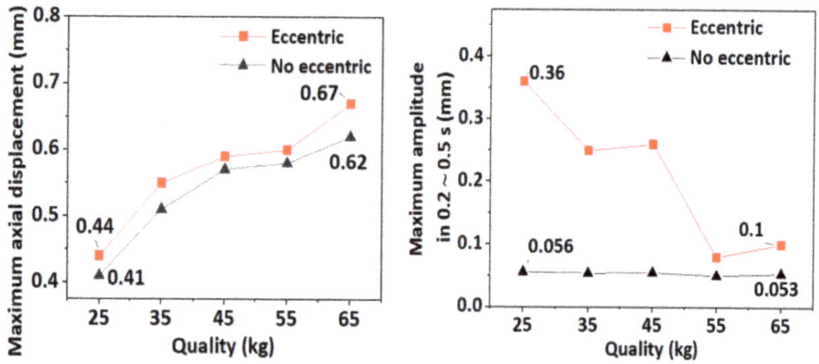

(a) (b)

Figure 9. Comparison of maximum axial displacement and maximum amplitude of different quality rotor drops without eccentricity and eccentricity 0.05° working condition. (**a**) Comparison of the maximum axial displacement; (**b**) Comparison of the maximum amplitude within the 0.2~0.5 s.

The rotor rotates non-eccentric. According to the impulse formula $I = FT$, the quality increases, and the depth of impact into the inner ring deepens. According to Equation (20), the ball displacement B_{za} (Y+) increases, and the contact angle α increases. Due to the fixed outer ring, the inner ring of the protective bearing generates a displacement in the axial direction of (Y+); that is, the axial displacement of the rotor decreases so that the axial displacement becomes stable rapidly.

Compared with the non-eccentric condition, the greater the quality, the greater the impact of the rotor drop on the bearing inner ring in the eccentric condition, the deeper the intrusion displacement, the smaller the amplitude of rebound depth, but the overall is still greater than the non-eccentric condition.

Figure 10a shows the 3D color mapping surface of the axial impact force affected by different qualities under the eccentricity angle increased by five gradients (0.01~0.09°).

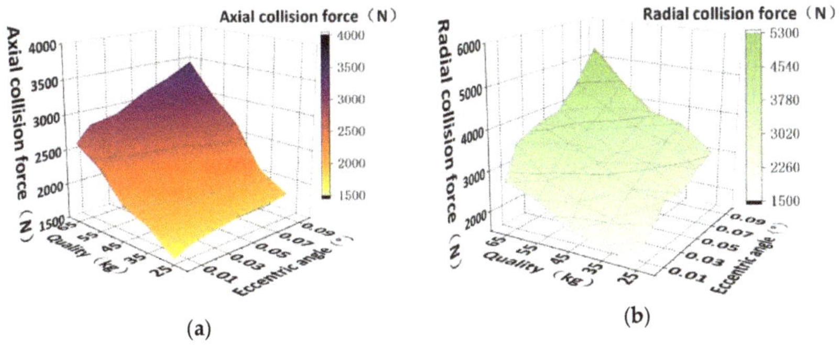

Figure 10. Axial and radial collision forces of quality on rotors with different speeds. (**a**) Maximum axial collision force; (**b**) Maximum radial collision force.

According to the gravity formula and the impulse formula, when other boundary conditions are fixed, the quality of the rotor increases, and the accumulated impulse of the rotor during the falling time will also increase, thus leading to the increase of the axial collision force between the rotor and the protective bearing. With the increase of eccentric angle, the maximum increase of axial impact force reaches 1790 N. According to Equation (8), with the increase of eccentric angle ζ, the axial protection clearance Δa between half of the rotor and the upper protective bearing increases, impulse increases, and the maximum axial impact force F_a increases. In other words, the increase in rotor eccentricity will affect the increase in axial impact force. The greater the rotor quality, the greater the impulse and the greater the axial impact force. As shown in Figure 11a, when the eccentric angle is 0.05° and the rotor quality increases by 150%, the axial impact force on the inner ring increases by 75% compared with the non-eccentric condition, which is similar, showing a nonlinear increase in general. Moreover, it is easy to see from Figure 11a that the eccentric drop impact force is greater than the non-eccentric condition.

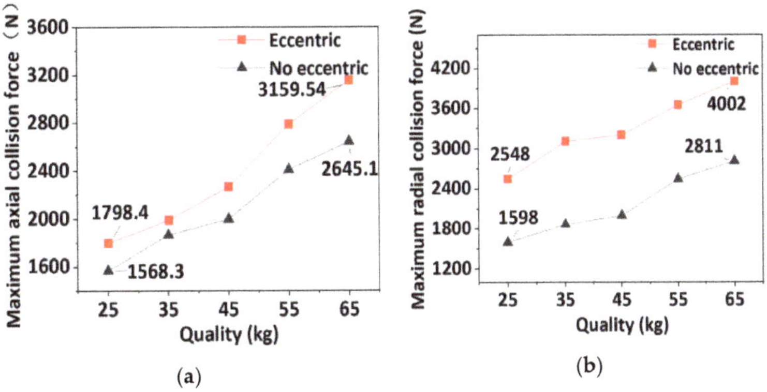

Figure 11. Comparison of maximum axial and radial collision forces. (**a**) Comparison of the maximum axial collision force of different quality rotor drops with an eccentricity of 0.05°; (**b**) Comparison of the maximum radial collision force of different quality rotor drops with an eccentricity of 0.05°.

Figure 10b is about the radial impact force, whose maximum increase reaches 3854 N, and the impact force also increases nonlinearly with the increase of quality and eccentricity angle. As shown in Figure 11b, under the condition of 0.05° eccentricity, the average value is 60% higher than that under the condition of non-eccentric eccentricity, the quality is 150% higher, and the radial impact force on the inner ring is 60% higher. Under eccentric conditions, along with the rotation and revolution of the rotor, the contact mode between the rotor and the inner ring changes from surface contact to local point contact, resulting in a local axial slip of the bearing inner ring and steel ball, constant changes in the rotation phase of the bearing inner ring, and the rotor falls to complete point contact collision. As the quality of the rotor increases, the component of gravity on the radial plane of the eccentric rotor also increases, leading to the component of the rotor on the radial plane of the inner ring. When the angle γ between the inner ring radial plane and the horizontal plane changes constantly, the high-frequency collision is repeated continuously; that is, the radial collision force between the two increases. With the increase of eccentric angle, it can be obtained from Equations (2) and (9) that the eccentric angle ζ increases, the transverse coordinate x_r of the rotor axis increases, the invasion depth τ_1 increases, and the normal velocity V_{n1} of the rotor relative to the inner ring of the bearing increases. According to Equation (8), the radial impact force F_{n1} between the rotor and the inner ring becomes larger if other fixed constants remain unchanged.

4.3. *Influence of Rotor Speed on Collision*

Table 3 shows the control parameters of different rotor speed conditions.

Table 3. Input boundary conditions.

Name	1	2	3	4	5
Rotor rotation speed (r/min)	5000	10,000	15,000	20,000	25,000
Rotor quality (kg)			45 kg		
Axial protection clearance (mm)			0.2		

In Figure 12a,b, it can be seen that the approximate fitting of the rotor axis displacement trajectory under non-eccentric working conditions is a circle with a diameter of about 0.155 mm, and there is no chaotic vibration trajectory in the trajectory, indicating that there is no collision with the bearing inner ring. However, the rotor speed increased from 5000 r/min to 25,000 r/min, and the degree of trajectory fitting was poor. In Figure 12c,d, it can be seen that, under eccentric conditions, the maximum displacement of the rotor's axial displacement track in both horizontal and vertical directions on the radial plane is wrong by a certain distance. This Figure approximates an ellipse with a major axis of 0.214 mm and a minor axis of 0.198 mm. As the speed reaches 25,000 r/min, the rotor collides with the inner ring of the bearing, resulting in a polygonal axis locus.

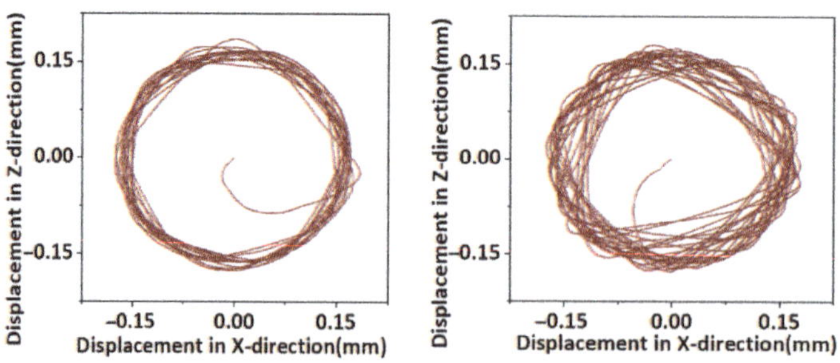

(**a**) (**b**)

Figure 12. *Cont.*

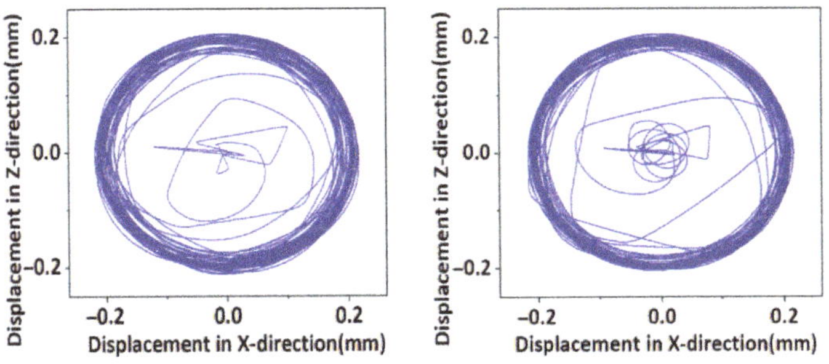

(c) (d)

Figure 12. Radial displacement trajectory of non-eccentric and eccentric rotor shaft centers at different speeds. (**a**) 5000 r non-eccentric drop axis displacement; (**b**) 25,000 r non-eccentric drop axis displacement; (**c**) 5000 r eccentric 0.05° drop axis displacement; (**d**) 25,000 r eccentric 0.05° drop axis displacement.

When the rotor falls on the protective bearing, the speed of the inner ring of the protective bearing increases rapidly in a short time. The radial displacement of the rotor is mainly caused by the relative rotation of the rotor and the inner ring and the centrifugal force of the rotor. As the rotational speed increases, the centrifugal force increases, and the relative motion velocity of the two increases. As can be seen from Figure 4, the radial force F_X generated by friction increases, leading to the change of rotor force time, resulting in trajectory disturbance. Under eccentric conditions, the radial force direction changes irregularly due to the collision between the rotor and the inner ring, which is the reason for the poor fitting degree of the rotor trajectory.

Figure 13 shows the trend diagram of axial impact forces of rotors at different speeds under non-eccentric and eccentric conditions. According to the mathematical model established by Equation (10) and indirectly verified by Figure 13a, there is no obvious correlation between the maximum axial impact force and the non-eccentric rotor speed [24,25], and the maximum axial impact force is maintained around 2250 N. Figure 13b shows that the rotor and bearing inner ring under eccentric conditions increase from 5000 r/min to 25,000 r/min, and the axial impact force increases from 1918.4 N to 2186.65 N.

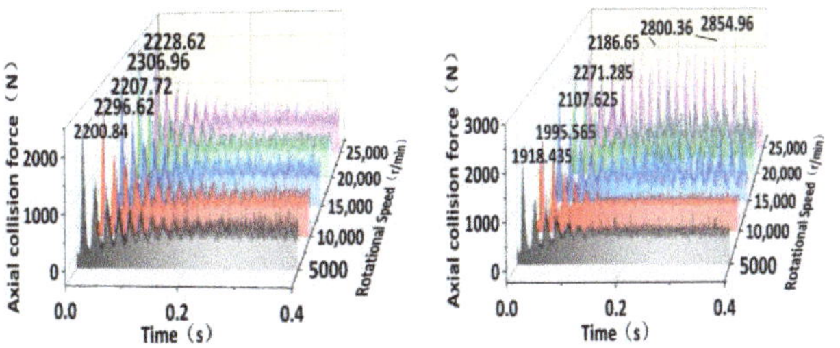

(a) (b)

Figure 13. Comparison of axial collision forces between non-eccentric and eccentric rotors at different speeds. (**a**) Non-eccentric fall axial collision force; (**b**) Eccentricity 0.05° drop axial collision force.

It can be seen that the axial collision force corresponding to each speed decreases slightly, and the stability degree of the collision force trend in the late period is not as good as that in the non-eccentric condition. The increase in the speed is accompanied by periodic high-value collision forces, such as the appearance of 2800.36 N and 2854.96 N.

As the speed increases, the kinetic energy of the rotor increases.

In non-eccentric working conditions, this part of the energy is transferred to the bearing or consumed only through contact friction. The friction factor μ, rotor quality m, axial protection clearance Δh, and other conditions do not change; that is, the friction force of the rotor and the gravitational potential energy of the rotor itself do not change, and the maximum impact force basically does not change with the change of rotor speed when the rotor starts to fall.

Under eccentric conditions, according to gravity formula $G = mg$ and impulse formula $I = FT$, part of the axial protection clearance decreases, and part of it increases. When the rotor quality is constant, the accumulated impulse of the part of the rotor with small protection clearance decreases, and the impact force decreases. However, when the protection gap is large, the cumulative impulse increases and the impact force increases, and the change trend of the impact force is periodic. As the rotational speed increases, the impact force does not change obviously; that is, it has little effect on the axial impact force.

Figure 14a shows the axial impact force generated by the rotor falls at different speeds under the condition of five declination gradients (0.01~0.09°), and its maximum increase is about 610 N. As can be seen from Figure 15a, with the increase in rotational speed, the maximum axial impact force under non-eccentric conditions is basically stable.

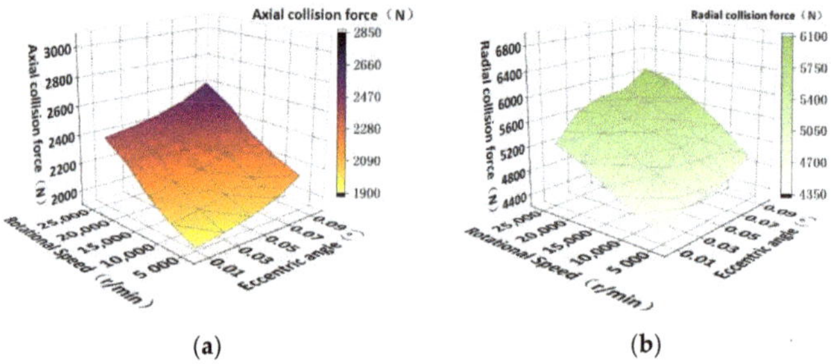

Figure 14. Eccentricity angle on the axial and radial collision force of rotor with different speeds. (**a**) Maximum axial collision force; (**b**) Maximum radial collision force.

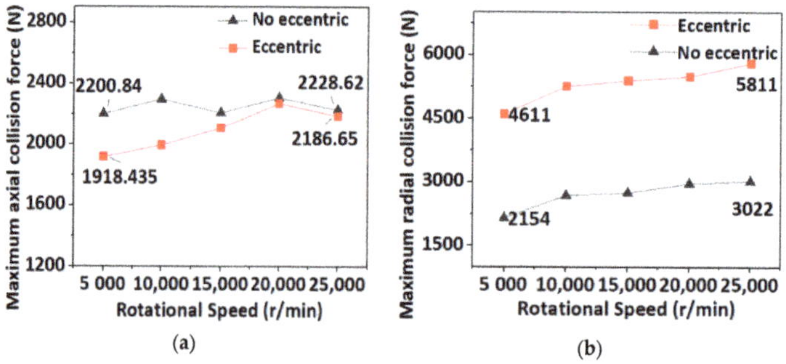

Figure 15. Comparison of maximum axial and radial collision forces under non-eccentric and eccentric working conditions. (**a**) Comparison of the maximum axial collision force of different speed rotor drops with eccentric 0.05°; (**b**) Comparison of the maximum radial collision force of different quality rotor drops with the eccentric 0.05°.

The corresponding maximum axial impact force is slightly less than the non-eccentric condition. With the increase in rotational speed, the nonlinear impact force increases slightly. The reasons have been explained above and will not be repeated here. As shown in Figure 14b, the maximum increase of radial impact force is about 1600 N. As shown in Figure 15b, the average maximum radial impact force under eccentric conditions is

110% higher than that under non-eccentric conditions, both of which are nonlinear and small increases.

According to Equations (2), (8)–(10), it can be seen that when the eccentric angle increases, the axial force F_a and radial force F_n both increase proportionally, and the radial impact force caused by the eccentric drop will increase the average force of the inner ring and steel ball, and the offset in the x and y directions of the inner ring will increase. According to Formula (3), with the increase of x_{b1} and y_{b1}, the relative compressive deformation between the inner ring and the steel ball increases τ_2; that is, the penetration depth of the channel between the protective bearing rolling body and the inner and outer ring increases, resulting in bearing pitting or fish-scale fatigue surface peeling, and reducing the life of the protective bearing [31,32].

4.4. Effect of Axial Protective Clearance on Collision

Table 4 shows the different control parameters of rotor speed conditions, with four sets of control simulations, and the eccentric angle is still five gradients.

Table 4. Input boundary conditions.

Name	1	2	3	4
Axial protection clearance (mm)	0.1	0.15	0.2	0.25
Rotor quality (kg)		45		
Rotor rotation speed (r/min)		20,000		

Figure 16 shows the axial displacement diagram of the rotor falling under non-eccentric conditions. The displacement of the rotor axis in the radial plane of the protective bearing shows a polygonal trend. With the increase of axial protection clearance, the rotor rebound height increases. The spatial displacement after rotation of the axis presents an oblique elliptic trajectory, while the one presented in the radial plane is a synthetic circular trajectory with a diameter of about 0.168 mm, which is less than the radial protection clearance of 0.2 mm.

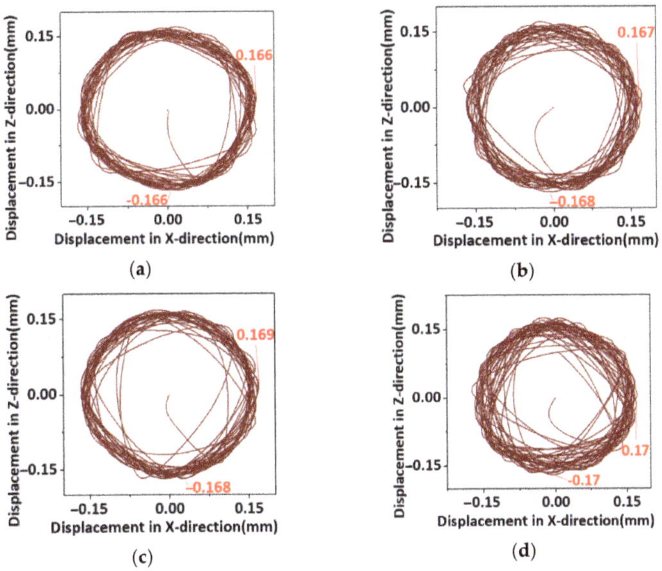

Figure 16. Radial axis displacement trajectory of the rotor under different axial protection gaps for non-eccentric working conditions. (**a**) 0.1 mm drop axis displacement; (**b**) 0.15 mm drop axis displacement; (**c**) 0.2 mm drop axis displacement; (**d**) 0.25 mm drop axis displacement.

As shown in Figure 17, the rotor axis presents a trend of collision rebound polygon variation in the radial plane of the protective bearing. The eccentricity of the rotor falls into the inner ring and superimposes its own rotation and revolution motion, lengthening the long axis of the space oblique elliptic trajectory. The direction of radial impact force changes constantly with the change of gyroscopic phase; the gyroscopic vibration of the rotor increases, and the variation amplitude of the radial displacement increases. With the increase of the axial protection clearance, the trajectory fitting growth axis is 0.21 mm, and the short axis is a 0.157 mm ellipse. This displacement has been greater than the radial protection clearance by 0.2 mm. With the increase of axial protection clearance, the radial collision between the rotor and the inner ring is severe, which will seriously affect the bearing life.

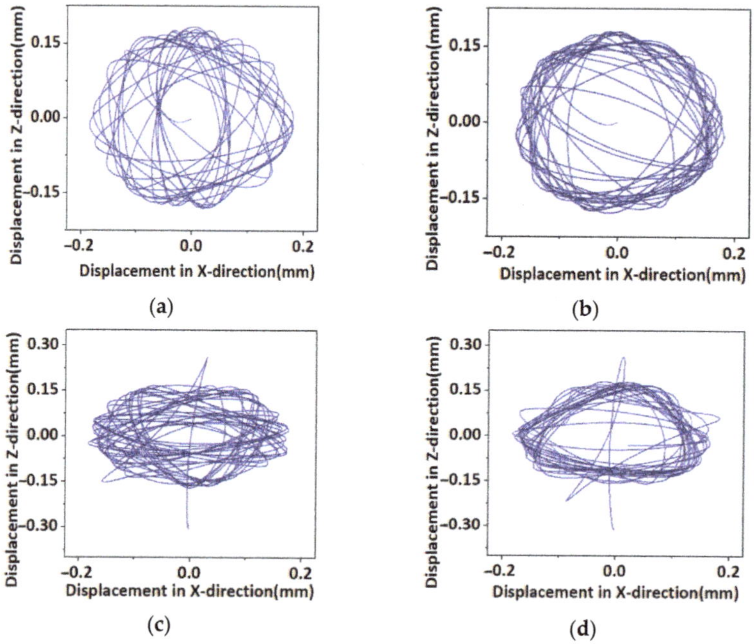

Figure 17. Radial axis displacement trajectory of eccentric 0.05° rotor under different axial protection clearance. (**a**) 0.1 mm Drop axis displacement; (**b**) 0.15 mm Drop axis displacement; (**c**) 0.2 mm Drop axis displacement; (**d**) 0.25 mm Drop axis displacement.

Figure 18a shows the axial impact force generated when the rotor with different axial protection clearance falls under the condition of five declination gradients (0.01°~0.09°). With the increase in the axial protection clearance and eccentricity angle, the maximum increase of axial impact force caused by rotor falling is 826 N. According to Figure 19a, with the increase of axial protection clearance, the average maximum axial impact force under eccentric conditions is 120 N larger than that under non-eccentric conditions, and the growth trend of both is similar. As the clearance Δh between the rotor and protective bearing increases, the initial energy of the rotor increases. According to the $h = 1/2\, gt^2$, $I = FT = mv$, the rotor drop time t increases, and the impulse and velocity are increased. Since the equivalent stiffness and damping of rotor and bearing are certain, the greater the speed at the moment of collision. The larger the eccentricity angle, the larger the contact pressure, the larger the deformation, and the larger the axial impact force.

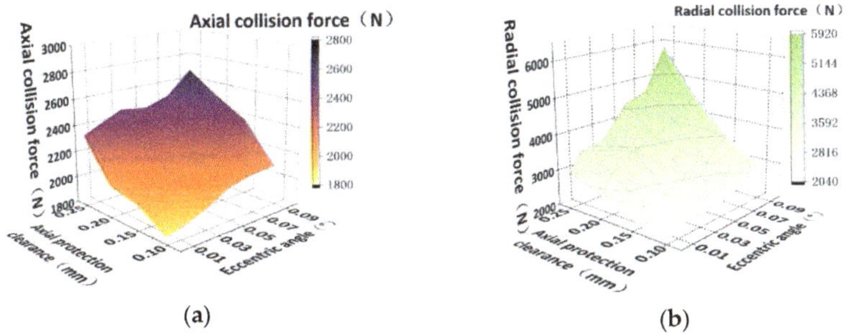

Figure 18. Eccentricity angle on the axial and radial collision forces of rotor drops with different axial clearances. (**a**) Maximum axial collision force; (**b**) Maximum radial collision force.

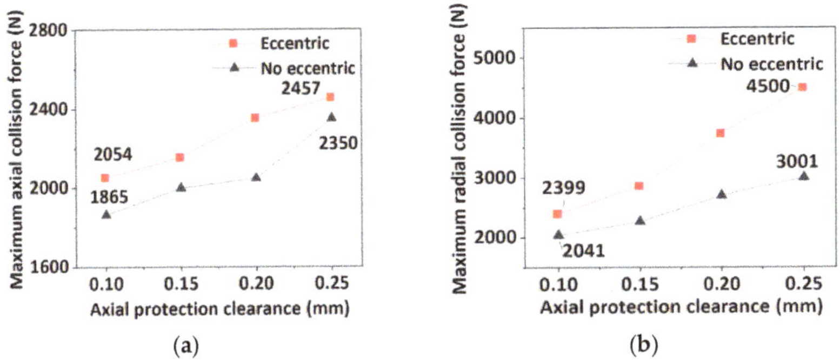

Figure 19. Comparison of maximum axial and radial collision forces under non-eccentric and eccentric working conditions. (**a**) Comparison of the maximum axial collision force of different axial protection clearance rotor drops under the 0.05° eccentric working condition; (**b**) Comparison of the maximum radial collision force of different axial protection gap rotor drops under the 0.05° eccentric working condition.

Figure 18b shows that the maximum radial impact force is 5896 N, twice the axial impact force. With the increase of eccentric angle, the maximum increase of radial impact force is 2806 N. As shown in Figure 19b, with the increase in axial protection clearance, the radial impact force increases by 1499 N and 1960 N in non-eccentric conditions as compared with that in non-eccentric conditions. Furthermore, the growth rate under eccentric conditions is obviously greater than that under non-eccentric conditions.

With the increase in the protection gap and impulse, it can be obtained from Equations (2) and (9) that the invasion depth τ_1 will deepen, and V_{n1} will increase. According to Equation (8), the radial impact force F_{n1} becomes larger. According to Figures 18 and 19, it can also be seen that the radial impact force increases with the increase of the axial protection clearance in the falling collision between the rotor and the bearing inner ring [28,33]. In other words, the fitting degree of the axial displacement trajectory under non-eccentric and eccentric conditions is also affected by the axial protection clearance.

5. Discussion of Drop Experiment

To simulate the drop performance of the eccentric rotor more in line with actual working conditions, a vertical rotor non-eccentric drop test was conducted using the protective bearing life test machine of the magnetic levitation bearing system shown in Figure 20. The 45 kg eccentric rotor speed rises with a 5, 10, 15, and 20 kr/min gradient. The protective bearing is a face-to-face mounted 71913C angular contact ball bearing, and the rest of the fall conditions are the same as the previous Table 3, and the rotor axis trajectory index is measured in real-time.

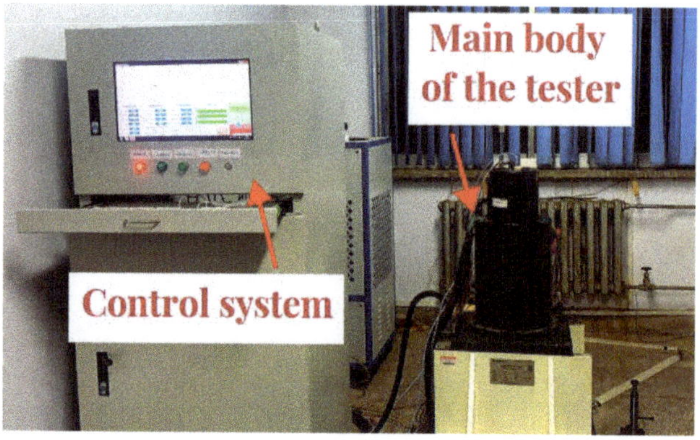

Figure 20. Life tester of auxiliary bearing for the active magnetic bearing system.

Figure 21 shows the radial axis displacement trajectory of the rotor set at different rotational speeds during the test; as the rotor speed increases, the degree of radial axis trajectory fitting into the circle gradually becomes worse, and the internal chaotic trajectory of the trajectory increases, and the rotor quality center trajectory in the radial direction has contact collision with the inner surface of the inner ring, resulting in the rotor in the process of axial collision rebound, while the velocity component of the rotor radial rebound exists. From the maximum diameter of the following four circular trajectory diagrams, it can be seen that as the rotor speed increases from 5 kr/min to 20 kr/min, the maximum displacement of the rotor radial center of quality increases from 0.15 mm to 0.18 mm, indicating that the rotor dynamic balance accuracy decreases as the speed increases. From the trajectory, this is consistent with the simulation results of the eccentric rotor above.

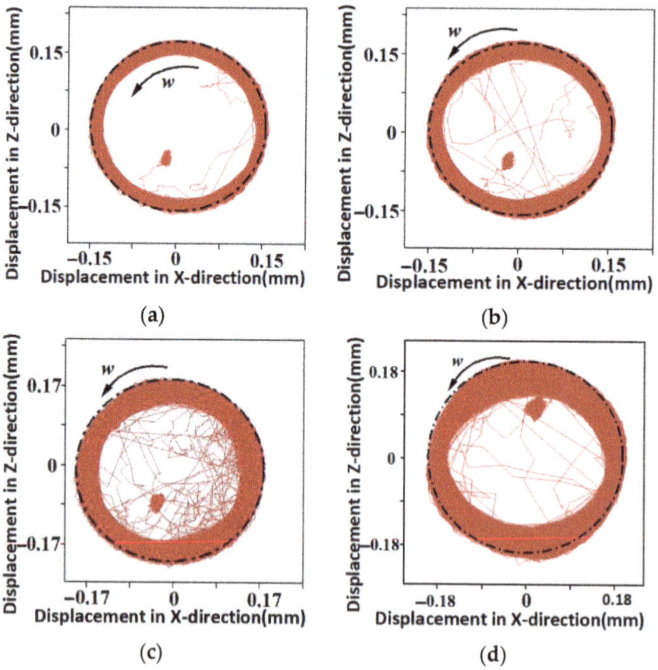

Figure 21. The centroid trajectory of rotors at different speeds in the process of falling collision. (**a**) 5 kr/min; (**b**) 10 kr/min; (**c**) 15 kr/min; (**d**) 20 kr/min.

In this paper, the impact force of the outer ring of the protective bearing is tested by the pressure sensor. As shown in Figure 22, sensor A and sensor B are symmetrically arranged at the bottom of the large end of the outer ring of the protective bearing. The outer ring is pressed and fixed by the pressing force exerted by the testing machine tooling, which conforms to the simulation model established above. The inner ring of the protective bearing is impacted by the rotor drop, and the axial impact force is transmitted to the outer ring through the rolling body so as to realize the change trend of the impact force indirectly. By comparison with Figures 6 and 23, it can be seen that, under the condition of the eccentric rotor, which is more in line with actual conditions, the maximum impact force of the outer ring is 979.477 N, and the average value of 212.71 N under stable conditions; both are about 1/2 of the corresponding force of the inner ring.

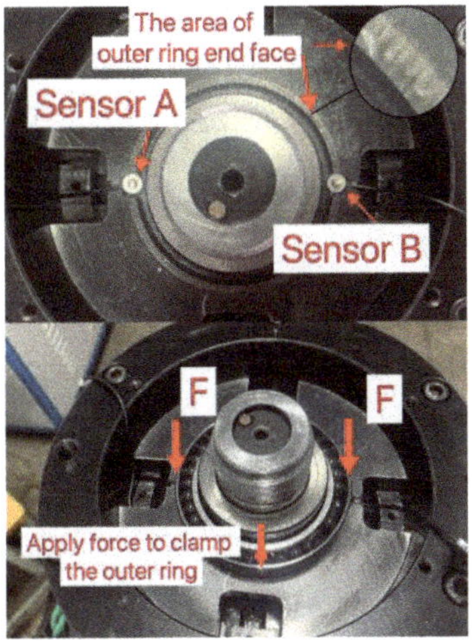

Figure 22. Installation diagram of axial collision force sensor.

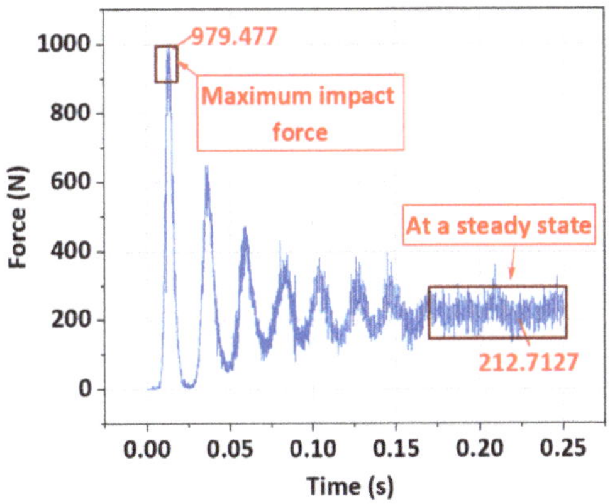

Figure 23. Simulation results of collision force on the outer ring.

As shown in Figure 24, according to the above experimental machine modification and simulation analysis results, the impact force of the outer ring of the protective bearing is tested. As shown in Figure 24a, the experimental data curve is divided into four stages: Rest, Rotor suspension acceleration, Falling collision stage, and Steady-state. The 389.1 N in the static stage in Figure 24a includes the outer ring force F= about 190 N and the dead weight of half of the rotor $G_{1/2}$ = 200 N. 192.4 N of the rotor suspension acceleration stage only includes the outer ring force $F = 190$ N of testing machine tooling. The drop collision stage of 1169.7 N includes the outer ring pressing force $F_O = 190$ N and the maximum impact force $F_{max} = 979.7$ N. The mean value of 372.4 in the stable stage corresponds to the force value in the static stage.

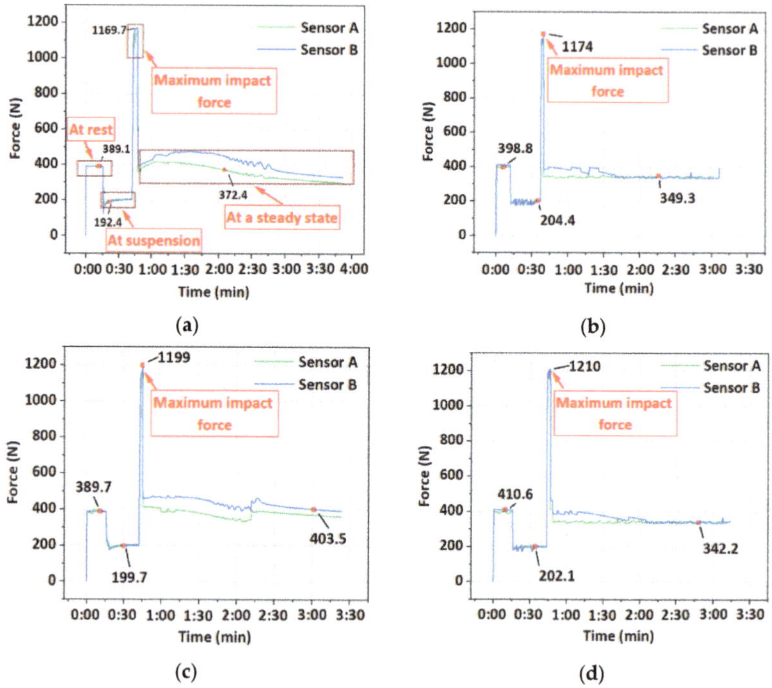

Figure 24. Force on the outer ring of the protective bearing with increasing speed. (**a**) 5 kr/min; (**b**) 10 kr/min; (**c**) 15 kr/min; (**d**) 20 kr/min.

Pay attention to the four figures in Figure 24. The rotor falls in the eccentric state, which is more consistent with the actual working condition. It can be seen that as the rotor speed increases from 5 kr/min to 20 kr/min, the force values of 1169.7 N, 1174 N, 1199 N, and 1210 N also rise gradient-wise. By comparison with Figure 13b above, it can be seen that the trend of simulation and experimental results is consistent and corresponding. In addition, after the four force values minus the outer ring force $F_O = 190$ N, compared with Figure 23, the maximum collision force F_{max} corresponds to the simulation, all of which are about 980 N. These are close to the simulation results above and also verify the accuracy of the self-programming simulation model above.

Figure 25 shows the appearance of the inner ring of the protective bearing after the collision of the eccentric rotor drop at different rotational speeds. 5 kr/min of the inner ring has no obvious changes; observe 20 kr/min of the inner ring, the ceramic ball, by the axial impact of running scratches the inner ring groove, indentation is obvious, the rotor drops to the inner ring end surface of the bearing, the speed is too high, deceleration time is extended, the inner ring end surface also reveals obvious scratches, resulting in the ceramic ball and groove, the ceramic ball and the groove, the friction between the rotor collision pad and the inner ring end surface increases, generating intense heat, affecting the protection of the bearing against falling performance and service life.

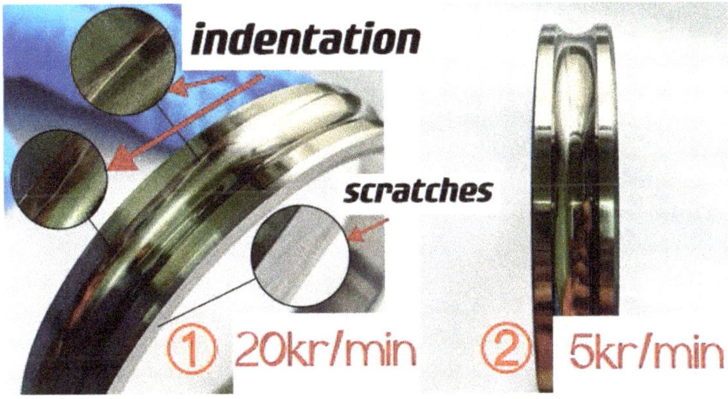

Figure 25. The appearance of the inner ring of the protective bearing at 20 kr/min and 5 kr/min operating conditions.

6. Conclusions

This paper first compares and discusses the rotor drop impact characteristics under eccentric and non-eccentric working conditions and then explores and analyzes the rotor axial displacement and axial and radial impact forces under three working parameters (quality, speed, axial protection clearance). The results are as follows:

(1) Under eccentric conditions, an ellipse with a long axis of 0.225 mm larger than the radial protection clearance is synthesized after the rotor falls. The axis locus under non-eccentric conditions is a circle with a diameter of 0.158 mm. Compared with the non-eccentric condition, the radial impact force increases by 140% on average, and the maximum axial impact force increases by 14%. Both of them produce high collision force, which is close to the maximum value continuously after 0.1 s. The results show that the eccentric condition has a great adverse effect on the stability of the axial displacement of the rotor, and both axial and radial impact forces are multiplied.

(2) With the increase in rotor quality, compared with the non-eccentric condition, the average increase of rotor axial drop displacement in the eccentric condition is 0.04 mm, and the amplitude in depth is increased by 350%. With the increase in eccentric angle, the axial impact force increases by 1.75 times, and the radial impact force increases by 60%. The results show that the change of quality can deepen the axial penetration depth after the rotor drops under eccentric conditions and has a great influence on the axial and radial impact force.

(3) With the increase in rotor speed, the maximum axial impact force decreases slightly, and the radial impact force increases by 110% compared with the non-eccentric condition. The maximum axial impact force changes slightly under non-eccentric conditions. An ellipse with a long axis of 0.214 mm, which is larger than the radial protection clearance, is synthesized from the rotor drop trajectory under eccentric conditions. The results show that the increase in rotational speed has little effect on the axial impact force, but a great effect on the radial impact force, and the collision between the rotor and the inner ring leads to the trajectory disorder.

(4) With the increase of the axial protection clearance Δh, the pseudo-circle degree of the rotor axis displacement trajectory becomes worse in both non-eccentric and eccentric conditions. The trajectory under eccentric conditions is an ellipse with a radial fitting growth axis of 0.21 mm larger than the radial protection clearance of 0.2 mm. With the increase of eccentric angle, the average increase of the maximum axial force is 120 N, and the average increase of the maximum radial force is two times. The results show that the increase of the axial protection clearance will cause the disturbance of

the radial displacement trajectory of the rotor, and the axial impact force will increase slightly, but the radial impact force will be greatly affected.

(5) After trial verification, the rotor speed increases the rotor radial axis trajectory to the degree of poor rounding and radial displacement from 0.15 mm to 0.18 mm. The force of the protective bearing outer ring increased from 1169.7 N to 1210 N. The inner ring groove and end face scratching are obvious, and frictional heat increases.

The eccentric working condition is 20~150% higher than the collision force without eccentric working conditions, and the shaft offset is larger than the radial protection gap of 0.2 mm, which indicates that the eccentric inclination of the rotor will cause greater impact damage to the bearing. Tsinghua University [15] studied the rotor drop of a heavy vertical magnetic bearing onto a protective bearing. In the experiment, the axial impact force and heating of vertical rotor falling were investigated. The change of axial impact force can only be measured indirectly by the sensor in the rotor drop experiment. In the future, intelligent, protective bearings may be used for temperature measurement or direct measurement of axial impact force so as to improve the impact resistance of protective bearings more effectively. Yang, G. also analyzed the inclination of the vertical rotor around the horizontal plane and proposed a detailed model of the vertical rotor falling process considering the inclination of rotation around the x and y axes [19]. He accurately predicted the dynamic behavior of the rotor in vertical descent. In his study on the axial impact force of rotor dropping, the influence of the friction coefficient is very weak, and the axial aerodynamic force tends to increase the peak axial contact force.

In view of the large impact force value in the case of eccentricity mentioned above, such high impact force exists objectively because the eccentricity error is unavoidable in actual manufacturing or the high-speed rotation of the motor-driven rotor [34]. This can reduce the axial impact force of the rotor by improving the rotor operating condition parameters and the structural parameters of the protective bearing, such as axial protection clearance, rotor dynamic balance accuracy, the number of ball fillings in the bearing, whether there is a cage or not, and changing the ring material, which will be a very meaningful study. The rotor of the vertical magnetic suspension bearing system will keep the eccentric condition when running. In this paper, the dynamic process of rotor eccentric drop is analyzed innovatively in the field of magnetic bearing research, and it is concluded that the actual impact of an eccentric rotor on the system is more serious than that of the non-eccentric rotor in all aspects. Therefore, before practical applications or tests, high requirements are put forward for the improvement of the overall impact resistance of protective bearings and the design of rotor dynamic balance precision. Future research in the field of impact protective bearings should focus on the rotor eccentricity drop conditions. This paper provides a reference for the research in this field. In addition, the research results of this paper will be helpful to the further improvement of protective bearing design and engineering application.

Author Contributions: Conceptualization, X.P. and D.Z.; methodology, X.P.; software, D.Z.; validation, D.Z. and X.W.; formal analysis, M.Q.; investigation, D.Z.; resources, D.W.; data curation, X.P. and D.Z.; writing—original draft preparation, X.P. and D.Z.; writing—review and editing, X.P. and D.Z.; visualization, D.W. and X.W.; supervision, X.P.; project administration, X.P. and D.Z.; funding acquisition, D.W. All authors have read and agreed to the published version of the manuscript.

Funding: This research received was founded by the National Natural Science Foundation of China (52205096).

Conflicts of Interest: The authors declare no conflict of interest.

References

1. Gao, H.; Xu, L. Real-time Feed-forward Force Compensation for Active Magnetic Bearings System Based on H∞ Controller. *Chin. J. Mech. Eng.* **2011**, *24*, 58–66. [CrossRef]
2. Zhang, W.; Zhu, H.; Yuan, Y. Study on Key Technologies and Applications of Magnetic Bearings. *Trans. China Electrotech. Soc.* **2015**, *30*, 12–20.

3. Xia, X. Dynamics of Magnetic Suspended Rotor Drop on Auto Eliminating Clearance Auxiliary Bearings. Master's Thesis, Nanjing University of Aeronautics and Astronautics, Nanjing, China, 2014.
4. Khanlo, H.M.; Ghayour, M.; Ebrahimi, R. Chaotic Vibration Analysis of a Coaxial Rotor System in Active Magnetic Bearings and Contact With Auxiliary Bearings. *J. Comput. Nonlinear Dyn.* **2016**, *12*, 031012.
5. Wu, G.; Lu, B.; He, D.; Song, C. The research and analysis on critical speed of auxiliary support system for magnetic suspension wind turbine. *Mod. Manuf. Eng.* **2018**, *458*, 14.
6. Kong, Y.; Xue, Y.; Ye, J.; Jiao, Y. Impact Characteristics of Vertical Rotor Dropping on Auxiliary Bearings. *Mach. Tool Hydraul.* **2021**, *49*, 6–11.
7. Zhu, Y.; Jin, C.; Xu, L. Effect of different elastic ring installation locations in catcher bearing system on rotor drop dynamic responses. *J. Aerosp. Power* **2012**, *27*, 1671–1680.
8. Wei, P.; Wang, Y.; Yang, Y.; Yan, L.; Xu, L. An analysis of the contact force between a floating rotor and a retainer bearing after rotor drop. *J. Vib. Shock* **2018**, *37*, 251–258.
9. Zhao, J.; Yang, G.; Li, Y.; Yu, S.; Xu, L. Numerical Analysis of Magnetically Suspended Rotor in HTR-10 Helium Circulator Being Dropped into Auxiliary Bearings. *Nucl. Power Eng.* **2012**, *33*, 61–64.
10. Keogh, P.S.; Yong, W.Y. Thermal Assessment of Dynamic Rotor/Auxiliary Bearing Contact Events. *J. Tribol.* **2006**, *129*, 143–152. [CrossRef]
11. Ma, Z.; Zhao, D.; Ni, Y. The Dynamic Response Analysis of Floating Rotor-Touchdown Bearing in AMB System. *Bearing* **2022**, *65*, 1–9.
12. Ma, L.; Lin, M.; Shijun, W.; Peng, W. Research on the Dynamic Characteristics of Magnetic Bearing-rotor System with Auxiliary Bearing. *J. Phys. Conf. Ser.* **2020**, *1637*, 012140. [CrossRef]
13. Yang, G.; Xu, Y.; Shi, Z.; Gu, H. Characteristic analysis of rotor dynamics and experiments of active magnetic bearing for HTR-10GT. *Nucl. Eng. Des.* **2006**, *237*, 1363–1371.
14. Zhu, C. Nonlinear dynamics of rotor drop on rolling element backup bearings after active magnetic bearings failure. *J. Mech. Eng.* **2006**, *7*, 196–202. [CrossRef]
15. Li, Z.; Shi, Z.; Tie, X.; Yang, G.; Ren, W.; Yao, J.; Wang, Y.; Wang, Z. Failure Mechanism on Touchdown Bearing of Heavy Maglev Rotor Drop. *China Mech. Eng.* **2023**, *34*, 1009–1018.
16. Altaf, S.; Mehmood, M.S.; Soomro, M.W. Advancement of Fault Diagnosis and Detection Process in Industrial Machine Environment. *J. Eng. Sci.* **2019**, *6*, D1–D8. [CrossRef]
17. Lyu, M.; Liu, T.; Wang, Z.; Yan, S.; Jia, X.; Wang, Y. A control method of the rotor re-levitation for different orbit responses during touchdowns in active magnetic bearings. *Mech. Syst. Signal Process.* **2018**, *105*, 241–260. [CrossRef]
18. Anna, T.; Enrico, M.; Andrea, R.; Alessandro, R.; Pierluca, D.A.; Amedeo, F.; Duccio, F.; Benjamin, D. Development and Experimental Validation of Auxiliary Rolling Bearing Models for Active Magnetic Bearings (AMBs) Applications. *Int. J. Rotat. Mach.* **2019**, *2019*, 4675286.
19. Wang, Z.; Ding, H.; Chen, L. Modeling of a magnetic levitation energy harvesting system attaching a linear oscillator under multi-frequency excitation and its output power's influencing parametric analysis. *J. Vib. Shock* **2018**, *37*, 225–229.
20. Fang, B.; Zhang, J.; Hong, J.; Yan, K. Research on the Nonlinear Stiffness Characteristics of Double-Row Angular Contact Ball Bearings under Different Working Conditions. *Lubricants* **2023**, *11*, 44. [CrossRef]
21. Fumagalli, M.A.; Schweitzer, G.; Ulbrich, H. Modelling and Measurement Analysis of the Contact Interaction between a High Speed Rotor and Its Stator. Ph.D. Dissertation, ETH Zürich, Zürich, Switzerland, 1997.
22. Neisi, N.; Sikanen, E.; Heikkinen, J.E.; Sopanen, J. Effect of off-sized balls on contact stresses in a touchdown bearing. *Tribol. Int.* **2018**, *120*, 340–349. [CrossRef]
23. Liu, T.; Lyu, M.; Wang, Z.; Yan, S. An identification method of orbit responses rooting in vibration analysis of rotor during touchdowns of active magnetic bearings. *J. Sound Vib.* **2018**, *414*, 241–260. [CrossRef]
24. Ma, S.; Yin, Y.; Chao, B.; Yan, K.; Fang, B.; Hong, J. A real-time coupling model of bearing-rotor system based on semi-flexible body element. *Int. J. Mech. Sci.* **2023**, *245*, 108098. [CrossRef]
25. Inayat-Hussain, J.I. Nonlinear Dynamics of a Magnetically Supported Flexible Rotor in Auxiliary Bearings. *J. Phys. Conf. Ser.* **2020**, *1489*, 012015. [CrossRef]
26. Yang, G.; Zhao, L.; Keogh, P.; Zhao, Y. Dynamic Analysis for the Rotor Drop Process and Its Application to a Vertically Levitated Rotor/Active Magnetic Bearing System. *J. Tribol.* **2017**, *139*, 041701.
27. Wang, J.; Jia, X.; Zhang, K.; Xu, Y. Research on Dynamics of Magnetic Bearing Rotor Drop on Back-up Bearings. *China Mech. Eng.* **2022**, *33*, 2403–2413.
28. Zhu, Y.; Jin, C.; Lian, C.; Zhen, Z. Dynamic Analysis of Vertical Rotor Drops Onto Deep Groove Ball Catcher Bearing. *Mach. Des. Res.* **2017**, *33*, 72–77.
29. Yu, C.; Sun, Y.; Wang, H.; Shan, W.; Chen, Y.; Qiu, R. Dynamic analysis of magnetic bearing rotor dropping on radial and axial integrated auxiliary bearing. *Mech. Mach. Theory* **2019**, *140*, 622–640. [CrossRef]
30. Heikkinen, J.E.; Ghalamchi, B.; Viitala, R.; Sopanen, J.; Juhanko, J.; Mikkola, A.; Kuosmanen, P. Vibration analysis of paper machine's asymmetric tube roll supported by spherical roller bearings. *Mech. Syst. Signal Process.* **2018**, *104*, 688–704. [CrossRef]
31. Neisi, N.; Heikkinen, J.E.; Sopanen, J. Influence of surface waviness in the heat generation and thermal expansion of the touchdown bearing. *Eur. J. Mech. A Solids* **2018**, *74*, 34–37. [CrossRef]

32. Liu, X.; Zhou, Y.; Yan, X.; Zhao, J.; Shi, Z.; Chen, J.; Zhao, Y.; Yang, G. Experimental study on the novel rolling-sliding integrated auxiliary bearing in active magnetic bearing during rotor drop. *Ann. Nucl. Energy* **2020**, *136*, 107044. [CrossRef]
33. Wang, D.; Wang, N.; Chen, K. Fixed-point Rubbing Characteristics of Magnetic Suspended Dual-rotor Systems. *China Mech. Eng.* **2021**, *32*, 1686–1699.
34. Witanowski, Ł.; Breńkacz, Ł.; Szewczuk-Krypa, N.; Dorosińska-Komor, M.; Puchalski, B. Comparable analysis of PID controller settings in order to ensure reliable operation of active foil bearings. *Eksploatacja i Niezawodność Maint. Reliab.* **2022**, *24*, 377–385. [CrossRef]

lubricants

Article

Bearing Non-Uniform Loading Condition Monitoring Based on Dual-Channel Fusion Improved DenseNet Network

Yanfei Zhang [1,2,*], Yang Liu [1], Lijie Wang [1], Dongya Li [2], Wenxue Zhang [3] and Lingfei Kong [1]

1 School of Mechanical and Precision Instrument Engineering, Xi'an University of Technology, Xi'an 710048, China; 2210221250@stu.xaut.edu.cn (Y.L.); wanglijie@xaut.edu.cn (L.W.); lingfeikong@xaut.edu.cn (L.K.)
2 Luoyang Bearing Science & Technology Co., Ltd., Luoyang 471039, China; zyszzbldy@163.com
3 Hangzhou Wren Hydraulic Equipment Manufacturing Co., Ltd., Hangzhou 311100, China; jsfwb@wrenchina.com
* Correspondence: yfzhang@xaut.edu.cn

Abstract: Misalignment or unbalanced loading of machine tool spindle bearings often results in skewed bearing operation, which makes the spindle more susceptible to failure. In addition, due to the weak impact signal of the bearing in skewed operation, a single feature information cannot accurately characterize the operation status of the bearing. To address the above problems, this paper proposes a method to monitor the uneven running state of bearing load based on a dual-channel fusion improved dense connection (DenseNet) network. First, the original signal is pre-processed by overlapping sampling method, and the dual-channel experimental data are obtained by frequency-domain and time-frequency-domain algorithms; then the processed data are input into the improved 1D-DenseNet and 2D-DenseNet models respectively for feature extraction; then the frequency-domain and time-frequency-domain features are fused by concat splicing operation, and the output belongs to each category The probability distribution is used to characterize the operating state of the bearings. Finally, the validity of the algorithm model is verified by using the Case Western Reserve University public rolling bearing data set, and an experimental bench is designed and built for experimental verification of the uneven bearing load operation. The comparative analysis of the experimental results in this paper shows that the algorithm can extract the features of the input signal more comprehensively and finally achieve 100% recognition accuracy.

Keywords: rolling bearing; condition monitoring; DenseNet network; dual channel; feature fusion

Citation: Zhang, Y.; Liu, Y.; Wang, L.; Li, D.; Zhang, W.; Kong, L. Bearing Non-Uniform Loading Condition Monitoring Based on Dual-Channel Fusion Improved DenseNet Network. *Lubricants* **2023**, *11*, 251. https://doi.org/10.3390/lubricants11060251

Received: 5 May 2023
Revised: 31 May 2023
Accepted: 1 June 2023
Published: 7 June 2023

1. Introduction

As the "workhorse" of the equipment manufacturing industry, machine tools are one of the most important tools in industrial production, with applications covering the mechanical industry, automotive industry, electric power equipment, railroad locomotives and aerospace. As the core component of machine tools, the operating condition of the spindle directly affects the machining accuracy and efficiency of machine tools, while the bearing, as the spindle support component of machine tools, directly affects the rotational accuracy of the spindle due to its assembly accuracy and the same of the enemy [1–3]. Due to long-term service in variable loads, high temperatures, shock, and other harsh environments, and by manufacturing errors, assembly accuracy and human operation errors, and other factors, will lead to bearing deflection in the service process, which is very easy to cause bearing failure. Therefore, accurate and efficient real-time condition monitoring of bearings is important to ensure the healthy operation of machine tools and improve productivity [4].

Traditional bearing condition monitoring methods are often studied with the help of time-domain features of the signal, and a few methods consider feature extraction in both frequency and time-frequency domains and use the extracted feature information for condition monitoring [5]. Previous mechanical fault diagnosis models based on deep learning have poor generalization ability and complex networks, Deng Mingyang et al. [6] combined frequency domain feature extraction self-encoder with variational self-encoder and proposed frequency domain feature variational self-encoder, which makes the extracted features more robust. Using the overall similarity of the same mode class vibration observation samples on the FFT amplitude spectrum feature waveform, Jiao Weidong et al. [7] proposed a fault diagnosis method based on the pattern matching of the frequency domain feature waveform. Kullbak-Leibler (KL) distance mutual parameter method can solve the blind deconvolution order uncertainty problem, Liu Feng [8] et al. proposed an improved time domain blind deconvolution order uncertainty method based on the combination of generalized morphological filtering and improved KL distance combination of improved time-domain blind deconvolution fault feature extraction algorithm, and the method can effectively extract rolling bearing fault features. Weimin Li [9] et al. proposed a diagnosis method based on a frequency-domain sparse classification algorithm, which effectively overcomes noise interference and avoids the problem of fault feature frequency estimation. For the cliffness, margin, and spectral cliffness, which are usually very sensitive to the data singularity of the signal due to chance factors, are easy to cause misjudgment in the condition monitoring of bearings, Wang Xiaoling [10] et al. proposed a frequency band entropy method based on time-frequency analysis and information entropy theory for rolling bearing fault monitoring. In order to make comprehensive use of the time-frequency domain information of vibration signal and the complexity characteristics of measuring time-frequency distribution, Jiaqi Li et al. [11] introduced two-dimensional multiscale entropy into the fault diagnosis of rolling bearings and proposed a rolling bearing fault diagnosis method based on two-dimensional time-frequency multiscale entropy and firefly algorithm optimized support vector machine. The above method only considers the fault feature extraction in the time domain, frequency domain, or time-frequency domain separately, which has certain limitations in bearing fault diagnosis and is difficult to reflect its fault state accurately, and it is difficult to guarantee the mapping relationship between its feature value and service state as the amount of data increases.

With the development of computer hardware, machine learning has become a very effective tool for classification, for which the classification problem is the basis and many applications have evolved from it. Machine learning is able to learn the laws and patterns of data with the help of computers in a large amount of data, and in the process of learning, the potential and valuable information within the data is mined deeply [12–14]. In order to improve the monitoring speed and monitoring accuracy, many scholars have introduced machine learning methods into the field of condition monitoring and achieved good results. Traditional shallow machine learning methods include support vector machines, decision trees, K-nearest neighbors, plain Bayes, artificial neural networks, etc., which, due to their need for large amounts of prior knowledge, lead to difficult feature extraction and selection [15,16].

In recent years, under the impact of the wave of artificial intelligence, people have started to introduce end-to-end deep learning methods into the fault diagnosis collar, and deep learning models have provided new ideas for fault diagnosis research by getting rid of the inevitable uncertainty of manual feature extraction methods [17]. A convolutional neural network (CNN) is a class of feedforward neural network that contains convolutional computation and has a deep structure, and is one of the representative algorithms of deep learning [18]. The algorithm has the capability of representational learning and is able to classify the input information in a translation-invariant manner according to its hierarchical structure. Janssens [19] proposed a three-layer CNN model for bearing fault detection using vibration signals, where a discrete Fourier transform is applied to the data before the model is trained and fed into the network model. Gu [20] proposed to feed the original vibration signals into 1-DCNN and Gu [21] proposed an improved convolutional neural network model with global mean pooling instead of the final fully connected layer of CNN for the purpose of reducing the number of parameters, which effectively improved the computational speed of the model. To address the problem that traditional fault diagnosis methods require manual feature extraction and feature information is difficult to be fully mined, Chen Ke et al. [22] proposed an end-to-end bearing fault diagnosis method based on CNN, LSTM, and attention mechanism. To address the problem that the traditional bearing fault diagnosis method does not sufficiently extract key features under strong noise and variable load, Yang Xianglan et al. [23] proposed an ECA_ResNet-based bearing fault diagnosis method.

As the layers of a neural network become deeper, the path from the output to the input becomes longer, which leads to the disappearance of the change gradient in the process of backpropagation back to the input. To address this problem, densely connected neural network (DenseNet), as an improved algorithm of the CNN network, solves this problem by establishing dense connections between all the previous layers and the later layers to reuse the features, and some research have been conducted by related scholars. Huang et al. [24] proposed a densely connected convolutional network (DenseNet), which improves the learning efficiency by feature reuse. enhance the learning efficiency, which is the most advanced convolutional neural network architecture. c Shi [25] proposed a wear-induced internal leakage fault diagnosis method based on intrinsic modal functions (IMFs) and weighted densely connected convolutional networks (WDenseNets), using the weighted optimal IMF components as inputs to WDenseNet for fault identification and classification. Yufeng Wang [26] proposed an improved one-dimensional DenseNet network structure capable of handling one-dimensional spectral sequences to achieve multi-scale for signal feature extraction. Rexiang Niu et al. [27] proposed an improved fault diagnosis method for densely connected convolutional networks, which extracts features through multi-scale convolutional layers and achieves weighting of multi-scale feature channels using an attention mechanism to improve the generalization performance of the model. Qingrong Wang et al. [28] proposed a dual-channel cross-dense connected fault diagnosis model incorporating parallel ECA modules, and designed a multi-convolutional residual module and a multi-scale dense connected network for fault feature extraction to achieve interaction and integration of fault information. To address the problem of insufficient ability of shallow features to characterize the fault information of vibration signals, K. Wang et al. [29] proposed an intelligent fault diagnosis neural network model combining a style recalibration module and a densely connected convolutional neural network. Some of the above methods do not give full play to the powerful feature extraction capability of the DenseNet network, the model network structure is shallow, the network generalization capability is weak, or the frequency domain and time-frequency domain signal feature extraction is not considered.

In summary, this paper proposes a method to evaluate the bearing load inhomogeneous operating state based on a dual-channel fusion improved DenseNet network, firstly, data enhancement is performed on the original sample data by overlapping sampling method; secondly, frequency domain and time-frequency domain transformations are performed on the processed data to obtain the experimental data set of two channels; then the processed experimental data are input into the improved 1D-DenseNet and 2D-DenseNet models for feature extraction; finally, the frequency-domain and time-frequency-domain features are fused by concat splicing operation to achieve the classification and recognition of the load inhomogeneity of bearings. The method uses densely connected networks to build the base model, which greatly increases the depth of the network structure and enhances the generalization ability of the model. The effectiveness of the proposed method is verified through experiments, which provides a new idea for rolling bearing fault diagnosis.

2. Fundamentals

2.1. Convolutional Neural Network

2.1.1. Principle of Convolutional Neural Network (CNN)

CNN is a supervised learning neural network with a multilayer convolutional structure, introduced by Hubel and Wiesel with the concept of Receptive fields [30]. First proposed by LeCun for image processing [31], it differs from the traditional neural network structure in that it consists of two main parts: a convolutional layer and a pooling layer. The convolutional layer is connected to the previous layer by local connectivity and weight sharing, which greatly reduces the number of required parameters; the downsampling layer is a method for feature dimensionality reduction, i.e., reducing the complexity of the network by reducing the input feature parameters, which not only improves the robustness of the neural network but also prevents the occurrence of overfitting [32,33].

The convolutional layer is the core of CNN, which mainly implements the feature extraction of the dataset, which is one of the most important differences from traditional neural networks. The features of each layer in the convolutional layer are obtained by convolving the convolutional kernel with the input features of the previous layer, and the parameters can be adjusted through training to obtain the optimal features. In practice, smaller convolutional kernels are used to reduce the amount of operations. Each convolutional kernel can be used as a tool for feature extraction, and a new feature map is generated by means of convolutional operations. The convolution operation is the process of letting the convolution kernel slide along the coordinate position of the input image or input signal horizontally or vertically for a certain number of steps to compute the data corresponding to it, and the computational equation is as follows [34]:

$$x_j^i = f(\sum_{i \in M_j} x_i^{l-1} * k_{ij}^l + b_i^l) \quad (1)$$

where M represents the set of input feature maps, $*$ represents the convolution operation, k represents the convolution kernel, b represents the bias term, x_j^i represents the jth feature map of the ith layer, and $f(-)$ represents the nonlinear activation function used to enhance the representation of the data.

Pooling layer, also known as downsampling layer, can reduce the dimensionality of the input feature set, reduce the computational effort of the neural network, and increase the perceptual field of the posterior neurons to achieve the effect of effective control of overfitting. The pooling operation can be divided into maximum pooling and average pooling, among which maximum pooling is more widely used. Maximum pooling plans the input features into several regions and outputs the maximum value of each region separately.

2.1.2. ResNet Network

As the depth of the network keeps increasing, CNN models begin to suffer from a series of problems such as gradient disappearance and explosion, which in turn lead to a decrease in the accuracy of the model. Based on this, He et al. [35] proposed ResNet by borrowing the idea of cross-layer linking of high-speed networks, the core of which is the residual block. Since this network structure utilizes the residual technique, it is also known as the residual network, and the specific structure is shown in Figure 1:

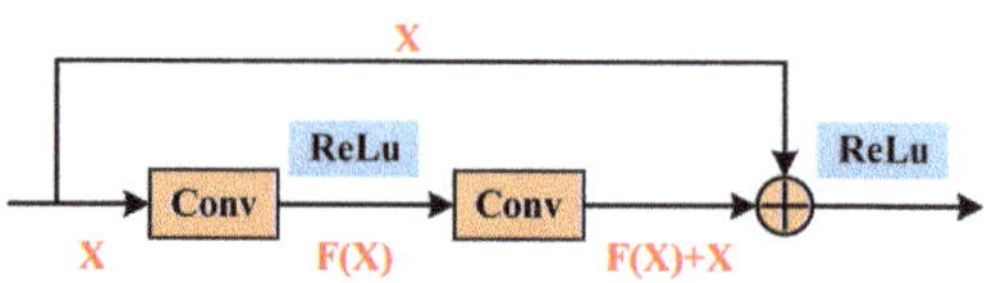

Figure 1. Structure of residual block.

The structural expression of the residual network is as follows:

$$H(x) = F(x) + x \quad (2)$$

where $H(x)$ denotes the output of the structure,x denotes the input, and $F(x)$ denotes the output obtained from the convolution layer. When the parameter of the convolution layer is 0, the formula is expressed as $H(x) = F(x)$. This is the core idea of the residual network, which achieves the transfer of feature information of each layer by constant connection, which ensures the depth of the network and improves the performance of the network model at the same time.

2.2. DenseNet Network

A dense convolutional neural network is an improved convolutional neural network algorithm based on the residual network (ResNet), which aims to alleviate the problem of gradient disappearance and model degradation by using fewer parameters. The core idea of a dense convolutional network as a neural network with dense connectivity is cross-layer connectivity, where each layer of input in the network model takes as input the feature information output from all previous layers, while the features of that layer are also directly passed to all subsequent layers as input to ensure maximum information transfer between layers, making the network perform similar deep supervision in an implicit way [24].

The DenseNet network proposes a new structure by multiplexing the features, which not only slows down the gradient disappearance, but also has a smaller number of parameters, and it is connected in the form of cross-channel with the formula:

$$x_l = H_l([x_0, x_1, \ldots, x_{l-1}]) \quad (3)$$

where: is the input to the x_0 network; x_l is the output of layer l in the network; x_{l-1} is the input to layer $l-1$ of the network. $H_l(\cdot)$ is the nonlinear transformation operation acting on layer l.

DenseNet mainly consists of convolutional layer, pooling layer, DenseBlock, TransitionLayer, and linear classification layer. As the network structure is based on dense connections between the layers, it is referred to as a densely connected network, as shown in Figure 2.

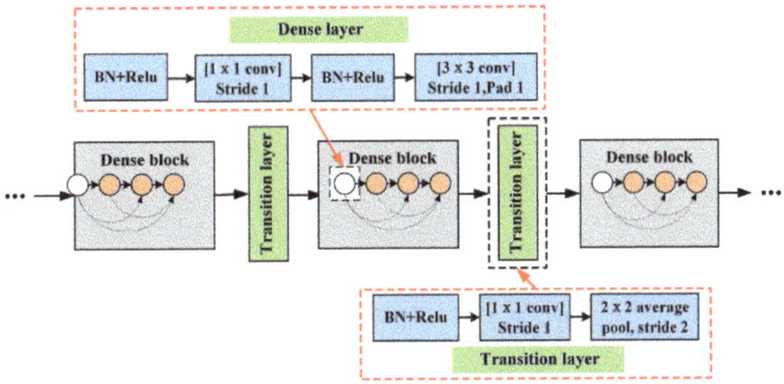

Figure 2. DenselNet model structure.

GrowthRate: The hyperparameter k is the network growth rate, which refers to the number of feature maps produced by each layer. An important feature of DenseNet is that k is very small for each layer, because each layer can be connected to all feature maps in the dense blocks that exist in it. The growth rate controls how much global information is added at each layer, and this information can be called anywhere in the network, which is the biggest difference between DenseNet and traditional neural networks.

DenseBlock: The network perceives the feature information locally through the first convolutional layer initially. Next, the data enters the dense block. A bottleneck layer structure is BN-Relu-Conv(1 × 1)-BN-Relu-Conv(3 × 3), which becomes Densenet-B. Each bottleneck layer generally contains a 1 × 1 convolution and a 3 × 3 convolution. The former serves to effectively reduce the number of feature maps, reduce the computational effort and achieve feature fusion for each channel, while the latter serves to perform feature extraction. A dense block can be composed of multiple bottleneck layers.

TransitionLayer: The network structure between two dense blocks is called the transition layer, and its structure is BN-Relu-Conv-Dropout-Pooling, which generally consistsa of 1×1 convolution and 2×2 pooling layer, the main role is to reduce the number of feature maps. θ denotes the compression factor, generally $\theta < 1$. If the forward thickening block generates n layers of feature maps, in order to compress the data, after the transition block, the number of feature maps as the input of the next thickening block becomes $\theta \times n$.

2.3. ECA-Net Module

The input time-frequency maps are learned by 2D-DenseNet dense network to obtain a large number of features, and the ECA-Net attention mechanism module is introduced to improve the classification efficiency of the fusion model, enhance the overall channel features, and improve the model performance [36].

The ECA-Net attention mechanism uses the global average pooling layer directly after the 1×1 convolutional layer, removing the fully connected layer. This module avoids dimensionality reduction and effectively captures cross-channel interactions. The module achieves good results with only a few parameters involved.

The ECA-Net module accomplishes cross-channel information interaction by one-dimensional convolution, and the size of the convolution kernel is adaptively varied by a function that allows more cross-channel interaction for layers with a larger number of channels, as shown in Figure 3.

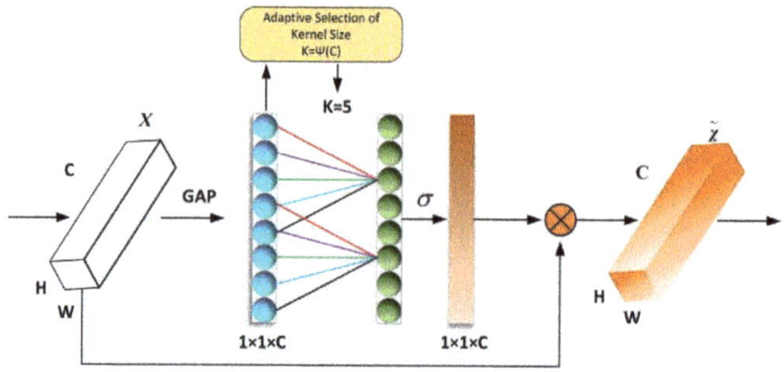

Figure 3. ECA-Net attention mechanism structure diagram.

The adaptive function is as follows (where $\gamma = 2,\ b = 1$):

$$k = \left| \frac{\log_2(c)}{\gamma} + \frac{b}{\gamma} \right| \tag{4}$$

The specific implementation process of the ECA-Net attention mechanism is as follows:

S1: Input feature maps with dimensions of H × W × C.

S2: Perform spatial feature compression on the input feature map. Implementation: in the spatial dimension, using global average pooling GAP to obtain the feature map of 1 × 1 × C.

S3: For the compressed feature map, channel feature learning is performed. Realization: through 1 × 1 convolution, learning the importance between different channels, at this time the output dimension is still 1 × 1 × C. The output dimension is still the same.

S4: Finally, the channel attention is combined with the feature map of channel attention 1 × 1 × C, the original input feature map H × W × C, perform channel-by-channel multiplication and finally outputs the feature map with channel attention.

According to the Efficient Channel Attention (ECA-Net) module shown in Figure 3. Considering the aggregated features obtained through the global average library (GAP), ECA-Net generates channel weights by performing a fast one-dimensional convolution of size k, where k is determined adaptively by mapping the channel dimension C.

2.4. LSTM-Attention Module

The addition of the LSTM-Attention module to the 1D-DenseNet densely connected network can effectively suppress gradient disappearance or explosion with good generalization ability.

The LSTM network is improved from the standard RNN. The LSTM effectively alleviates the long-term dependence problem of the standard RNN through its internal complex gate operation and the introduction of cellular states [37]. The unique feature of LSTM is that it introduces a memory cell and gate mechanism to solve the gradient disappearance and gradient explosion problems in the traditional RNN. LSTM is unique in that it introduces memory cell and gate mechanism to solve the gradient disappearance and gradient explosion problems in traditional RNNs, and enhances the ability to model long-term dependence.

The equations for the forgetting gate f_t, the input gate i_t, the output gate o_t, the cell state c_t and the output h_t are shown in the following equations:

$$f_t = \sigma\left(W_f \cdot [h_{t-1}, x_t] + b_f\right) \tag{5}$$

$$i_t = \sigma(W_i \cdot [h_{t-1}, x_t] + b_i) \tag{6}$$

$$o_t = \sigma(W_o \cdot [h_{t-1}, x_t] + b_o) \tag{7}$$

$$c_t = f_t \otimes c_{t-1} + i_t \otimes tanh(W_c \cdot [h_{t-1}, x_t] + b_c) \tag{8}$$

$$h_t = o_t \otimes \tanh(c_t) \tag{9}$$

where: x_t refers to the input at the current moment; h_{t-1} refers to the output at the previous moment; W refers to the weight matrix; b refers to the bias; $\sigma(x) = 1/(1 + e^x)$ is the sigmoid activation function; $\otimes$ refers to the dot product operation.

Self-Attention is an improvement of the attention mechanism, which not only can quickly filter out the key information and reduce the attention to other irrelevant information, but also can reduce the dependence on external information and be better at capturing the internal relevance of the input data [38]. By introducing the self-attention mechanism, the neural network solves the model information overload problem while also improving the accuracy and robustness of the network [39].

The computation of Self-Attention is divided into two steps. Step 1: Calculate the attention weights between any vectors of the input sequence; Step 2: Calculate the weighted average of the input sequence based on the attention weights. The specific operation is shown in the following equation:

$$Q = XW^q \tag{10}$$

$$K = XW^k \tag{11}$$

$$V = XW^v \tag{12}$$

$$Attention(Q, K, V) = softmax\left(\frac{QK^T}{\sqrt{dim}}\right)V \tag{13}$$

where: Q, K and V are the query matrix, key matrix and value matrix, respectively, obtained by multiplying the input X with the corresponding weight matrices W^q, W^k, W^v, respectively; dim denotes the dimensionality of Q, K and V.

In summary, vibration signal feature extraction by LSTM-Attention module can better capture the key features in the time series signal, and then improve the prediction accuracy of the model. This is very helpful for some application scenarios with high accuracy requirements, such as fault diagnosis and prediction.

3. Improved Evaluation Model of DenseNetnetwork Based Ontwo-Channel Fusion

3.1. Model Overview

In this paper, a dual-channel fusion DenseNet network model (Frequency and Time-Frequency domain fusion DenseNet, FTF-DFD) is constructed based on a densely connected neural network, and its structure is shown in Figure 4: Since the original vibration data samples are insufficient and cannot be directly used in the standard network model to obtain better evaluation results, this paper performs data enhancement on the original data by overlapping sampling method.

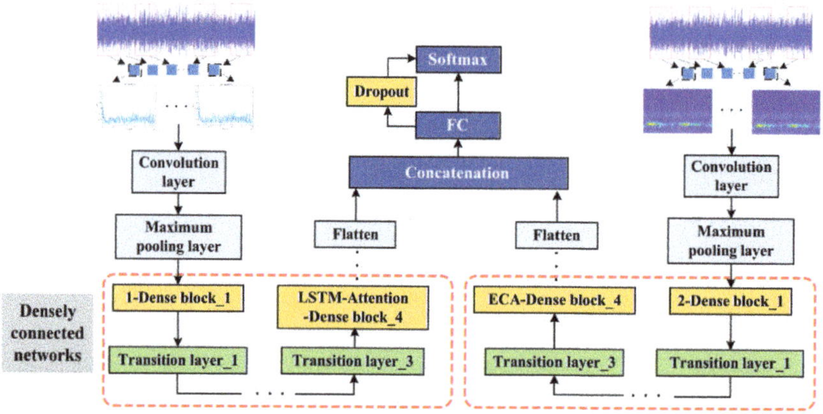

Figure 4. FTF-DenseNet model structure.

The model shown in Figure 4 is a two-channel DenseNet network structure consisting of the input layer, feature extraction module, and fault mode classification module. LSTM-Attention module for deep feature extraction; 1D-DenseNet model bottleneck layer structure is the same, both contain a combination of 1×1 and 1×3 size convolutional kernel, each dense block has a different number of bottleneck layers, this paper uses three groups of dense blocks, arranged according to the number of 3:2:1, adding a transition layer between every two dense blocks, which consists of a convolutional kernel size of 1×1 convolutional layer and a mean pooling layer of kernel size 1×2, which is used for downscaling and extracting global feature information, and finally, the output data of the network is linearized by squaring.

After the time domain signal is expanded by overlapping sampling, the wavelet time-frequency map is obtained by continuous wavelet transform, which is used as the input of the 2D-DenseNet model, and the ECA-Net attention mechanism is added after the last layer of dense blocks to effectively extract the model accuracy, strengthen the overall channel characteristics, and improve the model performance. The model structure of 2D-DenseNet is similar to that of 1D-DenseNet, which transforms the convolutional kernel from 1D to 2D, each bottleneck layer contains a combination of 1×1 and 3×3 size convolutional kernels, and the transition layer consists of a convolutional layer with 1×1 convolutional kernel size and an average pooling layer with 2×2 kernel size.

After the time-domain signal is overlapped sampled for data expansion, the wavelet time-frequency map is obtained by continuous wavelet transform, which is used as the input of the 2D-DenseNet model, and the ECA-Net attention mechanism is added after the last layer of dense blocks to effectively extract the model accuracy, strengthen the overall channel characteristics, and improve the model performance. 2D-DenseNet is similar to the model structure of 1D-DenseNet The model structure of 2D-DenseNet is similar to that of 1D-DenseNet, which transforms the convolutional kernel from 1D to 2D, each bottleneck layer contains a combination of 1×1 and 3×3 size convolutional kernels, and the transition layer consists of a convolutional layer with 1×1 convolutional kernel size and an average pooling layer with 2×2 kernel size.

The output feature data of the 1D-DenseNet model and 2D-DenseNet model are stretched into feature vectors, and the splicing operation is performed by concat, and the fused feature information is input into the fully connected network layer and SoftMax classifier, and the probability distribution belonging to each category is output to achieve the fault classification recognition of bearings. The frequency domain and time-frequency domain fusion method proposed in this paper has a good fusion effect and improves the classification accuracy of the model. The specific parameters of the model are shown in Table 1.

Table 1. FTF-DenselNet network parameters.

Model Name	1D-DenseNet		2D-DenseNet	
	Structure Type	**Convolution Kernel**	**Structure Type**	**Convolution Kernel**
Input layer	One-dimensional FFT spectrum	-	Two-dimensional time-frequency diagram	-
Convolutional layer	Conv	1×7	Conv	7×7
Pooling layer	Maxpooling	1×3	Maxpooling	3×3
Dense block 1	$\left\{\begin{matrix} BN-Relu-Conv \\ BN-Relu-Conv \end{matrix}\right\} \times 1$	$\left\{\begin{matrix} 1\times 1 \\ 1\times 3 \end{matrix}\right\} \times 1$	$\left\{\begin{matrix} BN-Relu-Conv \\ BN-Relu-Conv \end{matrix}\right\} \times 1$	$\left\{\begin{matrix} 1\times 1 \\ 3\times 3 \end{matrix}\right\} \times 1$
Transition layer	BN-Relu-Conv-Pooling	$\left\{\begin{matrix} 1\times 1 \\ 1\times 2 \end{matrix}\right\} \times 1$	BN-Relu-Conv-Pooling	$\left\{\begin{matrix} 1\times 1 \\ 2\times 2 \end{matrix}\right\} \times 1$
Dense block 2	$\left\{\begin{matrix} BN-Relu-Conv \\ BN-Relu-Conv \end{matrix}\right\} \times 3$	$\left\{\begin{matrix} 1\times 1 \\ 1\times 3 \end{matrix}\right\} \times 3$	$\left\{\begin{matrix} BN-Relu-Conv \\ BN-Relu-Conv \end{matrix}\right\} \times 3$	$\left\{\begin{matrix} 1\times 1 \\ 3\times 3 \end{matrix}\right\} \times 3$
Transition layer	BN-Relu-Conv-Pooling	$\left\{\begin{matrix} 1\times 1 \\ 1\times 2 \end{matrix}\right\} \times 1$	BN-Relu-Conv-Pooling	$\left\{\begin{matrix} 1\times 1 \\ 2\times 2 \end{matrix}\right\} \times 1$
Dense block 3	$\left\{\begin{matrix} BN-Relu-Conv \\ BN-Relu-Conv \end{matrix}\right\} \times 1$	$\left\{\begin{matrix} 1\times 1 \\ 1\times 3 \end{matrix}\right\} \times 1$	$\left\{\begin{matrix} BN-Relu-Conv \\ BN-Relu-Conv \end{matrix}\right\} \times 1$	$\left\{\begin{matrix} 1\times 1 \\ 3\times 3 \end{matrix}\right\} \times 1$
ECA-Net	None	-	$\times 1$	-
LSTM-Attention	$\times 1$	-	None	-
Fully connected layer	FC	-	FC	-
Fully connected layer	FC			
Output layer	SoftMax			

3.2. Data Pre-Processing

3.2.1. Normalization Process

In order to data reduce the effect of distribution changes, improve the convergence speed of the model and diagnostic accuracy, the data are normalized and preprocessed, and the results are mapped to the [0–1] interval through a linear transformation, assuming that the sample data $X = \{x_1, x_2, \ldots, x_n\}$, whose transformation equation is as follows.

$$y_i = \frac{x_i - min\{x_j\}}{max\{x_j\} - min\{x_j\}} \tag{14}$$

where, y_i is the normalized result, x_i is the i-th sample data, $max\{x_j\}$ is the maximum value of the sample data, and $max\{x_j\}$ is the minimum value of the sample data.

3.2.2. Data Enhancement

In the field of data-driven deep learning, having large enough training samples is the key to improve the accuracy of the model and effectively reduce the overfitting of the model. In this paper, we propose to increase the training samples by using overlapping sampling with moving sliding windows, as shown in Figure 5, which can effectively increase the training samples while maintaining the periodicity and continuity of the one-dimensional time-series vibration signals and avoiding problems such as signal loss caused by isometric sampling and sampling.

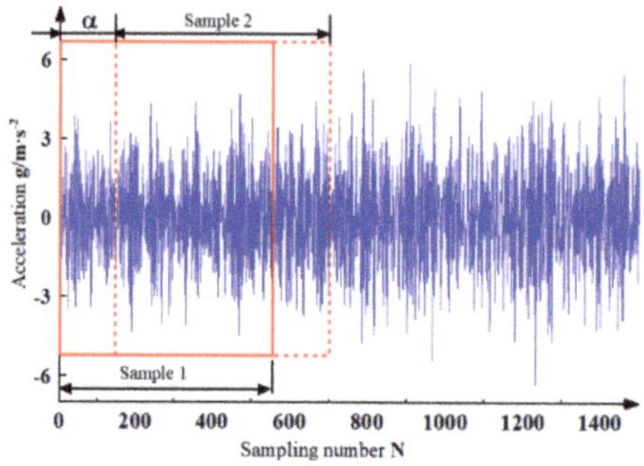

Figure 5. Schematic diagram of data enhancement.

From Figure 5, if the total length of data in a certain state is L = 245,759, the length of data for each sample is l = 1024. if no enhancement is applied, the number of samples A that can be segmented by the current vibration signal is:

$$A = [\frac{L}{l}] \tag{15}$$

Using the moving sliding window overlap method with offset α = 100 for data sampling, the length of the overlapping part of the data is 924; the number of samples obtained from the current signal that can be split is (the maximum total number of samples that can be split per group of data is 2448)

$$B = [\frac{L - l}{\alpha} + 1] \tag{16}$$

Then the multiplier of sample expansion after using data enhancement γ is

$$\gamma = \frac{B}{A} = \frac{l(L - l + \alpha)}{\alpha L} \tag{17}$$

The expansion of the original data samples is achieved by overlapping sampling to avoid the loss of detailed features. Different offsets α, data lengths l, expansion multipliers γ and sample numbers B are set to achieve the performance detection of the model under different sample numbers and prove the practicality of the model.

3.2.3. Data Conversion

FFT is an efficient algorithm of Discrete Fourier Transform (DFT), called Fast Fourier Transform (FFT), which improves the DFT algorithm according to the odd, even, imaginary and real characteristics of DFT, and its basic principle is still Fourier Transform, which will not be discussed here. By calling the fft function in python, the frequency domain characteristics of the signal can be obtained, and then the frequency distribution of different signals can be analyzed.

The signal after data enhancement is analyzed in the time-frequency domain, where the time-frequency map of the original signal is obtained by continuous wavelet variation

(CWT), which can clearly and accurately represent the time-frequency distribution of the vibration.

The continuous wavelet transform provides the best resolution results for non-periodic signals without leakage effects. Continuous wavelet transform $CWT(\alpha,\tau)$ can be calculated by the following equation:

$$CWT(\alpha,\tau) = \frac{1}{\sqrt{\alpha}}\int_{-\infty}^{+\infty} s(u)\psi(t) \times \left(\frac{u-\tau}{\alpha}\right)du \tag{18}$$

where, α is the scale, $s(u)$ is the original signal, τ is the translation, and $\psi(t)$ is the mother wavelet.

The continuous wavelet variation (Continuous Wavelet Transform, CWT) chooses Complex Morlet Wavelet (Cmor) as the wavelet basis, and the Cmor wavelet basis function is obtained by improving the Morlet wavelet basis function. It is a complex wavelet basis function with dual resolution properties in both frequency and time domains, which is widely used in the field of signal processing and wavelet analysis. The Cmor wavelet basis function has a similar shape to the Gaussian function but has better frequency localization properties in the frequency domain. It has better time-frequency localization properties in both time and frequency domains, and is suitable for processing non-stationary signals and analyzing transient phenomena in signals.

3.3. Model Training

The optimization algorithm used for the TADAT-based rolling bearing fault diagnosis model is chosen as Adam. The Adam algorithm adaptively adjusts the learning rate of each parameter, and different learning rates can be used for different parameters, thus making the training more efficient and stable.

The Adam algorithm dynamically corrects the training steps of each parameter using first-order moment estimation and second-order moment estimation of the gradient with the following update rules:

$$\theta_{t+1} = \theta_t - \frac{\eta}{\sqrt{v_t+\varepsilon}}m_t \tag{19}$$

where, θ_{t+1}, θ_t denotes the model parameters at step $t+1$ and step t, respectively; η is the learning rate; v_t denotes the value of the unbiased second-order moment estimate; m_t denotes the value of the unbiased first-order moment estimate; and ε is a very small positive number, generally taken as $\varepsilon = 10^{-8}$, preventing the denominator from being 0.

The TADAT-based rolling bearing fault diagnosis model diagnoses work conditions based on features, which belongs to the classification problem in supervised learning, so cross entropy is chosen as the loss function and optimized.

Cross entropy is mainly used to calculate the distance between the correct probability of labeling and the probability of prediction, and the smaller the value of cross entropy, the closer the prediction result is to the actual result, and the formula is as follows:

$$loss = -\sum_{\theta} p(\theta) lg q(\theta) \tag{20}$$

where, θ denotes the individual learning parameters; $p(\theta)$ denotes the probability of correct labeling; and $q(\theta)$ is the prediction probability.

For two probability distributions $p(\theta)$ and $q(\theta)$, define the K-L scatter of $p(\theta)$ and $q(\theta)$ as follows:

$$KLD = \sum_{\theta} p(\theta) lg \frac{p(\theta)}{q(\theta)} \tag{21}$$

When calculating the cross-entropy loss using KL scatter, the true labels need to be transformed into probability distributions, usually using methods such as one-hot encoding or smoothed labels. In this paper, a smoothed target label is used instead of the traditional one-hot encoded label, thus reducing the impact of the noise and uncertainty of the label on the model and obtaining a loss function *ce_loss*.

In *ce_loss* its L2 regularization penalty term is introduced to penalize the size of the model parameters to prevent overfitting. The strength of the regularization penalty can be controlled by adjusting the value of alpha to establish the regularized loss function as follows:

$$impro\_loss = ce\_loss + alpha * \sum_{\theta} \theta^2 \tag{22}$$

where, θ denotes each learning parameter; alpha is the regularization parameter.

After pre-processing, the data set is divided into training set, validation set and test set in the ratio of 7:2:1, the model uses the parameters with the highest training and validation accuracy as the final parameters, the optimizer uses the Adam optimizer with fast and stable convergence, the loss function is the regularized loss function, the initial learning rate is 0.01, and the learning rate decays by half for every 10 iterations. Normalization batch normalization is used to accelerate the convergence speed of the neural network, Dropout operation is added to prevent overfitting, and the number of training iterations of the model is set to 100; finally, the Softmax function is used to classify the target and output the probability distribution of each category; as shown in Table 2:

Table 2. Model parameters.

Parameter Category	Parameter Setting
Training set:Validation set:Test set	5600:1600:800
Optimizer	Adam
Number of training sessions	100
Learning Rate	0.02
Batch Size	64
alpha	0.05
smothing	0.1

In order to obtain the appropriate Batch Size parameter for the model, the mid-load experiment was used as the basis for comparison by setting different Batch Size parameter values, mainly setting three different sets of values of 32, 64, and 128, as shown in Figure 6, and it was found that the accuracy of the model improved the fastest when the Batch Size was set to 64, reaching 94% accuracy after 20 iterations, and After 60 iterations, the accuracy rate is stabilized at 100%, while the accuracy rate of the model test fluctuates more when the Batch Size is set to 32 and 128. It was concluded that the best iteration of model accuracy was achieved when the Batch Size was set to 64.

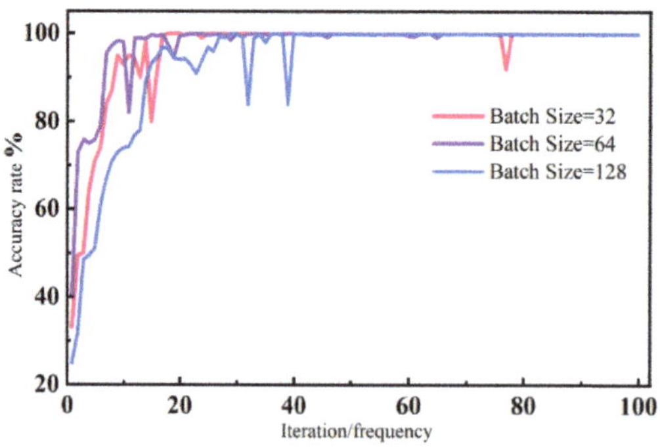

Figure 6. Comparison of Batch Size under medium load condition.

4. Experimental Verification

4.1. Environment Description

A non-equilibrium bearing load test stand was developed and built to further study the monitoring function of this technology in the process of double bearing operation, as shown in Figure 7. The test stand mainly consists of a motor, a precision spindle, a rolling bearing and an acceleration sensor, and the maximum speed read by the electric spindle is 10,000 r/min. The mechanical spindle is connected to the electric spindle through a flexible coupling, and the motor operation is controlled by a servo control system.

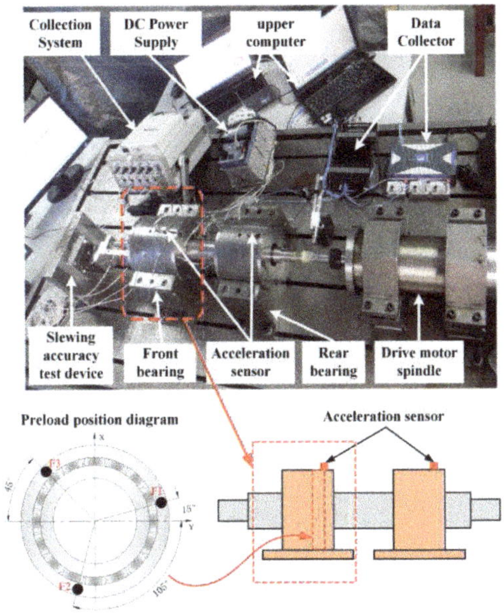

Figure 7. Structure of non-uniform preload test rig.

The test rig used four NSK7014C angular contact ball bearings, where F1, F2 and F3 were loaded 120° on the bearings, respectively, and the bearing bias operating condition was determined by setting different sizes of preload; the bearings were mounted back-to-back with a fixed speed of 4000 r/min, a sampling frequency of 8192 Hz and a sampling length of 512. Table 3 shows the bearing parameters.

Table 3. NSK7014C angular contact ball bearing parameters table.

Inner Ring Diameter/mm	Parameter Setting	Thickness/mm	Dynamic Load/KN	Static Load/mm
70	10	20	47	43

Software environment: The training and testing environment of this paper is 14 cores, 16 G memory, processor: 12th Gen Intel Core i7-12700H processor; programming environment Pytorch1.7.1.

The bearing non-uniform load test bench is designed to distinguish the operating condition of the bearing under unbalanced operation, so that the bearing failure caused by factors such as assembly or machining can be detected in time. Due to the limited conditions in the laboratory, the currently built test bench can only be used to verify the effectiveness and accuracy of the condition monitoring method, and cannot simulate the corresponding bearing failure state for verification experiments.

4.2. Example Analysis

4.2.1. Data Conversion

A total of twelve sets of data are collected through the bearing load non-uniform operation fault simulation test bench, including F1, F2, F3 loading and data under even load conditions, which are mainly divided into four conditions of light load, medium load, heavy load and even load, and there are 12 types of bearing vibration data. There are four sets of experiments in total, the first three sets of experiments with 1400 samples in the training set, 400 samples in the test set and 200 samples in the validation set; the last set of experiments with all types of data input; the specific experimental data are shown in Table 4.

Table 4. Experimental dataset.

Experiment Name	Signal Type	Training Set	Validation Set	Test Set
First group of experiments (Light load comparison)	F1 (C2) = 400 N	1400	400	200
	F2 (C2) = 400 N	1400	400	200
	F3 (C2) = 400 N	1400	400	200
	$F_{1.2.3}$(C1) = 200 N	1400	400	200
Second group of experiments (mid-load comparison)	F1 (C4) = 800 N	1400	400	200
	F2 (C4) = 800 N	1400	400	200
	F3 (C4) = 800 N	1400	400	200
	$F_{1.2.3}$(C3) = 400 N	1400	400	200
Third group of experiments (Heavy load comparison)	F1 (C6) = 1200 N	1400	400	200
	F2 (C6) = 1200 N	1400	400	200
	F3 (C6) = 1200 N	1400	400	200
	$F_{1.2.3}$(C5) = 600 N	1400	400	200
Fourth group of experiments	Enter 12 types of data	700–12	200–12	100–12

The vibration signals at F1 positions C2, C4, C6 and $F_{1.2.3}$ (C1) working conditions were subjected to signal analysis, and 1024 data points were taken as one sample and subjected to FFT transform with wavelet transform. The analysis in Figure 8 shows that the spectrograms of the four working conditions data at 3044 Hz have the maximum amplitude variation, which reflects the main frequency components of the signal in the frequency domain. As the load at the F1 position of the bearing gradually increases, the vibration characteristics of the bearing will be more intense, and more frequency components and amplitudes will be generated, and more noise and spurious frequencies may appear in the FFT analysis. Since the frequency and amplitude of the bearing vibration will change with the load, the amplitude of the main frequency components in the FFT spectrum is relatively small when the load is higher. Under the average load, the frequency and amplitude of bearing vibration are relatively stable, so the amplitude of the main frequency components in the FFT spectrum is relatively large. The spectrum analysis shows that the spectrum gradually increases with C6, C4, C2, and $F_{1.2.3}$ (C1). The signal analysis of the time-frequency diagram also proves this point. In the energy distribution in the frequency range of 2 to 3 kHz, the $F_{1.2.3}$ (C1) condition has the largest energy, while the C6 condition has the least energy.

4.2.2. Model Testing

The experiments were conducted with the input sample length of 1024, Batch Size of 64, training iterations of 100, learning rate of 0.002, and optimizer choice of Adam. The method was initially tested on the Case Western Reserve University bearing dataset.

The test selected the bearing data at the drive end, sampling frequency of 48 kHz, and load of 0 hp, the fault form contains outer ring fault, inner ring fault and rolling body fault three kinds of fault parts as shown in Figure 9, the fault type is specifically divided into 7 mils, 14 mils and 21 mils three kinds of fault diameter, plus the normal state, a total of ten kinds of bearing state data. The specific sample composition information is shown in Table 5. A total of 70% of the samples are selected as the training set, 20% as the validation set, and 10% as the test set.

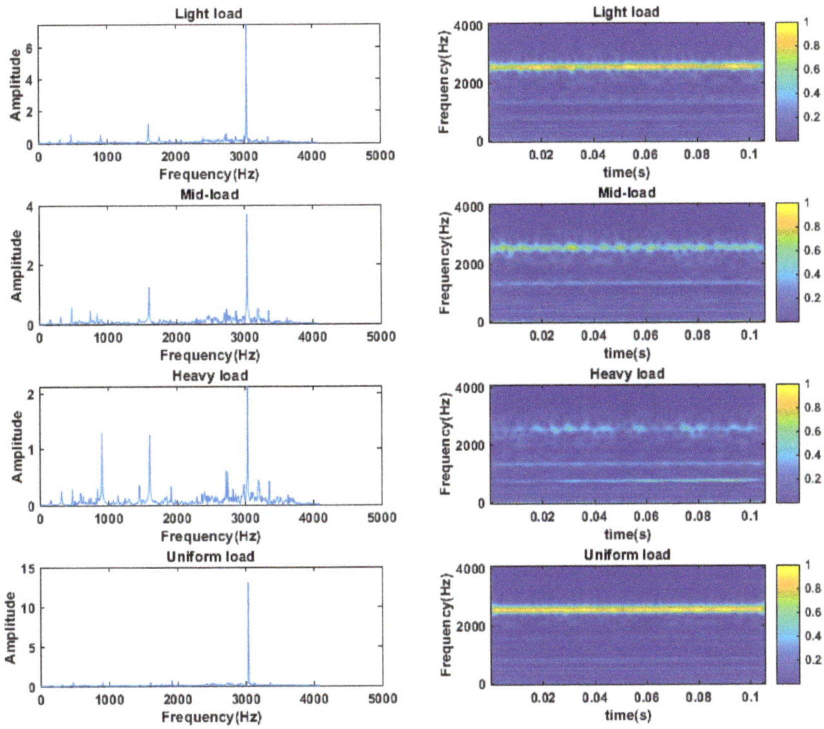

Figure 8. Signal analysis of three operating conditions at F1 position with equal load.

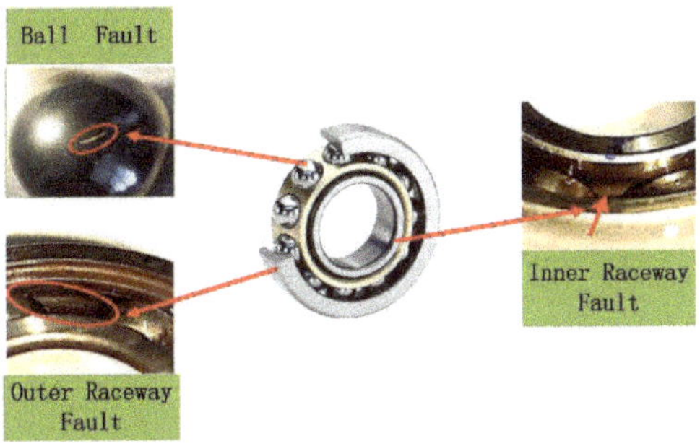

Figure 9. Bearing fault form distribution chart.

Table 5. Sample composition information.

Sample Type	Sample Length	Sample Size	Type Tags
Ball Fault (7 mils)	864	400	B007
Ball Fault (14 mils)	864	400	B014
Ball Fault (21 mils)	864	400	B021
Inner Raceway Fault (7 mils)	864	400	IR007
Inner Raceway Fault (14 mils)	864	400	IR014
Inner Raceway Fault (21 mils)	864	400	IR021
Outer Raceway Fault (7 mils)	864	400	OR007
Outer Raceway Fault (14 mils)	864	400	OR014
Outer Raceway Fault (21 mils)	864	400	OR021
Normal	864	400	normal

The time domain of the bearing vibration signal contains a large number of high and low-frequency components, which have different sensitivities for the diagnosis of bearing faults. Therefore, converting the time domain signal to the frequency domain signal for analysis can better capture the characteristics of bearing faults as shown in Figure 10. The fault signal with fault type 7 mils at 0 hp is taken for spectral analysis with the normal signal, and 1024 data points are taken as a sample for FFT transformation, and four states of the bearing, such as normal state, rolling element fault, inner ring fault, and outer ring fault, can be found. The difference of amplitude in the high-frequency band is large. Figure 11 shows that this mechanical vibration signal mainly contains energy in the frequency range of 0~5 kHz, in which the inner ring fault has obvious energy intensity transformation in both low and high-frequency bands, and its distribution is very dense because it is a different type of fault, which can effectively identify the frequency components and time domain features of the signal and provide useful information for applications such as signal feature extraction, classification, and diagnosis.

The ten bearing condition data in Table 5 were input into the model for condition diagnosis of the bearings. From the model output accuracy versus loss function curve in Figure 12, it can be seen that the model can reach 98% accuracy after 20 iterations, and after 50 iterations, the model can finally reach 100% accuracy.

In order to clearly represent the extraction ability of features in the model, we use the t-SNE technique to downscale and visualize the features in the input and output layers to indirectly represent the extraction ability of features in the model, where different colors and numbers indicate different fault categories and horizontal and vertical coordinates indicate different dimensions.

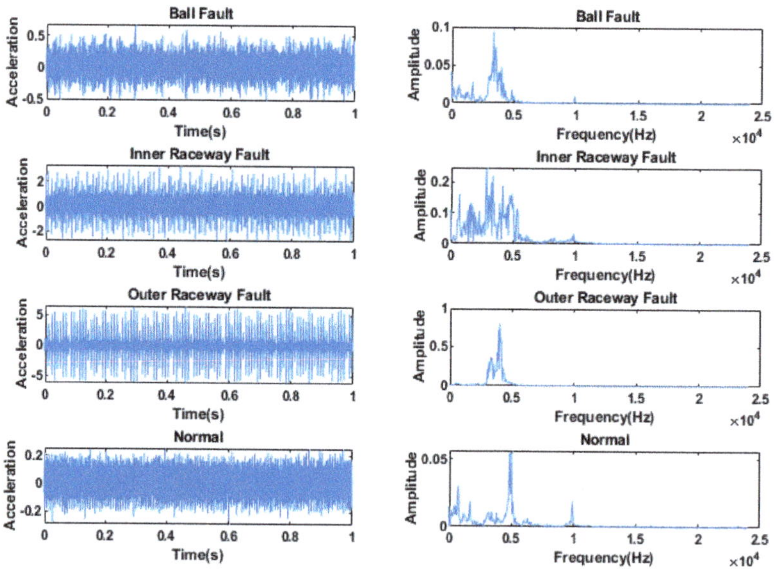

Figure 10. Time domain waveforms and frequency spectrum of CWRU bearing data.

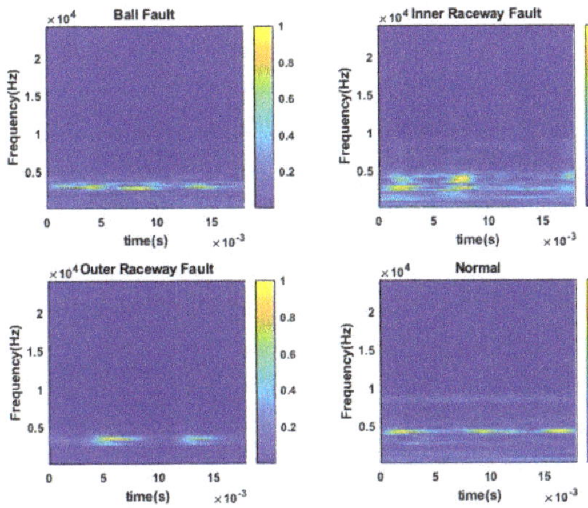

Figure 11. Wavelet time−frequency diagram of CWRU bearing data.

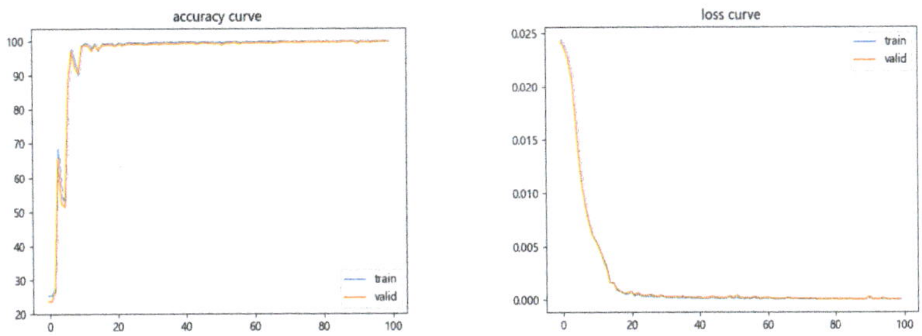

(a) (b)

Figure 12. (a) Accuracy curve; (b) Loss curve.

As shown in Figure 13, the input layer is disorganized and various features are mixed together. After the three-stage densely connected network and the two-domain feature fusion, the extraction of features by the model is basically completed, and the separation and convergence of all kinds of features are basically completed, and the visual classification results of the output layer show that the model has a good classification effect.

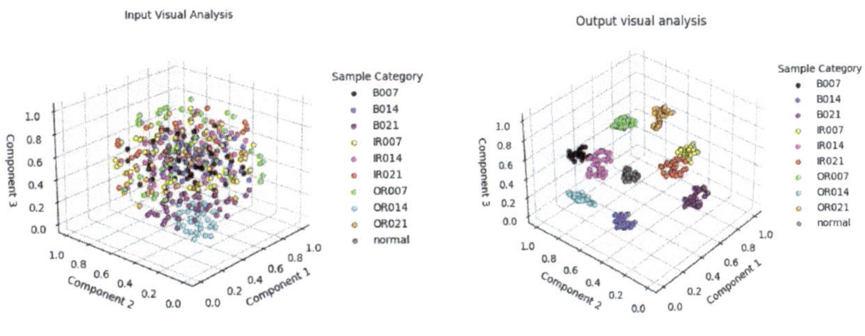

(a) (b)

Figure 13. (a) Input feature visualization; (b) Output feature visualization.

4.3. Comparison Experiments

At the position of bearing F1, three working conditions of light load (OC_1), medium load (OC_2) and heavy load (OC_3) were measured, and three data sets M1, M2, and M3 were established to contain the information of the above three working conditions, and 300 samples were taken for each working condition. In the network model, the Batch Size is set to 64, the number of training iterations is set to 100, the optimizer is Adam, the initial learning rate is 0.01, and the learning rate decays by half every 10 iterations, and the loss function is selected as *impro_loss*.

The diagnostic res CNN: The model structure is the input layer, Conv layer, MaxPool layer, ReLu activation function, BN layer, flat layer, Dropout layer, fully connected layer, and SoftMax output layer. The input data is a two-dimensional time-frequency map, and the middle layer is a two-layer convolutional pooling network, which is stretched by the flat layer and then passed through the fully-connected layer and the SoftMax output layer to achieve the classification of work conditions.

Improved-FTF-CNN: The model adopts the fusion of frequency domain and time-frequency domain, in which the 1D and 2D models, the three-level dense connection network is used, and the ratio of dense blocks are 3:2:1, relying on the feature fusion through concat, and finally the classification of working conditions through the SoftMax output layer.

DenseNet: The input layer of the model is fed with a two-dimensional time-frequency map, and the intermediate structure uses three groups of dense blocks, according to the number 3, 2, and 1. The output layer uses the SoftMax output layer for the classification of working conditions.

Improved-FTF-DenseNet: the base structure of the model is the same as the improved-FTF-CNN network structure, and the intermediate feature extraction structure replaces the CNN module in it with the DenseNet network structure, and the rest of the network results remain unchanged.

The diagnostic results of the above methods are shown in Figure 14, and the average value of five experiments is taken as the model evaluation result, and the average value of six groups of experiments is taken for model performance evaluation, which is shown in Table 6. the network structure of the CNN model is relatively simple and cannot extract accurate features, and the training time is the shortest, with an average accuracy rate of 87.43%; the improved-FTF-CNN model, compared with the simple CNN network, has an accuracy rate has significantly improved, and is 4.83% higher than the CNN model; the DenseNet model can improve the complexity of the model due to the dense connection structure, and after adjusting its parameters, the final accuracy can reach 92.57%; the improved-FTF-DenseNet can reach a final average accuracy of 93.88% through the model of dual-channel fusion, which is lower than the method of this paper by 3.18%.

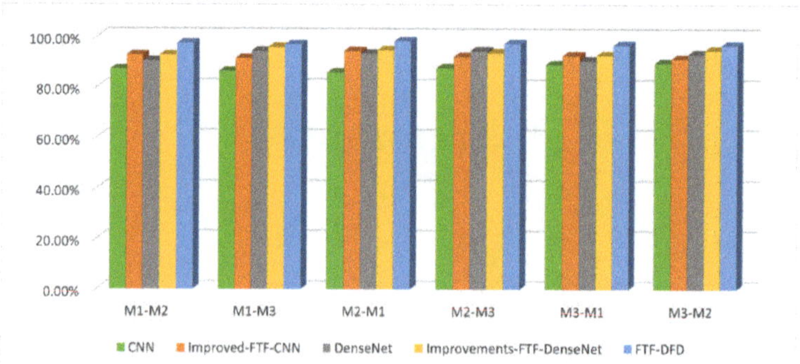

Figure 14. Model comparison chart.

Table 6. Diagnostic results of different methods.

Network Structure	Accuracy	Training Time/min
CNN	87.43%	0.5 min
Improved-FTF-CNN	92.26%	1.2 min
DenseNet	92.57%	3.5 min
Improvements-FTF-DenseNet	93.88%	4.2 min
FTF-DFD	97.06%	3.8 min

4.4. Uneven Bearing Load Experiment

The results of three sets of experiments of the FTF-DFD fault diagnosis model proposed in this paper are shown in Figure 15, which are the accuracy curves of the unbalanced experiments. The accuracy rates of the three different experimental conditions on the test set after 18 iterations of training all reach more than 98%, among which the accuracy curves of the light load experiments have a large abrupt change in the rising stage and the accuracy rate is not as fast as The accuracy curves of the medium-load and heavy-load experiments are not as fast as those of the medium-load and heavy-load experiments. The accuracy transformation curve is flatter under the medium-load experimental condition, and the accuracy can reach 100% on the test set after 50 iterations of training. The accuracy of all three unbalanced experiments reached 100% after 70 iterations of training, and it can be seen through the three sets of experiments that the model is more adaptable under medium-load and heavy-load working conditions.

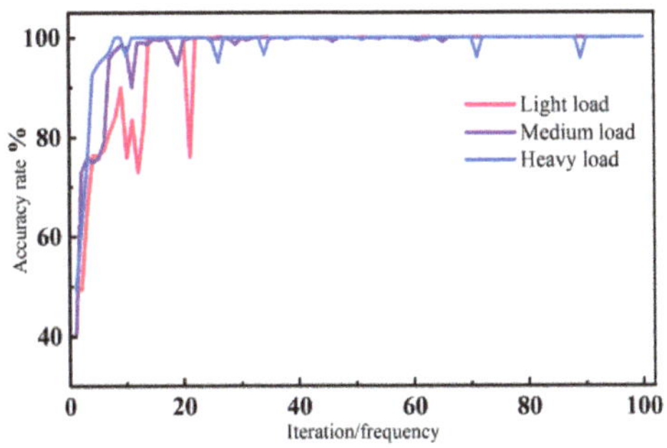

Figure 15. Comparison of three working conditions.

The deeper the layers of the neural network model, the better the extraction effect for signal features. In this paper, the complexity of the model is increased by the densely connected network, and the learning ability of the model is enhanced to extract the one- and two-dimensional features of the original signal, and the model is made to obtain more feature information through the mode of two-channel fusion, so as to improve the accuracy of the model for monitoring the load inhomogeneous state.

The comparison results of the first three sets of experiments are shown in Figure 16. Through the confusion matrix and classification result graphs of the three sets of experiments, it can be seen that the FTF-DFD model proposed in this paper achieves the recognition of four types of position information, F1, F2, F3 and $F_{1.2.3}$, respectively, under light load, medium load and heavy load conditions, and all of them achieve 100% recognition accuracy.

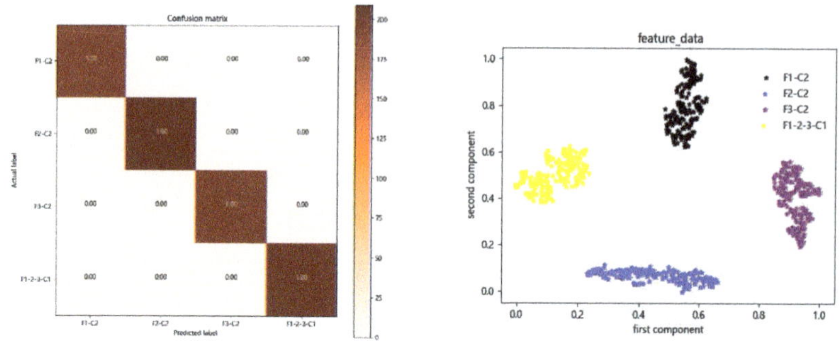

(**a**) Light load experiment

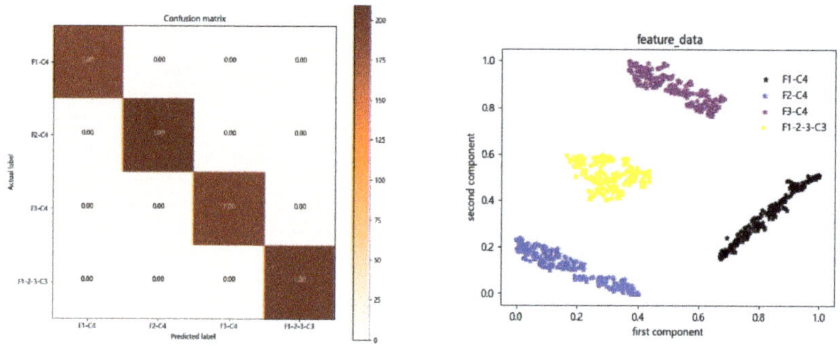

(**b**) Mid-load experiments

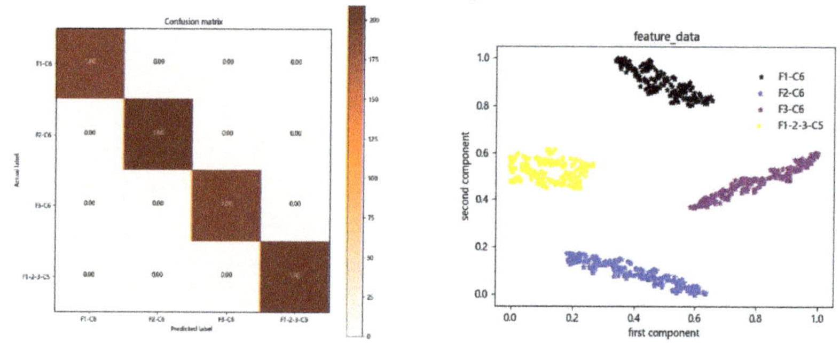

(**c**) Heavy load experiment

Figure 16. Comparison chart of the first three groups of experiments. Confusion matrix and visualization of classification results.

The fourth set of experiments took the 12 sets of work condition data involved in this paper and input them into the evaluation model after the data sample expansion of overlapping sampling, and the final output was divided into 12 clusters by t-SNE visualization. According to the input features in Figure 17a, it can be seen that compared with the Western Reserve University 10 classification task features are completely mixed, the original input data of this experiment are mainly divided into two parts, F1 and F3 positions at all the working condition data are mixed together, and F2 is mixed with all the working condition data at position $F_{1.2.3}$, indicating that the original feature distributions of these data are closer and cannot be easily distinguished from each other. The classification result graph of the output of Figure 17b shows that the experimental data of all kinds of working conditions of the model are improved from the chaotic state to the aggregated state, and the classification task of 12 working conditions is completed effectively, and all the working condition information is completely distinguished. The experiments are conducted by expanding different sample sizes, and it can be seen that the feature distributions in the two groups of working conditions, F2 (C2) and $F_{1.2.3}$ (C5), are always closer, indicating the similarity of the feature components in these two groups of working conditions. This proves that the algorithm in this paper can effectively realize the working condition recognition under the uneven bearing load.

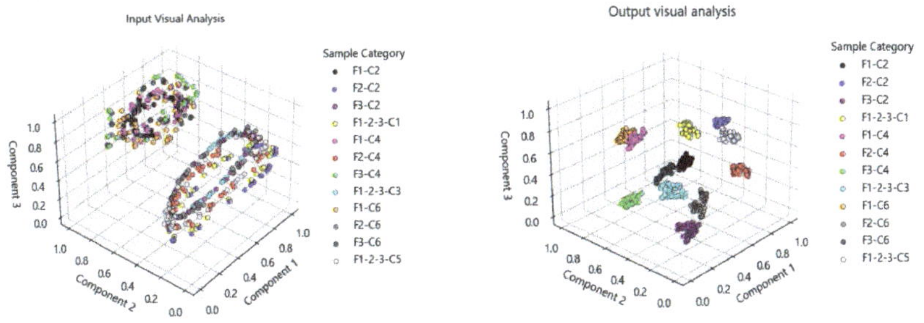

(**a**) Input characteristics (**b**) Output characteristics

Figure 17. Visual classification result chart.

5. Conclusions

In order to improve the condition monitoring performance of bearings with non-uniform loads, we propose a fault diagnosis method using the FTF-DFD model to identify the operating condition of spindles more accurately. First, the sample expansion of the original data is performed, and then the frequency domain and time-frequency domain conversion are performed. Then, the FTF-DFD model is constructed for the extraction of dual-domain feature information, and the overall iterative performance of the model is improved by the *Adam* dynamic adjustment strategy and the improved *ce_loss* loss function. Finally, the validity of the model was tested by the Case Western Reserve University data set, and an experimental bench for bearing non-uniform load operation was designed and built for validation, which was compared with the other four methods, and the following conclusions were drawn:

- Using the dual-channel model to extract the frequency domain features and time-frequency domain features of the original signal can reflect the vibration characteristics of the bearing more comprehensively and accurately, thus improving the accuracy of fault diagnosis.
- The condition monitoring model of the spindle bearing of FTF-DFD is established. The model has strong generalization performance, and the bearing condition under fault condition and variable load condition can be identified, and the condition detection rate is extremely high, reaching up to 100%.
- The overall iterative performance of the model is greatly improved, and the training time is reduced by using the Adam dynamic adjustment strategy in conjunction with the improved *ce_loss* loss function.
- This paper only validates the performance of the method for identifying non-uniform loads on bearings. In the future, the FTF-DFD model will be applied to other components of the spindle system, and the model will be migrated to other fields to complete further validation.

Author Contributions: Software, W.Z.; Formal analysis, D.L.; Resources, L.W.; Data curation, Y.L.; Writing—review & editing, Y.Z.; Supervision, L.K. All authors have read and agreed to the published version of the manuscript.

Funding: This research was funded by the "National Natural Science Foundation of China, grant number 52005405", "Major Scientific and Technological Project of China machinery industry group

Co., LTD, grant number ZDZX2021-2", "Shaanxi Provincial Key R&D Program, grant number 2023-YBGY-350" and "Shaanxi Provincial Key R&D Program (2022GY-211)".

Data Availability Statement: Not applicable.

Conflicts of Interest: The authors declare no conflict of interest.

References

1. Jing, L.; Zhao, M.; Li, P.; Xu, X. A convolutional neural network based feature learning and fault diagnosis method for the condition monitoring of gearbox. *Measurement* **2017**, *111*, 1–10. [CrossRef]
2. Jeschke, S.; Brecher, C.; Meisen, T.; Özdemir, D.; Eschert, T. *Industrial Internet of Things and Cyber Manufacturing Systems*; Springer International Publishing: Cham, Switzerland, 2017; pp. 3–19.
3. Yin, S.; Li, X.; Gao, H.; Kaynak, O. Data-based techniques focused on modern industry: An overview. *IEEE Trans. Ind. Electron.* **2014**, *62*, 657–667. [CrossRef]
4. Chen, Z. Research on Intelligent Diagnosis Method of Mechanical Equipment Based on Deep Migration Learning. Ph.D. Thesis, South China University of Technology, Guangzhou, China, 2020.
5. Wang, Y.J. Research on Rolling Bearing Vibration Signal Feature Extraction and State Evaluation Method. Ph.D. Thesis, Harbin Institute of Technology, Harbin, China, 2015.
6. Deng, M.Y.; Li, C.Z.; Yang, H. Research on bearing fault diagnosis based on frequency domain feature variational self-encoder. *Comput. Meas. Control* **2023**, *31*, 70–75.
7. Jiao, W.D.; Ding, X.M.; Yan, T.Y.; Yan, Y.Y. Research on fault diagnosis method based on frequency domain feature waveform pattern matching. *J. East China Jiaotong Univ.* **2021**, *38*, 73–81.
8. Liu, F.; Wu, X.; Pan, N.; Zhou, J. Application of improved time-domain blind deconvolution algorithm for bearing fault diagnosis. *Mech. Strength* **2016**, *38*, 207–214.
9. Li, W.; Ma, J.; Yu, F. A rolling bearing fault diagnosis method based on frequency domain sparse classification. *Bearing* **2016**, *33*, 58–61.
10. Wang, S.L.; Chen, J.; From, F. Application of time-frequency-based band entropy method in rolling bearing fault identification. *Vib. Shock* **2012**, *31*, 29–33.
11. Li, J.Q.; Zheng, J.D.; Pan, H.Y.; Tong, J.; Feng, K.; Ni, Q. A two-dimensional time-frequency multi-scale entropy method for rolling bearing fault diagnosis. *Mech. Sci. Technol.* **2023**, *111*, 1–10.
12. Zhou, J.; Zhu, J.W. Research on machine learning classification problems and algorithms. *Software* **2019**, *40*, 205–208.
13. Zhang, R.; Wang, Y.B. Research on machine learning and its algorithms and development. *J. Commun. Univ. China* **2016**, *23*, 10–18.
14. Pei, S. Research on Classification Algorithm Based on Machine Learning. Master's Thesis, North Central University, Taiyuan, China, 2016.
15. Wu, X.M.; Wu, Y.Y.; Wang, X.; Li, C.F.; Zhang, F.H. Application of machine learning in bearing fault diagnosis. *Equip. Manuf. Technol.* **2022**, *327*, 118–126.
16. Wang, J. Research and Application of Text Classification Algorithm Based on Machine Learning. Master's Thesis, University of Electronic Science and Technology, Chengdu, China, 2015.
17. Zhang, X.Y.; Luan, Z.Q.; Liu, X.L. A review of rolling bearing fault diagnosis research based on deep learning. *Equip. Manag. Maint.* **2017**, *414*, 130–133.
18. Lu, X.; Zhang, C.; Gao, J.; Xu, Y.; Shao, X. Bearing fault diagnosis algorithm based on convolutional neural network and CatBoost. *Mechatron. Eng.* **2023**, *40*, 1–10.
19. Janssens, O.; Slavkovikj, V.; Vervisch, B.; Stockman, K.; Loccufier, M.; Verstockt, S.; Van de Walle, R.; Van Hoecke, S. Convolutional neural network based fault detection for rotating machinery. *J. Sound Vib.* **2016**, *377*, 331–345. [CrossRef]
20. Gu, X.; Tang, X.H.; Lu, J.G.; Li, S.W. Adaptive Fault Diagnosis Method for Rolling Bearings Based on I-DCNN-LSTM. *Mach. Tool Hydraul.* **2020**, *48*, 107–113.
21. Gong, W.F.; Chen, H.; Zhang, Z.H.; Zhang, M.L.; Guan, C.; Wang, X. Intelligent fault diagnosis forrolling bearing based on improved convolutional neural network. *J. Vib. Eng.* **2020**, *33*, 400–413.
22. Chen, K.; Huang, M.; Li, Y. Bearing fault diagnosis method based on CNN-LSTM and attention mechanism. *J. Beijing Univ. Inf. Sci. Technol.* **2022**, *37*, 26–31.
23. Yang, X.; Sun, S.; Wang, G.; Shi, N.; Xie, Y. Bearing fault diagnosis method based on ECA_ResNet. *Bearing* **2023**, 1–8.
24. Huang, G.; Liu, Z.; Van Der Maaten, L.; Weinberger, K.Q. Densely connected convolutional networks. In Proceedings of the IEEE Conference on Computer Vision and Pattern Recognition, Honolulu, HI, USA, 21–26 July 2017; pp. 4700–4708.
25. Shi, C.; Ren, Y.; Tang, H.; Mupfukirei, L.R. A fault diagnosis method for electro-hydraulic directional valve based on intrinsic mode functions and weighted densely connected convolutional networks. *Meas. Sci. Technol.* **2021**, *32*, 084015. [CrossRef]
26. Wang, Y. Research on The Fault Diagnosis Method of Fusing Multiple Sensors. Master's Thesis, Beijing Jiaotong University, Beijing, China, 2019.
27. Niu, R.X.; Ding, H.; Shi, R.; Meng, X.L. Improved densely connected convolutional networks for rolling bearing fault diagnosis. *Vib. Shock* **2022**, *41*, 252–258.

28. Wang, Q.R.; Wang, Y.; Zhu, C.F.; Zhou, Y.T. Fault diagnosis of rolling bearings with dual-channel cross-dense connection. *Mech. Sci. Technol.* **2023**, 1–9.
29. Wang, K.; Liu, X.; Yang, J.Q.; Dong, Z.S. Fault diagnosis of rolling bearings with variable operating conditions based on improved DenseNet model. *Comb. Mach. Tools Autom. Mach. Technol.* **2022**, *580*, 78–81.
30. Hubel, D.H.; Wiesel, T.N. Receptive fields, binocular interaction and functional architecture in the cat's visual cortex. *J. Physiol.* **1962**, *160*, 106. [CrossRef] [PubMed]
31. LeCun, Y.; Boser, B.; Denker, J.; Henderson, D.; Howard, R.; Hubbard, W.; Jackel, L. Handwritten digit recognition with a back-propagation net-work. *Adv. Neural Inf. Process. Syst.* **1989**, *2*, 1–9.
32. Li, H.Y.; Su, T.B. A hand-drawn sketch recognition method based on Bayesian network and convolutional neural network. *J. Southwest Norm. Univ.* **2019**, *44*, 96–102.
33. Wu, D.H.; Ren, G.Q.; Wang, H.G.; Zhang, Y. A review of mechanical fault diagnosis methods based on convolutional neural networks. *Mech. Strength* **2020**, *42*, 1024–1032.
34. Wang, T.Y.; Gong, L.M.; Wang, P.; Qiao, H.U.; Ren, D. KD-DenseNet-based fault diagnosis model for rotating machinery. *Vib. Shock* **2020**, *39*, 39–45.
35. He, K.; Zhang, X.; Ren, S.; Sun, J. Deep residual learning for image recognition. In Proceedings of the IEEE Conference on Computer Vision and Pattern Recognition, Las Vegas, NV, USA, 26 June–1 July 2016; pp. 770–778.
36. Wang, Q.; Wu, B.; Zhu, P.; Li, P.; Zuo, W.; Hu, Q. ECA-Net: Efficient Channel Attention for Deep Convolutional Neural Networks. In Proceedings of the IEEE/CVF Conference on Computer Vision and Pattern Recognition, Long Beach, CA, USA, 15–20 June 2019.
37. Gers, F.A.; Schmidhuber, J.; Cummins, F. Learning to forget: Continual prediction with LSTM. *Neural Comput.* **2000**, *12*, 2451–2471. [CrossRef]
38. Zhao, H.; Jia, J.; Koltun, V. Exploring self-attention for image recognition. In Proceedings of the IEEE/CVF Conference on Computer Vision and Pattern Recognition, Virtual, 13–19 June 2020; pp. 10076–10085.
39. Zhang, H.; Zhang, Q.; Shao, S.Y.; Niu, T.; Yang, X. Attention-based LSTM network for rotatory machine remaining useful life prediction. *IEEE Access* **2020**, *8*, 188–199. [CrossRef]

lubricants

MDPI

Article

Experimental Study of the Rheology of Grease by the Example of CIATIM-221 and Identification of Its Behavior Model

Yuriy O. Nosov and Anna A. Kamenskikh *

Department of Computational Mathematics, Mechanics and Biomechanics, Perm National Research Polytechnic University, 614990 Perm, Russia; ura.4132@yandex.ru
* Correspondence: anna_kamenskih@mail.ru; Tel.: +7-(342)-2-39-15-64

Abstract: The study of the rheological properties of a lubricant allows for the assessment of the structure's durability in which they are used. Computer engineering enables the prediction of the structure performance using refined mathematical models of its materials. This paper presents an experimental investigation of the rheological behavior of a lubricant that is actively used in bridge structures. The paper proposed a methodology for determining the rheological characteristics of the lubricant using a rotational viscometer. Additionally, the article performed the task of identifying the mathematical model of the lubricant behavior based on the Maxwell body, using two approaches: the Anand model and the Prony series. The proposed models allow for numerical modeling of the structure's performance throughout their lifecycle within the scope of computer engineering.

Keywords: lubricant; viscoelastic; Prony series; Anand's model; viscosity; elastic; osciliration experiment; experimental data; numerical data

Citation: Nosov, Y.O.; Kamenskikh, A.A. Experimental Study of the Rheology of Grease by the Example of CIATIM-221 and Identification of Its Behavior Model. *Lubricants* **2023**, *11*, 295. https://doi.org/10.3390/lubricants11070295

Received: 16 May 2023
Revised: 4 July 2023
Accepted: 13 July 2023
Published: 15 July 2023

1. Introduction

1.1. Research Objectives

Research objective: conducting full-scale experiments and identifying a mathematical model of the viscoelastic lubricant behavior over a wide temperature range.

Research and development objectives:

1. Conducting a series of full-scale experiments to determine the viscoelastic lubricants properties over a wide range of temperatures;
2. Identifying a mathematical model of the viscoelastic lubricant behavior in the form of the Maxwell body based on two viscoelasticity models: the Prony series and the Anand's model;
3. Creating a unified numerical procedure to determine approaches to the parameters from item two based on the application of the multi-parameter Nelder–Mead optimization.

1.2. Problem Context and Description

Various lubricants are widely used in friction nodes [1–6]. Such structures operate within the framework of solid mechanics and contact mechanics. In operation, lubrication helps to reduce friction between contact surfaces [7,8], reduce the effect of surface roughness [9,10], control the temperature [11,12], etc. At the same time, many authors note the nonlinear lubricant behavior during operation [10,13–15]. Numerical models require a qualitative description of the material behavior as close to real constructions as possible. Lubricants are no exception; thus, it is necessary for the detailed study of the application of a mathematical model to describe the lubricants behavior.

At present, lubricants are divided into four groups: liquid, plastic, solid, and gaseous. Depending on their aggregate state, lubricants may be used in various areas of human activity: machine building [16,17], bridge building [18], hydraulic systems [19], etc. Lubricants work under different conditions in structures. Lubricant works in a state of constrained compression [18] and within the framework of thermomechanics and thermal application in bridge bearing structures. It increases the structure durability and improves the heat transfer efficiency of the condenser and evaporator [19]. Lubricants are an integral part of the actual tribology problem [20]. In this case, several works [21–23] point out the importance of conducting full-scale and numerical experiments to determine the properties and perform mathematical modeling of their behavior.

Computer engineering enables a numerical analysis of the structure to be conducted, assessing the possibility of using different materials in the structure, analyzing the influence of changes in geometry, and so on. Computer engineering allows for the rationalization of the structure and its elements relatively quickly. It is necessary to use material behavior models that are close to reality to obtain high-quality modeling results. Many authors emphasize the importance of constructing mathematical models of material behavior [24–28]. This allows for an approximation of the model to the actual structure.

One of the actual problems is the study of the dynamic characteristics of the lubricant behavior [21–23,29–31], including the viscoelasticity, viscoplasticity, etc. Many studies [32,33] describe lubrication by Maxwell-type equations. Maxwell's body is a viscoelastic fluid that can flow (relax) under any load. It is characterized by irreversible deformations [30]. The Prony series and Anand's model are the most common models for describing the Maxwell's body. The Prony series is widely used to describe the viscoelastic behavior of various materials [34–36]. This approach is applied to polymeric [37,38] and metallic materials [39]. The description of viscoelasticity by Anand's model enables the evaluation of the material plastic deformation. Initially, Anand's model was aimed at describing the behavior of metal melts [40,41]. Later, it was widely used to describe polymers, pure and alloyed glasses, composites, etc. [42,43]. An incomplete set of model parameters is often used when describing non-metallic materials. This is due to the peculiarities of the material behavior being described. At present, there are attempts to describe lubricant behavior based on the mathematics of Anand's model [44].

To describe the mathematics associated with the operation of lubricants, it is necessary to use methods of computer engineering and numerical algorithms. Thus, it is possible to automate the search for the coefficients of the defining relationships and conduct verification of the obtained data based on numerical and full-scale experiments. Such approaches are widespread and have proven their effectiveness in determining the materials characteristics [45].

The main directions of lubricant research can be distinguished as follows:

1. Study of tribological, electromechanical, and thermal characteristics of lubricants, including those with various additives and solutions;
2. Experimental research on a wide range of temperatures to identify dynamic and static characteristics;
3. Identification of mathematical models of material behavior and their implementation in numerical analogues of friction nodes.

Currently, there are various methods for researching the rheological characteristics of lubricants, including capillary viscometers, penetrometers, rotational viscometers, DWS technology, etc. The principle of capillary viscometers is based on determining the flow rate of the fluid under the influence of a pressure difference in the capillary [46,47]. Penetrometers determine the resistance of the lubricant by measuring the penetration of indenters with different geometrical, physical, and mechanical characteristics into the lubricant [48]. Rotational rheometers are characterized by coaxial cylindrical viscometers, which consist of two cylinders of the same size, one of which is fixed, and the other rotates around its axis at different frequencies [47–51]. DWS technology [52] is significantly different from other methods, requiring a minimal amount of sample. Spectroscopy allows for non-contact investigation of the sample, which eliminates the possibility of sample destruction. This type of technology allows for the evaluation of the rheological material properties over a wider range of temperatures and frequencies compared to analogs. The lubricant examined in this study is used in sliding bridge bearings. These structures work within a narrow temperature range of −60 to 60 °C and at higher shear rates. In the first approximation, it was decided to investigate the material on the Discovery HR2 rotational viscometer. DWS technology will be used for further research on the behavior of the lubricant.

The current research is aimed at obtaining an experimental basis for the deformation behavior of lubricants with refinement of the experimental research methodology. The second direction of the research is creating effective numerical algorithms to describe pasty lubricants using well-known models widely used in applied engineering analysis packages: ANSYS Mechanical APDL (ANSYS Inc., Canonsburg, PA, USA); ABAQUS (ABAQUS Inc., Velizy-Villacoublay, France), etc. Lubricants are used in the steel–polymer contact material pair structure, which can operate in a wide temperature range to ensure the durability. They include CIATIM-221, CIATIM-221F, TOMFLON SK 170 FH, TOMFLON SBS 240 FM, etc. CIATIM-221 and CIATIM-221F are produced all over the world. This paper presents the performance of full-scale experiments to determine the rheological properties of the lubricant CIATIM-221 (Center-Oil LLC, Polevskoy, Russia) over a wide temperature range. The identification of a mathematical lubricant model is required for further investigation of the design performance through numerical experiments.

2. Materials and Methods

2.1. Experimental Research

The lubricant CIATIM-221 was chosen as the research object. It is a frost- and heat-resistant lubricant, which allows its use in both hot and northern climatic zones.

Experimental studies are conducted at the premises of the PNIPU plastics laboratory. Thermophysical, rheological, and thermomechanical characteristics of the materials and products are tested in the laboratory [53,54]. A Discovery Hybrid Rheometer (TA Instruments—Waters LLC, New Castle, DE, USA) (Figure 1a) was used for experimental investigation of the rheological lubricant characteristics. The main characteristics are:

- minimum oscillation torque 2;
- minimum sustained shear torque 10;
- maximum torque 200;
- torque resolution 0.1;
- minimum frequency;
- maximum frequency 100;
- minimum angular frequency 0;
- maximum angular frequency 300;
- displacement resolution 10, etc.

An electrically heated plane is used for the active heating and cooling of the study samples. This design allows the specimens to be heated to 400 °C and cooled to −70 °C using liquid nitrogen (Figure 1b). The design allows for a uniform temperature distribution throughout the sample. As part of the operation, the temperature tolerance limit is ±1 °C.

To determine the true stresses and strains during the experiment, it is necessary to consider the measuring system. As part of the work, a comparison of two variants of measuring systems was performed: plane–plane and cone–plane. The basic relationships for determining the stress (1) and strain (2) are as follows:

$$\sigma = K_\sigma \cdot M, \tag{1}$$

$$\gamma = K_\gamma \cdot \varphi, \tag{2}$$

where M—rheometer torque; φ—offset angle; K_σ and K_γ—stress and strain constants, respectively, which depend on the geometry of the mating surfaces. According to [55], dynamic tests are run in the framework of linear viscoelasticity. This allows the specimens to be examined without destroying the structure. To determine the viscoelastic characteristics of the material, we will use a technique that consists of 4 stages.

(a) (b)

Figure 1. Experimental study on the Discovery HR2 rheometer. (**a**) General view of the setup; (**b**) application of liquid nitrogen.

In the first stage of the studies, it is necessary to determine the shear value of the sample [56]. To determine the linear viscoelasticity domain, a strain sweep experiment is conducted in which the value of the complex shear modulus $G^*=$ const is monitored.

The second stage of the study is the selection of the measuring system for dynamic tests. Two options can be considered with the presented equipment: cone–plane [57] and plane–plane [58]. The main advantage of the cone–plane measuring system is the uniform shear rate distribution. However, due to the small gap size (50–100 times smaller than for the second variant), there are errors in the investigation of temperature dependence.

The third stage of the study involves a series of experiments to determine the dependence of the material properties on temperature [59]. The sample is dynamically deformed at a small displacement angle, which is determined at stage 1 of the study, with a constant shear rate and cooling/heating of the sample at a constant rate.

The final stage of the study consists of determining the dependence of the material stresses on the shear rate [60]. Similar to stage 3, the material is deformed, the temperature remains constant, and the shear rate changes.

2.2. Identification of a Mathematical Model of Lubricant Behavior

In [32,33], the lubricant is described as a Maxwell body. It is a sequential connection of an elastic spring (3) and a viscous element (4):

$$\tau_e = G\gamma_e, \tag{3}$$

$$\tau_v = \eta\dot{\gamma}_v, \tag{4}$$

where G—shear modulus; η—dynamic viscosity. Thus, the elastic and viscous element tangential stresses are equal to each other (5), and the shear strain is the sum of the elastic and viscous parts (6) in the case of a series connection:

$$\tau = \tau_e = \tau_v, \tag{5}$$

$$\gamma = \gamma_e + \gamma_v. \tag{6}$$

Total shear strain as a function of time:

$$\gamma = \tau/G + 1/\eta\left(\int_0^t \tau(q)dq\right). \tag{7}$$

Further research is based on Equations (3)–(7).

The search for unknown parameters of the presented models is performed using the multi-parameter Nelder–Mead optimization algorithm with experimental data. The problem of minimization of the functional is as follows:

$$F = \left|\left(\tau_{\exp} - \tau_{\text{num}}(\overline{\mathbf{x}})\right)/\tau_{\exp}\right| \times 100\% \to \min, \tag{8}$$

where $\overline{\mathbf{x}}$—vector of unknowns. $\overline{\mathbf{x}}$ has a different number of optimization parameters for different viscoelastic models. The search continues until the functional is less than 5%.

2.2.1. Prony Series

The Prony series in conjunction with the Williams–Landela–Ferry (*WLF*) model allow the behavior of a material to be described over a wide temperature range. The stress–strain relationship has the form:

$$\tau(t) = \int_0^t 2\left[G_\infty + G_0\sum_{i=1}^{k} \alpha_i \exp\left(-(t-q)/\beta_i'\right)\right]d\gamma(q), \tag{9}$$

where G_0—shear modulus at t = 0; k—the number of Prony series variables; q—relaxation time; α_i—shear modulus coefficients. The experimental data are given in the form of a shear modulus distribution as a function of temperature. The temperature–time analogy *WLF* (10) is used to identify the model:

$$\beta_i' = \beta_i/A_{WLF}(T), \tag{10}$$

where $A_{WLF}(T)$—the shift function used in the *WLF*, which is of the form:

$$A_{WLF}(T) = (C_1(T - T_r))/(C_2 + (T - T_r)), \tag{11}$$

where T—current temperature; T_r—constant baseline temperature; C_1, C_2—empirical material constants.

Based on Equations (9)–(11), the unknowns vector $\overline{\mathbf{x}} = \{\beta_i, \alpha_i, T_r, C_1, C_2\}$ is constructed, which is retrieved by minimizing the functional (8).

2.2.2. Anand's Model

Anand's model allows for the description of not only the viscoelastic but also the plastic behavior of the materials. It includes the viscous shear rate:

$$\dot{\gamma}_v = Ae^{-U/RT}[\sinh(\xi\tau/S)]^{1/m}, \tag{12}$$

and the evolutionary equation:

$$\dot{S} = \{h_0(|B|)^a B/|B|\}\dot{\gamma}_v, \tag{13}$$

where $B = 1 - S/S^*$; $S^* = S_1\left[\dot{\gamma}_v e^{U/RT}/A\right]^n$; S—strain resistance; S^*—the saturation value of the hardening function; h_0—material curing constant; R—universal constant gas; U—activation energy; T—absolute temperature; n—sample saturation as a function of shear rate. Based on Equations (12) and (13), the unknowns vector $\bar{\mathbf{x}} = \{S_0,\ A,\ U/R,\ \xi,\ m,\ h_0, S_1,\ n,\ a\}$ is constructed.

To describe the Maxwell body, transform Equation (4) to the form (12):

$$\dot{\gamma}_v = \tau/\eta = Ae^{-U/RT}[\sinh(\xi\tau/S)]^{1/m}, \tag{14}$$

by adopting a number of simplifications ($m = 1$; $\xi << 1$; $S = 1$), Equation (14) takes the form:

$$\dot{\gamma}_v = \tau/\eta = A\xi e^{-U/RT}\tau, \tag{15}$$

where a number of empirical constants and dynamic viscosity $\eta = e^{U/RT}/(A\xi)$ are related, which can be determined from the experimental data. With $S = 1$, Equation (13) is zeroed. Then, $h_0 = 0$, and the vector of the unknowns takes the form $\bar{\mathbf{x}} = \left\{S_0,\ A,\ U/R,\ e^{U/RT}/(A\eta),\ 1,\ 0, S_1,\ n,\ a\right\}$. Transforming $\bar{\mathbf{x}}$ by excluding the known constants and relationships, the final vector of unknowns takes the form $\bar{\mathbf{x}} = \{S_0,\ A,\ U/R, S_1,\ n,\ a\}$.

2.3. Mathematical Model Identification Procedure

The identification procedure is built on the synergy of ANSYS and Python. A pure shear numerical experiment is simulated in ANSYS Mechanical APDL (Figure 2).

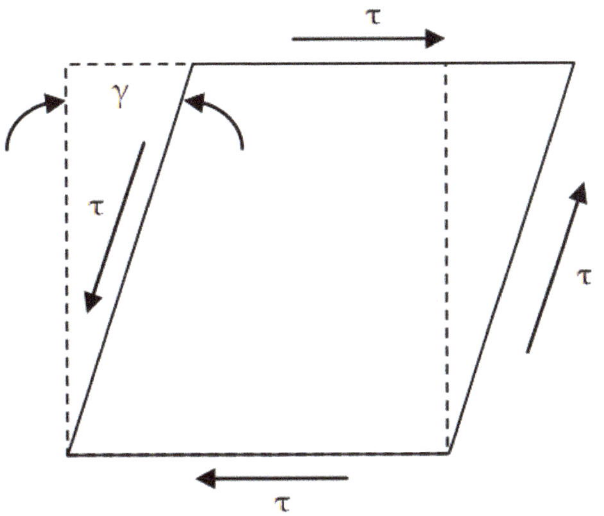

Figure 2. Numerical experiment for pure shear.

The shear value of the sample corresponds to the experimental data of 0.1%. This repeats the cyclic shear strain of the sample over a wide temperature range from 80 to −40 °C a rate of 2 °C per minute. A simplified scheme of the mathematical model identification procedure is shown in Figure 3.

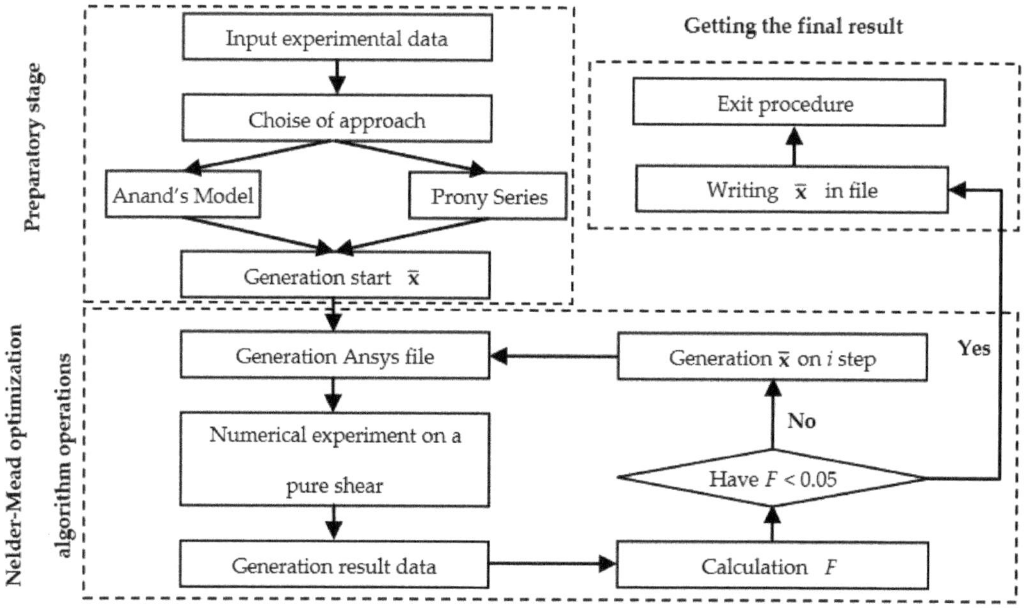

Figure 3. Simplified scheme of the mathematical model identification procedure.

The numerical procedure consists of 3 main parts:

- Preliminary step. It consists in the form of experimental data into the procedure, the choice of a mathematical model, and the setting of initial values of the unknowns vector $\bar{\mathbf{x}}$;
- Nelder–Mead multi-parameter optimization operations. An ANSYS file is generated with a sequence of commands to build the numerical model. The pure shift problem is solved with the generation of a results file. The function F is calculated and comparing to the required error. If the condition is not met, a new vector of unknowns is generated. Then, the optimization procedure is repeated;
- Obtaining a result file. If the error condition is met, the final value of the unknowns vector $\bar{\mathbf{x}}$ is written down and the procedure is exited.

3. Results

3.1. Results of Full-Scale Experiments

Friction nodes often operate in aggressive environments, under increased loads, in temperature zones with elevated/reduced temperatures and frequent temperature fluctuations. CIATIM-221 is often used in the sliding bearings of bridges [61–63]. These structures operate in the temperature range −60 to 60 °C. Research has been performed on the temperature range of −40 to 80 °C. This range includes most of the operating ranges of plain bearings.

The shear strain limit of the lubricant has to be estimated in the first phase of the study. This is necessary to determine the linear viscoelasticity interval (Figure 4). According to [56], the limit of the linear viscoelasticity interval decreases as the temperature increases. Consequently, the study will be conducted at elevated temperatures close to the maximum values: 50 °C and 80 °C. The study was performed using two measuring systems: plane–plane and cone–plane.

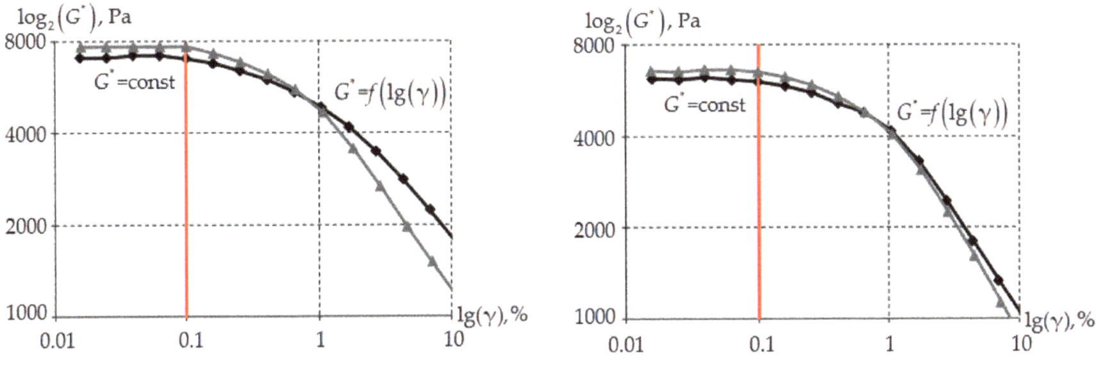

a b

Figure 4. Linear viscoelasticity limit analysis: (**a**) 50 °C, (**b**) 80 °C; measuring systems: black line—plane–plane, gray line—cone–plane.

The distribution of the complex modulus has a nonlinear characteristic after the shear strain of the sample reaches more than 0.1%. Consequently, in the next stages of the viscoelasticity study, the lubricant will be subjected to a shear strain value of 0.1%. In addition, at the linear viscoelasticity interval, the cone–plane measuring system has values higher by an average of 3% than the plane–plane measuring system.

Next, tangential stresses are investigated as a function of shear rate. Experiments are conducted at temperatures of 50 and 80 °C. Within the study, a shear rate of 100 Hz is the maximum possible value. The minimum value of the shear rate was chosen out of consideration of the time cost of the experiment. To obtain the value, it is necessary to perform three to five stress measurements [32,33] at each value of the shear rate (Figure 5).

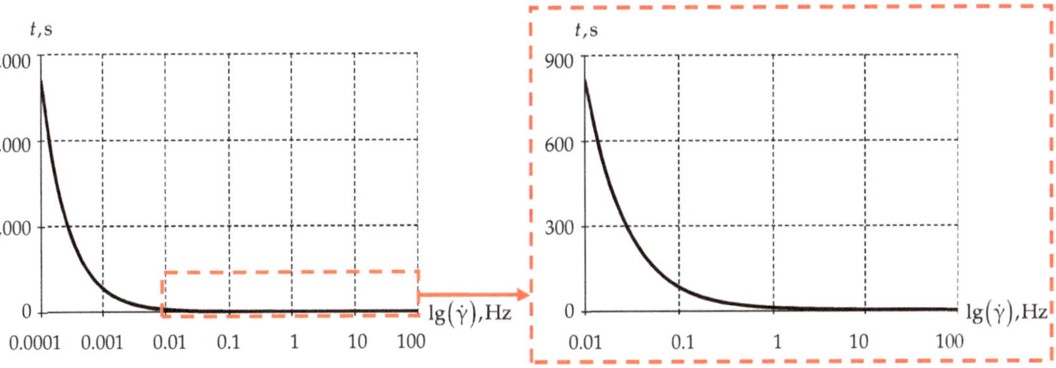

Figure 5. Dependence of the experiment time on the choice of the minimum shear rate.

An increase in the experiment time is observed with a decrease in the minimum value of the shear rate. In this case, the experiment time is presented without taking into account the inertial loads. The elimination of the inertia leads to an increase in the experiment time by an average of 20–30%. It was decided to use a range of shear rates (0.01 Hz to 100 Hz) to minimize the time of the experiment.

In the second stage of the study, it is necessary to determine the measuring system. For this purpose, we will conduct experiments on the dependence of tangential stresses on the shear rate at temperatures of 50 °C and 80 °C (Figure 6).

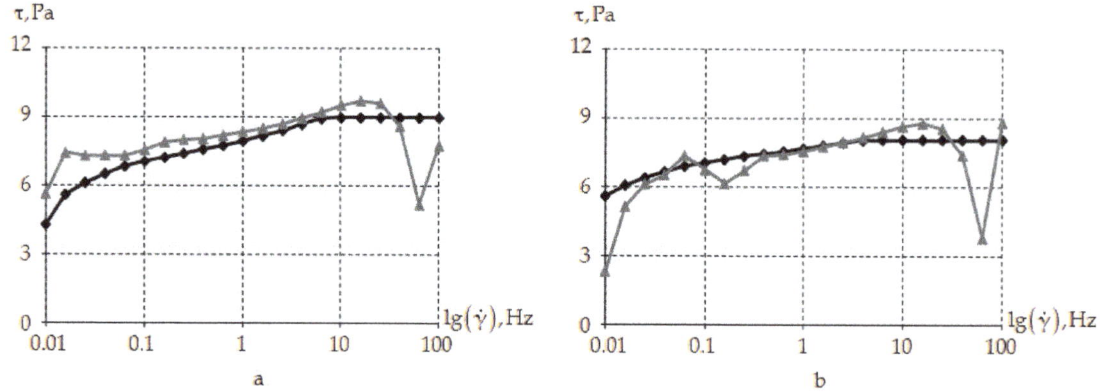

Figure 6. Comparison of measuring systems: (**a**) 50 °C, (**b**) 80 °C; measuring systems: black line—plane–plane, gray line—cone–plane.

The value of the tangential stresses when using the cone–plane measuring system is non-uniform. With the plane–plane system, the dependence $\tau(\lg(\dot{\gamma}))$ is uniform. It becomes constant at a certain shear rate. The cone–plane measuring system introduces a significant error at temperatures other than room temperature.

The third stage of the study consists of conducting experiments to determine the dependence of the material properties on temperature. For this purpose, a sample shear strain value of 0.1% and an average share rate of 1 Hz are chosen based on stages 1 and 2. The sample is cooled from 80 °C to −40 °C a rate of 2 °C per minute (Figure 7).

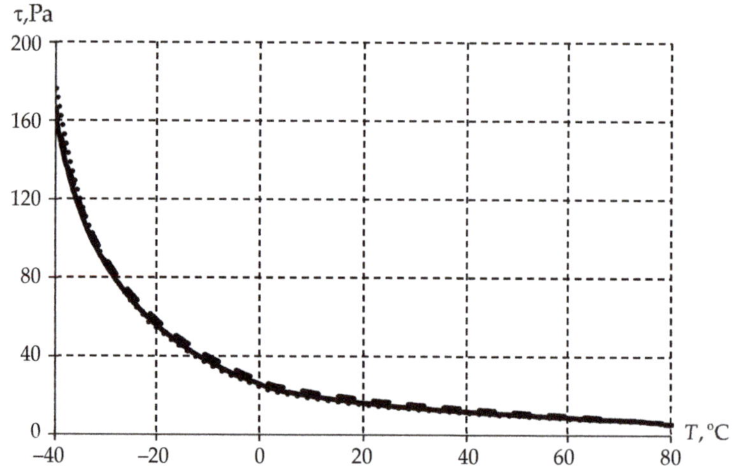

Figure 7. Dependence of tangential stress on temperature in a series of experiments: solid line—experiment 1; dashed line—experiment 2; dots—experiment 3.

The distribution of tangential stresses as a function of temperature is exponential. A significant increase in tangential stresses occurs at temperatures less than 0 °C. The difference does not exceed 5% in the experimental data.

In addition, temperature dependence was obtained for the following lubricant characteristics: complex shear modulus, accumulation modulus, and loss modulus. These characteristics serve us to assess its aggregate state (Figure 8).

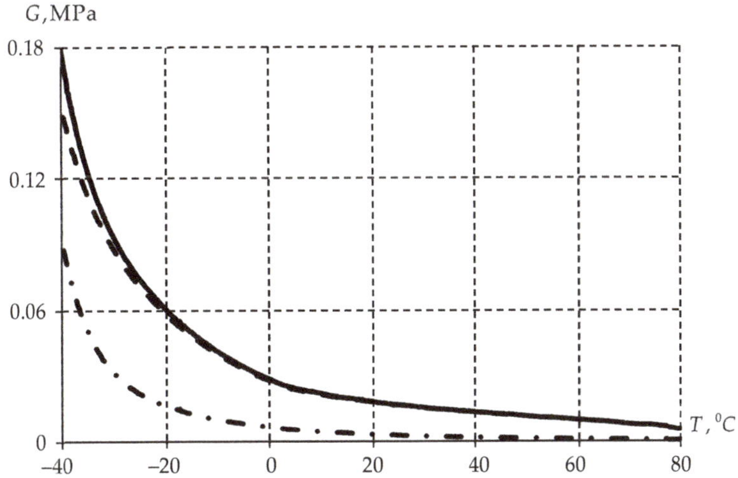

Figure 8. Dependence of physical and mechanical characteristics of a lubricant on temperature: solid line—complex shear modulus; dashed line—accumulation modulus; dashed line—loss modulus.

The distributions of the complex modulus, accumulation modulus, and loss modulus have similar characteristics. The distribution is exponential with a decreasing sample temperature. Over the entire temperature interval, the accumulation modulus is higher than the loss modulus. According to [55,56], this behavior indicates a constant aggregate state of the material as an elastic or gel-like body.

As part of the lubricant research methodology, the last step is to analyze the behavior of the sample when the shear rate changes. The experiment is conducted at a number of temperatures: −40 °C, −20 °C, 0 °C, 20 °C, 50 °C, and 80 °C (Figure 9).

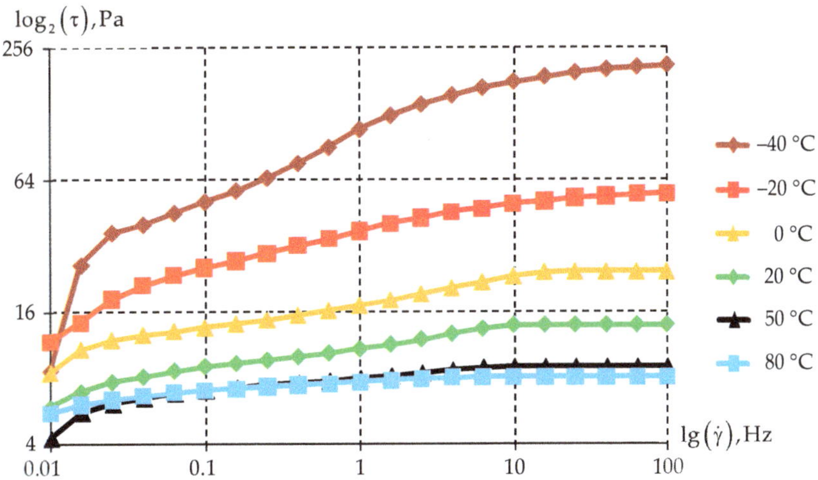

Figure 9. Tangential stress distribution as a function of shear rate.

The behavior of tangential stresses as a function of shear rate for non-negative temperatures minimally differs. The values increase by gradually reaching a constant value when a certain shear rate is obtained. The boundary of constant tangential stress increases as the temperature decreases. In the case of negative temperatures, there is no such interval.

3.2. Identification of a Mathematical Model of the Lubricant as a Function of Temperature

A pure shear numerical experiment is simulated as part of the identification of the mathematical model of the lubricant. The numerical experiment corresponds to the laboratory experiment: shear strain value at 0.1% with a shear rate of 1 Hz and a temperature rate of 2 °C per minute.

The experimental data were approximated over the range of −40 to 80 °C, with a rate of 2 °C, to compare the values of tangential stress. Tangential stress values corresponding to the experimental data will be collected every 60 s (Figure 10), in the process of solving the dynamic problem in a numerical experiment.

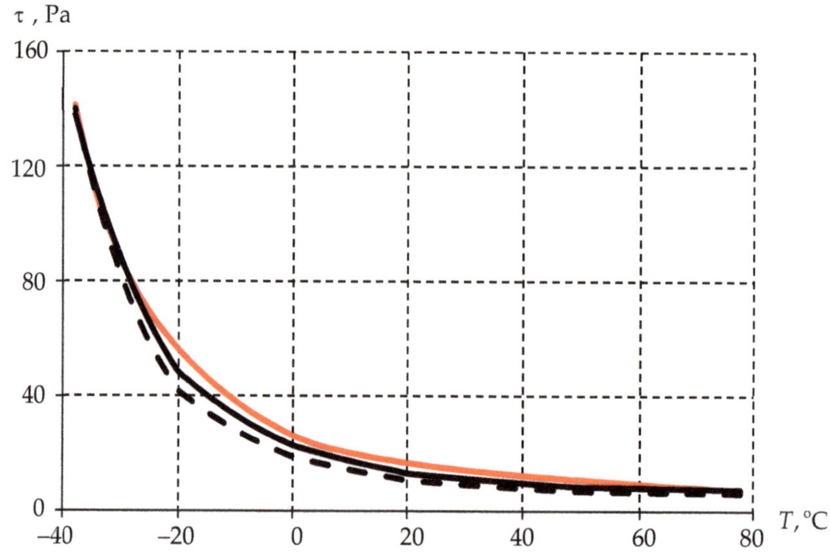

Figure 10. Tangential stress distribution as a function of temperature: red line—approximation of experimental data; solid black line—Anand's model; dashed black line—Prony series.

The experimental dependence $\tau(T)$ is obtained on the basis of statistical data processing. An approximation of the function of voltage dependence on temperature was carried out. Further studies are based on this curve. The scatter of experimental results does not exceed 5%. The identification procedure with the use of two mathematical approaches allows us to obtain $\tau(T)$ at a sufficiently good level.

The relaxation times β_i of the Prony series include 21 coefficients in the range of 10^{-10} to 10^{10}. The initial distribution of the weight coefficients, α_i, was set as constant, equal to 0.01. In the course of the algorithm's work, the value of the weight coefficients changed significantly (Figure 11).

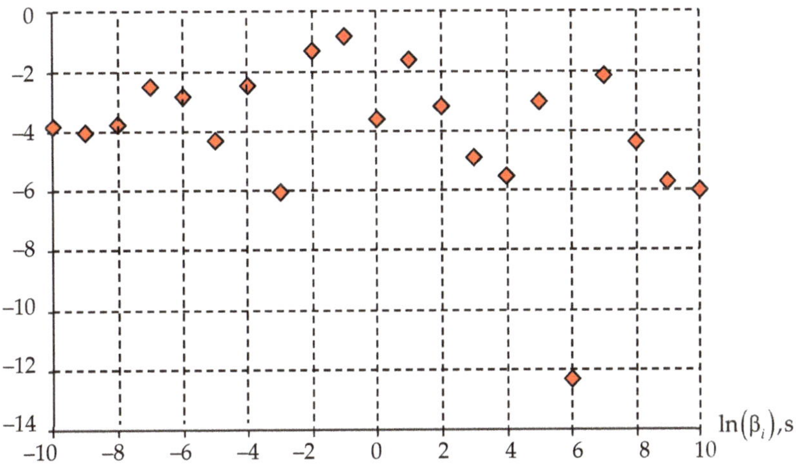

Figure 11. Final distribution of the Prony model coefficients for CIATIM-221.

It can be seen that the weight coefficient at a relaxation time of 10 s has a significant effect. In addition, the values of the coefficients of Equation (11) have the form: T_r = 12.1513 K, C_1 = 4.1469, C_2 = 20.5555.

The numerical procedure for identifying Anand's model converges much faster. This is due to a smaller set of material constants. Table 1 shows the initial and final values of the coefficients of Anand's model.

Table 1. Value of Anand's model coefficients for CIATIM-221.

Values	S_0, MPa	A, 1/s	U/R, K	S_1, MPa	n	a
Initial	1	10^{10}	1000	1	1	1
Final	15.6218	3.518×10^9	985.0634	7.023×10^{-7}	4.498×10^{-4}	2.8112

The initial shear strain resistance value, S_0, is related to the significant viscosity of the lubricant. The shear strain resistance saturation coefficient, S_1, decreases with exposure time. The decrease, n, can be attributed to the shear yield characteristic of the lubricant.

3.3. Dependence of Rheological Properties of the Lubricant on the Shear Rate

A characteristic feature of pasty lubricants is the predominance of the viscous component at low shear rates and the elastic component at higher shear rates. It is necessary to investigate the lubricant's dependence on the shear rate. Figure 12 shows the dependence $\tau(\lg(\dot{\gamma}))$ at different temperatures.

The nature of the tangential stress distribution as a function of shear rate at different temperatures has minor differences; there is a coincidence in the values at shear rates greater than 10 Hz but a significant error at lower shear rates. The Prony series description of grease has a higher error than Anand's model. This is related to the fact that Anand's model accumulates plastic deformation. An extension of the mathematical description of the defining relationships is required in other viscoelastic models for pasty greases.

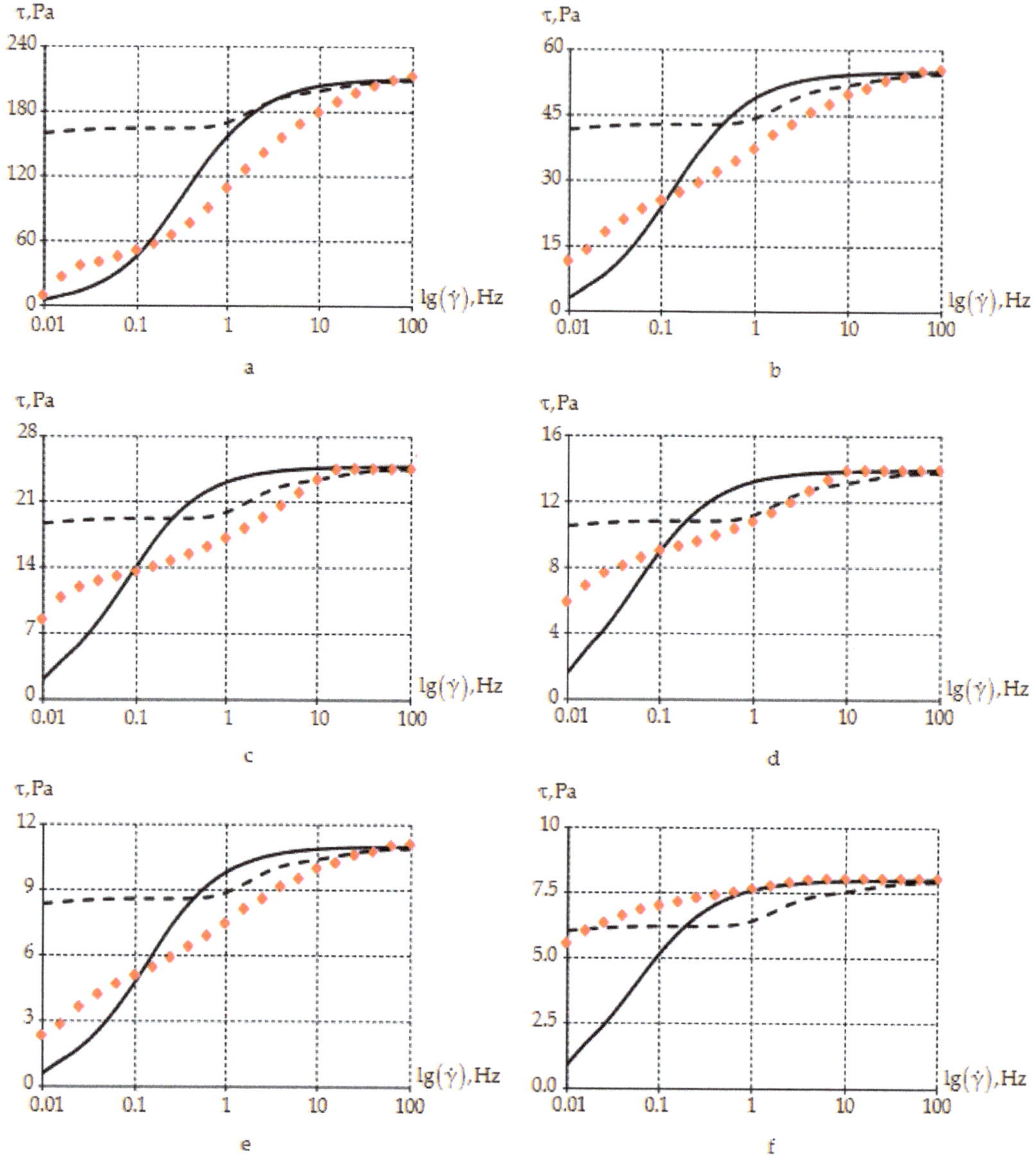

Figure 12. Tangential stress distribution as a function of shear rate over a wide range of temperatures: (**a**) −40 °C, (**b**) −20 °C, (**c**) 0 °C, (**d**) 20 °C, (**e**) 50 °C, and (**f**) 80 °C; red markers—experimental data; solid black line—Anand's model; dashed black line—Prony series.

4. Discussion

4.1. Limitation Statement

The presented study has several limitations:

- The full-scale experiment is conducted in a small range of shear rates, from 0.01 to 100 Hz, which does not give a complete picture of the viscous and elastic components at small and large share rates, respectively;
- The behavior of the lubricant was investigated using the Discovery HR2 rotational viscometer with a limited range of temperatures and share rates;

- The lubricant is capable of operating in the temperature range of −60 to +150 °C, but the equipment allows evaluation of behavior at temperatures of −40 to +80 °C;
- Lubricant is treated as a Maxwell body, in fact, the object of study has a more complex pattern of behavior;
- Lubricant is considered within the problem of deformable solid mechanics; the problem of fluid and gas mechanics is not set.

In the future, many tasks will be set before the researchers:

- Consideration of other linear viscoelastic models (Kelvin model, Voigt model, etc.);
- Using a temperature–time superposition to be able to describe and analyze decreased and increased shear rates and temperatures;
- Numerical simulation of the structure as a whole with the use of a lubricant, using the example of a spherical sliding bearing of a bridge span;
- The lubricant will be examining using the DWS technology.

4.2. Prediction of Structure Behavior Based on Computer Engineering

Computer engineering allows for predicting the behavior of a structure. To do this, refined mathematical models of the behavior of lubricants should be used [64–66]. Chong et al. conducted research to determine the mathematical model of lubricant behavior, which allows for the accurate prediction of the frictional characteristics of lubrication systems. Mukutadze et al. also conducted research on the mathematical model of liquid lubricant behavior, which allows for predicting the influence of the non-stationary profile on the shear rates, pressure, and friction force of the lubricant. In addition, the authors [66] actively describe the mathematical model of the rheological behavior of lubrication at higher frequencies. Two modes were identified: at high temperature and low pressure, the response is adapted to the Kappo model, and as the viscosity increases, the response changes and is consistent with the theory of Airy. This, in turn, allows for predicting the overall behavior of the structure at higher frequencies of its operation.

This work considers the lubricant CIATIM-221, which is widely used in bridge structures. Identifying the mathematical model allows for predicting the behavior of the structure over a wide range of temperatures. The model can be used under higher frequency loads on the structure. To describe the low-frequency effect on the structure, more complex models, as described above, should be used.

4.3. On the Choice of a Mathematical Model

Many researchers consider a lubricant as a more complex viscoelastic or viscoelastic-plastic body [67,68]. Ivins et al. describe the rheological behavior of a lubricant with the Burger model, which, unlike the Maxwell model, includes two springs and two viscous elements. The description of the rheological behavior of a lubricant by a more complex models leads to refinement of the results on the dependence on shear rates. Kvarda et al. [69] propose to describe the lubricant, at higher shear rates, by Newton's law of viscosity, which is related to the output of a constant stress value, as presented in the experiments in Section 3.1.

Larson [44] offers several mathematical models to describe a complete model of the rheological behavior of a lubricant: the model of Mujumdar et al. [70], the de Souza Mendes and Thompson model [71], the Radhakrishnan model [72], etc.

The considered mathematical models and relationships for describing the lubricant viscoelastic behavior are presented in Table 2.

The description of pasty greases, such as CIATIM-221, based only on the Maxwell equations for a wide range of temperatures and shear rates is impossible. Selection of a mathematical model is required, which will allow a better description of the complex behavior of greases at a wide range of temperatures and shear rates.

Table 2. Mathematical models for describing the lubricant viscoelastic behavior.

Matehematical Model	Scheme of the Model	Equation
Kelvin–Voigt model	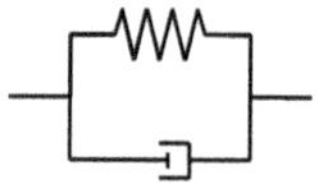	$\sigma(t) = E\varepsilon(t) + \eta\frac{d\varepsilon(t)}{dt}$
Burgers Material (Maxwell representation)	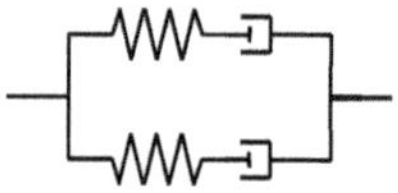	$\sigma + \left(\frac{\eta_1}{E_1} + \frac{\eta_2}{E_2}\right)\dot{\sigma} + \frac{\eta_1\eta_2}{E_1E_2}\ddot{\sigma} = (\eta_1 + \eta_2)\dot{\varepsilon} + \frac{\eta_1\eta_2(E_1+E_2)}{E_1E_2}\ddot{\varepsilon}$
Burgers Material (Kelvin representation)	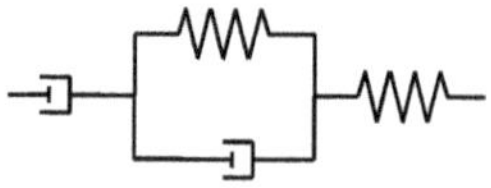	$\sigma + \left(\frac{\eta_1}{E_1} + \frac{\eta_2}{E_1} + \frac{\eta_2}{E_2}\right)\dot{\sigma} + \frac{\eta_1\eta_2}{E_1E_2}\ddot{\sigma} = \eta_2\dot{\varepsilon} + \frac{\eta_1\eta_2}{E_1}\ddot{\varepsilon}$

4.4. Scope of Application Results

CIATIM-221 is widely used in bridge bearings. The lubricant must operate over a wide temperature range and retain its structure. Increasing the durability of structural elements is one of the urgent tasks of bridge building [73]. A large amount of research is related to numerical modeling and computer engineering (Section 4.2.). However, the dynamic characteristics of materials must be taken into account to analyze the behavior of the structure over time (Section 4.3.). Using the approaches described above will enable:

- Numerical experiments to be conducted on the operation of structural elements during the life cycle;
- Extension of the presented study to determine the rheological properties of polymeric materials [74];
- Consideration of the possibility of using a set of materials in structural elements at the decision-making stage [75,76];
- The reduction of material and time costs for field research, etc.

5. Conclusions

The main goal of the research is to conduct natural experiments and identify a mathematical model of the lubricant. The object of the study is the lubricant CIATIM-221, which is widely used in bridge structures with a working temperature range of −60 °C to +60 °C.

For the experimental study, the rotational viscometer Discovery HR2 was chosen. The experimental investigations were carried out in the temperature range of −40 to +80 °C. In the course of the work, a methodology for determining rheological characteristics based on oscillatory research was proposed. During the experiment, it was found that the lubricant is in a gel-like or elastic state over the entire temperature range.

The second stage of the research is aimed at identifying a mathematical model of the behavior of the lubricant. In the course of the work, it was assumed that the lubricant behaves like a Maxwell body. Two mathematical models of viscoelastic behavior were chosen: the Anand model and the Prony series. It was found that the mathematical model describes the behavior of the material well depending on the temperature. However, depending on the frequency of the sample's action, there are differences:

- At higher frequencies, the error is minimal;
- At low frequencies, there is a significant error.

Therefore, the lubricant does not behave like a Maxwell body at low frequencies. The presented model can only be applied at higher frequencies. For low frequencies, more complex models of viscoelastic behavior of the material are required.

Author Contributions: Conceptualization, A.A.K. and Y.O.N.; methodology, A.A.K.; software, A.A.K. and Y.O.N.; validation, A.A.K.; writing—original draft preparation, A.A.K. and Y.O.N.;

writing—review and editing, A.A.K. and Y.O.N.; visualization, A.A.K. and Y.O.N.; funding acquisition, A.A.K. All authors have read and agreed to the published version of the manuscript.

Funding: The study supported by a grant of Russian Science Foundation (project No. 22-29-01313).

Data Availability Statement: Not applicable.

Conflicts of Interest: The authors declare no conflict of interest.

Nomenclature

φ	angular displacement of the rheometer;
G^*	complex shear modulus;
γ	shear strain;
τ_e	tangential stress of an elastic element;
γ_e	shear strain of an elastic element;
τ_v	tangential stress of a viscous element;
η	viscosity;
$\dot{\gamma}_v$	the rate of viscous shear strain;
f	functional;
$\tau_{\exp}$	experimental tangential stress;
τ_{num}	numerical tangential stress;
$\overline{\mathbf{x}}$	vector of unknowns;
G_∞	long shear modulus;
G_0	initial shear modulus;
α_i	weight coefficients;
β_i'	reduced time;
k	number of relaxation times;
$A_{WLF}(T)$	temperature-time analogy shift function;
T	absolute temperature;
C_1, C_2	empirical material constants;
T_r	base temperature;
A	pre-exponential multiplier;
U	activation energy;
R	universal gas constant;
ξ	stress multiplier;
S	shear strain resistance;
h_0	material hardening constant;
S^*	saturation value of the hardening function;
n	sample saturation as a function of shear rate.

References

1. Wen, C.; Meng, X.; Gu, J.; Xiao, L.; Jiang, S.; Bi, H. Starved lubrication analysis of angular contact ball bearing based on a multi-degree-of-freedom tribo-dynamic model. *Friction* **2023**, *11*, 1395–1418. [CrossRef]
2. Stahl, L.; Müller, M.; Ostermeyer, G.P. On the experimental characterization of the fluid volume influence on the friction between rough surfaces. *Friction* **2023**, *11*, 1334–1348. [CrossRef]
3. Jiang, S.; Liu, J.; Yang, Y.; Lin, Y.; Zhao, M. Experimental research and dynamics analysis of multi-link rigid–flexible coupling mechanism with multiple lubrication clearances. *Arch. Appl. Mech.* **2023**, *93*, 2749–2780. [CrossRef]
4. Sokolov, N.V.; Khadiev, M.B.; Fedotov, P.E.; Fedotov, E.M. Influence of the Lubricant's Supply Temperature on the Operation of a Fluid Film Thrust Bearing. *Russ. Eng. Res.* **2023**, *43*, 264–271. [CrossRef]
5. Li, Y.; Shi, L.; Liu, Z.; Wang, X.; Qiao, X.; Zhang, Z.; Yan, S. Study on the lubrication state and pitting damage of spur gear using a 3D mixed EHL model with fractal surface roughness. *J. Mech. Sci. Technol.* **2022**, *36*, 5947–5957. [CrossRef]
6. Cao, W.; He, T.; Pu, W.; Xiao, K. Dynamics of lubricated spiral bevel gears under different contact paths. *Friction* **2022**, *10*, 247–267. [CrossRef]
7. Zhang, Y.; Shakil, A.; Humood, M.; Polycarpou, A.A. Finite element simulations of sliding contact of the head-disk interface in magnetic storage with lubricant effects. *Appl. Surf. Sci. Adv.* **2021**, *6*, 100155. [CrossRef]
8. Smith, E.H. On the Design and Lubrication of Water-Lubricated, Rubber, Cutlass Bearings Operating in the Soft EHL Regime. *Lubricants* **2020**, *8*, 75. [CrossRef]
9. Wang, P.; Qiu, J.; Gao, P.; Dong, R.; Han, Y.; Fan, M. The tribological behaviors and anti-corrosion performances of 5-phenyltetrazole ionic liquid additives for water lubricants. *Wear* **2023**, *516–517*, 204621. [CrossRef]

10. Sharma, S.C.; Singh, A. A study of double layer conical porous hybrid journal bearing operated with non-Newtonian lubricant. *Tribol. Int.* **2023**, *179*, 108183. [CrossRef]
11. Qiu, Y.L.; Zhou, C.G.; Ou, Y.; Feng, F.T. Theoretical and experimental analysis of the temperature rise of a ball screw. *Int. J. Adv. Manuf. Technol.* **2023**, *127*, 703–715. [CrossRef]
12. Li, Z.; Shen, H.; Liang, K.; Chen, X.; Zhu, Z. A numerical study on the effect of oil lubricant on the heat transfer and efficiency of a vapour compression refrigeration system. *Int. Commun. Heat Mass Transf.* **2022**, *134*, 106016. [CrossRef]
13. Torabi, A.; Alidousti, M.H. Numerical and experimental study of elastohydrodynamic grease lubrication of dimple textured surfaces. *Acta Mech.* **2023**, *234*, 2919–2931. [CrossRef]
14. Dang, R.K.; Goyal, D.; Chauhan, A.; Dhami, S.S. Numerical and Experimental Studies on Performance Enhancement of Journal Bearings Using Nanoparticles Based Lubricants. *Arch. Comput. Methods Eng.* **2021**, *28*, 3887–3915. [CrossRef]
15. Jadhav, S.; Thakre, G.D.; Sharma, S.C. Numerical modeling of elastohydrodynamic lubrication of line contact lubricated with micropolar fluid. *J. Braz. Soc. Mech. Sci. Eng.* **2018**, *40*, 326. [CrossRef]
16. Summer, F.; Bergmann, P.; Grün, F. On the Wear Behaviour of Bush Drive Chains: Part II—Performance Screening of Pin Materials and Lubricant Effects. *Lubricants* **2023**, *11*, 157. [CrossRef]
17. Bouchehit, B.; Bou-Saïd, B.; Tichy, J. Towards Ecological Alternatives in Bearing Lubrication. *Lubricants* **2021**, *9*, 62. [CrossRef]
18. Nosov, Y.O.; Kamenskikh, A.A. Influence Analysis of Lubricant Recesses on the Working Capacity of the Bridge Span Spherical Bearing. *Lubricants* **2022**, *10*, 283. [CrossRef]
19. Wu, P.; Chen, J.; Sojka, P.E.; Li, Y.; Cao, H. Experimental study of the lubricant-refrigerant flow in a rotary compressor. *Int. J. Refrig.* **2023**. [CrossRef]
20. Singh, N.; Sinha, S.K. Tribological and mechanical analysis of hybrid epoxy based polymer composites with different in situ liquid lubricants (silicone oil, PAO and SN150 base oil). *Wear* **2022**, *504–505*, 204404. [CrossRef]
21. Katsaros, K.P.; Nikolakopoulos, P.G. Performance Prediction Model for Hydrodynamically Lubricated Tilting Pad Thrust Bearings Operating under Incomplete Oil Film with the Combination of Numerical and Machine-Learning Techniques. *Lubricants* **2023**, *11*, 113. [CrossRef]
22. Tomanik, E.; Aubanel, L.; Bussas, M.; Delloro, F.; Lampke, T. Tribological Performance of a Composite Cold Spray for Coated Bores. *Lubricants* **2023**, *11*, 127. [CrossRef]
23. Qiang, H.; Gao, G.; Ye, S.; Cheng, L.; Wang, Q. Effect of Characteristic Parameters and Distribution of Friction Pair Surface Texture on Lubrication Properties. *Lubricants* **2023**, *11*, 139. [CrossRef]
24. Kodnyanko, V.; Shatokhin, S.; Kurzakov, A.; Pikalov, Y. Mathematical Modeling on Statics and Dynamics of Aerostatic Thrust Bearing with External Combined Throttling and Elastic Orifice Fluid Flow Regulation. *Lubricants* **2020**, *8*, 57. [CrossRef]
25. Orozco Lozano, W.; Fonseca-Vigoya, M.D.S.; Pabón-León, J. Study of the Kinematics and Dynamics of the Ring Pack of a Diesel Engine by Means of the Construction of CFD Model in Conjunction with Mathematical Models. *Lubricants* **2021**, *9*, 116. [CrossRef]
26. Yang, H.; Majeed, A.; Al-Khaled, K.; Abbas, T.; Naeem, M.; Khan, S.U.; Saeed, M. Significance of Melting Heat Transfer and Brownian Motion on Flow of Powell–Eyring Fluid Conveying Nano-Sized Particles with Improved Energy Systems. *Lubricants* **2023**, *11*, 32. [CrossRef]
27. Valigi, M.C.; Malvezzi, M.; Logozzo, S. A Numerical Procedure Based on Orowan's Theory for Predicting the Behavior of the Cold Rolling Mill Process in Full Film Lubrication. *Lubricants* **2020**, *8*, 2. [CrossRef]
28. Poliakov, A.M.; Pakhaliuk, V.I. Predictive Estimates of Short-Term and Long-Term Results for Regenerative Rehabilitation of Local Articular Cartilage Defects in Synovial Joints. *Lubricants* **2023**, *11*, 116. [CrossRef]
29. Slabka, I.; Henniger, S.; Kücükkaya, D.; Dawoud, M.; Schwarze, H. Influence of Rheological Properties of Lithium Greases on Operating Behavior in Oscillating Rolling Bearings at a Small Swivel Angle. *Lubricants* **2022**, *10*, 163. [CrossRef]
30. Farré-Lladós, J.; Westerberg, L.G.; Casals-Terré, J.; Leckner, J.; Westbroek, R. On the Flow Dynamics of Polymer Greases. *Lubricants* **2022**, *10*, 66. [CrossRef]
31. Conrad, A.; Hodapp, A.; Hochstein, B.; Willenbacher, N.; Jacob, K.-H. Low-Temperature Rheology and Thermoanalytical Investigation of Lubricating Greases: Influence of Thickener Type and Concentration on Melting, Crystallization and Glass Transition. *Lubricants* **2022**, *10*, 1. [CrossRef]
32. Li, X.K.; Luo, Y.; Qi, Y.; Zhang, R. On non-Newtonian lubrication with the upper convected Maxwell model. *Appl. Math. Model.* **2011**, *35*, 2309–2323. [CrossRef]
33. Madsen, E.; Rosenlund, O.S.; Brandt, D.; Zhang, X. Adaptive feedforward control of a collaborative industrial robot manipulator using a novel extension of the Generalized Maxwell-Slip friction model. *Mech. Mach. Theory* **2021**, *155*, 104109. [CrossRef]
34. Suarez-Afanador, C.A.; Cornaggia, R.; Lahellec, N.; Maurel-Pantel, A.; Boussaa, D.; Moulinec, H.; Bordas, S.P.A. Effective thermo-viscoelastic behavior of short fiber reinforced thermo-rheologically simple polymers: An application to high temperature fiber reinforced additive manufacturing. *Eur. J. Mech. A/Solids* **2022**, *96*, 104701. [CrossRef]
35. Sun, Y.; Huang, B.; Chen, J.; Jia, X.; Ding, Y. Characterizing rheological behavior of asphalt binder over a complete range of pavement service loading frequency and temperature. *Constr. Build. Mater.* **2016**, *123*, 661–672. [CrossRef]
36. Zhang, X.; Gu, X.; Lv, J.; Zou, X. 3D numerical model to investigate the rheological properties of basalt fiber reinforced asphalt-like materials. *Constr. Build. Mater.* **2017**, *138*, 185–194. [CrossRef]
37. Lesnikova, Y.I.; Trufanov, A.N.; Kamenskikh, A.A. Analysis of the Polymer Two-Layer Protective Coating Impact on Panda-Type Optical Fiber under Bending. *Polymers* **2022**, *14*, 3840. [CrossRef]

38. Sakhabutdinova, L.R.; Smetannikov, O.Y.; Il'inykh, G.V. Numerical Simulation of the Process Manufacture of Large-Scale Composite Shell Taking Into Account Thermo Viscoelastic. *Tomsk. State Univ. J. Math. Mech.* **2022**, *76*, 165–181. [CrossRef]
39. Nuwayer, H.M.; Newaz, G.M. Flexural Creep Behavior of Adhesively Bonded Metal and Composite Laminates. *Int. J. Adhes. Adhes.* **2018**, *84*, 220–226. [CrossRef]
40. Anand, L. Constitutive Equations for Hot-Working of Metals. *Int. J. Plast.* **1985**, *3*, 213–231. [CrossRef]
41. Puchi-Cabrera, E.S.; Guérin, J.D.; Barbera-Sosa, J.G.; Dubar, M.; Dubar, L. Plausible Extension of Anand's Model to Metals Exhibiting Dynamic Recrystallization and its Experimental Validation. *Int. J. Plast.* **2018**, *108*, 70–87. [CrossRef]
42. Liu, E.; Bhogaraju, S.K.; Wunderle, B.; Elger, G. Investigation of stress relaxation in SAC305 with micro-Raman spectroscopy. *Microelectron. Reliab.* **2022**, *138*, 114664. [CrossRef]
43. Liang, Z.; Yonghuan, G.; Lei, S.; Chengwen, H. Reliability of SnAgCuFe Solder Joints in WLCSP30 Device. *Rare Met. Mater. Eng.* **2016**, *45*, 2823–2826. [CrossRef]
44. Larson, R.G.; Wei, Y. A Review of Thixotropy and its Rheological Modeling. *J. Rheol.* **2019**, *63*, 477–501. [CrossRef]
45. Hu, Y.; Wang, X.; Gao, Y.; Xu, J.; Ding, Y. Numerical Simulation of Effect of Glass Lubricant on Hot Extrusion of Inconel 625 Alloy Tubes. *Procedia Manuf.* **2019**, *37*, 119–126. [CrossRef]
46. Wolak, A.; Zając, G.; Słowik, T. Measuring Kinematic Viscosity of Engine Oils: A Comparison of Data Obtained from Four Different Devices. *Sensors* **2021**, *21*, 2530. [CrossRef]
47. Cerpa-Naranjo, A.; Pérez-Piñeiro, J.; Navajas-Chocarro, P.; Arce, M.P.; Lado-Touriño, I.; Barrios-Bermúdez, N.; Moreno, R.; Rojas-Cervantes, M.L. Rheological Properties of Different Graphene Nanomaterials in Biological Media. *Materials* **2022**, *15*, 3593. [CrossRef]
48. Kaushik, S.; Sonebi, M.; Amato, G.; Das, U.K.; Perrot, A. Optimisation of Mix Proportion of 3D Printable Mortar Based on Rheological Properties and Material Strength Using Factorial Design of Experiment. *Materials* **2023**, *16*, 1748. [CrossRef]
49. Xue, X.; Gao, J.; Wang, J.; Chen, Y. Evaluation of High-Temperature and Low-Temperature Performances of Lignin–Waste Engine Oil Modified Asphalt Binder and Its Mixture. *Materials* **2022**, *15*, 52. [CrossRef]
50. Peng, Y.; Via, B. The Effect of Cellulose Nanocrystal Suspension Treatment on Suspension Viscosity and Casted Film Property. *Polymers* **2021**, *13*, 2168. [CrossRef]
51. Quan, L.; Kalyon, D.M. Parallel-Disk Viscometry of a Viscoplastic Hydrogel: Yield Stress and Other Parameters of Shear Viscosity and Wall Slip. *Gels* **2022**, *8*, 230. [CrossRef]
52. Kozdrach, R. The Innovative Research Methodology of Tribological and Rheological Properties of Lubricating Grease. *Tribol. Ind.* **2021**, *43*, 117–130. [CrossRef]
53. Trufanova, N.M.; Ershov, S.V. Comparative Analysis of Heat and Mass Transfer Processes in the Extruder Dosing Zone with the Use Different Spatial Mathematical Model and Rheological Law. *J. Appl. Mech. Tech. Phys.* **2017**, *2*, 153–163. [CrossRef]
54. Davydova, V.A.; Shcherbinin, A.G.; Naumiv, M.D.; Ershov, S.V. A Numerical Study on Induction-Resistive Electric-Heating Processes of Pipelines. *Russ. Electr. Eng.* **2021**, *11*, 668–671. [CrossRef]
55. Schramm, G. *A Practical Approach to Rheology and Rheometry*; Gebrueder HAAKE GmbH: Karlsruhe, Germany, 1994; 290p.
56. Barnes, A.H. *A Handbook of Elementary Rheology*; University of Wales, Institute of Non-Newtonian Fluid Mechanics: Cardiff, Wales, 2000; 200p.
57. Zhai, M.; Zhou, K.; Sun, Z.; Xiong, Z.; Du, Q.; Zhang, Y.; Shi, L.; Hou, J. Rheological Characterization and Shear Viscosity Prediction of Heavy Oil-in-Water Emulsions. *J. Mol. Liq.* **2023**, *381*, 121782. [CrossRef]
58. Song, Y.; Won, C.; Kang, S.; Lee, H.; Park, S.; Park, S.H.; Yoon, J. Characterization of Gglass Viscosity with Parallel Plate and Rotational Viscometry. *J. Non-Cryst. Solids* **2018**, *486*, 27–35. [CrossRef]
59. Chereches, E.I.; Ibanescu, C.; Danu, M.; Minea, A.A. Studies on Rheological Properties and Isobaric Heat Capacity of ZnO-[C4mim][BF4] Nanoparticle Enhanced Ionic Liquid. *J. Mol. Liq.* **2023**, *380*, 121759. [CrossRef]
60. Romanova, Y.N.; Koroleva, M.Y.; Musina, N.S.; Maryutina, T.A. Rheology of Gel-Containing Water-in-Crude Oil Emulsions. *Geoenergy Sci. Eng.* **2023**, *226*, 211757. [CrossRef]
61. Adamov, A.A.; Kamenskikh, A.A.; Pankova, A.P.; Strukova, V.I. Comparative Analysis of the Work of Bridge Spherical Bearing at Different Antifriction Layer Locations. *Lubricants* **2022**, *10*, 207. [CrossRef]
62. Kim, J.-H.; Kim, W.S.; Yoo, Y. Friction Properties of Solid Lubricants with Different Multiwalled Carbon Nanotube Contents. *Materials* **2022**, *15*, 4054. [CrossRef]
63. Adamov, A.A.; Kamenskikh, A.A.; Pankova, A.P. Influence Analysis of the Antifriction Layer Materials and Thickness on the Contact Interaction of Spherical Bearings Elements. *Lubricants* **2022**, *10*, 30. [CrossRef]
64. Chong, W.W.F.; Hamdan, S.H.; Wong, K.J.; Yusup, S. Modelling Transitions in Regimes of Lubrication for Rough Surface Contact. *Lubricants* **2019**, *7*, 77. [CrossRef]
65. Mukutadze, M.A.; Lagunova, E.O. Mathematical Model of Flow of Lubricant and Molten Coating with Micropolar Rheological Properties in Running Clearance of Journal Bearing with Non-circular Bearing Surface Profile, Considering Pressure Dependence of Viscosity. In *Proceedings of the 8th International Conference on Industrial Engineering, Belgrade, Serbia, 29–30 September 2022*; ICIE, Lecture Notes in Mechanical Engineering; Springer: Cham, Switzerland, 2022; pp. 587–597. [CrossRef]
66. Jadhao, V.; Robbins, M.O. Rheological Properties of Liquids Under Conditions of Elastohydrodynamic Lubrication. *Tribol. Lett.* **2019**, *67*, 66. [CrossRef]

67. Gamaniel, S.S.; Dini, D.; Biancofiore, L. The Effect of Fluid Viscoelasticity in Lubricated Contacts in the Presence of Cavitation. *Tribol. Int.* **2021**, *160*, 107011. [CrossRef]
68. Ivins, E.R.; Caron, L.; Adhikari, S.; Larour, E. Notes on a Compressible Extended Burgers Model of Rheology. *Geophys. J. Int.* **2022**, *228*, 1975–1991. [CrossRef]
69. Kvarda, D.; Skurka, S.; Galas, R.; Omasta, M.; Shi, L.; Ding, H.; Wang, W.; Krupka, I.; Hartl, M. The Effect of Top of Rail Lubricant Composition on Adhesion and Rheological Behaviour. *Eng. Sci. Technol. Int. J.* **2022**, *35*, 101100. [CrossRef]
70. Mujumdar, A.; Beris, A.N.; Metzner, A.B. Transient Phenomena in Thixotropic Systems. *J. Non-Newton. Fluid Mech.* **2002**, *102*, 157–178. [CrossRef]
71. Mendes, P.R.; Thompson, R.L. A Critical Overview of Elasto-Viscoplastic Thixotropic Modeling. *J. Non-Newton. Fluid Mech.* **2012**, *187–188*, 8–15. [CrossRef]
72. Radhakrishnan, R.; Divous, T.; Manneville, S.; Fielding, S.M. Understanding Rheological Hysteresis in Soft Glassy Materials. *Soft Matter* **2017**, *9*, 1834–1852. [CrossRef]
73. Niemierko, A. Modern Bridge Bearings and Expansion Joints for Road Bridges. *Trans. Res. Procedia* **2016**, *14*, 4040–4049. [CrossRef]
74. Yi, X.; Du, S.; Zhang, L. *Composite Materials Engineering, Volume 1: Fundamentals of Composite Materials*; Springer: Singapore, 2018. [CrossRef]
75. Adamov, A.A.; Kamenskih, A.A.; Pankova, A.P. Numerical analysis of the spherical bearing geometric configuration with antifriction layer made of different materials. *PNRPU Mech. Bull.* **2020**, *4*, 15–26. [CrossRef]
76. Ono, K. Structural materials: Metallurgy of bridges. In *Metallurgical Design and Industry*; Springer: Cham, Switzerland, 2018; pp. 193–269. [CrossRef]

lubricants

MDPI

Article

Dynamic Temperature Prediction on High-Speed Angular Contact Ball Bearings of Machine Tool Spindles Based on CNN and Informer

Hongyu Li [1], Chunyang Liu [1,2,*], Fang Yang [1,3], Xiqiang Ma [1,2], Nan Guo [1,3], Xin Sui [1,3] and Xiao Wang [1,3]

1 School of Mechatronics Engineering, Henan University of Science and Technology, Luoyang 471003, China; lihy_026@163.com (H.L.); yangfanghkd@haust.edu.cn (F.Y.); maxiqiang@haust.edu.cn (X.M.); guonan@haust.edu.cn (N.G.); sxwd@163.com (X.S.); 9906416@haust.edu.cn (X.W.)

2 Longmen Laboratory, Luoyang 471003, China

3 Henan Key Laboratory for Machinery Design and Transmission System, Henan University of Science and Technology, Luoyang 471003, China

* Correspondence: chunyangliu@haust.edu.cn; Tel.: +86-187-3905-8729

Abstract: This study addressed the issues related to the difficulty of determining the operating status of machine tool spindle bearings due to the high rotational speeds and rapid temperature fluctuations. This paper presents an optimized model that combines Convolutional Neural Networks (CNNs) and Informer to dynamically predict the temperature rise process of bearings. Taking the H7006C angular contact ball bearing as the research object, a combination of experimental data and simulations was used to obtain the training dataset. Next, a model for predicting the temperature rise of the bearing was constructed using CNN + Informer and the structural parameters were optimized. Finally, the model's generalization ability was then verified by predicting the bearing temperature rise process under various working conditions. The results show that the error of the simulation data source model was less than 1 °C at steady state; the temperature error of the bearing temperature rise prediction model was less than 0.5 °C at both the temperature rise and steady-state stages under variable rotational speeds and variable load conditions compared to Informer and Long Short Term Memory (LSTM) models; the maximum prediction error of the operating conditions outside the dataset was less than 0.5 °C, and the temperature rise prediction model has a high accuracy, robustness, and generalization capability.

Keywords: ball bearing; temperature prediction; parameter optimization; CNN; informer

Citation: Li, H.; Liu, C.; Yang, F.; Ma, X.; Guo, N.; Sui, X.; Wang, X. Dynamic Temperature Prediction on High-Speed Angular Contact Ball Bearings of Machine Tool Spindles Based on CNN and Informer. *Lubricants* **2023**, *11*, 343. https://doi.org/10.3390/lubricants11080343

Received: 2 July 2023
Revised: 3 August 2023
Accepted: 8 August 2023
Published: 11 August 2023

1. Introduction

Angular contact ball bearings for machine tool spindles operate in a complex environment and inevitably generate a variety of problems that are difficult to detect promptly [1,2]. These problems can significantly impact the machining accuracy of the machine tool and even result in damage to the machine. Temperature is a critical parameter for monitoring the operational state of the bearing system, and extended exposure to abnormally high temperatures can result in bearing failure. Under the operating conditions of machine tool spindle bearings, predicting the dynamic temperature rise of bearings is a pressing research issue. By comparing the measured values with the predicted values, potential abnormal conditions of the bearings can be promptly identified, which holds significant importance for ensuring the precision and longevity of the machine tool equipment.

In recent years, numerous scholars have been researching methods to solve the temperature field of high-speed rolling bearings. Popescu et al. [3] proposed four methods for analyze the motion of angular contact ball bearings and calculated the internal friction torque and power loss of the bearings. Kim et al. [4] proposed a numerical method to estimate the steady-state temperature of spindle bearings. They constructed a finite element analysis model and compared the measured data to validate the effectiveness of the finite element analysis method. Xu et al. [5] developed a heat transfer model for high-speed railway bearings, taking into account the bearing characteristics. They conducted simulations and analyses of the temperature fields in the inner ring, outer ring, and roller using a finite element model. Additionally, they proposed a method for distributing the heat source in a reasonable manner. Deng et al. [6] developed a mathematical model using the heat source method and performed a comparison between the calculated results of the temperature field and a finite element model. The comparison was carried out for various parameters, including heat generation, heat transfer coefficient, heat source location, and bearing size. Their study confirmed the accuracy of the calculations and demonstrated improved efficiency of their mathematical model. Wu et al. [7] developed an analytical model of the spindle bearing system by incorporating Hertz and contact theory. This model took into account factors such as preload and centrifugal force. Additionally, they established a mathematical model of the temperature field based on heat transfer theory to analyze how the cooling system affects the temperature distribution within the system. Zheng et al. [8] integrated the effect of the contact angle on thermal deformation into the force equilibrium equation of angular contact ball bearings. They computed the bearing load and resolved it to determine the heat generation. Additionally, they furnished a comprehensive exposition of heat generation and its transmission from individual heat sources. Li et al. [9] developed a computational method to predict the thermodynamic properties of high-speed spindle bearings. The method was based on thermodynamic and quasi-static models. To accurately predict the spindle bearing, they employed a Monte Carlo optimization algorithm to invert the experimentally measured temperature data. Zhang et al. [10] developed a local frictional heat generation model for grease-lubricated angular contact ball bearings. This model enables the calculation of heat generation in each contact region of the bearing and predicts the bearing temperature values under high-speed operating conditions. These methods are effective in calculating the steady-state temperature field of the bearing. During the modeling process, it is common for the ambient temperature to be considered fixed, which neglects the variations in the lubrication state during bearing operation. This limitation can lead to errors that vary with the operating conditions.

To achieve real-time prediction of bearing temperature, Yan et al. [11] proposed a hybrid model for real-time prediction of bearing temperature. This model decomposes the plain bearing temperature data and optimizes the weights of the subseries to obtain the final prediction results. Liu et al. [12] compared and analyzed the temperature characteristics of bearings with different temporal distributions. They proposed a two-way long short-term memory (BILSTM)-based model for predicting abnormal bearing temperatures, which enables the diagnosis of bearing status in both temporal and spatial dimensions. This model achieves more accurate detection of abnormal states and provides effective early warning capabilities. Chen et al. [13] proposed a long short-term memory neural network that incorporates multi-task learning and attention mechanisms to accurately predict bearing temperature in complex environments. This model takes into consideration the impact of current working conditions and historical data on bearing temperature, leading to effective temperature predictions. Xiao et al. [14] introduced a novel deep learning algorithm called Stacked Sparse Self-Encoder Multilayer Perceptron (SSAE-MLP) for predicting wind turbine spindle temperatures. This algorithm utilizes multiple stacked sparse self-encoders to extract intricate features from the input data. Additionally, a regression predictor was added to the top layer of the model for supervised learning. The experimental results demonstrated the method's effectiveness in accurately predicting wind turbine spindle temperatures.

Traditional time series prediction methods rely heavily on mathematical and statistical principles. These methods establish a connection between predicted data and historical data by utilizing techniques such as linear regression or least squares regression analysis. One commonly used model in this category is the Autoregressive Integrated Moving Average (ARIMA) model [15]. Traditional prediction analysis methods, while having simple models and precise theories, have limitations in accurately predicting complex, nonlinear, nonstationary time series. The advancement and refinement of neural network theory have brought about significant applications in engineering. In the field of wind power prediction, Zhang [16] utilized support vector regression (SVR) as an external neural network. On the other hand, Cui [17] employed BP neural networks for predicting geotechnical engineering parameters. The primary limitation of external neural networks is their simplicity, which poses a constraint on prediction accuracy in practical applications. A specific type of recurrent feedback neural network framework known as the recurrent neural network (RNN) [18] addresses this issue by considering the initial inertia in time series data and constructing a comprehensive time series model through analysis of historical information. However, during usage, RNNs are susceptible to challenges like gradient vanishing and gradient explosion. On the other hand, the Long Short-Term Memory (LSTM) model [19], also a type of RNN, effectively mitigates these problems and facilitates the utilization of accurate historical information. It is worth noting though that the convergence time of such recurrent neural networks is relatively long, and they still exhibit inherent limitations when dealing with longer sequences.

Due to the increasing recognition and effectiveness of Transformer models, researchers have begun to favor Transformers over traditional RNN structures. Fan et al. [20] employed a multimodal attention mechanism to enhance the integration of historical information from various phases and employed it for predicting future time steps. However, the development of this approach in the field of time series prediction is constrained by challenges related to space and time complexity, as well as memory occupation rates, especially when dealing with long sequence inputs. Zhou et al. [21] addressed the challenge of capturing long-range dependencies in long sequences by introducing the Informer model, which utilizes a Transformer architecture. The Informer model replaces the original attention mechanism with sparse self-attention, resulting in reduced time–space complexity. Additionally, its generative decoder accomplishes long sequence output with a single forward step, effectively avoiding cumulative error expansion during the inference stage. Gong et al. [22] applied the Informer model to predict regional thermal loads and conducted a comparative study with several other models using room temperature, wind speed, and air quality as input features. The results showed that the prediction model using Informer performed better than other models in terms of accuracy and stability. Yang et al. [23] introduced the Informer model as a solution to address the error accumulation problem associated with conventional time series prediction methods when applied to motor bearing vibration data. They conducted a comparative analysis of the prediction results using a publicly available dataset, aiming to verify the superior performance of the Informer model in handling long time series data.

To tackle the aforementioned issues, this study built a composite dataset comprising both experimental and simulation data on bearing temperature rise. The H7006C angular contact ball bearing was selected as the research object for investigation. Furthermore, a novel approach combining a Convolutional Neural Network (CNN) and Informer method was proposed for dynamically predicting the bearing temperature rise process. The model parameters were optimized to enhance performance. Moreover, the accuracy and generalization ability of the bearing temperature rise prediction model were evaluated by using prediction data generated under various operating conditions.

2. Temperature Rise Prediction Model Training Set Data Sources

Given the limited load conditions that can be applied to the experimental equipment, a hybrid dataset comprising both simulation and experimental data is necessary to acquire more diversified information, encompassing various operating conditions and scenarios. This approach aims to enhance the accuracy and generalization capability of the bearing temperature rise prediction model.

2.1. Experimental Data Sources

The experimental equipment, which includes an infrared thermal imager, bearing axial loading device, and PC-based temperature data acquisition and analysis software, is shown in Figure 1. The experimental bearing was arranged with the end face facing upwards, and the loading was applied as an axial load. The axial force generated by the loading mechanism was first applied to the bearing housing cover and then transferred from the bearing housing cover to the test bearing.

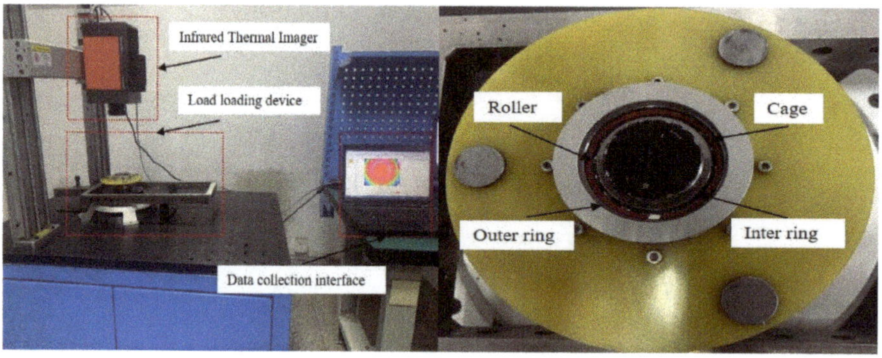

Figure 1. Angular contact ball bearing temperature rise experiment equipment.

The infrared thermal imager is an ImagelR8355 type; its main technical indicators are temperature measurement range −10~+175 °C, thermal image sampling frequency 10~110 Hz, measurement accuracy ±0.5 °C, thermal sensitivity 20 mK, and infrared image resolution 640 × 512. The rotational speed range of the experimental machine is 0 to 6000 rpm, and the rotational speed control function can be achieved through control software. The axial loading device applies loads ranging from 0 to 30 N. The experimental machine is capable of real-time monitoring of test parameters such as inner ring and cage rotation speed, axial loading load, and contact area temperature between the rolling elements and the inner/outer rings of the experimental bearing and can provide real-time feedback to the display.

The bearing temperature rise experiment process is as follows:

1. Preliminary preparation: Assess the condition of the spindle drive device, axial loading device, infrared thermal imager, and other equipment to ensure the safety, reliability, and clear image display in the experimental process.
2. Experimental bearing installation: Identify the type of the target bearing for the experiment and proceed with the installation of the experimental bearing.
3. Determine the test condition: Establish the preload axial force and motor rotational speed based on the specific objectives of the experiment.
4. Data acquisition: Adjust the parameters, such as the emissivity of the infrared thermal imager, set the sampling frequency, and complete the experimental data acquisition.

To ensure the reliability and stability of the experimental results, as well as to eliminate potential chance factors and errors, it was necessary to verify the universal applicability of the obtained data. The experiment was repeated three times at 5000 rpm and with a load of 30 N. The errors between the test results were within 0.5 °C, indicating the reliability of the experiment data.

Dynamic load rating of the bearing C = 200 N was performed according to ISO 281 (C/P) with different axial loads (10 N, 20 N, and 30 N), i.e., load factor C/P (20, 10, and 6.67), combined with various rotational speeds (2000 rpm, 3000 rpm, 4000 rpm, 5000 rpm, and 6000 rpm). Temperature data were collected from the surface of the inner ring of the bearing. Each operating condition was sampled every minute for a duration of one hour.

2.2. Simulation Data Sources

In this paper, a training set was constructed using both simulation and experimental data to predict the temperature rise process of an H7006C angular contact ball bearing under various operating conditions. The main structural parameters of the model object were as follows: inner race diameter (d) of 30 mm, outer race diameter (D) of 55 mm, width (B) of 13 mm, rolling diameter (Dw) of 5.556 mm, number of rolling elements (Z) of 16, contact angle (α) of 15 °, and grease lubrication as the lubrication method. This approach aims to enhance the accuracy and generalization capability of the bearing temperature rise prediction model.

(1) Simulation model

Based on the structural parameters described in the previous section, a three-dimensional model of the H7006C angular contact ball bearing was established, as shown in Figure 2. The material parameters for this bearing are listed in Table 1.

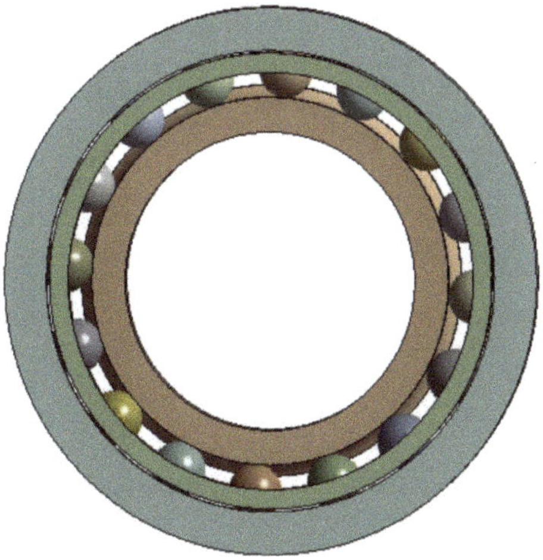

Figure 2. H7006C bearing three-dimensional model.

Table 1. H7006C material parameters.

Parameters	GCr15 (Rings)	Si3N4 (Rollers)	Pi (Cage)
Density	7800	3200	1120
Modulus of elasticity	208	300	300
Poisson's ratio	0.3	0.26	0.34
Thermal conductivity	40	11	0.15
Specific heat capacity	450	800	1250

Note: Pi represents polyimide, the material utilized for the bearing cage in this study.

The transient temperature field simulation process was conducted as follows:

1. Model building: Establish a 3D simulation model based on bearing geometry information and material parameters.
2. Heat generation calculation: Determine the test conditions and each bearing component's heat generation.
3. Pre-processing: Given the boundary conditions, such as heat convection, heat flow, etc., set the time step and initial temperature.
4. Post-processing: Start the transient temperature field simulation, save, and analyze the simulation results.

(2) Validation

To ensure the accuracy of the simulation data for the bearing temperature rise process in this paper, the experimental results were compared under two operating conditions: Operating Condition 1 (2000 rpm, 30 N) and Operating Condition 2 (6000 rpm, 20 N). The comparison results are illustrated in Figure 3.

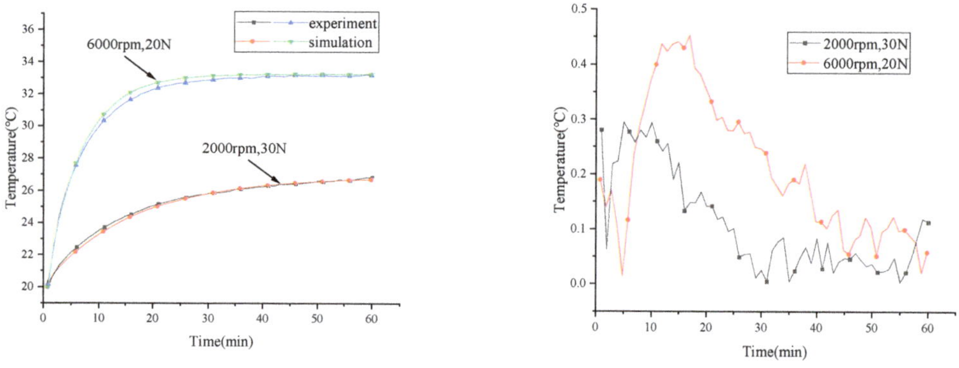

(a) Comparison of results **(b)** Error variation

Figure 3. Simulation results verification.

Figure 3 demonstrates that both operating conditions reached a steady-state stage within 30 min, with a temperature change of less than 1 °C. The simulation results align with the experimental results in terms of the temperature rise trend. The slight error in the temperature rise process may be attributed to the neglect of frictional power loss from the cage when constructing the thermal simulation model of the angular contact ball bearing. As a result, the raw heat input to the model was slightly lower than the actual value. However, the temperature error at steady-state was below 0.3 °C in both cases. Generally, a deviation of less than 1 °C under steady-state conditions is considered negligible for the dimensional changes of ball-bearing components. Therefore, the simulation data utilized in this paper can be employed as the training set for the prediction model, enabling the prediction of the bearing temperature rise process.

Dynamic load rating of the bearing C = 200 N was performed according to ISO 281 (C/P) with different axial loads (40 N, 50 N, and 60 N), i.e., load factor C/P (5, 4, and 3.34), combined with various rotational speeds (2000 rpm, 3000 rpm, 4000 rpm, 5000 rpm, and 6000 rpm). Each operating condition was sampled every minute for a duration of one hour in the simulation.

The experimental and simulation data consisted of a total of 30 operating conditions. Among these, the test set was composed of different load factors C/P (20, 10, 6.67, 5, 4, and 3.34) corresponding to a rotational speed of 6000 rpm, and different rotational speeds (2000 rpm, 3000 rpm, 4000 rpm, 5000 rpm, and 6000 rpm) corresponding to a load of 60 N. This test set was specifically utilized for evaluating the performance of the model, while the remaining data were used for pre-training the model.

3. Bearing Temperature Rise Prediction Based on CNN and Informer Combination Method

3.1. Convolutional Neural Networks

Compared to traditional feature extraction methods, CNN usage offers a more effective approach in extracting relevant information from data. In the temperature rise prediction model described in this paper, we employed convolution by taking nested sequence features as input. The key objective of the convolution process involves extracting input feature values through the utilization of a convolution kernel, specifically creating an $M \times N$ array.

By sliding the kernel with a specific stride, a local matrix is obtained. The convolution kernel transforms this local matrix, resulting in the output matrix for the convolution layer. This convolution process discards certain eigenvalue points while preserving the order relationship among the original input matrix features. Ultimately, it reduces the computational burden on the neural network. The process of convolution operation is as follows:

$$x_l^i = f(x_{l-1}^r \times K_l^i + b_l^i) = f(\sum_r x_{l-1}^r \times K_{l,r}^i + b_l^i) \tag{1}$$

where x_l^i is the feature map output by the i-th convolutional layer l, x_{l-1}^i is the r-th convolution region of the feature map generated by convolution layer $l-1$, K_l^i is the weight matrix of the i-th convolution kernel of convolution layer l, and b_l^i is the bias value.

Typically, a pooling layer is added after each convolutional layer in order to generate a lower-dimensional feature map and decrease computational complexity. The maximum pooling layer takes the maximum value of features within a certain region as the output, achieving further feature extraction.

3.2. *Informer Model*

The Informer model is a network architecture built upon an attention mechanism. It primarily enhances the computational efficiency of several components, including the self-attention mechanism, stacked layers of the network, and the incremental decoding method.

The model consists of two parts, an encoder and a decoder, which accept different input data. The encoder is responsible for receiving long sequence data as input. To replace the traditional self-attentive mechanism, it employs a sparse self-attentive mechanism. This modification effectively reduces the network size and enhances the model's robustness when multiple layers are stacked. On the other hand, the decoder receives input data in the form of long sequences. It fills the target elements with zeros and utilizes these all-zero sequences as part of the weighted attention for the feature map. Subsequently, it utilizes a generative approach to predict the sequences. The calculation of the sparse self-attention mechanism is as follows:

$$A(Q, K, V) = Softmax(\frac{\overline{Q}K^T}{\sqrt{D_K}})V \tag{2}$$

where $\overline{Q}$ is the matrix obtained by probabilistic sparse of Q and *Softmax* is the normalized activation function.

The encoder employs a "distillation" operation to prioritize the prominent high-level features and generate a concentrated self-attentive feature map in the lower layer, thereby reducing the input length. The "distillation" operation from layer j to layer $j+1$ at time t can be described as follows:

$$X_{j+1}^t = MaxPool(ELU(Conv1d\left[X_j^t\right]_{AB})) \tag{3}$$

where $[*]_{AB}$ basic operations contain attention blocks and sparse attention mechanisms; *Conv1d* denotes the one-dimensional convolution operation; *ELU* is the activation function; and *MaxPool* is the maximum pooling operation.

The decoder design aims to generate long sequence predictions through a single forward process. The model adopts a traditional Decoder structure, which includes two identical multi-headed attention layers, in order to address the issue of high time complexity involved in generative prediction for long sequential data. The input vector of the decoder is represented as follows:

$$X_{de}^t = Concat(X_{token}^t, X_0^t) \in R^{(L_{token}+L_y)d_{model}} \tag{4}$$

where X^t_{token} is the start token, and X^t_0 is a placeholder for the target sequence, which is set to 0.

3.3. CNN + Informer Bearing Temperature Rise Prediction Model

A CNN and Informer fusion approach was proposed to enhance the accuracy of predicting the bearing temperature rise process. The constructed dataset was utilized as input for the CNN model, employing the same parameters across different regions to calculate convolutional features. This technique reduces the number of parameters in the training process, mitigating the risk of overfitting. Subsequently, through multi-level convolution and pooling operations, the model extracts multi-scale features from the data. These extracted features are then fed into the Informer model via a fully connected layer. The Informer network model is iteratively trained to achieve precise prediction of bearing temperature rise. The structure of the prediction model for bearing temperature rise based on the CNN + Informer fusion is depicted in Figure 4.

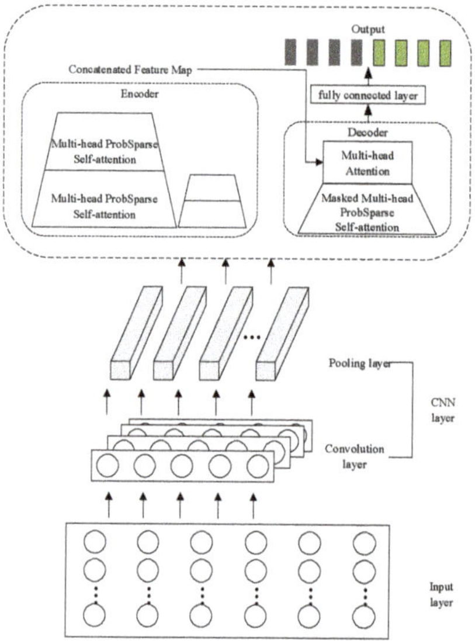

Figure 4. Basic structure of CNN + Informer network model.

The combined prediction model consists of two main parts: two-dimensional convolutional feature extraction and prediction. Initially, the input features of the variables are extracted using two-dimensional convolution, constructing a high-dimensional mapping feature vector. To enhance the feature extraction capability, this study adopted two layers of two-dimensional convolutional layers. The output of the convolutional module is received by both the encoder and decoder parts of the Informer model. The encoder handles long time series data, while the decoder processes short sequences and vectors of equal length comprising zero values as placeholders for predicted values. As the data pass through the encoder, intermediate results are generated through the multi-head sparse self-attentive module and the "distillation" mechanism module, which are computed multiple times in succession. The decoder takes the encoded input data and performs a multi-head sparse self-attentiveness operation with a mask. Subsequently, the intermediate result from the encoder undergoes a multi-head self-attentiveness operation. Finally, a fully connected layer adjusts the dimensionality of the output data and produces the prediction results.

The mean absolute error (*MAE*) and mean squared error (MSE) are common evaluation metrics for model performance. *MAE* represents the average error between predicted and true values, calculated as the average of the absolute differences. A smaller *MAE* indicates a higher accuracy. On the other hand, MSE represents the average of the squared differences between predicted and true values, which assesses the effectiveness of the model. A smaller MSE signifies better performance. In the case of the prediction model for the overall temperature rise process in different bearing operating conditions, the focus lies on the differences between predicted and actual values at each time point. Therefore, this paper utilized *MAE* as the evaluation metric for accuracy. The specific formula for calculating *MAE* is as follows:

$$MAE = \frac{1}{n}\sum_{i=1}^{n}|y_i - y_i^*| \quad (5)$$

where n is the number of sample data points; y_i is the true value; and y_i^* is the predicted value.

4. Results and Discussion

4.1. Parameter Optimization

The experiments were conducted on a Windows 10 system, equipped with an Intel Core i5 processor and an NVIDIA GeForce 930 graphics card.

In the temperature rise prediction model using CNN + Informer, there are several hyperparameters that need to be set separately. In this paper, the parameter selection range is shown in Table 2.

Table 2. Model parameters.

Parameter Name	Parameter Value
Number of convolution layers	2
Convolution kernel size	$3 \times 1/2 \times 1$
Number of convolution kernels	1~10/1~20
Number of encoder layers	3, 4, 6
Number of decoder layers	2
Head number of multi-head attention	8, 16

First, the number of encoder layers and the head number of the multi-head attention in Informer were selected as 2 and 8, respectively, to determine the convolution kernel parameters. The convolution kernels for convolution layer 1 range from 1 to 10, and the convolution kernels for convolution layer 2 range from 1 to 20, and partial error data are shown in Table 3.

Table 3. Comparison of partial errors with different numbers of convolution kernels.

Model	Convolutional Layer 1 Number of Convolutional Kernels	Convolutional Layer 2 Number of Convolutional Kernels	MAE/°C
CNN	1	20	0.6263
	3	1	0.6125
	1	2	0.6005
	2	15	0.5963
	4	1	0.4775

Table 3 displays combinations of convolutional kernel counts with relatively smaller errors, and it can be observed that our proposed model achieved the minimum testing error on the dataset when the numbers of convolutional kernels were set to 4 and 1. After determining the convolution kernel parameters, the head number of the multi-head attention and the number of encoder layers for the Informer model were set accordingly. The resulting error outcomes are illustrated in Figure 5.

Based on the findings in Figure 5, when the head number of the multi-head attention was set to 16, there were smaller errors across different encoder layers compared to when it was set to 8. Notably, the model's error was minimized when the number of encoder layers was specified as 4. Consequently, the structure of the temperature rise prediction model using the CNN + Informer approach was determined, and the specific parameters are outlined in Table 4.

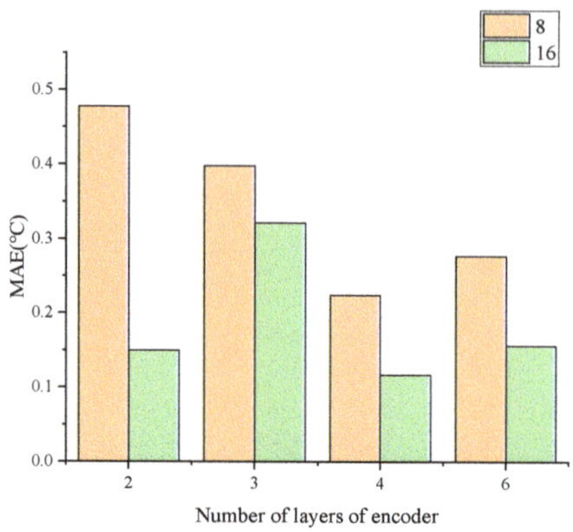

Figure 5. Prediction errors for different Informer parameters.

Table 4. CNN + Informer model parameters.

Model	Parameter Name	Parameter Value
CNN	Number of convolution layers	2
	Convolution kernel size	$3 \times 1/2 \times 1$
	Number of convolution kernels	4/1
Informer	Number of encoder layers	4
	Number of decoder layers	2
	Head number of multi-head attention	16

4.2. Comparison and Analysis of Prediction Results

Based on the previous analysis, it was observed that the bearing temperature, under different working conditions, generally reached a steady-state at around the 30 min mark. To evaluate the prediction capability of the proposed CNN + Informer model, we constructed models using LSTM, Informer, and CNN + Informer. The errors during both the temperature rise stage and the steady-state stage were compared among these models.

4.2.1. Model Prediction Results with Varying Rotational Speeds

The predictions of the multiple models under a rotational speed of 5000 rpm and a load of 60 N are presented in Figure 6. Furthermore, Figure 7 showcases the variations in prediction errors throughout the temperature rise and steady-state stages at various rotational speeds (2000 rpm, 3000 rpm, 4000 rpm, 5000 rpm, and 6000 rpm).

Based on the comparison of the model prediction results presented in Figure 6a, it can be observed that both the Informer and LSTM models exhibited varying degrees of fluctuation in their prediction results. In contrast, the model used in this paper demonstrated a consistent upward trend that closely aligned with the actual values, resembling the real curve more closely. Furthermore, Figure 6b illustrates a comparison of the prediction errors. It is evident that the maximum error of the model proposed in this paper remained within 0.5 °C, whereas the maximum error of the other two models exceeded 1 °C.

From Figure 7, it is evident that the model used in this paper exhibited lower prediction errors compared to the other two models at the various rotational speeds. The average absolute error of the predictions made by this model consistently remained below 0.2 °C. Furthermore, there is no discernible trend in the change of prediction errors as the rotational speed increased. This indicates that the model demonstrates good stability and is capable of accurately predicting temperature rise under varying rotational speeds.

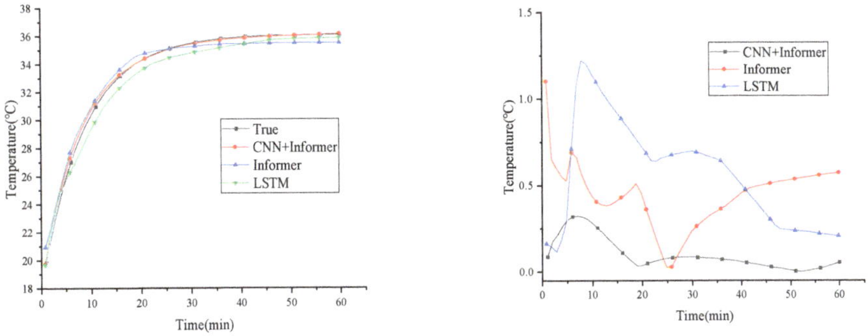

(**a**) Comparison of results (**b**) Comparison of errors

Figure 6. Model prediction results at 5000 rpm rotational speed and 60 N load.

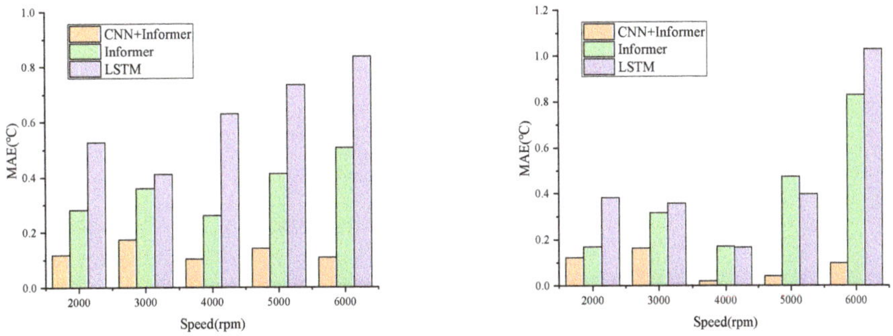

(**a**) Temperature rise stage (**b**) Steady-state stage

Figure 7. Prediction errors of each model at varying rotational speeds with a load of 60 N.

4.2.2. Model Prediction Results with Varying Load

When the bearing rotational speed is 6000 rpm, Figure 8 depicts the prediction results of the multiple models for a load of 50 N. Meanwhile, Figure 9 demonstrates the prediction errors during both the temperature rise phase and steady-state phase under various loads (10 N, 20 N, 30 N, 40 N, and 50 N).

The comparison of the prediction results in Figure 8a reveals a gradual deviation between the predictions of the Informer and LSTM models from the actual values over time. Figure 8b shows the error between predictions from Informer and LSTM models and the actual values gradually grows over time. In contrast, the model utilized in this paper consistently maintained an error within 0.5 °C without any discernible trend over time.

The prediction errors of the model under different loads in Figure 9 indicate that its error consistently remained within 0.5 °C, which is lower compared to the other two models. Moreover, as the load increased, there is no clear upward trend observed in the prediction errors of this model during both the temperature rise and steady-state stages. This suggests that the model exhibits high prediction accuracy and good stability.

4.2.3. Experimental Data Prediction Results and Analysis

Figure 10 depicts the evaluation of the model using the predicted results of the experimental data. In Figure 10a, it is evident that the error in the model's prediction results was more pronounced during the temperature rise stage. However, it can be observed that the error gradually decreased over time. Additionally, Figure 10b displays the error under various working conditions, indicating that the maximum error remained within 0.2 °C. This demonstrates the high accuracy of the model and verifies its prediction capability and reliability as presented in this paper.

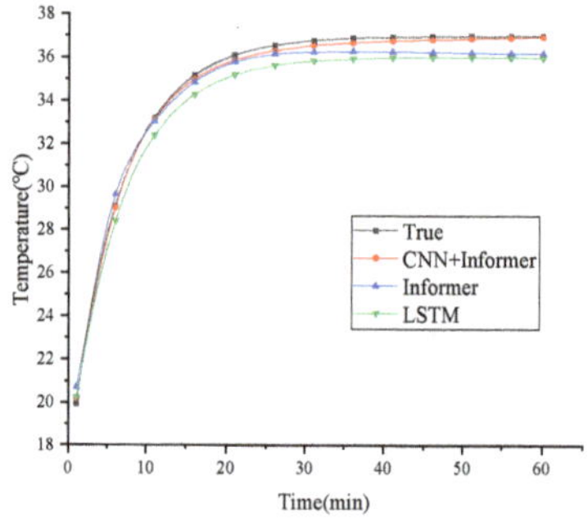

(**a**) Comparison of results

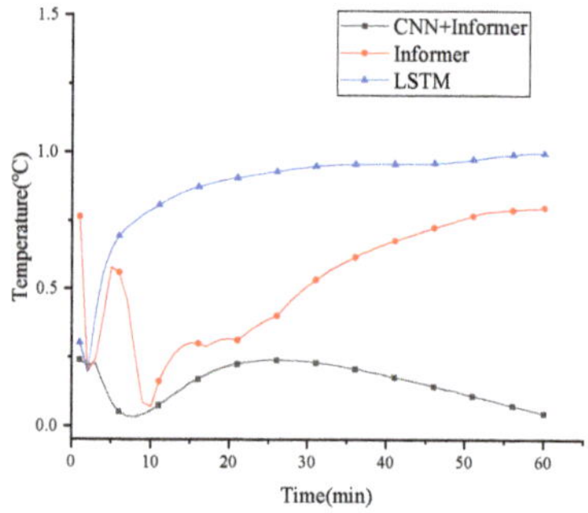

(**b**) Comparison of errors

Figure 8. Model prediction results at 6000 rpm and 50 N load.

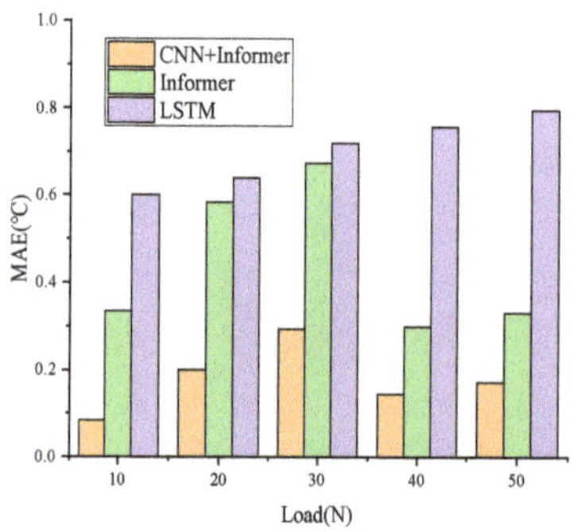

(**a**) Temperature rise stage

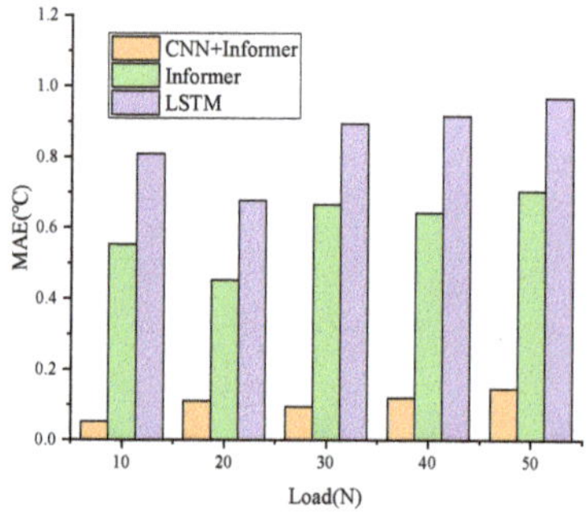

(**b**) Steady-state stage

Figure 9. Prediction errors of each model at varying load with a rotational speed of 6000 rpm.

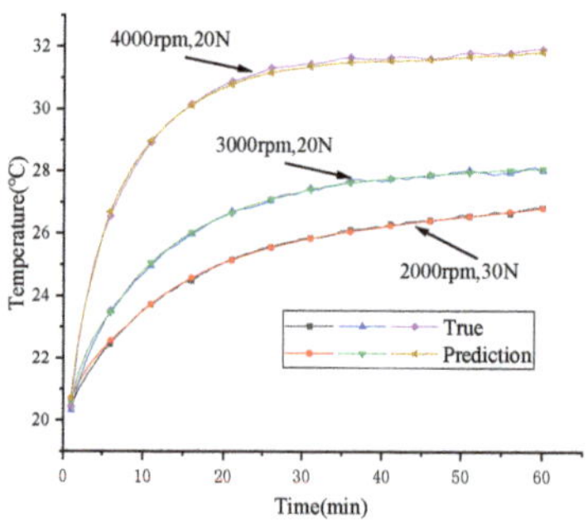

(**a**) Comparison of results

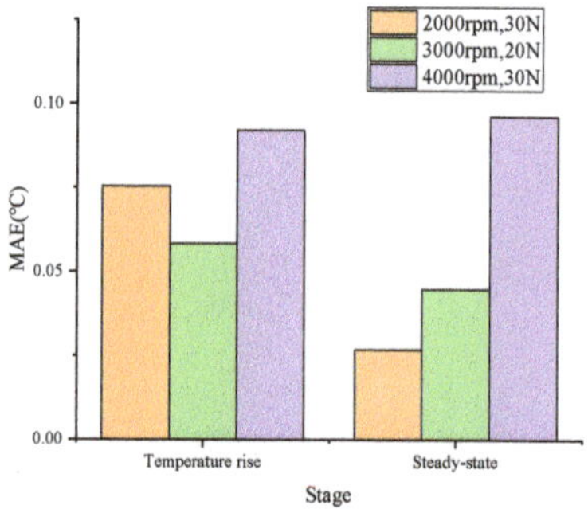

(**b**) Comparison of errors

Figure 10. Experimental data prediction results and errors.

4.3. Generalization Ability Experiment

To evaluate the generalization ability of the model, its prediction results were observed and evaluated under various operating conditions that were not included in the original dataset. The dataset includes simulation data under different room temperature conditions. As depicted in Figure 11, the model demonstrated good prediction performance when tested with new conditions through both empirical testing and simulations. The predicted curves gradually approached the true curve, indicating accurate predictions. When reaching the steady state, the deviation between the prediction curve and the true value was less than 0.5 °C, and the error during the steady-state stage was lower compared to that during the temperature rise stage. This could be attributed to the stabilization of the heat generation and dissipation processes during the steady-state stage, whereby this equilibrium state contributes to a reduction in prediction errors.

The results of the prediction and error analysis under different operating conditions demonstrated the accuracy of the temperature rise prediction model used in this paper for various rotational speed and load conditions. The proposed model exhibited lower errors compared to two other models, both during the temperature rise stage and the steady-state stage, confirming its validity and performance. Moreover, as the rotational speed and load increased, there is no noticeable trend in the prediction error, indicating the excellent stability of the model. Furthermore, the model demonstrated favorable prediction results for operating conditions not included in the original dataset, providing further evidence of its generalization ability. In summary, our research demonstrated that the predictive model exhibits a consistent trend with the true values, and there is no delay issue in its predictions. This enables real-time bearing condition monitoring, facilitating proactive maintenance strategies and ensuring efficient and reliable operation of rotating machinery.

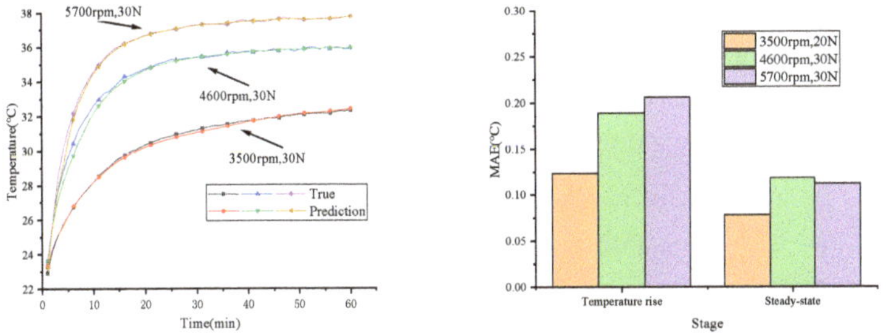

(**a**) Experimental data

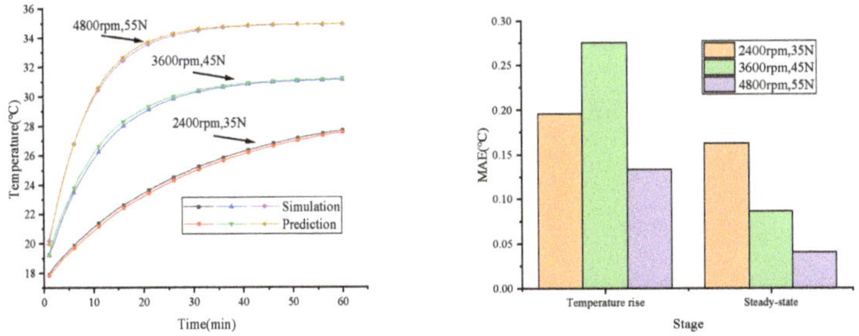

(**b**) Simulation data

Figure 11. Prediction results and errors of working conditions outside the dataset.

5. Conclusions

This study proposed a temperature rise prediction method based on deep learning models for spindle bearings in the high-speed operating conditions of machine tools. The method was developed to address challenges associated with timely assessment of bearing conditions due to rapid temperature changes. The specific conclusions are as follows:

(1) The proposed simulation model for the temperature field of angular contact ball bearings demonstrated an error of less than 0.5 °C when compared to the experimental results. This finding suggests that the simulation data exhibit high reliability and fulfill the requirements for training samples in the prediction model.

(2) Compared to LSTM and Informer, the optimized CNN + Informer model achieved higher accuracy and prediction stability. It achieved errors within 0.5 °C for multiple operating conditions and showed no variation trend with changing operating conditions.

(3) For operating conditions outside the dataset, the model predicted errors within 0.5 °C and 0.2 °C during the temperature rise and steady-state stages, respectively. This suggests that the prediction error decreases over time, providing further evidence for the model's generalization ability and effectiveness.

Author Contributions: Conceptualization, H.L. and C.L.; methodology, H.L.; software, H.L. and C.L.; validation, H.L. and F.Y.; formal analysis, X.S.; writing—original draft preparation, H.L.; writing—review and editing, H.L., C.L. and N.G.; supervision, X.M. and X.W.; project administration, N.G. and X.M.; funding acquisition, F.Y. All authors have read and agreed to the published version of the manuscript.

Funding: This research was funded by the National Key R&D Program of China (No. 2021YFB2011000, No. 2020YFB2009602); Major Science and Technology Projects of Longmen Laboratory (No. 231100220 500); Major Science and Technology Project of Henan Province (No. 221100220100); and Natural Science Foundation of Henan Province of China (No. 232300421336).

Institutional Review Board Statement: Not applicable.

Informed Consent Statement: Not applicable.

Data Availability Statement: All data are contained within the article.

Conflicts of Interest: The authors declare no conflict of interest.

References

1. Ma, S.; Yin, Y.; Chao, B.; Yan, K.; Fang, B.; Hong, J. A Real-time Coupling Model of Bearing-Rotor System Based on Semi-flexible Body Element. *Int. J. Mech. Sci.* **2023**, *245*, 108098. [CrossRef]
2. Fang, B.; Zhang, J.; Hong, J.; Yan, K. Research on the nonlinear stiffness characteristics of double-row angular contact ball bearings under different working conditions. *Lubricants* **2023**, *11*, 44. [CrossRef]
3. Popescu, A.; Houpert, L.; Olaru, D.N. Four approaches for calculating power losses in an angular contact ball bearing. *Mech. Mach. Theory* **2020**, *144*, 103669. [CrossRef]
4. Kim, K.-S.; Lee, D.-W.; Lee, S.-M.; Hwang, J.-H. A numerical approach to determine the frictional torque and temperature of an angular contact ball bearing in a spindle system. *Int. J. Precis. Eng. Manuf.* **2015**, *16*, 135–142. [CrossRef]
5. Xu, J.; Zhang, J.; Huang, Z.; Wang, L. Calculation and finite element analysis of the temperature field for high-speed rail bearing based on vibrational characteristics. *J. Vibroeng.* **2015**, *17*, 720–732.
6. Deng, X.; Fu, J.; Zhang, Y. A predictive model for temperature rise of spindle–bearing integrated system. *J. Manuf. Sci. Eng.* **2015**, *137*, 021014. [CrossRef]
7. Wu, L.; Tan, Q. Thermal characteristic analysis and experimental study of a spindle-bearing system. *Entropy* **2016**, *18*, 271. [CrossRef]
8. Zheng, D.; Chen, W.F. Effect of structure and assembly constraints on temperature of high-speed angular contact ball bearings with thermal network method. *Mech. Syst. Signal Process.* **2020**, *145*, 106929. [CrossRef]
9. Li, Z.; Zhao, C.; Lu, Z.; Liu, F. Thermal Performances Prediction Analysis of High Speed Feed Shaft Bearings Under Actual Working Condition. *IEEE Access* **2019**, *7*, 168011–168019. [CrossRef]
10. Zhang, C.; Guo, D.; Tian, J.; Niu, Q. Research on the influencing factors of thermal characteristics of high-speed grease lubricated angular contact ball bearing. *Adv. Mech. Eng.* **2021**, *13*, 16878140211027398. [CrossRef]

11. Yan, G.; Yu, C.; Bai, Y. Wind turbine bearing temperature forecasting using a new data-driven ensemble approach. *Machines* **2021**, *9*, 248. [CrossRef]
12. Liu, Y.Z.; Zou, Y.S.; Wu, Y. A novel abnormal detection method for bearing temperature based on spatiotemporal fusion. *Proc. Inst. Mech. Eng. Part F J. Rail Rapid Transit* **2022**, *236*, 317–333. [CrossRef]
13. Chen, Y.; Zhang, C.; Zhang, N.; Chen, Y. Multi-task learning and attention mechanism based long short-term memory for temperature prediction of EMU bearing. In Proceedings of the 2019 Prognostics and System Health Management Conference (PHM-Qingdao), Qingdao, China, 25–27 October 2019; pp. 1–7.
14. Xiao, X.; Liu, J.; Liu, D.; Tang, Y.; Dai, J.; Zhang, F. SSAE—MLP: Stacked sparse autoencoders-based multi-layer perceptron for main bearing temperature prediction of large-scale wind turbines. *Concurr. Comput. Pract. Exp.* **2021**, *33*, e6315. [CrossRef]
15. Ariyo, A.A.; Adewumi, A.O.; Ayo, C.K. Stock price prediction using the ARIMA model. In Proceedings of the 2014 UKSim-AMSS 16th International Conference on Computer Modelling and Simulation, Cambridge, UK, 26–28 March 2014; pp. 106–112.
16. Zhang, Y.; Sun, H.; Guo, Y. Wind power prediction based on PSO-SVR and grey combination model. *IEEE Access* **2019**, *7*, 136254–136267. [CrossRef]
17. Cui, K.; Jing, X. Research on prediction model of geotechnical parameters based on BP neural network. *Neural Comput. Appl.* **2019**, *31*, 8205–8215. [CrossRef]
18. Lukoševičius, M.; Jaeger, H. Reservoir computing approaches to recurrent neural network training. *Comput. Sci. Rev.* **2009**, *3*, 127–149. [CrossRef]
19. Yu, Y.; Si, X.; Hu, C.; Zhang, J. A review of recurrent neural networks: LSTM cells and network architectures. *Neural Comput.* **2019**, *31*, 1235–1270. [CrossRef]
20. Fan, C.; Zhang, Y.; Pan, Y.; Li, X.; Zhang, C.; Yuan, R.; Wu, D.; Wang, W.; Pei, J.; Huang, H. Multi-horizon time series forecasting with temporal attention learning. In Proceedings of the 25th ACM SIGKDD International Conference on Knowledge Discovery & Data Mining, Anchorage, AK, USA, 4–8 August 2019; pp. 2527–2535.
21. Zhou, H.; Zhang, S.; Peng, J.; Zhang, S.; Li, J.; Xiong, H.; Zhang, W. Informer: Beyond efficient transformer for long sequence time-series forecasting. In Proceedings of the AAAI Conference on Artificial Intelligence, Virtual, 2–9 February 2021; Volume 35, pp. 11106–11115.
22. Gong, M.; Zhao, Y.; Sun, J.; Han, C.; Sun, G.; Yan, B. Load forecasting of district heating system based on Informer. *Energy* **2022**, *253*, 124179. [CrossRef]
23. Yang, Z.; Liu, L.; Li, N.; Tian, J. Time series forecasting of motor bearing vibration based on informer. *Sensors* **2022**, *22*, 5858. [CrossRef] [PubMed]

lubricants

MDPI

Article

Numerical Investigation of Bump Foil Configurations Effect on Gas Foil Thrust Bearing Performance Based on a Thermos-Elastic-Hydrodynamic Model

Bin Hu [1], Anping Hou [1,*], Rui Deng [1], Rui Wang [1], Zhiyong Wu [2], Qifeng Ni [3] and Zhong Li [4]

1 School of Energy and Power Engineering, Beihang University, Beijing 100191, China; bh1204hb@163.com (B.H.); drds12138@163.com (R.D.); wangrbuaa@126.com (R.W.)
2 The Key Laboratory of Solar Thermal Energy and Photovoltaic System, IEE-CAS, Beijing 100190, China; wuzhiyong@mail.iee.ac.cn
3 Ningbo Hudu Energy Technology Co., Ltd., Ningbo 315000, China; nqf888@163.com
4 Zhengzhou Aerotropolis Institute of Artificial Intelligence, Zhengzhou 451162, China; 17888821559@163.com
* Correspondence: houap@buaa.edu.cn

Abstract: The performance of gas foil thrust bearings is critical to the successful design and operation of the high axial load rotatory machines that employ gas foil bearings. However, our understanding of gas foil thrust bearings remains incomplete. To enhance our understanding and predict the performance of gas foil thrust bearings, we have established a detailed three-dimensional thermo-elastic-hydrodynamic model of a gas foil thrust bearing based on a fluid-thermal-structure interaction approach in this study. To validate the accuracy of our model, a gas foil thrust bearing test rig was developed. Moreover, we present a numerical investigation of the influence of bump foil configurations on gas foil thrust bearing performance. The results show that the gas foil thrust bearing that fixes the bump foil at the trailing edge and splits the bump foil into several strips exhibits a 36.4% increase in load capacity compared to the gas foil thrust bearing that fixes a whole piece of bump foil at the leading edge. Fixing the bump foil at the trailing edge and splitting it into several strips effectively decreases power loss and reduces the risk of bearing thermal failure.

Keywords: thermo-elastic-hydrodynamic; gas foil thrust bearing; load capacity; thermal characteristic

Citation: Hu, B.; Hou, A.; Deng, R.; Wang, R.; Wu, Z.; Ni, Q.; Li, Z. Numerical Investigation of Bump Foil Configurations Effect on Gas Foil Thrust Bearing Performance Based on a Thermos-Elastic-Hydrodynamic Model. *Lubricants* **2023**, *11*, 417. https://doi.org/10.3390/lubricants11100417

Received: 4 September 2023
Revised: 19 September 2023
Accepted: 21 September 2023
Published: 22 September 2023

1. Introduction

With the increasing demands of high-speed rotating machinery, the application of conventional oil-lubricated bearings is increasingly limited. In comparison, gas foil bearings (GFBs) utilize ambient air as a lubricant, offering numerous advantages. GFBs require minimal maintenance, can operate at high speeds and temperatures [1], and have been reported to have a DN (shaft diameter in millimeters multiplied by shaft rotational speed in rev/min) limit of about 4.56×10^6 [2,3]. As a result, GFBs have gained significant popularity in recent years and are widely utilized in various high-speed rotating shaft systems, including air cycle machines, turboexpanders, compressors, and small microturbine systems [4–7]. After decades of development, GFBs have made great progress in design and application. However, the understanding of gas foil thrust bearing (GFTB) supporting axial load remains incomplete, as most of the interest and development has been attracted to gas foil journal bearing (GFJB) supporting radial load and controlling the rotor orbit [8]. Then, the purpose of the current study is to promote the research of GFTB.

Heshmat et al. [9] conducted numerical research on GFTB performance and obtained the optimum geometry through parametric studies of structural variables. However, they simplified the bump structure as a uniform elastic foundation and overlooked the coulomb friction and the effect of the top foil. Iordanoff [10] developed a more comprehensive GFTB model that considered the coulomb friction of the bump structure in order to predict GFTB performance. Experimental research was also conducted on the load capacity of GFTBs. The results indicated that free-fixed-end bumps had a higher stiffness coefficient and the predictions for heavy loads agreed well with experimental data. However, the stiffness distribution between the welded bump and the free bump was simplified as the linear distribution, and the effect of the top foil was still not considered. Heshmat et al. [11] conducted numerical investigations on the load performance of GFTBs by coupling Finite Element and Finite Difference methods. The top foil was regarded as a part of bearing compliance; it was proved that it had an essential effect on the performance of GFTB. However, the complex bump structure was assumed as an independent spring support. Park et al. [12] also coupled Finite Element and Finite Difference methods to predict the static and dynamic performance of tilted GFTB. To improve the accuracy of the bump structure, the interaction between bumps was considered. However, the effect of the top foil was ignored. Gad et al. [13] introduced a more comprehensive structural stiffness model for the foil structure of generation II GFTB. The model considered not only the interaction between bumps, but also the possibility of the rigid bearing surface and the flat segment between bumps. The effect of two different bump strip arrangements on the performance of generation II GFTB was investigated. However, the deflection of the top foil relative to bumps was simplified and the top foil sagging was neglected. Xu et al. [14] focused on the effect of top foil sagging on GFTB performance. The simulation results revealed that large top foil sagging caused high power loss and small minimum film thickness. The thicker top foil was proved to decrease the effect of top foil sagging and achieved better performance of GFTB. However, each bump was simplified, as many springs attached to the top foil computational node and the interaction of bumps was ignored. In order to analyze arbitrarily shaped foils, Xu et al. [15] presented a comprehensive model of GFTBs that considered both bump interaction and top foil sagging. They adopted the assumption of isothermal ideal gas in the compressible Reynolds equation. The numerical results demonstrated that the configuration of the bump foils significantly influenced the load capacity of GFTBs. However, the aforementioned models assumed a constant temperature, resulting in inaccurate predictions of GFTB performance.

To investigate the temperature effect on GFTB performance, various simulated thermal field models were developed. Bruckner [16] presented a simple thermohydrodynamic (THD) model of GFTB to study the thermal behavior. However, the thrust runner and foil structure were lumped together and treated as a single heat transfer mechanism, and the foil structure was considered as a compliant foundation with a linear stiffness distribution. Lee et al. [17] developed a three-dimensional THD model of GFTB with radially arranged bump foils. The temperature distribution of the thrust runner, top foil, bump foils, and other bearing parts was investigated using the heat balance equation with surroundings. However, each individual bump was simplified as a lumped thermal resistance in the temperature field and an independent support spring in the foil structural field. Gad et al. [18], on the other hand, utilized a 2D compressible Reynolds equation that considered the effects of centrifugal forces to model fluid flow. They employed the Couette Approximation to analyze the temperature distribution. The simulation results revealed that the side-leakage heat transfer had relatively poor effects. However, the inlet temperature was assumed to be constant due to the adoption of the Couette Approximation. Xu et al. [19] presented a 3D THD model that considered real gas effects and flow turbulence. They calculated and compared the performance of hydrostatic thrust foil bearings using two different types of lubricants. However, the temperature of each individual half bump arc was assumed to be consistent, as each half bump arc was simplified as a lumped thermal resistance in the temperature field. Lehn et al. [20] derived a comprehensive 3D thermo-elastic-hydrodynamic (TEHD) model for GFTB and a detailed shell model for the foil structure. They also adopted the assumption of a thermal resistance for a half bump arc. An exhaustive thermal analysis was conducted for GFTB and the thrust disk. The results demonstrated that the deformation of the thrust disk was the source of thermal runaway under high load and high-speed conditions. However, these aforementioned models used the 2D Reynolds equation to simulate the flow field and assumed constant fluid properties throughout the film thickness.

Most recently, the application of the fluid-thermal-structure interaction (FTSI) approach in TEHD has significantly improved the accuracy of predicting GFTB performance. This approach involves solving the full Navier–Stokes equations to model fluid flow, considering frictional contact, bump interaction, and top foil sagging in the foil structure model, and solving the 3D energy equation to investigate heat transfer. Liu et al. [21] utilized the FTSI approach to examine the impact of boundary pressure on GFTB performance and elucidate the heat flux transfer path. The same approach was employed to study the mechanism of thermal seizures and failure behavior in GFTB [22], demonstrating that an additional forced cooling flow can effectively prevent thermal failure. However, these studies focused only on the effects of boundary pressure and additional forced cooling flow on fluid-thermal-structure performance. Comprehensive parametric studies on GFTB performance are yet to be completed for obtaining more accurate results.

In this study, we present a comprehensive 3D TEHD model of GFTB based on the FTSI approach to investigate the impact of bump foil configurations on GFTB performance. Variation in fluid properties of gas film, bump interaction and top foil sagging are taken into consideration. The paper is organized as follows. Firstly, a description of the computational domains and boundary conditions used in the model is presented. Secondly, the experimental test and model validation is conducted to ensure its accuracy and reliability. Finally, load capacity and thermal characteristic of GFTBs with four different bump foil configurations are analyzed.

2. Numerical Method and Experimental Validation

2.1. Description of GFTB and Bump Foil Configurations

The general GFTB structure used in this study is shown in Figure 1a. The bearing consists of six pads, with a pad angle of 60° and a top foil angle of 45°. The remaining space is utilized for welding and accommodating manufacturing errors. The specific structure of GFTB is shown in Figure 1b. Each pad comprises a top foil, a single bump foil and a spacer. The spacer is responsible for controlling the ramp height of the GFTB. The top foil is securely attached to the spacer through spot welding and further connected to the back plate through spot welding as well. The bump foil is also spot-welded to the back plate. The installation position of the bump foil is determined through bearing design, and its geometry is based on the specifications outlined in [23]. The geometric details and material properties of the GFTB are shown in Figure 1c and summarized in Table 1. Due to the deformation of the foil structure, the film thickness varies across the flat surface of the bearing. For the sake of convenience in research, the nominal clearance is used to describe the bearing capacity. The nominal clearance is defined as the distance between the thrust collar surface and the top foil (undeformed).

Table 1. Dimensions and material properties of GFTB.

GFTB Parameters	Value
GFTB outer radium (mm), Ro	40
GFTB inner radium (mm), Ri	20
Pad angle (deg), α	60
Foil angle (deg), β	45
Ramp height (mm), hRamp	0.11
Ramp area ratio	0.32
Top foil thickness(mm), TF	0.10
Bump foil thickness (mm), tBump	0.10
Bump foil pitch (mm), SBump	3.17
Bump foil height (mm), hBump	0.51
Foil Poisson's ratio, ν	0.29
Foil Young's modulus, E	209 GPa

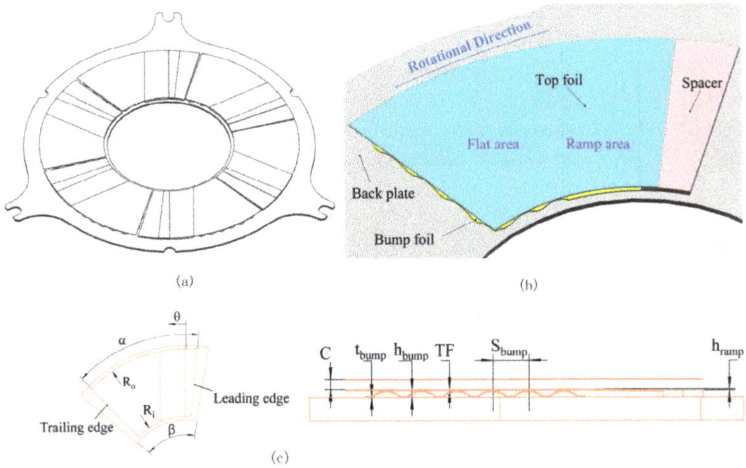

Figure 1. Schematic diagram of GFTB structure. (**a**) GFTB structure; (**b**) one GFTB pad; (**c**) variables describing of one GFTB pad.

In order to investigate the impact of bump foil configuration on GFTB performance, four different bump foils are designed and labeled as bearing 1 to 4, as illustrated in Figure 2. All four configurations share the same bump foil shape parameters, as detailed in Table 1. In bearing 1 and 3, the bump foil is fixed on the leading edge of the flat area, while in bearing 2 and 4, it is fixed on the trailing edge. Additionally, bearing 1 and 2 have a single-piece bump foil, whereas in bearing 3 and 4, the bump foil is divided into three independent, same-width strips.

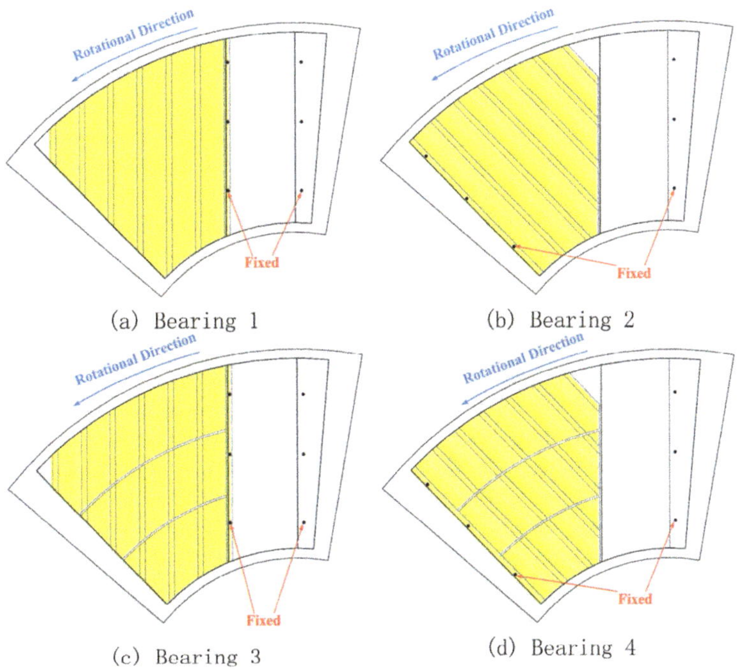

Figure 2. Four bump foil configurations of GFTB.

2.2. Computational Mode and Boundary Conditions

In this study, the commercial software ANSYS is utilized for simulation purposes. Figure 3 illustrates the information exchange process between the solvers of the TFSI approach. The calculation procedure is as follows: Firstly, the Computational Fluid Dynamics (CFD) solver calculates the hydrodynamic pressure and temperature data on the coupling surface. These data are then transferred to the structural model. Next, the Finite Element Method (FEM) solver evaluates the node displacement and heat flux of the foil structure. These results are sent back to the fluid model. Within each coupling iteration steps, both the CFD solver and FEM solver undergo multiple sub-iterations to achieve a convergence in their respective domains. Once the convergence requirements are met, the calculation proceeds to the next coupling iteration step. This process continues until the desired results are obtained.

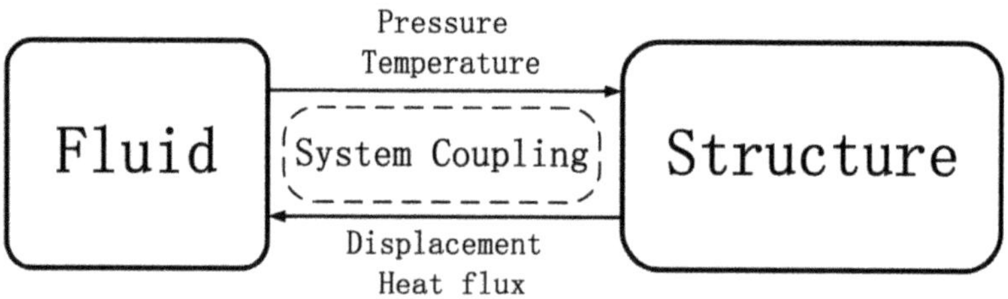

Figure 3. Computer solution flow of TFSI.

The governing equations are presented as follows. The continuity equation is written as:

$$\frac{\partial \rho}{\partial t} + \nabla \cdot (\rho U) = 0 \tag{1}$$

where the ρ and U are the density and velocity vector, respectively.

The momentum equation is written as:

$$\frac{\partial (\rho U)}{\partial t} + \rho U \cdot (\nabla U) = -\nabla P - \nabla \cdot \tau + F_e \tag{2}$$

where P is the pressure, τ is the stress tensor and F_e is the external force.

The energy equation in fluid domain is written as:

$$\frac{\partial (\rho c_p T)}{\partial t} + \nabla \cdot (\rho U c_p T) = -\nabla \cdot \left(\lambda_f \nabla T\right) + Q_e \tag{3}$$

where c_p is the specific heat capacity, T is the absolute temperature, λ_f is the fluid thermal conductivity and Q_e is the external heat source.

The energy equation in the structure domain is written as:

$$\nabla \cdot (\lambda_s \nabla T) = 0 \tag{4}$$

where λ_s is the structure thermal conductivity.

The fluid is treated as the ideal gas, and the physics of hydrodynamic film in the fluid domain is described by the Reynolds-averaged Navier–Stokes equations. Second-order upwind space discretization is employed for the spatial discretization, and second-order backward Euler discretization is applied for the temporal discretization. The k-ω based SST turbulent model is employed and validated to achieve more accurate simulated results [24]. As shown in Figure 4, the boundary condition for the inner and outer diameter surfaces of the fluid domain are set to "opening" to allow the bearing to entrain the ambient gas from the outside. A no-slip and moving wall boundary at a constant angular velocity is applied to the surface of the thrust disk. The thrust disk is treated as a thermal convection boundary, but its thermal convection coefficient is determined using a forced convection model [22]. Additionally, a stationary no-slip wall boundary condition is adopted for the coupling surface. The model incorporates rotational periodicity boundary conditions at both ends of the domain sector to simulate the entire bearing in the fluid domain.

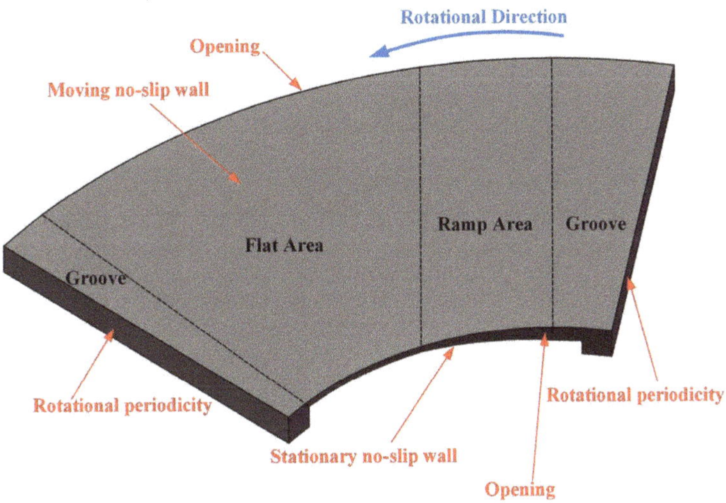

Figure 4. Fluid computational domain (the film thickness is stretched 10 times).

In the structure model, hexahedron solid element 185 is used to accurately represent the GTFB foils, with each foil consisting of two layers of solid elements. To establish the contact model within the foil structure, contact 174 elements and target 170 elements are employed. The augmented Lagrange method is utilized to determine the contact status and frictional forces between the foils and back plate. The contact and friction equations are written as:

$$F_N = FKN \cdot K_N \cdot \Delta + \lambda_0 \tag{5}$$

$$f_T = \mu_f \cdot F_N \tag{6}$$

where F_N is the normal contact force, FKN is the normal contact stiffness factor, K_N is the normal contact stiffness, Δ is the penetration tolerance, λ_0 is an additional term reducing the sensitivity of the normal contact force to the normal contact stiffness, f_T is the tangential friction force and μ_f is the friction coefficient.

The normal contact stiffness factor is set to 0.1 for better convergence, as a large deformation effect is active in the simulation. The friction coefficient between the foils and back plate is set to 0.1 according to previous studies [25,26]. Since the simulation considers the contact status and frictional forces, the structure model exhibits highly nonlinear behavior. Therefore, the full Newton–Raphson solution is employed to solve the structure model. As Figures 1b, 2 and 5 showed, the spacer and back plate are treated as fixed rigid bodies, while the fixed edges of the foils are set as fixed boundaries. A natural convection model is used to calculate the thermal convection coefficient of the foil surface. The contact region is set as a thermal contact boundation. The remaining surfaces of the foil structure, the spacer and the back plate are treated as thermal convection boundaries, and a natural convection model is utilized to calculate the thermal convection coefficient. The thrust disk is also treated as a thermal convection boundary, but its thermal convection coefficient is determined using a forced convection model [22]. The ambient gas is the ideal gas of 25 °C and 1 atm. The model also incorporates rotational periodicity boundary conditions at both ends of the domain sector to simulate the entire bearing in sturcture domain. The detailed boundary conditions are listed in Table 2. In additional, the initial temperature of both the fluid domain and structure domain is 25 °C, and the initial gas pressure is 1 atm.

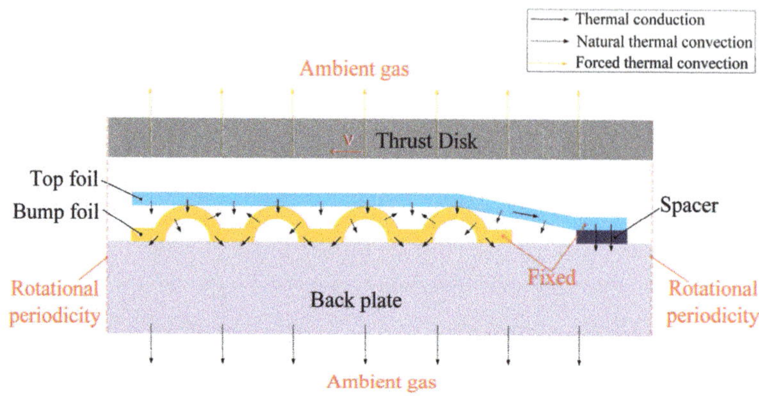

Figure 5. Structure computational domain of bearing 1.

Table 2. Detailed boundary conditions.

Boundary Type	Part	Boundary Condition
Opening	Inner and outer diameter surfaces of the fluid domain	Ambient temperature $T_0 = 25$ °C, ambient press $p_0 = 1$ atm
Moving no-slip wall	Thrust disk	Rotational speed $v = 30,000$ rpm, forced thermal convection coefficient $\lambda_{fc} = 84\ \mathrm{W/(K{\cdot}m^2)}$
Rotational periodicity	Both ends of the fluid and structure domain sector	
Fluid–solid interface	Top foil surface Top side of back plate	Transfer pressure, displacement, temperature and flux
Solid–solid interface	Top foil and bump foil bump foil and back plate	Contact and friction, solid thermal conductivity coefficient $\lambda_s = 16.9\ \mathrm{W/(K{\cdot}m)}$
Fixed	Fixed edges of top foil and bump foil	Fixed, solid thermal conductivity coefficient $\lambda_s = 16.9\ \mathrm{W/(K{\cdot}m)}$
Free wall	Underside of top foil bump foil	Unconstrained, natural thermal convection coefficient $\lambda_{fn} = 12\ \mathrm{W/(K{\cdot}m^2)}$, ambient temperature $T_0 = 25$ °C, ambient press $p_0 = 1$ atm
Fixed wall	Back plate Spacer	Fixed, rigid body, natural thermal convection coefficient $\lambda_{fn} = 12\ \mathrm{W/(K{\cdot}m^2)}$, ambient temperature $T_0 = 25$ °C, ambient press $p_0 = 1$ atm

In this paper, the structured meshes are used for all fluid and structure domains. The determination of grid independence is based on the load capacity as the criterion. As Figure 6 showed, the number of meshes is increased until the load capacity converges. For the fluid domain, a mesh number of 1.2 million is selected, while for the structure domain, it ranges from 30,000 to 33,000 depending on the specific bump foil configuration, and Figure 7 illustrates the part of structure mesh for bearing 1. To ensure convergence, the RMS residual value in the fluid domain is set to 1×10^{-5}. Additionally, a coupling data transfer convergence target of 0.01 is set to guarantee accurate calculations within each coupling iteration step. A workstation with 64 cores is used for computations, and each simulation case requires approximately 40 h of computational time.

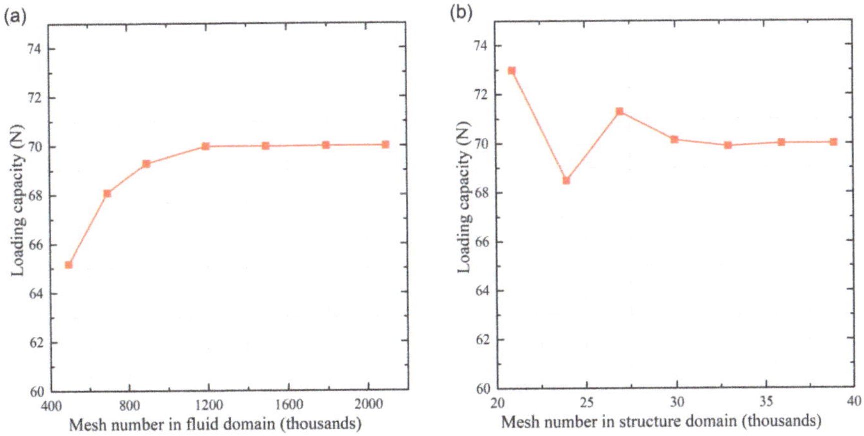

Figure 6. Mesh independence test of bearing 1. (**a**) Fluid domain. (**b**) Structure domain.

Figure 7. Part of structure mesh for bearing 1.

2.3. Experimental Validation

To validate the efficacy of the computational model, a GFTB test rig was designed and experiments were conducted. As illustrated in Figure 8a, the left side of the rig accommodates the drive unit. The thrust disk is mounted on a high-speed motor, and the speed is adjusted using a frequency converter. On the right side of the test rig, the loading and measurement components are situated. The test GTFB is installed on the bearing housing and connected to the bearing housing shaft. The shaft is supported by two gas hydrostatic bearings with different diameters, which facilitate minimal resistance in shaft movement and rotation. This setup allows for highly accurate drag torque measurements. The drag torque is measured using a force transducer, and a torque measuring rod is installed on the side of the bearing housing to connect with the force transducer. Figure 8b illustrates that the load is adjusted by regulating the pressure in the pressure chamber. The adjustment of the load is expressed by high-pressure air ejected from gas hydrostatic bearing nozzles, which form hydrostaic gas film between the gas hydrostatic bearing surface and the bearing housing shaft surface. Some of the air escapes to the external atmosphere through the bearing clearance, while the rest enters the pressure chamber along the bearing clearance. The pressure chamber is connected to the external atmosphere via a pressure regulator. The piston surface of the bearing housing shaft is designed with two gas hydrostatic bearings of different diameters. Consequently, an increase in pressure within the pressure chamber leads to an increase in the load acting on the piston surface.

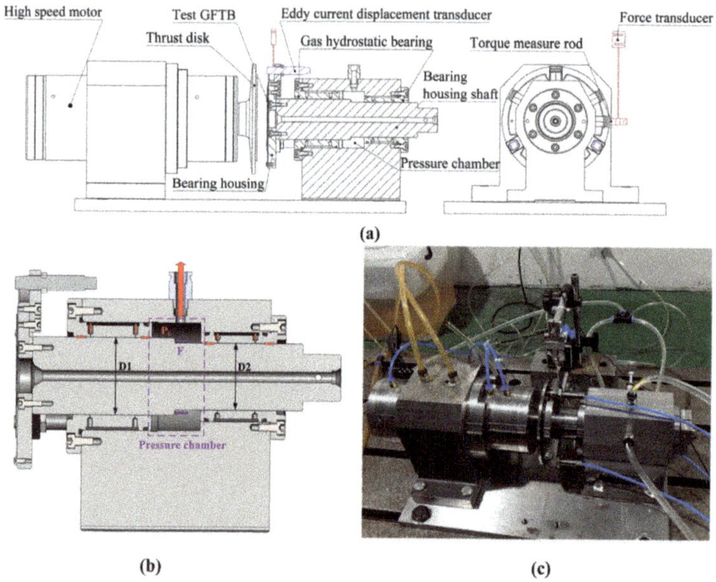

Figure 8. Test system for GFTB. (**a**) 3D model; (**b**) load part; (**c**) the photo of test rig.

Bearing 1 was selected for the model validation, and the actual bearing 1 is depicted in Figure 9. The numerical results of the drag torque, obtained at a rotational speed of 30,000 revolutions per minute (rpm), were compared with the experimental results in Figure 10. The numerical results exhibit a good agreement with the experimental data for thrust loads below 60 N. However, a notable disparity in drag torque suddenly emerges at higher thrust loads, which can be attributed to the rubbing contact between the top foil and the thrust disk during the test [10,27]. The maximum thrust load tested can be considered as 72 N, since the test drag torque increases significantly when the thrust load exceeds this value. Additionally, the maximum-tested thrust load closely approximates the predicted thrust load at the nominal thickness of 5 μm. This observation demonstrates the reasonableness of predicting the maximum load at a minimum film thickness of 5 μm [5,28,29]. Hence, the model is deemed valid for predicting the performance of the GFTB.

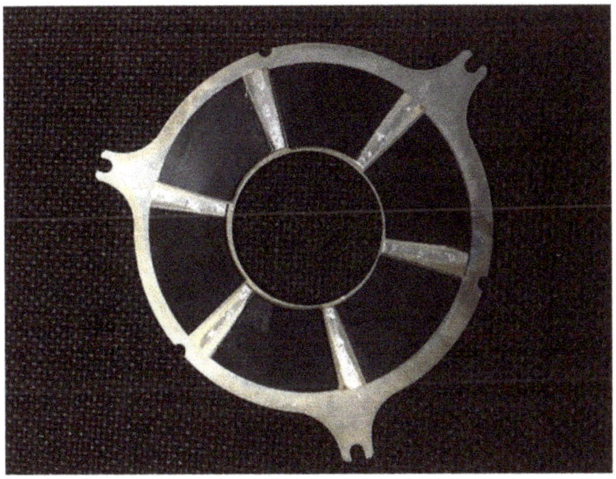

Figure 9. The photo of test GTFB.

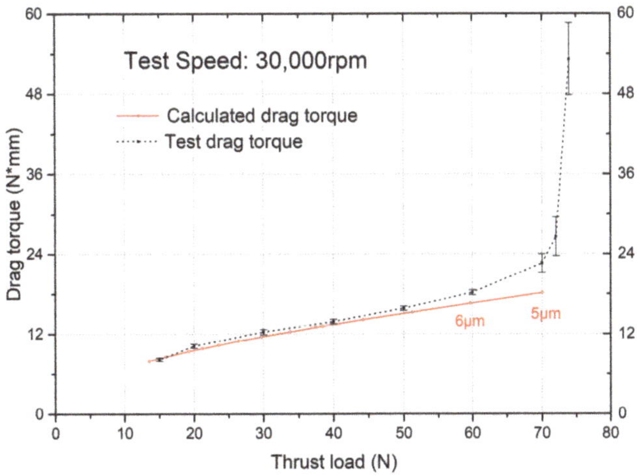

Figure 10. Comparison between numerical and experimental results of drag torque with different thrust loads.

3. Results and Discussion

3.1. Load Capacity

Figure 11 illustrates the load capacity of GFTBs at a rotational speed of 30,000 rpm for four different bump foil configurations. It is apparent that the chosen bump foil configurations have a significant impact on the bearing performance. Specifically, when the nominal clearance is below 10μm, bearings with the bump foil fixed at the pad trailing edge (bearing 2 and 4) exhibit a higher load capacity compared to bearing 1 and 3. According to the above section, the minimum nominal clearance of the gas film in this study is set at 5 μm. Table 3 presents the maximum values of bearing capacity loss at the nominal clearance of 5 μm. The results demonstrate a 20.8% increase in the maximum load capacity of bearing 3 compared to bearing 1. Moreover, bearing 2 shows an even more substantial improvement with a 33.5% increase in maximum load capacity. In contrast, when compared with bearing 4, bearing 3 experiences a decrease of 11.4% in the maximum load capacity, whereas the decrease in bearing 2 is minimal at only 2.1%. Thus, it can be concluded that the fixed position of the bump foil significantly influences the load performance of GFTBs. Furthermore, the impact of independent bump foil strips on static performance is closely related to the fixed position of the bump foil. For GTFBs with the bump foil fixed at the leading edge (bearing 1 and 3), the use of independent bump foil strips leads to a substantial improvement in load capacity. However, in the case of the bump foil fixed at the trailing edge (bearing 2 and 4), splitting the bump foil into several strips results in a weaker enhancement of the load capacity.

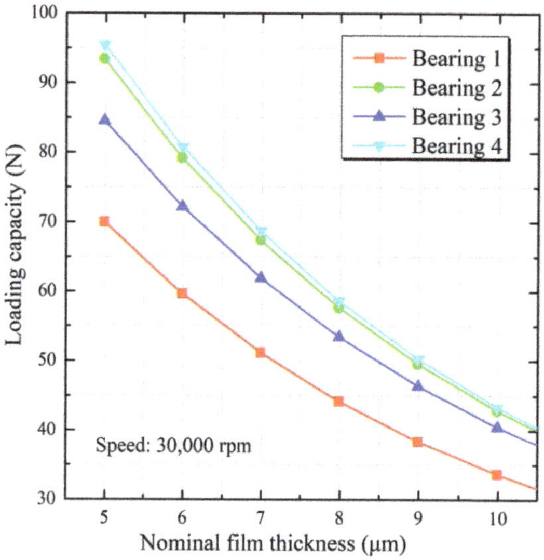

Figure 11. Load capacity of bearing 1–4 with different nominal film thickness.

Table 3. Maximum load capacity of bearing 1–4 at the nominal clearance of 5 μm.

Bearing Number	Maximum Load Capacity (N)	Difference from Bearing 1	Difference from Bearing 4
Bearing 1	70.0		−26.7%
Bearing 2	93.5	33.5%	−2.1%
Bearing 3	84.6	20.8%	−11.4%
Bearing 4	95.5	36.4%	

To further explore the underlying mechanism behind these findings, Figure 12 showcases the pressure distribution on the thrust disk at the minimum nominal clearance for the four bearings. A consistent color scale is used for the pressure distribution across all four configurations, with the maximum pressure value of the gas film serving as the upper limit. It is observed that all four bearings exhibit pressure blocks resembling "tiger stripes", which are caused by the sagging of the top foil between adjacent bumps. Bearing 1 and bearing 3 display a spiky pressure distribution, while bearing 2 and bearing 4 feature a plateau-like pressure distribution. This leads to a substantially larger high-pressure area in bearing 2 and bearing 4 compared to bearing 1 and bearing 3, resulting in higher load capacity. Regarding the impact of independent bump foil strips, it is notable that the maximum pressure in bearing 1 is significantly smaller than that in bearing 3. On the other hand, the maximum pressure and high-pressure area in bearing 2 exhibit minimal differences compared to bearing 4. This clearly indicates that the effect of independent bump foil strips on bearing capacity is contingent upon the fixed position of the bump foil.

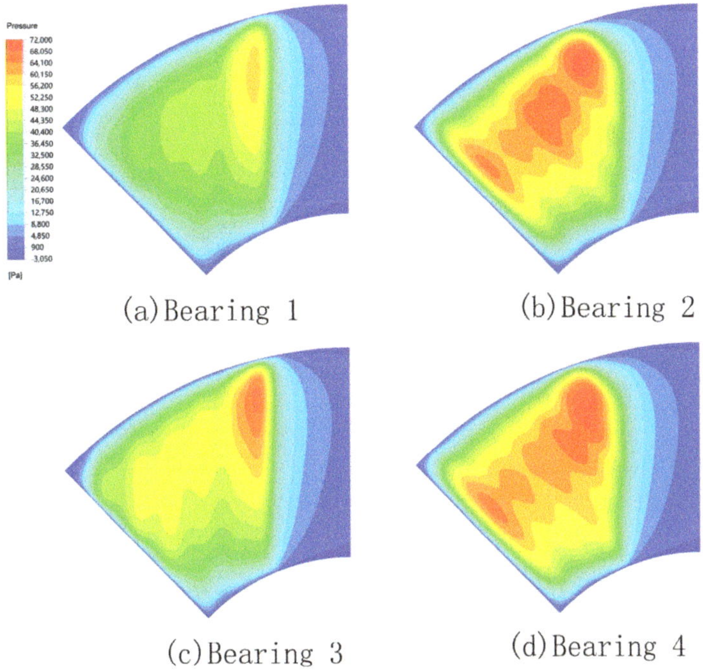

Figure 12. Pressure distribution at the minimum nominal clearance on the thrust disk of bearing 1–4.

The gas film thickness distribution along the circumferential direction at the minimum nominal clearance is presented in Figure 13. The analyzed section radius within the high-pressure area is 32 mm. Based on extensive literature and practical knowledge, it is widely acknowledged that the bump stiffness at the fixed end is greater than at the free end [10,30]. When the bump foil is fixed at the pad leading edge, as observed in bearing 1 and bearing 3, the thinnest film occurs at the beginning of the flat area. Furthermore, the film thickness of the flat area exhibits an overall increasing trend with the angle. Consequently, the high-pressure region in bearing 1 and 3 is primarily concentrated at the leading portion of the flat area due to the diverging gas film. In contrast, bearing 2 and bearing 4 display a converging gas film towards the trailing edge. This configuration causes the high-pressure area to extend and cover most of the flat area. Therefore, for optimal load capacity and a converging gas film, it is recommended to fix the bump foil at the trailing edge [28,30,31].

3.2. Thermal Characteristic

The power loss resulting from shear stress in the fluid film serves as the heat source in the computational model and significantly impacts the temperature distribution. Figure 14 shows the power loss of GFTBs with four distinct bump foil configurations. As the nominal film thickness decreases, the power loss of all bearings increases, with a more drastic increase when the thickness approaches the minimum film clearance value. This can be attributed to the intensified hydrodynamic effect of the fluid film when the nominal film thickness decreases, leading to an increase in shear stress. The relationship between load capacity and power loss is shown to be linear in Figure 14b. Bearing 1 exhibits the highest slope, while bearing 4 displays the lowest slope. Under the same load capacity, bearing 2 and bearing 4 experience significantly smaller power loss compared to bearing 1 and bearing 3. This indicates that fixing the bump foil at the trailing edge, along with splitting it into multiple strips, can effectively reduce power loss and enhance GFTB efficiency.

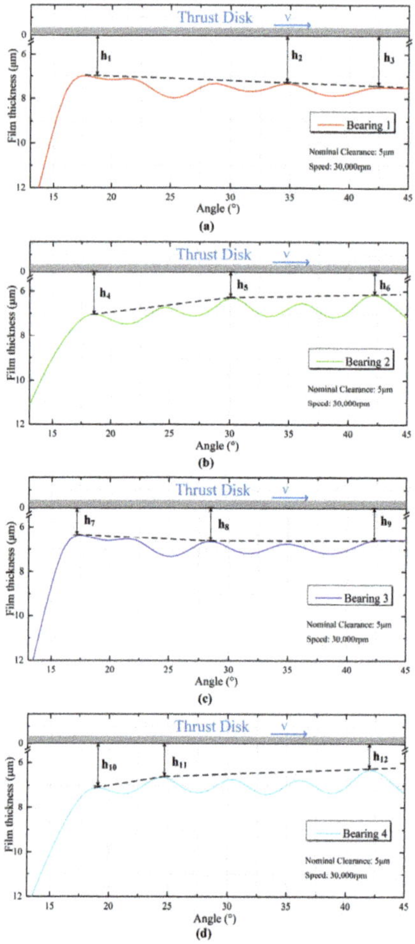

Figure 13. Gas film thickness distribution at a radius of 32 mm of bearing 1–4. (**a**) Bearing 1. (**b**) Bearing 2. (**c**) Bearing 3. (**d**) Bearing 4.

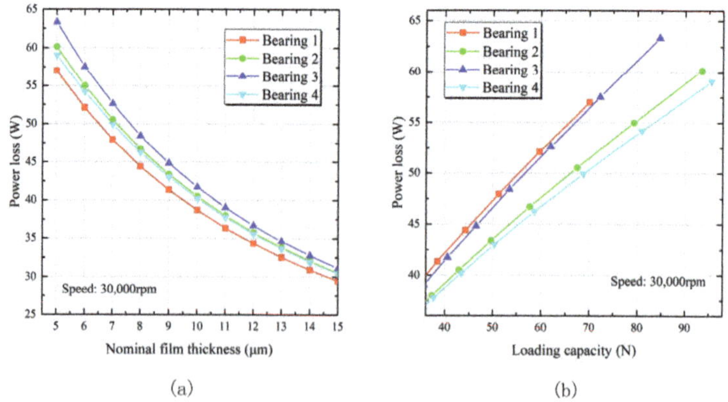

Figure 14. Power loss of bearing 1–4. (**a**) Relation to the nominal film thickness. (**b**) Relation to the load capacity.

Figure 15 showcases the foil temperature distributions of GFTBs featuring four different bump foil configurations at the minimum nominal clearance. In the radial direction, all bearings exhibit a gradient distribution of the temperature. The outer diameter position, characterized by a higher linear speed, experiences significantly higher temperatures compared to the inner diameter position, owing to the positive correlation between shear stress and speed.

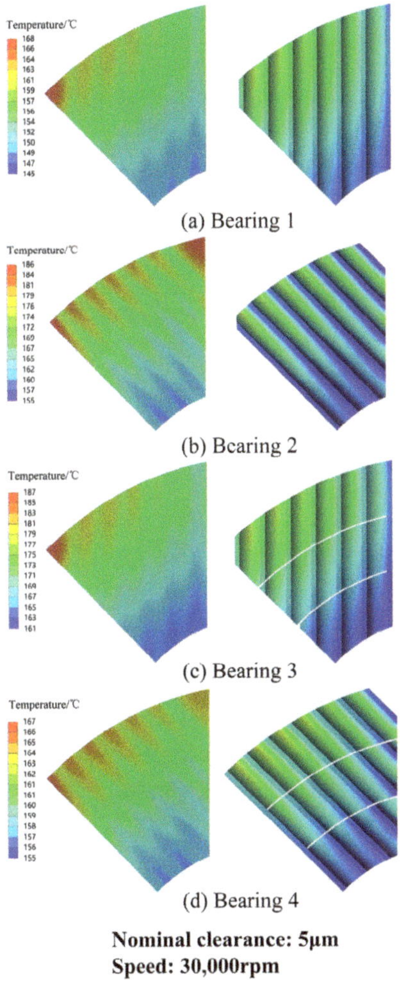

Figure 15. Temperature distribution of bearing 1–4.

In the circumferential direction, all four bearings display "tiger strip" high-temperature blocks, resulting from thermal contact between the top foil and bump foil. The highest temperature regions are located at the outer diameter positions of the trailing edge, consistent with previous research findings [21,32]. Table 4 presents the thermal characteristics of the four bearings. The average temperature of the top foils shows a positive correlation with power loss among all bearings. Bearing 1, with the lowest power loss, exhibits the lowest average temperature of the top foil. However, when considering the maximum temperature of the top foil, there are some differences. Bearing 4, despite not having the lowest power loss, experiences the lowest maximum temperature of the top foil.

Table 4. Thermal characteristic at the minimum nominal clearance of bearing 1–4.

Bearing Number	Maximum Temperature of Top Foil (°C)	Average Temperature of Top Foil (°C)	Power Loss (W)	Leakage (mg/s)
Bearing 1	167.16	153.87	56.99	57.59
Bearing 2	185.37	168.26	60.15	59.01
Bearing 3	186.42	170.34	63.33	57.69
Bearing 4	166.14	160.14	59.04	59.48

To explain this phenomenon, the leakage of the fluid film is analyzed. Figure 16 illustrates that most ambient air enters the fluid film through the inner and outer diameter surfaces in the groove region, with outflow occurring mostly in the ramp area and flat area. As both ends of the fluid domain are set as rotational periodicity boundaries, the inflow and outflow solely take place through the inner and outer diameter surfaces. Thus, the outflow is considered as leakage since it matches the inflow. Comparing bearing 4 to bearing 1 in Table 4, it is observed that bearing 4 exhibits more significant leakage, leading to more heat transfer from the fluid film to the ambient air. As a result, it displays a lower maximum temperature of the top foil. This finding suggests that fixing the bump foil at the trailing edge and splitting it into several strips can reduce the risk of bearing thermal failure.

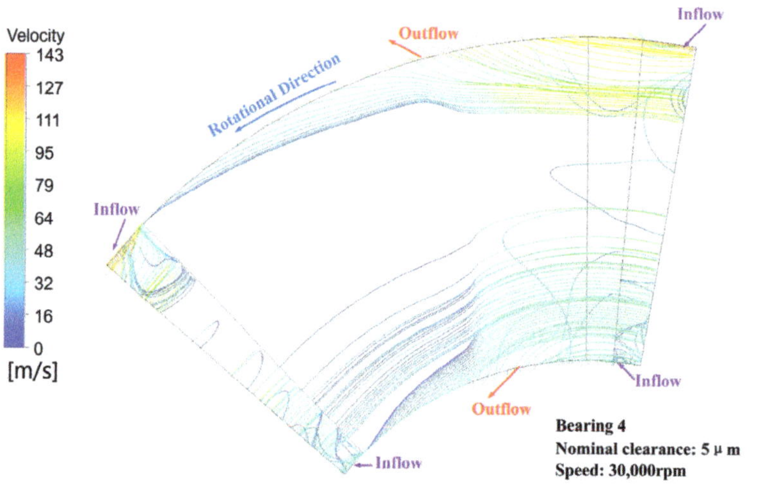

Figure 16. Flow trend of opening boundary.

4. Conclusions

In this study, a comprehensive 3D THED model of GFTB is established based on the FTSI approach to investigate the impact of bump foil configurations on GFTB performance. The model is validated using experimental results, considering factors such as frictional contact, bump interaction, tap foil sagging, and temperature effects. The load capacity and thermal characteristics of GFTBs with four different bump foil configurations are examined, leading to the following main conclusions:

(1) The 3D THED model of GFTB based on the FTSI approach is valid for the GFTB performance prediction.
(2) Fixing the bump foil at the trailing edge improves the load capacity.
(3) The influence of independent bump foil strips on load depends on the position where the bump foil is fixed. When the bump foil is fixed at the trailing edge, splitting it into several strips only slightly enhances the load capacity.

(4) To reduce power loss and minimize the risk of bearing thermal failure, it is recommended to fix the bump foil at the trailing edge and split it into multiple strips.

This study not only provides design guidance for the present GFTB, but also demonstrates an advanced tool to design and predict the performance of future generations of GFTB.

Author Contributions: Conceptualization, B.H. and A.H.; methodology, B.H.; software, R.D.; validation, B.H. and R.W.; formal analysis, B.H.; investigation, B.H; data curation, B.H.; writing—original draft preparation, B.H.; writing—review and editing, B.H. and Z.W.; resources, Q.N. and Z.L.; visualization, B.H.; supervision, A.H.; funding acquisition, A.H. All authors have read and agreed to the published version of the manuscript.

Funding: This research was funded by the National Science and Technology Major Project of China, grant number J2019-V-0017-0112.

Data Availability Statement: The data presented in this study are available on reasonable request from the corresponding author.

Conflicts of Interest: The authors declare no conflict of interest.

References

1. Feng, K.; Kaneko, S. Analytical Model of Bump-Type Foil Bearings Using a Link-Spring Structure and a Finite-Element Shell Model. *J. Tribol.* **2010**, *132*, 021706. [CrossRef]
2. DellaCorte, C.; Bruckner, R.J. Remaining Technical Challenges and Future Plans for Oil-Free Turbomachinery. *J. Eng. Gas Turbines Power* **2011**, *133*, 042502. [CrossRef]
3. Somaya, K.; Yamashita, T.; Yoshimoto, S. Experimental and Numerical Investigation of the High-Speed Instability of Aerodynamic Foil Journal Bearings for Micro Turbomachinery. In Proceedings of the ASME/STLE 2012 International Joint Tribology Conference, Denver, CO, USA, 7–10 October 2012.
4. Agrawal, G.L. Foil Air/Gas Bearing Technology—An Overview. In Proceedings of the ASME 1997 International Gas Turbine and Aeroengine Congress and Exhibition, Orlando, FL, USA, 2–5 June 1997.
5. Kim, T.H.; Lee, Y.-B.; Kim, T.Y.; Jeong, K.H. Rotordynamic Performance of an Oil-Free Turbo Blower Focusing on Load Capacity of Gas Foil Thrust Bearings. *J. Eng. Gas Turbines Power* **2011**, *134*, 022501. [CrossRef]
6. Samanta, P.; Murmu, N.C.; Khonsari, M.M. The evolution of foil bearing technology. *Tribol. Int.* **2019**, *135*, 305–323. [CrossRef]
7. Walton, J.F., II; Tomaszewski, M.J.; Heshmat, H. The Role of High Performance Foil Bearings in Advanced, Oil-Free, High-Speed Motor Driven Compressors. In Proceedings of the ASME 2003 1st International Conference on Fuel Cell Science, Engineering and Technology, Rochester, NY, USA, 21–23 April 2003.
8. Fu, G.; Untaroiu, A.; Swanson, E. Effect of Foil Geometry on the Static Performance of Thrust Foil Bearings. *J. Eng. Gas Turbines Power* **2018**, *140*, 082502. [CrossRef]
9. Heshmat, H.; Walowit, J.A.; Pinkus, O. Analysis of Gas Lubricated Compliant Thrust Bearings. *J. Lubr. Technol.* **1983**, *105*, 638–646. [CrossRef]
10. Iordanoff, I. Analysis of an Aerodynamic Compliant Foil Thrust Bearing: Method for a Rapid Design. *J. Tribol.* **1999**, *121*, 816–822. [CrossRef]
11. Heshmat, C.A.; Xu, D.S.; Heshmat, H. Analysis of Gas Lubricated Foil Thrust Bearings Using Coupled Finite Element and Finite Difference Methods. *J. Tribol.* **1999**, *122*, 199–204. [CrossRef]
12. Park, D.-J.; Kim, C.-H.; Jang, G.-H.; Lee, Y.-B. Theoretical considerations of static and dynamic characteristics of air foil thrust bearing with tilt and slip flow. *Tribol. Int.* **2008**, *41*, 282–295. [CrossRef]
13. Gad, A.M.; Kaneko, S. A New Structural Stiffness Model for Bump-Type Foil Bearings: Application to Generation II Gas Lubricated Foil Thrust Bearing. *J. Tribol.* **2014**, *136*, 041701. [CrossRef]
14. Xu, F.; Kim, D.; Zamanian Yazdi, B. Theoretical Study of Top Foil Sagging Effect on the Performance of Air Thrust Foil Bearing. In Proceedings of the ASME Turbo Expo 2016: Turbomachinery Technical Conference and Exposition, Seoul, Republic of Korea, 13–17 June 2016.
15. Xu, Z.; Li, C.; Du, J.; Li, J.; Wang, Y. Load-carrying characteristics of bump-type gas foil thrust bearings. *Int. J. Mech. Sci.* **2023**, *244*, 108080. [CrossRef]
16. Bruckner, R.J. Simulation and Modeling of the Hydrodynamic, Thermal, and Structural Behavior of Foil Thrust Bearings. Ph.D. Thesis, Case Western Reserve University, Cleveland, OH, USA, 2004.
17. Lee, D.; Kim, D. Three-Dimensional Thermohydrodynamic Analyses of Rayleigh Step Air Foil Thrust Bearing with Radially Arranged Bump Foils. *Tribol. Trans.* **2011**, *54*, 432–448. [CrossRef]
18. Gad, A.M.; Kaneko, S. Fluid Flow and Thermal Features of Gas Foil Thrust Bearings at Moderate Operating Temperatures. In Proceedings of the 9th IFToMM International Conference on Rotor Dynamics, Cham, Switzerland, 22–25 September 2015; pp. 1223–1233.

19. Xu, F.; Kim, D. Three-Dimensional Turbulent Thermo-Elastohydrodynamic Analyses of Hybrid Thrust Foil Bearings Using Real Gas Model. In Proceedings of the ASME Turbo Expo 2016: Turbomachinery Technical Conference and Exposition, Seoul, Republic of Korea, 13–17 June 2016.
20. Lehn, A.; Mahner, M.; Schweizer, B. A thermo-elasto-hydrodynamic model for air foil thrust bearings including self-induced convective cooling of the rotor disk and thermal runaway. *Tribol. Int.* **2018**, *119*, 281–298. [CrossRef]
21. Liu, X.; Li, C.; Du, J. The Fluid-Structure-Thermal Performance Analysis of Gas Foil Thrust Bearing by Using Computational Fluid Dynamics. *Lubricants* **2022**, *10*, 294. [CrossRef]
22. Xiong, C.; Xu, B.; Yu, H.; Huang, Z.; Chen, Z. A thermo-elastic-hydrodynamic model for air foil thrust bearings considering thermal seizure and failure analyses. *Tribol. Int.* **2023**, *183*, 108373. [CrossRef]
23. DellaCorte, C.; Radil, K.C.; Bruckner, R.J.; Howard, S.A. Design, Fabrication, and Performance of Open Source Generation I and II Compliant Hydrodynamic Gas Foil Bearings. *Tribol. Trans.* **2008**, *51*, 254–264. [CrossRef]
24. Qin, K.; Li, D.; Huang, C.; Luo, K. Comparative analysis of turbulence models for gas bearings flowfield simulations. *Fluid Dyn. Res.* **2019**, *51*, 045505. [CrossRef]
25. Hoffmann, R.; Liebich, R. Experimental and numerical analysis of the dynamic behaviour of a foil bearing structure affected by metal shims. *Tribol. Int.* **2017**, *115*, 378–388. [CrossRef]
26. Zywica, G.; Baginski, P.; Bogulicz, M.; Martowicz, A.; Roemer, J.; Kantor, S. Numerical identification of the dynamic characteristics of a nonlinear foil bearing structure: Effect of the excitation force amplitude and the assembly preload. *J. Sound Vib.* **2022**, *520*, 116663. [CrossRef]
27. San Andrés, L.; Ryu, K.; Diemer, P. Prediction of Gas Thrust Foil Bearing Performance for Oil-Free Automotive Turbochargers. *J. Eng. Gas Turbines Power* **2014**, *137*, 032502. [CrossRef]
28. Dykas, B. Factors Influencing the Performance of Foil Gas Thrust Bearings for Oil-Free Turbomachinery Applications. Ph.D. Thesis, Case Western Reserve University, Cleveland, OH, USA, 2006.
29. Conboy, T.M. Real-Gas Effects in Foil Thrust Bearings Operating in the Turbulent Regime. *J. Tribol.* **2013**, *135*, 031703. [CrossRef]
30. Gad, A.M.; Kaneko, S. Tailoring of the bearing stiffness to enhance the performance of gas-lubricated bump-type foil thrust bearing. *Proc. Inst. Mech. Eng. Part J J. Eng. Tribol.* **2016**, *230*, 541–560. [CrossRef]
31. Dickman, J.R. An investigation of gas foil thrust bearing performance and its influencing factors. Master's Thesis, Case Western Reserve University, Cleveland, OH, USA, 2010.
32. Rieken, M.; Mahner, M.; Schweizer, B. Thermal Optimization of Air Foil Thrust Bearings Using Different Foil Materials. *J. Turbomach.* **2020**, *142*, 101003. [CrossRef]

lubricants

MDPI

Article

Research on the Service Condition Monitoring Method of Rolling Bearings Based on Isomorphic Data Fusion

Yanfei Zhang [1,2,*], Yang Liu [1], Mingqi Yang [2], Xiaoyang Feng [3], Qianxiang Zhu [4] and Lingfei Kong [1]

1 School of Mechanical and Precision Instrument Engineering, Xi'an University of Technology, Xi'an 710048, China; 2210221250@stu.xaut.edu.cn (Y.L.); lingfeikong@xaut.edu.cn (L.K.)
2 Luoyang Bearing Science & Technology Co., Ltd., Luoyang 471039, China; 15937904981@163.com
3 Shaanxi Robot Automation Technology Co., Ltd., Xi'an 710061, China; fxy@xknc.net
4 Xi'an Research Institute Co., Ltd., China Coal Technology and Engineering Group Corp, Xi'an 710048, China; zhuqianxiang@cctegxian.com
* Correspondence: yfzhang@xaut.edu.cn

Abstract: In order to solve the problem that it is difficult for a single sensor to accurately characterize the running state of rotating bearings under complex working conditions, this paper proposes a data-level fusion method based on multi-source isomorphic sensors to monitor spindle bearings. First, new vibration signals in the X,Y,Z direction were obtained through the process of decomposing, de-noise, and reconstructing. Second, the PCA algorithm was used to select the time-domain and frequency-domain features of the vibration signals, construct the feature matrix, and perform dimensionality reduction in the feature matrix. Finally, the entropy weight method was introduced to obtain the initial weights of the three directions as the inputs of the adaptive function. The chaotic particle swarm optimization algorithm proposed in this paper helps particles jump out of the local optimum. Chaotic mapping is used to initialize the velocity and position of the particles, which calculates globally optimal weights in three directions. In order to extract bearing signal features more accurately and efficiently, a DenseNet and Transformer (DAT) feature extraction model is proposed to deal with the complex changes and noise interference of bearing signals. Through the open data set of Jiangnan University and the data collected by our own experimental platform, the maximum accuracy of the DAT model was verified to be 100%.

Keywords: multi-source; entropy weighting; chaotic particle swarm optimization algorithm; data-level fusion; feature extraction

Citation: Zhang, Y.; Liu, Y.; Yang, M.; Feng, X.; Zhu, Q.; Kong, L. Research on the Service Condition Monitoring Method of Rolling Bearings Based on Isomorphic Data Fusion. *Lubricants* **2023**, *11*, 429. https://doi.org/10.3390/lubricants11100429

Received: 26 July 2023
Revised: 28 September 2023
Accepted: 29 September 2023
Published: 4 October 2023

1. Introduction

Machine tools are important and necessary instruments in the equipment manufacturing sector. The spindle, which is the machine tool's core component, determines its precision and productivity. Bearing assembly precision and performance are critical because they determine the spindle's running condition and performance, which influences the machine tool's overall machining quality and effectiveness. [1–3]. To forecast the dynamic performance of the bearing-rotor system, Ma S et al. [4] developed a dynamic model based on SFBE. By describing SFBE-specific physical properties, this model provides real-time coupling and the synchronous solution of bearing and rotor models. Fang B et al. [5] proposed a generalized mathematical model of DR-ACBB under three different configurations to study the variation rule of nonlinear stiffness. It resulted in the skewed running of bearings in the service, which can very easily cause bearing failures. This is due to long-term service in harsh environments such as variable loads, high temperatures, and impacts, as well as under the influence of factors like manufacturing errors, assembly accuracy, and human operation errors. Relying on mathematical models alone is limited and no longer allows for the complete condition characterization of bearings. To ensure the safe operation of machine tools and to boost production, the reliable and effective real-time condition monitoring of bearings is crucial [6]. Machine learning has developed into a very powerful classification tool with the advancement of computer technology. Machine learning can delve deeper into potentially advantageous information in data since computers can handle enormous amounts of data [7,8]. Traditional shallow machine-learning techniques do have some limitations. These techniques usually require a lot of prior knowledge, which makes selecting and extracting features challenging [9,10].

End-to-end deep learning methodologies have been increasingly brought into the field of fault diagnosis in recent years as a result of the impact of the artificial intelligence (AI) wave. Deep learning techniques, as opposed to conventional approaches, offer fresh perspectives and opportunities for defect detection research by automatically learning feature representations and eliminating inevitable ambiguities in the manual feature extraction process [11]. One of these representative deep learning methods, the convolutional neural network (CNN), is a subclass of the feed-forward neural network that includes convolutional computation and has a deep structure [12]. This algorithm is capable of representational learning and can categorize input data according to its hierarchical structure in a translation-invariant manner. Janssens [13] suggested a three-layer CNN model for bearing defect identification based on vibration signals. Prior to training the model and feeding data into the network model, these data were discretely Fourier-converted. Gu [14] suggested that the 1-DCNN and LSTM two channels be fed raw vibration signals to fuse feature information in both the temporal and spatial dimensions, thereby classifying the bearing problems. Zhang et al. [15] proposed a method to monitor the uneven operating conditions of bearings based on a two-channel fusion of the improved DenseNet network, which realizes the fusion of features in the frequency domain and the time-frequency domain. This method addresses the issue that traditional bearing fault diagnosis methods are insufficient to extract key features under strong noise and variable loads. Aiming at the above references, the feature extraction model in this article introduces DenseNet and Transformer modules to improve this model's ability to deal with complex working conditions.

The multi-sensor measurement and sensing system is a complex information processing system that integrates target measurements, data processing, and information fusion. It is widely used in the fields of industrial system monitoring [16], fault diagnosis [17], spatial localization [18], and environmental observation [19]. Among them, at the data level, the fusion of homologous and homomorphic multi-sensor sensing sequences is one of the key elements of this system. By fusing data from multiple sensors, the accuracy and reliability of information can be improved, and then more accurate measurement and sensing results can be realized, which provides important support for achieving efficient data processing and decision making [20]. In the field of intelligent manufacturing, data fusion technology effectively improves people's processing ability and utilization efficiency of industrial big data, where multi-source data have the characteristics of comprehensively describing the target alongside complementary data, and its fusion operation can improve the decision-making credibility and anti-interference ability of the model, reduce the redundancy existing in multi-source data, and reduce the waste of storage resources [21].

Data fusion methods of multi-source homogeneous sensors have been introduced to comprehensively characterize the operating state of rolling bearings because it is challenging for a single sensor to accurately characterize the operating state of machine tool spindle bearings under complex working conditions. Common fusion methods for homogeneous- and homogeneous-type multi-sensor sensing sequences include the weighted averaging method [22,23], Bayesian estimation [24], maximum likelihood estimation [25], Kalman filtering [26], neural network [27] and fuzzy logic [28]. Among these, the weighted average method is suitable for data layer fusion, but the distribution of weights has a significant impact on the fusion effect [22]. Bayesian estimation, maximum likelihood estimation, and other statistical methods require a priori knowledge of the target object. Kalman filtering requires that the system model and statistical characteristics of noise are known, and it cannot deal with the problem of adding new sensors. Neural network-based methods require training and learning, and their applicability is affected by the number of input dimensions and the number of neurons and cannot handle input source variations. Fuzzy C-mean clustering-based methods are computationally straightforward [28], do not require a priori knowledge and model limitations, and can be applied online, but their results depend on the precise estimation of the number of clusters. Neural network-based methods require training and learning, and their applicability is affected by the number of input dimensions and neurons and cannot handle input source variations.

Su [29] proposed a homogeneous multi-sensor online fusion method based on improved fuzzy clustering and the aforementioned analysis. This method uses a robust fuzzy clustering method that introduces noise classes to analyze multiple sources of data simultaneously and does not depend on the number of clusters set in the traditional fuzzy clustering fusion method. A multi-sensor data fusion approach based on an adaptive weighting algorithm was proposed by Tang [30]. This method fuses signals from several sensors using an adaptive weighting algorithm and then uses a Kalman filtering algorithm to decrease noise in the output. In their algorithm, Cai et al. [31] combined measurement data preprocessing with improved batch estimation adaptive and weighted data fusion, introducing environmental factors and enhancing the batch estimation algorithm to determine the ideal monitoring value of individual sensors, and realizing adaptive weighted data fusion in accordance with the principle of optimal weight allocation. Zhu et al. [32] proposed a multi-sensor data fusion algorithm based on wavelet noise reduction and adaptive weighting to address the issues of large error, conflict, and redundancy in the multi-node data acquisition of greenhouse environmental information. The proposed algorithm processed the collected data through wavelet noise reduction to make it have good smoothness and stability and fused multi-sensor data using the adaptive weighting algorithm. With the guidance of the above literature, this article constructs a fusion algorithm based on weighted fusion, which is based on the entropy weighting method and chaotic particle swarm search.

In conclusion, this paper proposes a data-level fusion method based on multi-source isomorphic sensors to monitor the running state of rolling bearings and constructs a DAT feature extraction model for the deep feature extraction of fused data to detect this bearing's service state. It is challenging for a single sensor to accurately characterize the service state of machine tool spindle bearings under complex working conditions. In order to improve the accuracy, reliability, coverage, time-domain continuity, and consistency of data, as well as the fault tolerance and robustness of the system, a data-level fusion method with multi-source isomorphic sensors is proposed to monitor the operational status of rolling bearings. We created the DAT deep learning model (DenseNet and Transformer, DAT), which introduces the serial combination of DenseNet and Transformer modules to enable feature reuse and improve the model's capacity for handling time-series data, enabling more sophisticated feature extraction and transformation.

2. Relevant Theoretical Approaches

2.1. Wavelet Packet Denoising

Wavelet packet decomposition is substantially more effective than wavelet analysis's capacity to analyze signals since it decomposes both the high- and low-frequency portions of the signal. As seen in Figure 1, the following is an illustration of a four-layer wavelet packet decomposition, with a denoting the low-frequency portion and b denoting the high-frequency portion [33].

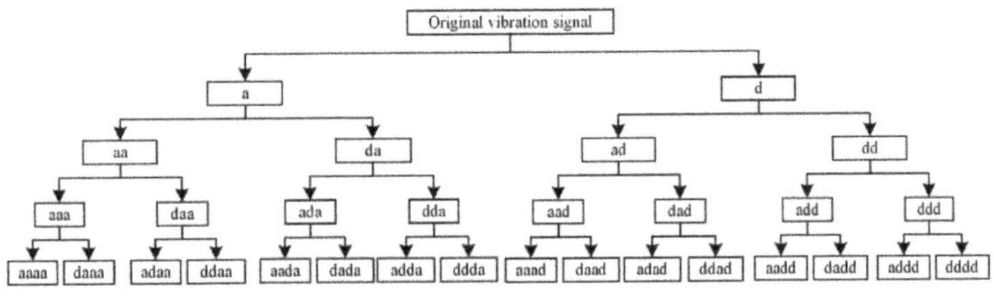

Figure 1. Four-layer wavelet packet decomposition.

In wavelet packet decomposition, $\mu_0 = \varphi(t)$, $\mu_1 = \psi(t)$ where $\varphi(t)$ and $\psi(t)$ denote the scale function and wavelet function, respectively, and $\{h_n\}_{n\in Z}$ and $\{g_n\}_{n\in Z}$ denotes the low-pass and high-pass filters, respectively; a set of functions known as wavelet packets can be defined by $\mu_0,\ \mu_1,\ h,\ g$ at a fixed scale:

$$\mu_{2n}(t) = \sqrt{2}\sum_k h_k \mu_n(2t-k) \tag{1}$$

$$\mu_{2n+1}(t) = \sqrt{2}\sum_k g_k \mu_n(2t-k) \tag{2}$$

where $\mu_n, n = 0,\ 1,\ 2, \ldots n.$ is called the wavelet packet and is determined by the orthogonal scale function $\mu_0 = \varphi(t)$.

A noisy signal is provided as follows:

$$s = x + n \tag{3}$$

where s is the measured noise signal, x is the original signal, n is the noise, and the essence of signal denoising is to estimate the original signal x based on the detected noise signals. The corresponding wavelet packet threshold denoising steps are as follows.

$$\begin{cases} y = W(s) \\ \tilde{y} = D(y,\lambda) \\ \tilde{x} = \overline{W}\left(\tilde{y}\right) \end{cases} \tag{4}$$

where W and $\overline{W}$ denote the wavelet packet transform and its inverse, respectively, λ is the threshold, and D is the signal thresholding denoising process.

$$SNR = 10lg\left[\frac{\sum_n x^2(n)}{\sum_n(\hat{x}(n) - x(n))^2}\right] \tag{5}$$

$$RMSE = \sqrt{\frac{1}{n}\sum_n [\hat{x}(n) - x(n)]^2} \tag{6}$$

In this formula, SNR represents the signal-to-noise ratio, which is the ratio of the energy of the useful signal to the energy of noise; the larger the SNR is, the smaller the noise mixed in the measured signal is. RMSE represents the root mean square error, which is the root mean square error between the signal after reconstruction and the original signal, and the smaller this value is, the better the de-noising effect is; $\hat{x}(n)$ is the signal with noise, and $x(n)$ is the original signal.

The original vibration signal can be decomposed to a maximum of 15 layers. After comparing the signal-to-noise ratio and the root mean square error of the various layers, the decomposition of the wavelet packet after four layers was selected. Too many layers result in the loss of actual useful information, and too few layers are not able to play a role in improving the signal-to-noise ratio.

2.2. Entropy Weighting Method

The entropy weight method is a multi-criteria decision analysis method that realizes the comprehensive evaluation of each indicator by calculating the entropy value and weight of the indicator. This method does not need to standardize the data and is suitable for situations where each indicator has a different scale and a different direction. The core idea of the entropy weighting method is that the smaller the entropy value of an indicator is, the more informative it is, and the greater the impact on the comprehensive evaluation it has; the weight of the indicator is calculated according to the proportion of the entropy value of each indicator [34], and the features extracted in this paper are shown in Table 1.

Table 1. Feature extraction.

Feature Classification	Feature Extraction
Time domain features	Maximum value, minimum value, peak-to-peak value, average value, absolute average value, root mean square, variance, standard deviation, steepness, skewness, peak factor, waveform factor, pulse factor, margin factor
Frequency domain features	Mean frequency, mean square frequency, root mean square frequency, frequency variance, frequency standard deviation

By extracting the time-domain and frequency-domain features of the original signal, the matrix can be defined $X = \{X_{ij}\}$, where i is the number of sensors, $i = 1, 2, \ldots, n$; j is the number of feature indicators, $j = 1, 2, \ldots, m$; and then the feature matrix can be expressed as:

$$X = \begin{bmatrix} X_{11} & X_{12} & \ldots & X_{1m} \\ X_{21} & X_{22} & \ldots & X_{2m} \\ \vdots & \vdots & \ddots & \vdots \\ X_{n1} & X_{n2} & \ldots & X_{nm} \end{bmatrix} \quad (7)$$

The weight of the j-th feature indicator under the i-th sensor is as follows:

$$P_{ij} = \frac{X_{ij}}{\sum_{i=1}^{N} x_{ij}} \quad (8)$$

The entropy value under the i-th sensor can be calculated as shown below:

$$H_i = -\frac{1}{\ln N} \sum_{i=1}^{N} P_{ij} \ln P_{ij} \quad (9)$$

where $P_{ij} \ln P_{ij}$ is considered to be 0 if $P_{ij} = 0$.

The weight of the i-th sensor can be calculated as follows:

$$W_i = \frac{1 - H_i}{n - \sum_{i=1}^{n} H_i} \quad (10)$$

2.3. Principle of the PCA Downscaling Algorithm

PCA (Principal Component Analysis) is a commonly used data dimensionality reduction method, which can reduce high-dimensional data into a low-dimensional space while trying to retain the information of the data [35]. The PCA dimensionality reduction algorithm flow of this paper is shown in Figure 2.

2.4. Chaos Mapping

The basic idea of the chaotic optimization algorithm is to map chaotic variables from a chaotic space to a solution space and then search for this using the characteristics of chaotic variables with traversability, randomness, and regularity. The chaotic optimization algorithm has the characteristics of not appearing sensitive to the initial value, it is easy to jump out of the local minima, and has a fast search speed, high computational accuracy, and global asymptotic convergence. Chaotic sequences commonly used in the field of group intelligence mainly include Logistic mapping, PWLCM mapping, Singer mapping, Sine mapping, etc., and in this method, Logistic mapping was chosen to optimize the particle swarm algorithm [36].

Logistic mapping is implemented as follows:

$$z_{k+1} = \mu z_k (1 - z_k) \quad (11)$$

where $z_0 \notin \{0, 0.25, 0.5, 0.75, 1.0\}, \mu \in [0, 4]$.

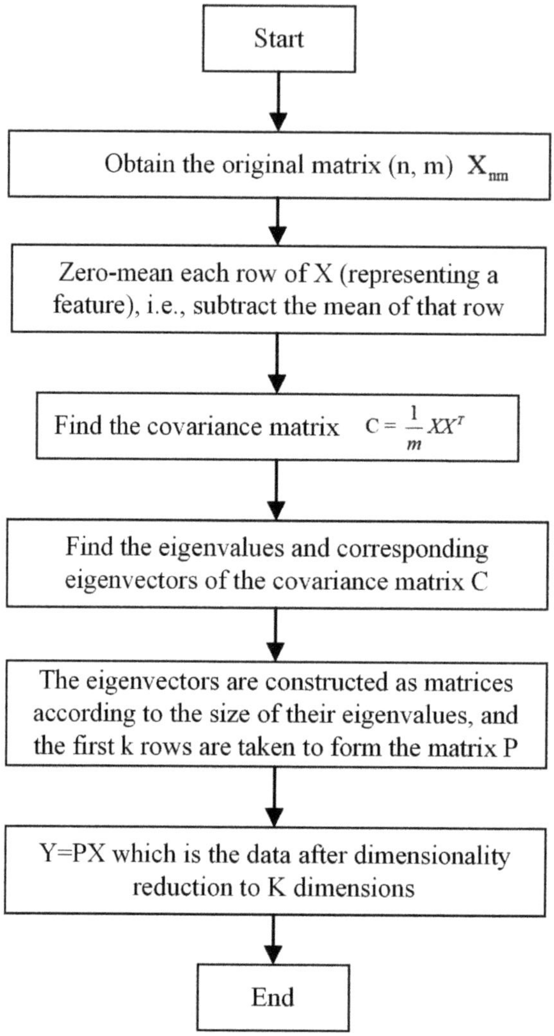

Figure 2. Flowchart of PCA dimensionality reduction algorithm.

2.5. Particle Swarm Optimization Algorithm

Weighted data fusion refers to the statistical analysis of multi-sensor data in different times and spaces and then using relevant mathematical methods or practical experience to assign different weights to different sensing data and obtain data fusion values. This includes the weighted average method, Kalman filter, and artificial neural network method. Weighted data fusion aims to obtain a better representation of the features of multi-source data. Weighted data fusion is the fusion of sensor data according to a certain weight, which sets the sensor observation value at $a_i, i = 1, 2, \ldots, n$, where each sensor weighting coefficient is set to $w_{a_i}, i = 1, 2, \ldots, n$ to obtain fused data:

$$\overline{Y} = \sum_{i=1}^{n} w_{a_i} a_i \quad (12)$$

In order to implement the distribution of adaptive weight coefficients among sensor observations, the particle swarm optimization technique and the entropy weight method were introduced in this research. The chaotic mapping algorithm is used to optimize the particle swarm algorithm because it helps particles jump out of the local optimum and speeds up convergence, which addresses the issue that the particle swarm optimization algorithm is prone to premature convergence to the local optimum and slow convergence at later stages of iteration. This particle's fundamental formula for updating its position and velocity is:

$$X_i = (x_{i1}, x_{i2}, \ldots, x_{iD}), i = 1,\ 2, \ldots,\ N \tag{13}$$

$$V_i = (v_{i1}, v_{i2}, \ldots, v_{iD}), i = 1,\ 2, \ldots,\ N \tag{14}$$

$$v_{ij}(t+1) = wv_{ij}(t) + c_1 r_1(t)\left[p_{ij}(t) - x_{ij}(t)\right] + c_2 r_2(t)\left[p_{gi}(t) - x_{ij}(t)\right] \tag{15}$$

$$x_{ij}(t+1) = x_{ij}(t) + v_{ij}(t+1) \tag{16}$$

where $i = 1,\ 2, \ldots,\ N$ denotes the number of particle swarms; j denotes the dimension, P_{ij} denotes the j-th dimension of the individual extreme value of the ith particle; P_{gj} denotes the j-th dimension of the global optimal solution; t denotes the number of iterations of the particle swarms; w is the inertia factor, which is generally taken as the value of 0.5–0.8, and denotes the strength of the algorithm's global optimization seeking ability; and c_1, c_2 is the learning factor, which is generally taken as the value of 0–4.

2.6. Comparison of Fusion Effect

2.6.1. Improved Chaotic Particle Swarm Optimization Algorithm

Mainly in the particle swarm optimization algorithm, chaotic mapping and the entropy weighting method are introduced to achieve the adaptive weighted fusion of vibration signals in the X, Y, and Z directions so that fused signals can characterize more features, the specific process of which is shown in Figure 3.

The specific implementation steps of the improved chaotic particle swarm optimization algorithm are as follows: (where the observation of the sensor is defined as $a_i, i = 1,\ 2, \ldots,\ n$. The weighting coefficient of the sensor is also set to $w_{a_i}, i = 1,\ 2, \ldots,\ n$.)

(1) Obtaining the original vibration signals: load the original vibration signals in the X, Y, and Z directions to obtain three vectors of length N.
(2) Wavelet packet denoising: 4-layer wavelet packet denoising is performed on the loaded signal to obtain the reconstructed signal in the x, y, and z directions. $a_i, i = 1,\ 2, \ldots,\ n$.
(3) Divide the samples: each vector is randomly divided into 200 samples of length 1024 to obtain 600 samples, which are then stored in a 600×1024 matrix.
(4) Entropy weighting method to extract time domain and frequency domain features: for each sample, 14 time domain features and 5 frequency domain features are calculated to obtain a 19-dimensional feature vector. For all 600 samples, a 600×19 feature matrix is formed.
(5) The obtained feature matrix is downscaled using the PCA downscaling algorithm, and the first three principal components are selected according to the contribution rate, constituting a brand new feature matrix of 600×3. This matrix is normalized, and the weight of the feature matrix is calculated using the entropy weight method.
(6) Chaotic particle swarm optimization algorithm: the initial positions and velocities of the particles are optimized using the Logistic chaotic mapping search algorithm, and the fusion weights are iteratively updated using Shannon's direct as the fitness function. According to the optimization results, the optimal fusion weights of the vibration signals in three directions are obtained, $w_{a_i}, i = 1,\ 2, \ldots,\ n$. The number of particles is set to 20, the maximum number of iterations for the particle swarm optimization algorithm is set to 50, and the number of iterations of the chaotic mapping is set to 30.

(7) Data fusion: according to the optimal weights, the vibration signals in the three directions are weighted and fused, and the fused data are obtained and saved as a new data set, which is biased and calculated afterward.

$$\overline{Q} = \overline{w}_{a1}\overline{a}_1 + \overline{w}_{a2}\overline{a}_2 + \cdots \overline{w}_{an}\overline{a}_n \tag{17}$$

where Q is the fused vibration signal.

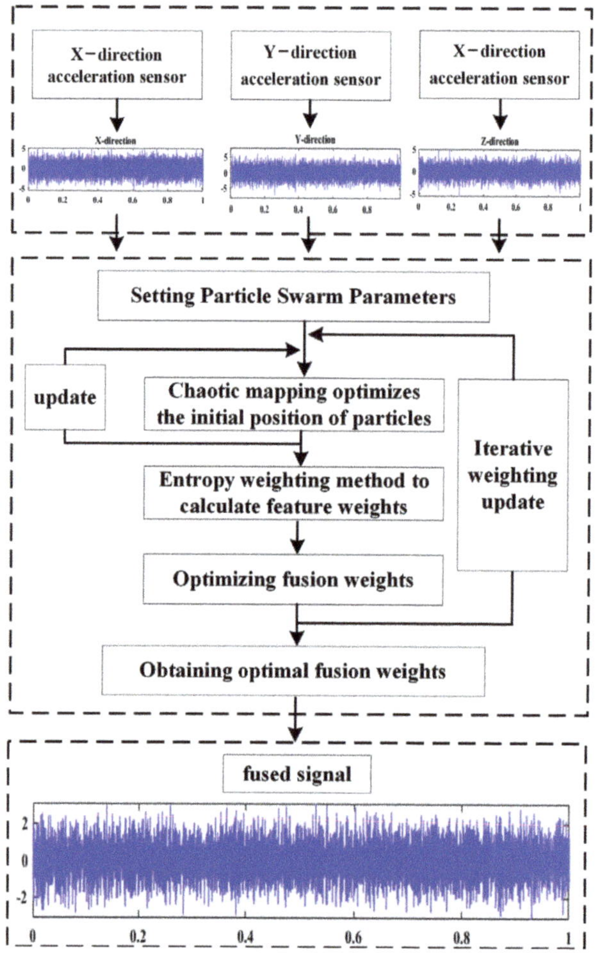

Figure 3. Improved chaotic particle swarm optimization algorithm.

2.6.2. Comparison of Algorithm Fusion Effects

The particle swarm optimization algorithm is primarily used for the adaptive weighted fusion of vibration sensor data in the X, Y, and Z directions to assign the weight coefficients in three directions; therefore, the fused data can more thoroughly and better characterize the effective features of bearings in different states. The comprehensive indexes are primarily established for the analysis of the data-level fusion effects of the following four schemes:

Scheme I: particle swarm optimization (PSO)
Scheme II: particle swarm optimization + entropy weight method (CPSO-EWM)
Scheme III: Chaos mapping + particle swarm optimization (CPSO)

Scheme IV: Chaos mapping + particle swarm optimization + entropy weight method (CPSO-EWM)

The establishment of comprehensive indexes: comprehensive indexes are established based on the characteristics of three indexes: signal-to-noise ratio (SNR), root mean square error (RMSE), and correlation coefficient (Corrcoef).

$$CI = \sum_{n=1}^{3} \left(\frac{SNR}{RMSE} + Corrcoef \right), n = [x,\ y,\ z] \tag{18}$$

Table 2 shows that the fused dataset produced by the improved chaotic particle swarm search technique has a superior fusion effect, a greater correlation with the original signal, and a superior fusion of useful aspects for the vibration signals in the *X*, *Y*, and *Z* directions.

Table 2. Comparison of data fusion effects for four schemes.

	PSO	PSO-EWM	CPSO	CPSO-EWM
CI	9.3194	13.9624	11.5777	18.0705

Except for the above comparison on particle swarm algorithms, Table 3 shows that there are some other similar population intelligence optimization algorithms, such as Ant Colony Optimization (ACO), the Artificial Fish Swarm Algorithm (AFSA), and Fish School Search (FSS). The adaptive weighted fusion of isomorphic signals can be achieved by these population intelligence optimization algorithms.

Table 3. Comparison of the effectiveness of common adaptive optimization algorithms.

	PSO	PSO-EWM	CPSO	CPSO-EWM
CI	9.3194	13.9624	11.5777	18.0705

From Table 3, it can be analyzed that the CPSO-EWM algorithm proposed in this paper has some advantages over the other three optimization algorithms. Among them, the fusion effect of ACO and AFSA is close to that of AFSA, and both AFSA and FSS are optimization algorithms based on the behavior of fish populations, though clearly, the fusion effect of AFSA is slightly better.

2.7. Deep Learning Related Modules Introduction

2.7.1. DenseNet Module

DenseNet is a densely connected convolutional neural network whose main purpose is to solve the problems of gradient vanishing and feature repetition during deep network training. The main function of the DenseNet module is to extract effective feature information from input data, and after each convolutional layer, its output is spliced with the output of previous convolutional layers to form a dense connection.

The DenseNet module contains several dense blocks, where each dense block consists of several convolutional layers and pooling layers. In each dense block, all the convolutional layers accept the outputs of all previous convolutional layers and are used as inputs, thus enhancing feature multiplexing and information transfer.

By multiplexing these features, the DenseNet network presents a new structure that not only slows the occurrence of gradient vanishing but also has fewer parameters and is coupled via cross-channel formulas like:

$$x_l = H_l([x_0, x_1, \ldots, x_{l-1}]) \tag{19}$$

where x_0 and x_l denote inputs to the network and the outputs of layer l, respectively, x_{l-1} is the input to layer $l-1$ of the network, and $H_l(\cdot)$ is the nonlinear transformation operation that acts on layer l.

2.7.2. Transformer Module

The transformer is a neural network model based on the self-attention mechanism for processing sequence data. It mainly consists of two parts: the encoder and the decoder. The Transformer module contains a multi-head self-attention layer, a feed-forward neural network layer, and a residual connection. In signal processing tasks, the Transformer module can be used to extract the important features of elements in the sequence, thus improving the performance of the model. The specific module structure involved is as follows.

(1) Position Encoding: with the introduction of positional encoding, the Transformer model, in order to obtain better parallel computing power, is added to the embedding vector (embedding) of an element as an overall vector by encoding the position of the element in the sequence. Positional coding uses the following functions:

$$PE_{(pos,2i)} = sin\left(pos/10,000^{2i/d\_model}\right) \tag{20}$$

$$PE_{(pos,2i+1)} = cos\left(pos/10,000^{2i/d\_model}\right) \tag{21}$$

where P is the position matrix whose parameters can be updated with the model training process, $P \in R^{n \times d}$.

(2) Multi-head attention: the multi-head attention mechanism used inside the encoder and decoder structures in the Transformer model is obtained by extending the dimensions based on the Scaled Dot-product Attention mechanism.

Scaled Dot-product Attention is a kind of self-attention mechanism, i.e., its own vectors, including Q(query), K(key), and V(value), participate in the computation, and its specific computation is as follows:

$$S_A(Q, K, V) = Softmax(A) \cdot V = Softmax\left(\frac{Q \cdot K^T}{\sqrt{d}}\right) \cdot V \tag{22}$$

where d is the number of dimensions; $S_A(\cdot)$ is the self-attention computation operation; A is the self-attention matrix, $A \in R^{n \times n}$, and n is the sequence length.

The multi-head attention mechanism extends the scaling dot product attention algorithm to multiple dimensions; that is, for the multi-head, after calculating the scaling dot product attention used for multiple information, each result is spliced. The calculation process is as follows:

$$S_{MHA}\left(\overline{Q}, \overline{K}, \overline{V}\right) = concat(H_1, H_2, \ldots, H_h) \cdot W \tag{23}$$

$$H_i = S_A(Q_i, K_i, V_i), i = 1,\ 2, \ldots,\ h \tag{24}$$

where, Q, K, V is the Q_i, K_i, V_i splicing composition, respectively; W is the linear transformation matrix, $W \in R^{d \times d}$; and $concat(\cdot)$ is the splicing operation.

(3) Residual Connection: the Transformer uses residual connection to enhance the flow of information to improve performance and optimize the training process in combination with the layer normalization operation as follows:

$$R_c = L_N[X + S_{MHA}(x)] \tag{25}$$

where $R_c(\cdot)$ is the residual join operation; $L_N(\cdot)$ is the layer normalization operation; X is the input sequence; and $S_{MHA}(\cdot)$ is the use of multiple attention mechanisms.

(4) Data enter a fully connected network made up of two linear transformation layers and one nonlinear activation layer after being output from the multi-attention layer. The activation function in this network uses a linear rectification function.

$$Net_{FFN} = W_2 \cdot f(W_1 \cdot X) \tag{26}$$

where $Net_{FFN}(\cdot)$ is the feedforward network; W_1, W_2 is the parameter of each of the 2 linear layers; and $f(\cdot)$ is the nonlinear activation function.

(5) Max-pooling: the Transformer module's encoder ends with the introduction of the pooling layer downsampling function. The pooling procedure, in which the pooling layer adopts the maximum pooling can lower the size of the feature vectors and the danger of overfitting.

$$y_a = max\left(r_a^{n \times n} u(n,n)\right) \tag{27}$$

where y_a is the output feature of region a; $r_a^{n \times n}$ represents the $\alpha - th$ region of size $n \times n$; and $u(n,n)$ represents the window function of size $n \times n$.

2.7.3. Introduction to the DAT feature extraction model

The DAT deep learning model (DenseNet and Transformer, DAT) is a tandem combination of DenseNet and Transformer modules, which can realize more complex feature extraction and transformation; this structure is shown in Figure 4. The deep feature extraction model uses a tandem combination of DenseNet and Transformer for one-dimensional signal feature extraction and has the following roles and advantages:

(1) Increase feature extraction's effectiveness: the Transformer, on the one hand, uses the self-attention mechanism, which is able to better capture key information in the sequence and improve the accuracy and efficiency of feature extraction. DenseNet, on the other hand, has the characteristic of dense connection, which can more fully utilize low-level features for classification and improve the efficiency of feature extraction.

(2) Increase the model's capacity for generalization: Transformer and DenseNet both possess excellent feature extraction and generalization capabilities, enabling them to deal with complicated changes and noise interference in the bearing signal and increase the model's capacity for generalization.

(3) Make the most of the bearing vibration signal's time series properties. The bearing signal is a type of time series signal and contains a few time series features. Both DenseNet and Transformer can fully utilize time series features to extract more thorough and precise feature representations, resulting in better classification of the signal.

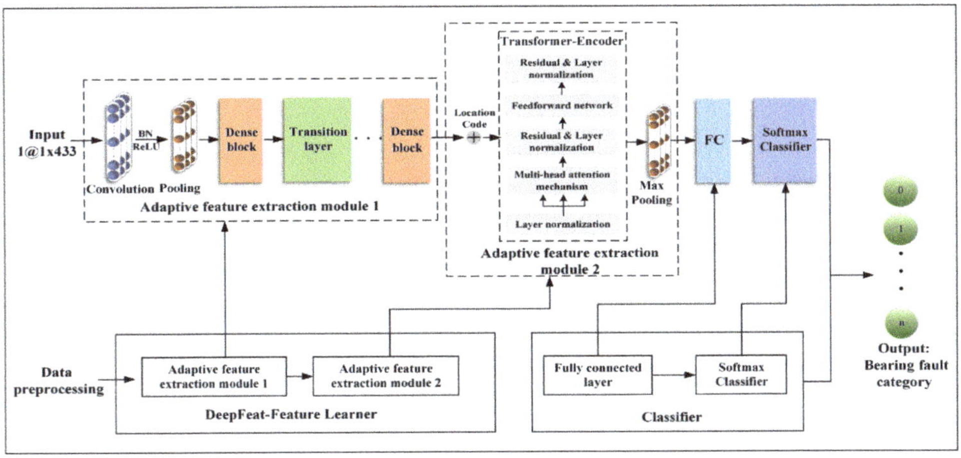

Figure 4. Overall structure of the DAT-based diagnostic model.

In conclusion, the feature extraction of bearing one-dimensional signals using DenseNet and Transformer in tandem can significantly increase the model's classification accuracy, generalizability, and interpretability, making it ideal for challenging signal classification applications.

Figure 4 depicts the overall structure of the DAT-based fault detection model, which is composed of a feature learner and a deep learning classifier and is implemented as follows: (1) preprocessing and the sample expansion of sensor measurement data is performed (see Section 2.3); (2) Fourier variation is used to transform the preprocessed one-dimensional vectors into a spectrum of 1433 before being input into the DAT model; (3) in the learning feature, two deep feature extraction processes are used. the convolutional layer, pooling layer, dense block (DenseBlock), transition layer, etc., make up the main components of the DenseNet dense connected network (deep feature extraction module 1), and the multi-head self-attention layer, feed-forward neural network layer, residual connection, and other model structures make up the main components of the Transformer module (deep feature extraction module 2). (4) The extracted features are input into the feature classifier (the fully connected network layer and residual connection) and other model structures; Table 4 displays the unique DAT diagnostic model parameters.

Table 4. DAT network parameters.

Model Name	**1D-DAT**	
	Structure Type	**Convolution Kernel**
Input layer	1D FFT spectrum	—
Convolution layer	Conv	1×7
Pooling layer	Max-pooling	1×3
Adaptive Features Extraction Module 1	$\text{Dense block} - 1 : \left\{\begin{matrix}\text{BN} - \text{Relu} - \text{Conv} \\ \text{BN} - \text{Relu} - \text{Conv}\end{matrix}\right\} \times 1$	$\left\{\begin{matrix}1 \times 1 \\ 1 \times 3\end{matrix}\right\} \times 1$
	Transition layer-1:BN-Relu-Conv-Pooling	$\left\{\begin{matrix}1 \times 1 \\ 1 \times 2\end{matrix}\right\} \times 1$
	……	……
	Transition layer-2:BN-Relu-Conv-Pooling	$\left\{\begin{matrix}1 \times 1 \\ 1 \times 2\end{matrix}\right\} \times 1$
	$\text{Dense block} - 3 : \left\{\begin{matrix}\text{BN} - \text{Relu} - \text{Conv} \\ \text{BN} - \text{Relu} - \text{Conv}\end{matrix}\right\} \times 1$	$\left\{\begin{matrix}1 \times 1 \\ 1 \times 3\end{matrix}\right\} \times 1$
Adaptive Features Extraction Module 2	Position code$\times$1	—
	Encoder$\times$1	—
	Encoder$\times$1	
	Max-pooling	$\times$1
Fully connected layer	FC	—
Output layer	SoftMax	—

3. Model Setup and Training

3.1. Condition Monitoring Model Introduction

A data-level fusion method is proposed in this paper based on multi-source isomorphic sensors to monitor the operational status of rolling bearings, as shown in Figure 5. This procedure includes the decomposition of wavelet packets, denoising, the reconstruction of the vibration signal, feature extraction in the time and frequency domains, and dimension reduction using the PCA algorithm. The chaotic particle swarm algorithm is used with the entropy weighting method to produce the initial weights and global optimal weights. Last but not least, the Transformer module is shown in order to build the DAT feature extraction model and enhance the precision and effectiveness of feature extraction. This technique can successfully handle the requirement for tracking the operational status of machine tool spindle bearings under challenging operating conditions.

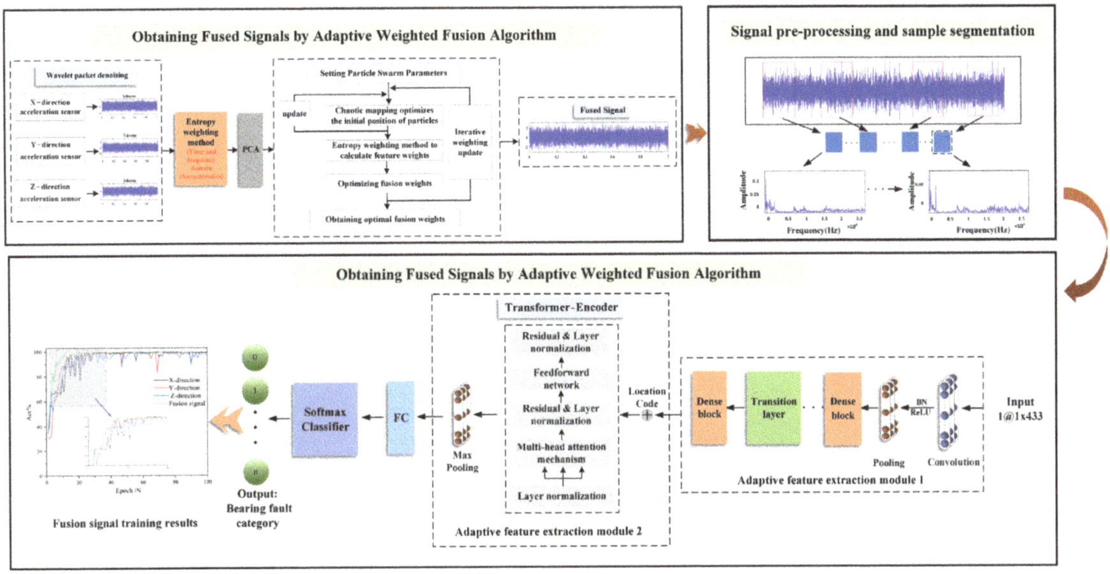

Figure 5. Overall structure of the condition monitoring model.

3.2. Data Pre-Processing

Experimental dataset S1: Open-source data on rolling bearing faults were published by Jiangnan University's experimental platform to evaluate and study the model for identifying rolling bearing faults. The experimental rolling bearing fault diagnosis system for wind turbines at Jiangnan University is depicted in Figure 6. Rolling bearing vibration signals were collected at speeds of 600, 800, and 1000 rpm with a constant rotational speed of 1 krpm, a sampling frequency of 50 kHz, and a sampling time of 10s [37]. Bearing failure was man-made through the wire cutting technology, respectively, in the bearing inner ring, outer ring, rolling body, and the processing of 0.3 * 0.05 mm (width * depth) tiny wounds, as can be seen from the waveform diagram. The data of various states are difficult to distinguish directly from this waveform diagram.

Figure 6. Rolling bearing failure test platform of Jiangnan University.

Each set of experiments involved three different fault conditions (rolling body damage, inner ring damage, and outer ring damage) as well as one type of normal condition. Table 5 for the speed of 600 r/min in the experimental dataset describes the type of fault conditions, as shown in Figure 7, for the experiments according to fan speed.

Table 5. Experiment 1 (600 r/min) data set.

Experimental Conditions	Training Set: Validation Set: Test Set	Labels
Inner ring failure	280:80:40	IF
Outer ring failure	280:80:40	OF
Ball Failure	280:80:40	BF
Normal	280:80:40	Normal

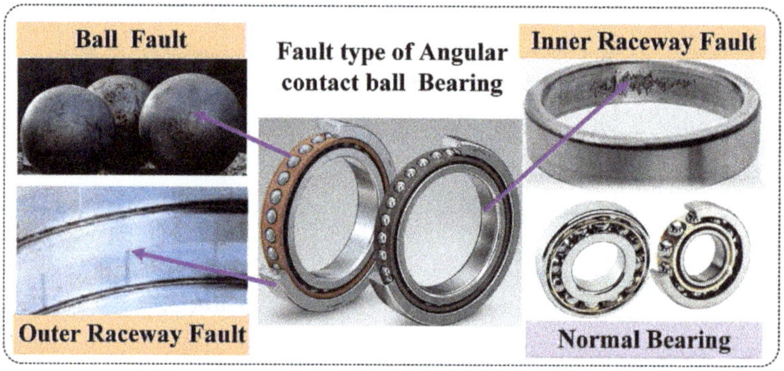

Figure 7. Distribution of bearing condition types.

Numerous high-frequency and low-frequency components with varying sensitivity levels can be found in the time domain of the vibration signal, which is used to diagnose bearing defects. In order to analyze the time-domain signals more effectively, it is necessary to convert them into frequency-domain signals. As depicted in Figure 8, four bearing signals—BF, IF, OF, and normal—under a rotation speed of 600 r/min were taken for spectrum analysis, and 1024 data points were taken as a sample for FFT transformation. It can be found that the normal state of the bearing, the fault of the rolling element, the fault of the inner ring, and the fault of the outer ring have obvious differences in the amplitude of the whole stage of the spectrum, which can effectively identify the frequency component of the signal and provide useful information for the application of signal feature extraction, classification, and diagnosis.

Experimental dataset S2: To further investigate the monitoring function of this method during the double-bearing operation, an unbalanced bearing load test rig was developed, as shown in Figure 9. A non-balanced bearing load test platform was designed and manufactured, and the monitoring function of this technology was investigated during the bearing operation. The test platform, as illustrated in Figure 9, consists of a motor, precision spindle, roll bearing, and acceleration sensor with a maximum speed of 10,000 r/min. A flexible coupling connects the mechanical spindle to the electric spindle, and the motor action is regulated by a servo control system. The hardware consists of a motorized spindle, a rotational accuracy test device, a data collector, a computer, and other components.

The platform employed four NSK 7014C angular contact ball bearings, with the positions of F1, F2, and F3 evenly spaced at 120°. Preloads of different sizes were set to determine the operating conditions of bearings, including light (C2), medium (C4), and heavy (C6) loads. The bearings were mounted back-to-back, and the fixed speed of the test platform was set at 4000 r/min with a sampling frequency of 8192 Hz. The parameters of the bearings are shown in Table 6.

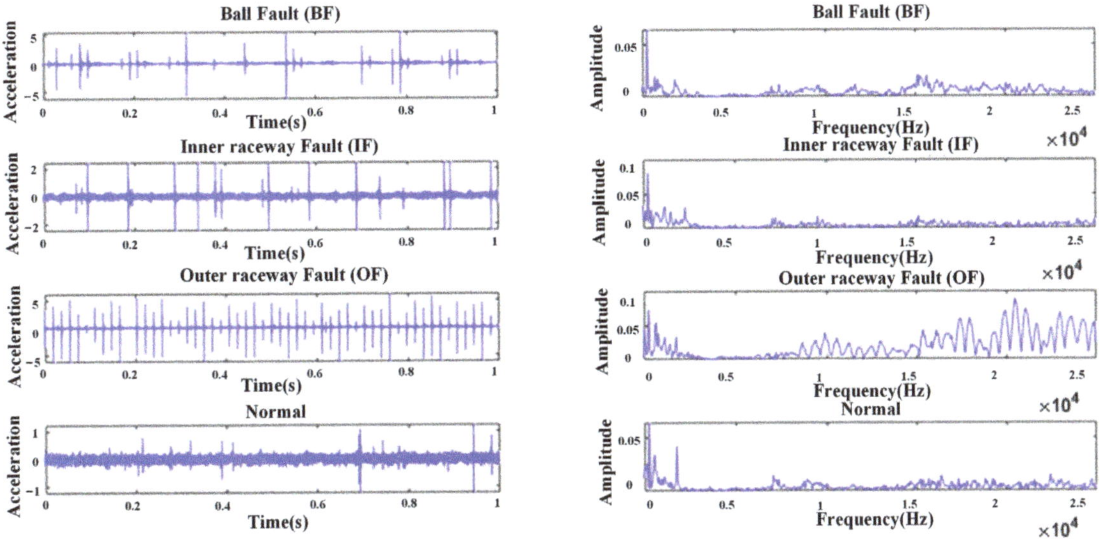

Figure 8. Time-domain waveforms and spectrograms of bearing data at Jiangnan University.

The purpose of building the test platform was to distinguish the working condition of bearings under an unbalanced operation so as to detect the bearing failures caused by wrong assembly or processing in real-time. Due to the limited conditions of the laboratory, the current test platform can only be used to verify the effectiveness and accuracy of the condition monitoring method and cannot simulate the corresponding bearing fault state under different loads.

3.2.1. Data Normalization

Normalized preprocessing is a commonly used data processing method. These data are mapped to a specific interval by applying a linear transformation to the data and commonly mapping the data to the interval [0, 1]. If sample data are supposed to be $X = \{x_1, x_2, \ldots, x_n\}$, the normalized transformation formula is as follows:

$$y_i = \frac{x_i - min\{x_j\}}{max\{x_j\} - min\{x_j\}} \quad (28)$$

where y_i is the result of normalization, x_i is the $i - th$ sample data, $max\{x_j\}$ is the maximum value of the sample data, and $min\{x_j\}$ is the minimum value of the sample data.

3.2.2. Overlapping Sampling

In the realm of data-driven deep learning, having enough big training samples is essential to increase model accuracy and significantly lower overfitting. As illustrated in Figure 10, we used overlapping sampling with a moving sliding window to increase the number of training samples, which can better capture changes and patterns in time series data. This method can avoid the problem of signal loss caused by equidistant sampling and sampling, thus improving the training effect and generalization ability of the model.

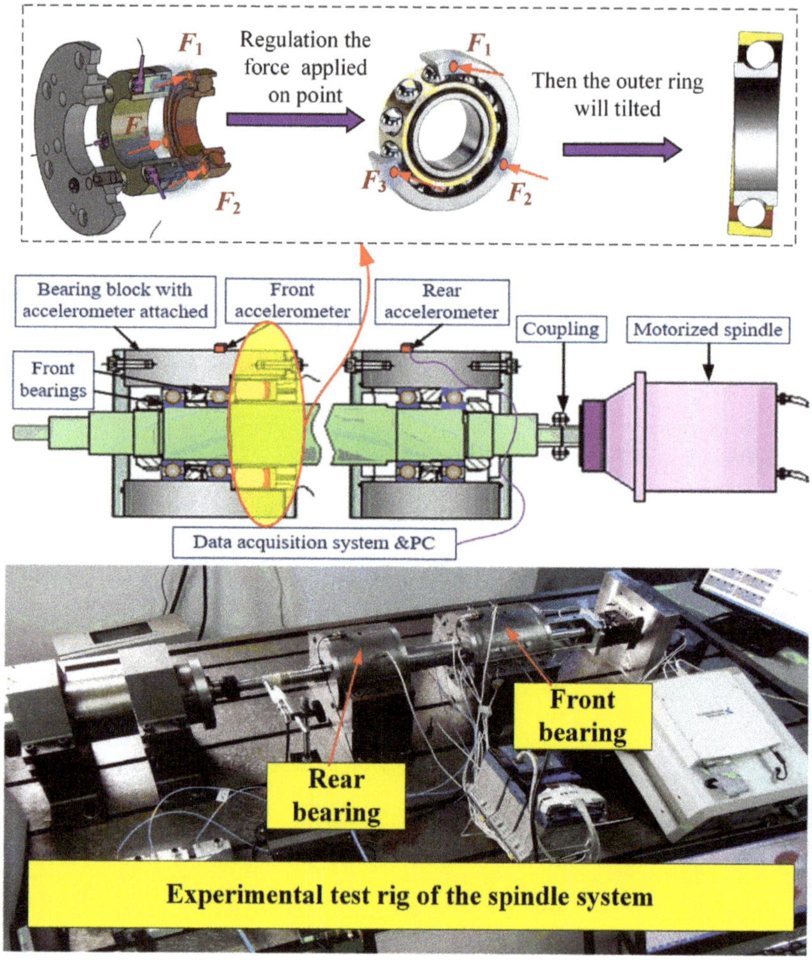

Figure 9. Structure of non-uniform preloading test stand.

Table 6. Parameters of NSK 7014C angular contact ball bearings.

Inner Ring Diameter/mm	Outer Ring Diameter/mm	Thickness/mm	Dynamic Load/KN	Static Load/mm
70	100	20	47	43

In order to avoid the loss of detailed features, overlap sampling is used to extend the original data samples. By adjusting the parameters such as offset, data length, expansion multiplier, and the number of samples, the detection of the model performance under different numbers of samples can be achieved. This can be achieved by flexibly adjusting the sampling parameters so as to optimize the training and prediction ability of the model.

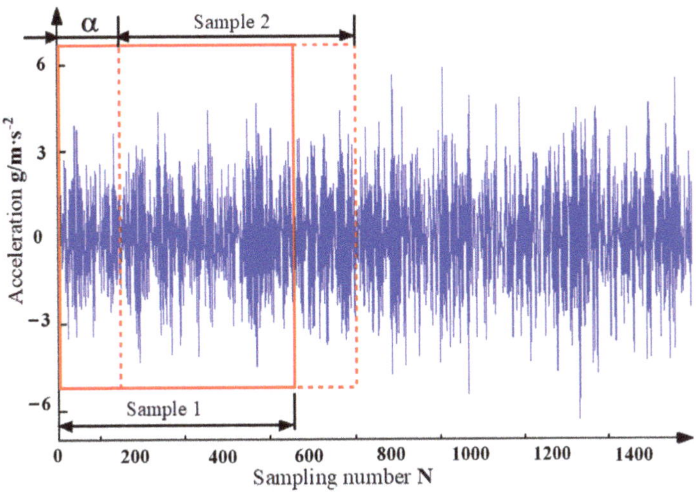

Figure 10. Schematic diagram of data enhancement.

3.3. DAT Model Hyperparameter Settings

The AdamP backpropagation algorithm proposed by ByeonghoHeo et al. [38] was selected as the optimization method, which improved the performance of small-batch training, reduced the risk of overfitting, and delayed the attenuation of effective step size, thus training the model at a barrier-free speed, retaining many advantages of the Adam algorithm, such as adaptive learning rate adjustment and momentum term.

The rolling bearing service condition monitoring model based on DAT is based on features that diagnose working conditions, with data under different working conditions mainly classified and recognized. The Cross-entropy loss function was chosen as the base function, and some improvements were made.

The Cross-entropy loss function is formulated as follows:

$$loss = -\sum_{\theta} p(\theta) \log q(\theta) \tag{29}$$

where θ denotes the learning parameter; $p(\theta)$ and $q(\theta)$ are the correct probability and prediction probability of the label.

The K-L scatter of $p(\theta)$ and $q(\theta)$ is as follows:

$$KLD = \sum_{\theta} p(\theta) \log \frac{p(\theta)}{q(\theta)} \tag{30}$$

When calculating Cross-entropy loss using KL scatter, it is often necessary to transform the true labels into probability distributions. The traditional approach is to use one-hot coding, where only one element is 1, and the rest are 0, indicating the category to which the true label belongs. However, this approach may lead to the overconfidence or over-sensitivity of the model in the presence of noise and uncertainty.

In order to reduce the impact of label noise and uncertainty in the model, this paper uses smoothed target labeling. By smoothing the target labels, we can better adapt to complex data distributions and noise situations and, thus, obtain a more accurate loss function of Ce.

Figure 11 shows that the iteration effect of the experiment is significantly improved when using Ce_loss compared to the iteration curve of the model under Cross-entropy loss. The accuracy of the training set of this paper's method reached 99% after 10 iterations, while the accuracy of Cross-entropy loss reached 99% after 85 iterations. The experiment at 600 r/min under the S1 dataset was chosen for analysis. The experimental results demonstrate that the strategy suggested in this research can significantly enhance the model's training performance while shortening the training period. Through the experimental analysis, the parameters of the model were set, as shown in Table 7.

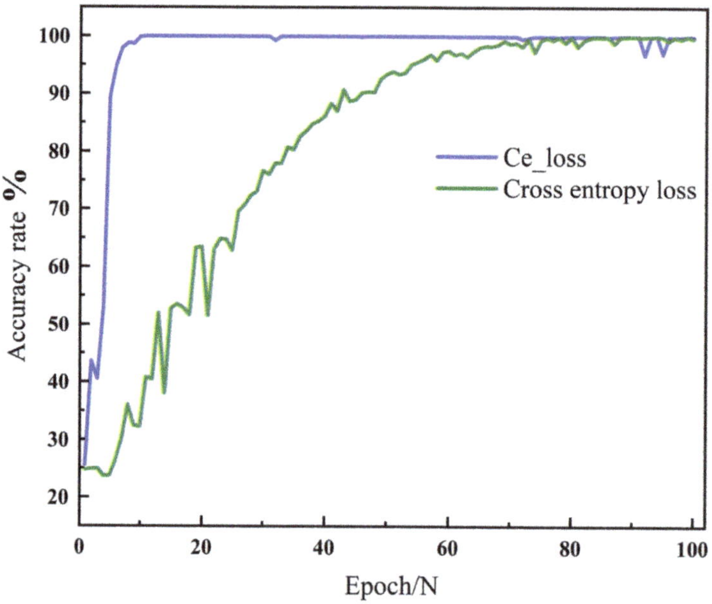

Figure 11. Model iteration curves under 600 r/min experiment with different loss functions.

Table 7. Model parameters (experimental data set S1).

Parameter Category	Parameter Setting
Optimizer	AdamP
Loss function	Ce_loss
Number of iterations	100
Initial learning rate	0.001
Smoothing	0.1
Batch Size	64

3.4. Model Training

Each group of tests was repeated five times to assess the accuracy and stability of the proposed model for rolling bearing failure diagnosis with the average accuracy and greatest accuracy of the experiments provided in Table 8. On the rolling bearing fault data collected by Jiangnan University's rolling bearing fault diagnostic platform, fault diagnosis was performed using the deep learning framework PyTorch (see Table 5 for details).

Table 8. Fault diagnosis results of rolling bearing based on DAT model.

Data Sets	Experimental Setup	Number of Iterations	Training Time/s	Average Acc/%	Maximum Acc/%
Jiangnan University	600 r/min	100	57	99.375%	99.756%
	800 r/min	100	62	99.583%	99.625%
	1000 r/min	100	58	99.285%	99.423%

As shown in Figure 12, the DAT model's greatest accuracy on the validation set at 600 r/min was 99.8%, and after five iterations, training accuracy was 90%. The model was more stable throughout the training process, and there was no abrupt change in accuracy, which suggests that this model has a strong ability to generalize, is highly robust, is able to capture data in real patterns, and is not easily disturbed by noise and outliers. This can be seen by analyzing the iteration curves in the Figure below.

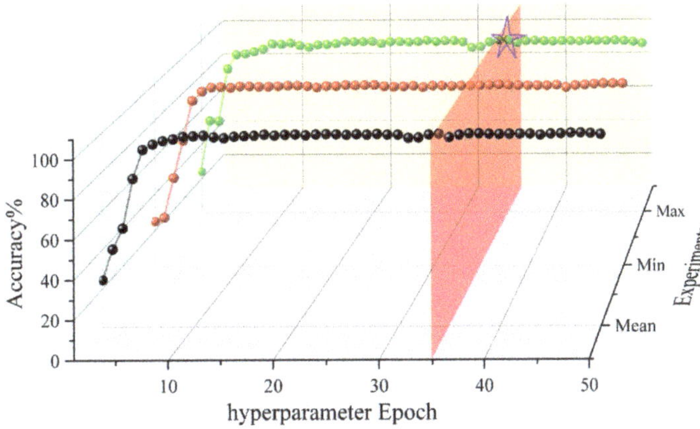

Figure 12. Model training at 600 r/min rotational speed.

Figure 13 shows the confusion matrix of the model under the rolling bearing fault diagnosis platform of Jiangnan University, which was used to evaluate the performance of the classification model and helped us gain a more comprehensive understanding of the classification effect of the model in different categories.

Figure 13 shows that the DAT model achieves the highest accuracy of 99.5% in five experiments on the experimental dataset of 600 r/min at Jiangnan University. By analyzing the confusion matrix, it was found that 2% of the IF600 (Inner ring failure) were incorrectly predicted as BF600 (Ball failure), 2% of the Normal600 (Normal) were incorrectly predicted as IF600 (Inner ring failure), and the rest were correctly classified.

Under the same experimental conditions, the proposed DAT fault diagnosis model was compared with Transformer, DenseNet-LSTM, CNN-LSTM, DenseNet, and other models in the comparison experiments. For the experimental data of rolling bearing faults in Jiangnan University, the average accuracy comparison results of the above different models are shown in Figure 14.

On the basis of DenseNet, the DAT model introduces the Transformer module, in which the Transformer employs the self-attention mechanism, which can better capture key information in the sequence, improve the accuracy and efficiency of feature extraction, and make the DAT model have a better fault diagnosis effect. Table 9 shows that the average accuracy of the DAT model on the experimental data of Jiangnan University's rolling bearing defects is higher than that of the other four models. DenseNet-LSTM adds the LATM layer to the DenseNet network, whereas the DAT model adds the Transformer module to the DenseNet network where both models have higher diagnostic accuracy, but the DAT model has a slightly higher diagnostic accuracy. In terms of learning features, the DAT model is slightly inferior to the DenseNet-LSTM model, implying that the self-attention mechanism is superior to LSTM, which is suited for temporal signal prediction.

In summary, it can be seen that the DAT fault diagnosis model has higher fault diagnosis accuracy and better stability than Transformer, DenseNet-LSTM, CNN-LSTM, DenseNet and other models.

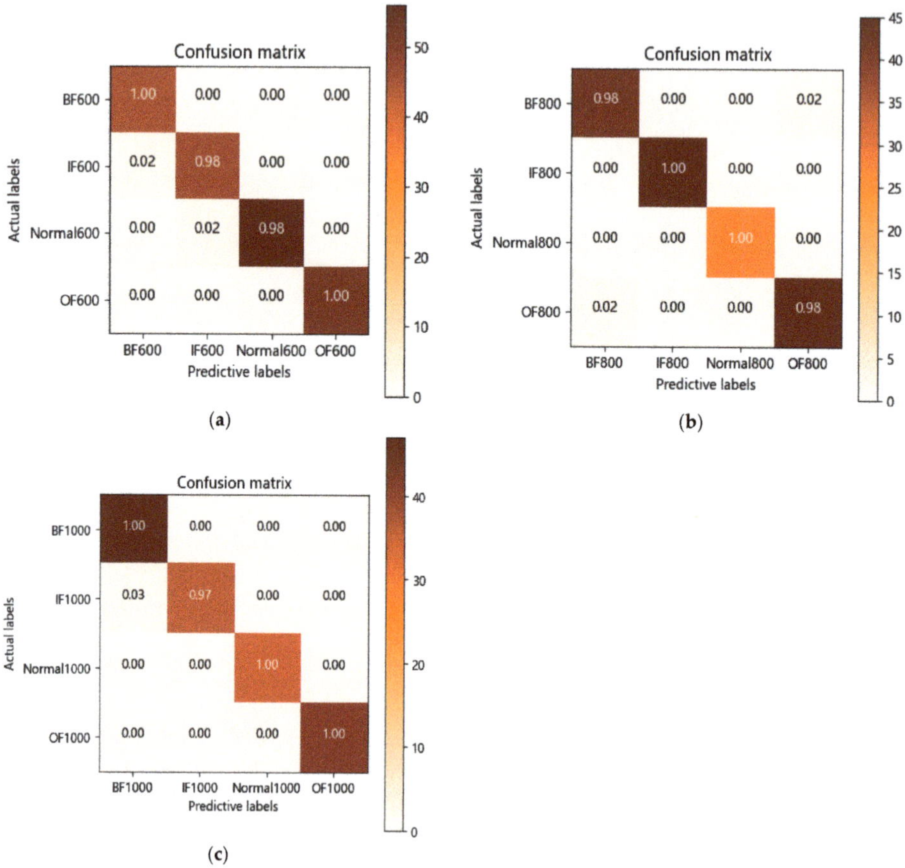

Figure 13. Confusion matrix based on DAT modeling. (**a**) 600 r/min (**b**) 800 r/min (**c**) 1000 r/min.

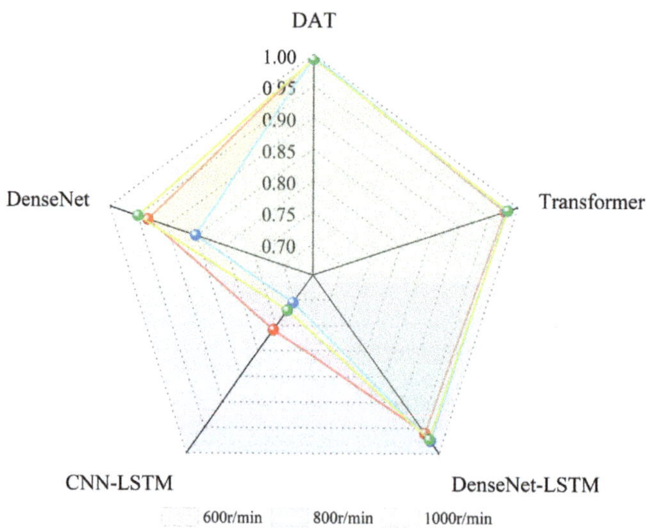

Figure 14. DAT model and its comparison model with average accuracy at different rotational speeds.

Table 9. Average accuracy of DAT model and its comparison model under 5 experiments.

Experiments	DAT	Transformer	DenseNet-LSTM	CNN-LSTM	DenseNet
600 r/min	99.275%	97.725%	96.768%	75.833%	93.333%
800 r/min	99.583%	98.583%	97.525%	70.512%	85.233%
1000 r/min	99.158%	98.375%	97.245%	72.012%	95.076%

4. Fusion Data Testing

The experimental data set S2 was measured using the non-uniform load operation fault simulation test platform. The data of the light load (C2), medium load (C4), and heavy load (C6) under the positions of F1, F2, and F3 were collected, and the experimental data sets were divided as shown in Table 10. Bearing vibration data can be classified into nine types. The experiment was divided into three groups, each of which included 840 samples in the training set, 240 samples in the test set, and 120 samples in the verification set.

Table 10. Experimental data set (experimental data set S2).

Experimental Setup	Signal Type	Training Set	Validation Set	Test Set
Experiment 1 (F1 position)	F1(C2) = 400 N	840	240	120
	F1(C4) = 800 N	840	240	120
	F1(C6) = 1200 N	840	240	120
Experiment 2 (F2 position)	F2(C2) = 400 N	840	240	120
	F2(C4) = 800 N	840	240	120
	F2(C6) = 1200 N	840	240	120
Experiment 3 (F3 position)	F3(C2) = 400 N	840	240	120
	F3(C4) = 800 N	840	240	120
	F3(C6) = 1200 N	840	240	120

In the time domain, by converting the signal to the frequency domain, we could decompose this signal into components of different frequencies and analyze the amplitude of each frequency component. In Figure 15, we selected the bearing signal under the C2 operating condition at the F2 position for spectral analysis. The 1024 data points were taken as a sample, and FFT (Fast Fourier Transform) was transformed to obtain the corresponding spectrogram. From the spectrograms, it can be observed that the spectrograms of all four sets of vibration data had the maximum amplitude variation at 3044 Hz. This reflects the major frequency components of the signal in the frequency domain. In isomorphic data fusion, the low-frequency component often represents low-frequency noise or a slow vibration, which is usually as small as possible. By analyzing the spectrogram, it can be seen that the fused signal had a smaller low-frequency component amplitude at 3044 Hz, indicating that the fusion effect of the isomorphic data fusion method proposed in this paper is clearer.

The experiment's input sample length was set to 1×1024, its Fourier transform was run to choose a 1×433 spectrum as the model's input, the batch size was 64, the number of training iterations was 100, the learning rate was 0.05, and Adam was chosen as the optimizer.

As can be seen from Figure 16, the accuracy curve of the fused signal is relatively stable, reaching 94% after 10 iterations, and the accuracy of the model is stable at 99.34% after 50 iterations. The local zoom-in graph shows that the iterative process of the fused signal is more stable, indicating that the fused signal has more effective features.

According to Table 11, it can be seen that the accuracy of the fused data for the fault identification on the DAT model is slightly higher than that of using a single-direction vibration signal, and the accuracy of the fused data is more stable compared to the unisex signal. In order to judge the classification performance of the training results intuitively, a confusion matrix was used to present them visually, as shown in Figure 17, where the horizontal and vertical axes labels represent the three working conditions of light load (C2), medium load (C4) and heavy load (C6).

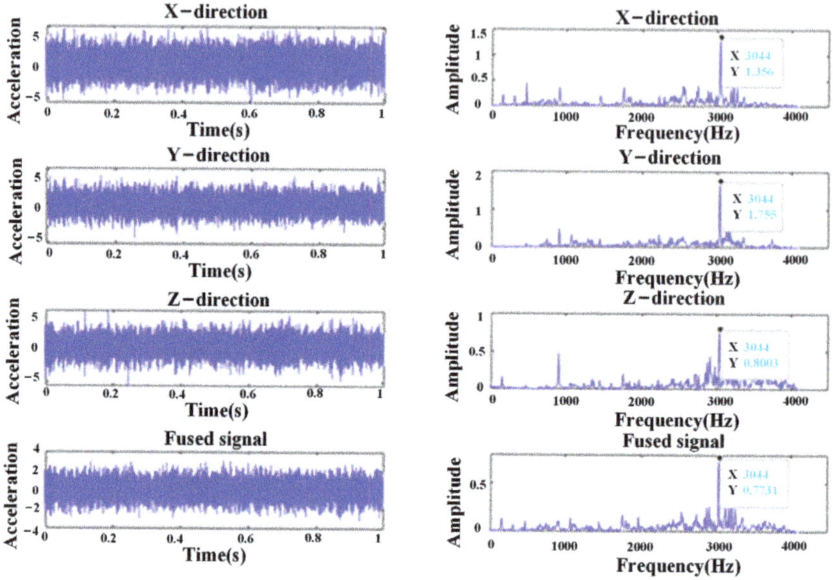

Figure 15. Signal analysis of X, Y, Z and fused signal at F1 position.

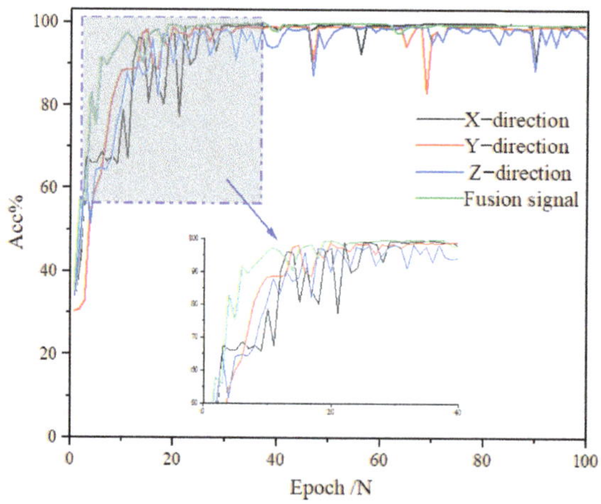

Figure 16. Accuracy curve of fused signal with X, Y and Z signals.

Table 11. Results of control experiments (50 iterations and average accuracy over 5 experiments).

Experimental Setup	Fusion Data		X-Direction Average/Acc%	Y-Direction Average/Acc%	Z-Direction Average/Acc%
	Avg/Acc%	Max/Acc%			
F1 position	99.76%	100%	99.15%	98.56%	97.72%
F2 position	99.82%	100%	98.66%	97.66%	99.17%
F3 position	99.92%	100%	98.52%	99.23%	99.40%

For visualization, the high-dimensional features collected from the DAT model's input layer and final hidden layer were mapped into three-dimensional feature vectors. Figure 16 depicts the visualization results of the features with the best accuracy in five experiments, where the numerical point reflects the fault diagnosis efficiency of the algorithm utilized in this study. Figure 18 shows that among X, Y, Z, and the fusion signal, the fusion signal had the best fault classification impact, showing that the DAT model's classification effect was superior.

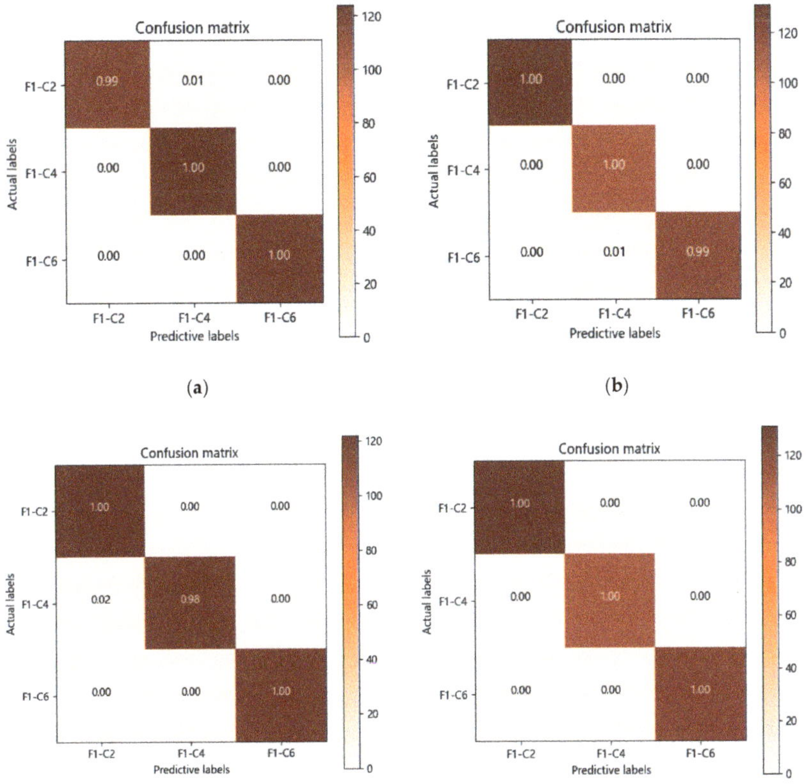

(**a**) (**b**)

(**c**) (**d**)

Figure 17. Confusion matrix for fault classification of the first set of experiments. (**a**) X direction confusion matrix (**b**) Y direction confusion matrix. (**c**) Z direction confusion matrix (**d**) Fusion signal confusion matrix.

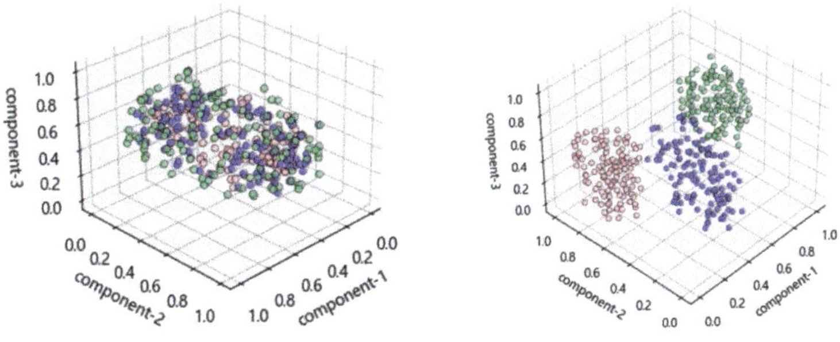

(**a**)

Figure 18. *Cont.*

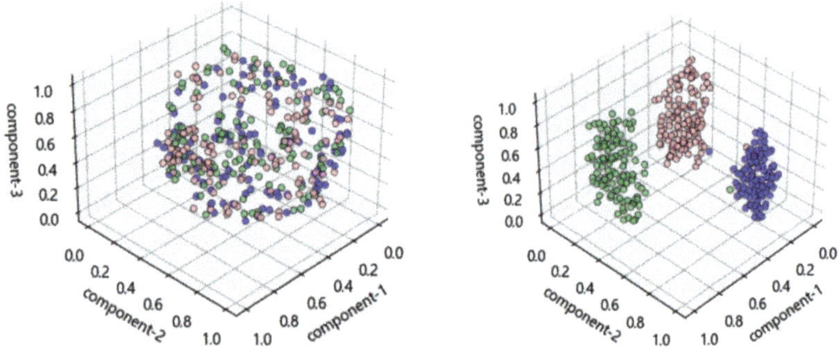

(b)

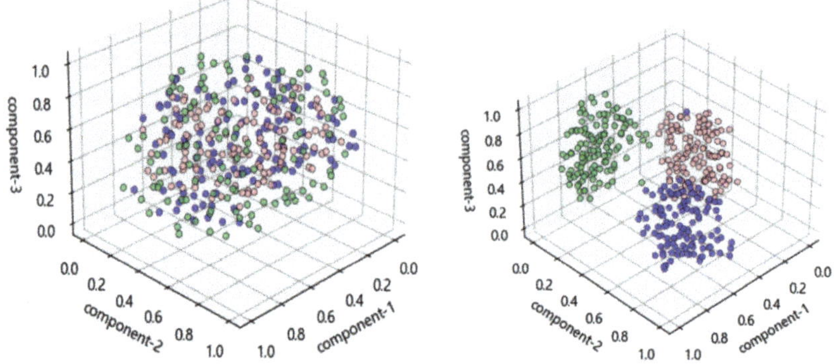

(c)

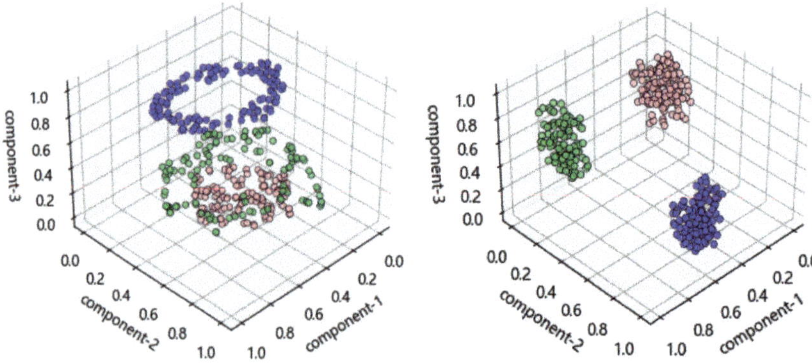

(d)

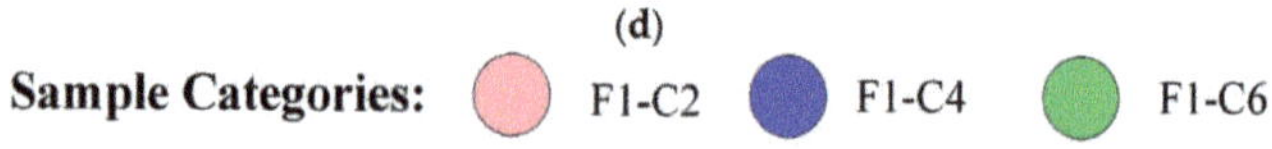

Figure 18. Visualization of input and output layer features for the first set of experiments. (**a**) X direction (Max-99.72%). (**b**) Y direction (Max-99.72%). (**c**) Z direction (Max-99.58%). (**d**) Fusion signal (Max-100%).

Figures 17 and 18 show that in five tests from the first set of experimental datasets of S2 obtained from the bearing non-uniform load test platform, the DAT model similarly attained a minimum accuracy of more than 99%. In total, 1% of the samples from F1-C2 (low load) in X-direction tests were wrongly projected as F1-C4 (medium load), according to the confusion matrix and the output feature downscaling visualization plot, while the remaining samples were correctly classified. Combining information from various isomorphic acceleration sensors can boost redundancy and boost the accuracy of defect finding. Data from other sensors are still available in the event of a sensor failure or abnormality, maintaining the stability of the system and the accuracy of fault identification. By examining the fused signals' X, Y, and Z directions as well as the classification outcomes, it was discovered that the fused signals' highest degree of classification accuracy is 100%, their original features are more distinct from one another, and there was no classification abnormality when compared to the other three signal groups.

5. Conclusions

A data-level fusion method based on multi-source isomorphic sensors is proposed in this paper to monitor the working condition of rolling bearings. The vibration data in the X, Y, and Z directions of raw data were fused firstly using a chaotic particle swarm optimization algorithm. Then, a DAT feature extraction model was built to extract the deep features of the fused signals. Finally, the overall iterative performance of the model was improved using the AdamP optimization algorithm and the improved Ce_loss loss function, reaching the following conclusion.

- The data-level fusion method of multi-source homogeneous sensors is proposed by fusing data from different sensors. Information of multiple dimensions can be obtained, which makes the perception of the target object or the environment more comprehensive and accurate and enhances time-domain continuity alongside the consistency of data, which can be enhanced as well as the fault tolerance and robustness of the system.
- A DAT deep feature extraction model can be constructed to monitor the working condition of spindle bearings, which can recognize the bearing faults and unbalanced loads.
- Through the AdamP optimization algorithm and the improved Ce_loss loss function, the iterative performance of the proposed model can be drastically improved, and the steady state can be reached faster.
- This study validates the fusion performance of isomorphic signals and the diagnostic performance of the model. In the future, we plan to apply the DAT model to other components of the spindle system and migrate it to other fields for validation. This could expand the applicability of the model and increase its value in practical engineering applications.

Author Contributions: Conceptualization, Y.Z.; methodology, Y.Z., Y.L.; software, Y.L.; validation, Y.L., M.Y. and X.F.; formal analysis, Q.Z.; investigation, Y.L.; resources, L.K.; data curation, Y.L.; writing—original draft preparation, Y.L.; writing—review and editing, Y.Z.; visualization, Q.Z.; supervision, Y.Z.; project administration, L.K.; funding acquisition, Y.Z. All authors have read and agreed to the published version of the manuscript.

Funding: This research was funded by National Natural Science Foundation of China (52005405), Shaanxi Provincial Key R&D Program (2022GY-211&No.2023-YBGY-098), Science Foundation of The Tian Di Science & Technology Co., Ltd. (2021-TD-MS006).

Data Availability Statement: Not applicable.

Conflicts of Interest: The authors declare no conflict of interest.

References

1. Jing, L.; Zhao, M.; Li, P.; Xu, X. A convolutional neural network based feature learning and fault diagnosis method for the condition monitoring of gearbox. *Measurement* **2017**, *111*, 2–4. [CrossRef]
2. Jeschke, S.; Brecher, C.; Meisen, T.; Özdemir, D.; Eschert, T. *Industrial Internet of Things and Cyber Manufacturing System*; Springer: Cham, Switzerland, 2017; pp. 3–19.
3. Yin, S.; Li, X.; Gao, H.; Kaynak, O. Data-based techniques focused on modern industry: An overview. *IEEE Trans. Ind. Electron.* **2014**, *62*, 657–667. [CrossRef]

4. Ma, S.; Yin, Y.; Chao, B.; Yan, K.; Fang, B.; Hong, J. A Real-time Coupling Model of Bearing-Rotor System Based on Semi-flexible Body Element. *Int. J. Mech. Sci.* **2023**, *245*, 108098. [CrossRef]
5. Fang, B.; Zhang, J.; Hong, J.; Yan, K. Research on the Nonlinear Stiffness Characteristics of Double-Row Angular Contact Ball Bearings under Different Working Conditions. *Lubricants* **2023**, *11*, 44. [CrossRef]
6. Zhang, Y.F.; Li, Y.H.; Kong, L.F.; Li, W.C.; Yi, Y.J. Rolling bearing condition monitoring method based on multi-feature information fusion. *J. Adv. Manuf. Sci. Technol.* **2022**, *3*, 2022020. [CrossRef]
7. Zhou, J.; Zhu, J.W. Research on machine learning classification problems and algorithms. *Software* **2019**, *40*, 205–208.
8. Zhang, R.; Wang, Y.B. Research on machine learning and its algorithms and development. *J. Commun. Univ. China Nat. Sci. Ed.* **2016**, *23*, 10–18.
9. Wu, X.M.; Wu, Y.Y.; Wang, X.; Li, C.F.; Zhang, F.H. Application of machine learning in bearing fault diagnosis. *Equip. Manuf. Technol.* **2022**, *03*, 03,118–126.
10. Zhang, Y.F.; Li, Y.H.; Wang, D.F.; Kong, L.F. A rolling bearing fault monitoring method based on multi-source information fusion. *Bearing* **2022**, *12*, 12,59–65.
11. Zhang, X.Y.; Luan, Z.Q.; Liu, X.L. A review of deep learning based rolling bearing fault diagnosis research. *Equip. Manag. Maint.* **2017**, *18*, 130–133.
12. Lu, X.Y.; Zhang, C.B.; Gao, J.; Xu, Y.P.; Shao, X. Bearing fault diagnosis algorithm based on convolutional neural network and CatBoost. *Electromechanical Eng.* **2023**, *01*, 1–10.
13. Janssens, O.; Slavkovikj, V.; Vervisch, B.; Stockman, K.; Loccufier, M.; Verstockt, S.; Van de Walle, R.; Van Hoecke, S. Convolutional neural network based fault detection for rotating machinery. *J. Sound Vib.* **2016**, *377*, 331–345. [CrossRef]
14. Gu, X.; Tang, X.H.; Lu, J.G.; Li, S.W. Adaptive Fault Diagnosis Method for Rolling Bearings Based on I-DCNN-LSTM. *Mach. Tool. Hydraul.* **2020**, *48*, 107–113.
15. Zhang, Y.; Liu, Y.; Wang, L.; Li, D.; Zhang, W.; Kong, L. Bearing Non-Uniform Loading Condition Monitoring Based on Dual-Channel Fusion Improved DenseNet Network. *Lubricants* **2023**, *11*, 251. [CrossRef]
16. Jin, X.B.; Lin, Y.S.; Zhang, H. Multisensor fusion estimation in state monitoring. *Control. Theory Appl.* **2009**, *26*, 296–298.
17. Okatan, A.; Hajiyev, C.; Hajiyeva, U. Fault detection in sensor information fusion Kalman filter. *Int. J. Electron. Commun.* **2009**, *63*, 762–768. [CrossRef]
18. Bath, W.G.; Boswell, C.M.; Sommerer, S.; Wang, I. Detection systems information fusion. *Johns Hopkins APL Tech. Dig.* **2005**, *26*, 306–313.
19. Sung, W.T.; Tsai, M.H. Multi-sensor wireless signal aggregation for environmental monitoring system via multi-bit data fusion. *Appl. Math. Inf. Sci.* **2011**, *5*, 589–603.
20. Li, Z.M.; Chen, R.Z.; Zhang, B.M. Study of adaptive weighted estimate algorithm of congeneric multi-sensor data fusion. *J. Lanzhou Univ. Technol.* **2006**, *32*, 78–82.
21. Yan, J.; Hu, Y.; Guo, C. Rotor unbalance fault diagnosis using DBN based on multi-source heterogeneous information fusion. *Procedia Manuf.* **2019**, *35*, 1184–1189. [CrossRef]
22. Duan, Z.S.; Han, C.Z.; Tao, T.F. Consistent multi-sensor data fusion based on nearest statistical distance. *Chin. J. Sci. Instrum.* **2005**, *26*, 478–481.
23. Wang, J.Q.; Zhou, H.Y.; Wu, Y. The theory of data fusion based on state optimal estimation. *Math. Appl.* **2007**, *20*, 392–399.
24. Zheng, Y.J.; Niu, R.X.; Varshney, P.K. Sequential Bayesian estimation with censored data for multi-sensor systems. *IEEE Trans. Signal Process.* **2014**, *62*, 2626–2641. [CrossRef]
25. Fan, S.L.; Li, D.G.; Zhao, J.M. An mine multi-sensor maximum likelihood estimation data fusion algorithm. *J. Inf. Comput. Sci.* **2013**, *10*, 3809–3814. [CrossRef]
26. Lei, X.F.; Zhu, B. Perception of microburst based on multi-sensor data fusion. *Inf. Control.* **2011**, *40*, 296–301.
27. Fincher, D.W.; Mix, D.F. Multi-sensor data fusion using neural networks. In Proceedings of the IEEE International Conference on Systems, Man and Cybernetics, Los Angeles, CA, USA, 4–7 November 1990; pp. 835–838.
28. Tang, A.H.; Zhang, Y.M. Application of fuzzy clustering in multi-sensor information fusion. *J. Theor. Appl. Inf. Technol.* **2012**, *45*, 661–667.
29. Su, W.X.; Zhu, Y.L.; Liu, F.; Ma, L.B. A homogeneous multi-sensor online data fusion method based on improved fuzzy clustering. *Inf. Control.* **2015**, *44*, 557–563.
30. Tang, Z.Y.; Cai, Y.; Wang, H. A multi-sensor data fusion method based on adaptive weighting algorithm. *Command. Inf. Syst. Technol.* **2022**, *13*, 66–70.
31. Cai, B.L.; Su, G.D. Research on improved batch estimation and adaptive weighted fusion method. *Meas. Control. Technol.* **2019**, *38*, 122–126.
32. Zhu, K.; Song, X.; He, J.X.; Yang, L. Greenhouse data fusion based on wavelet noise reduction and adaptive weighting method. *Jiangsu Agric. Sci.* **2021**, *49*, 180–186.
33. Feng, A.A.; Yue, J.H.; Zheng, Y.; Guo, X.Y. Simulation analysis of wavelet packet denoising optimized by thought evolution algorithm. *Comput. Simul.* **2020**, *37*, 285–290.
34. Guo, C.J.; Gong, C.Y.; Rong, F.; Song, Y.Q. Real-time vibration signal storage management technology based on entropy weight method. *J. Tianjin Polytech. Univ.* **2015**, *34*, 67–71.

35. Chen, L.; Zhang, C.L. Rolling bearing fault diagnosis based on EMD envelope spectral features and PCA-PNN. *Coal Mine Mach.* **2022**, *43*, 173–176.
36. He, X.S.; Yang, X.R. Analytical comparison of particle swarm algorithms for chaotic mapping. *J. Basic Sci. Text. Coll. Univ.* **2023**, *36*, 86–93.
37. Li, K.; Xiong, M.; Su Lei Lu, L.X.; Chen, S. Fault diagnosis method based on improved deep limit learning machine. *Vib. Test Diagn.* **2020**, *40*, 1120–1127.
38. Heo, B.; Chun, S.; Oh, S.J.; Han, D.; Yun, S.; Kim, G.; Uh, Y.; Ha, J. Adamp: Slowing down the slowdown for momentum optimizers on scale-invariant weights. *arXiv* **2020**, arXiv:2006.08217.

lubricants

MDPI

Article

Nonlinear Dynamic Responses of Rigid Rotor Supported by Thick Top Foil Bearings

Bin Hu [1], Xiaodong Yang [2,*], Anping Hou [1], Rui Wang [1], Zhiyong Wu [3], Qifeng Ni [4] and Zhong Li [5]

1 School of Energy and Power Engineering, Beihang University, Beijing 100191, China; bh1204hb@163.com (B.H.); houap@buaa.edu.cn (A.H.); wangrbuaa@126.com (R.W.)
2 Institute of Artificial Intelligence, Beihang University, Beijing 100191, China
3 The Key Laboratory of Solar Thermal Energy and Photovoltaic System, Institute of Electrical Engineering, Chinese Academy of Sciences (IEE-CAS), Beijing 100190, China; wuzhiyong@mail.iee.ac.cn
4 Ningbo Hudu Energy Technology Co., Ltd., Ningbo 315000, China; nqf888@163.com
5 Zhengzhou Aerotropolis Institute of Artificial Intelligence, Zhengzhou 451162, China; 17888821559@163.com
* Correspondence: yangxiaodong@buaa.edu.cn

Abstract: This study focuses on thick top foil bearings (TTFBs), which can prevent top foil from sagging and significantly reduce the load capacity of gas foil bearings (GFBs). However, the limited research on the dynamic responses of TTFB-rotor systems has hindered their wide application of TTFBs with high load capacity. To address this, an integrated nonlinear dynamic model is developed to analyze the linear dynamic responses of a rigid rotor supported on TTFBs. The model incorporates time domain orbit simulation, considering unsteady Reynolds equations, foil deformation equations, thick top foil motion equations, and rotor motion equations. A symmetrical test rig is used to validate the model, and three types of TTFBs with different bump foil stiffness are tested, with experimental results aligning with the model predictions. This study also investigates the effects of nominal clearance, static load, and unbalance on TTFB-rotor systems. The results indicate that unbalance has minimal impact on subsynchronous vibrations. However, larger bump foil stiffness, increased normal clearance, and higher static load contribute to improved stability and higher maximum stable speed of the TTFB-rotor system. Moreover, other relevant parameters reducing the bearing attitude angle can further enhance the system's stability.

Keywords: nonlinear numerical prediction; thick top foil; subsynchronous vibrations; experimental investigation

Citation: Hu, B.; Yang, X.; Hou, A.; Wang, R.; Wu, Z.; Ni, Q.; Li, Z. Nonlinear Dynamic Responses of Rigid Rotor Supported by Thick Top Foil Bearings. *Lubricants* **2023**, *11*, 453. https://doi.org/10.3390/lubricants11100453

Received: 12 September 2023
Revised: 2 October 2023
Accepted: 12 October 2023
Published: 20 October 2023

1. Introduction

Increasing the rotational speed of rotating machinery can significantly improve energy density and efficiency. Gas foil bearings (GFBs) are oil-free bearings that meet the high-speed requirement due to their low friction and high DN (shaft diameter in millimeters multiplied by shaft rotational speed in rev/min) limit [1–3]. In contrast with traditional oil-lubricated bearings, GFBs eliminate the need for an oil-lubrication system, resulting in a compact structure and reduced energy loss [4,5]. These advantageous characteristics have greatly facilitated the successful application of GFBs in high-speed turbomachinery systems, including cryogenic turboexpanders, Brayton air refrigerator systems, high-speed machining centers, and micro-power generator systems [6–9].

However, GFB-rotor systems often experience severe subsynchronous vibrations at high rotating speeds, limiting the range of their applications [10]. As a result, numerous studies have focused on investigating the dynamic responses of GFB-rotor systems. Several researchers have used a perturbation method to calculate the dynamic stiffness and damping coefficients for analyzing the dynamic responses of GFB-rotor systems. Ku et al. [11] introduced this method to predict the structural stiffness and damping of foil when a rotor is in its static equilibrium position. Lee et al. [12] compared the dynamic responses of a super-critical rotor supported by conventional bump foil bearings and viscoelastic bearings using experimental and theoretical results. The theoretical orbit calculations, based on dynamic stiffness and damping coefficients, showed a good agreement with the experimental results when the vibration amplitude of the rotating speed remained below the bending critical speed. The theoretical findings indicated that enhancing the structural damping of the foil structure reduced vibration. Vleugels et al. [13] conducted a stability analysis of the GFB-rotor system using dynamic stiffness and damping coefficients. They assumed the foil structure acted as a uniform elastic foundation to calculate the dynamic stiffness and damping properties. Numerical results indicated that bearing compliance significantly influenced the dynamic stiffness, and increasing the bearing load and reducing the ratio of the nominal bearing clearance to the rotor radius enhanced stability. Kim et al. [14] performed theoretical and experimental research on the effects of mechanical preload. Metal shims were inserted beneath the bump-strip layers of shimmed GFBs to introduce mechanical preload. Compared with the original GFBs, the shimmed GFBs exhibited a noticeable increase in dynamic direct stiffness and direct damping. The multiple lobe film clearance profile created by the metal shims reduced the amplitude of subsynchronous whirl motions and increased the natural frequency of the GFB-rotor system, as confirmed by experimental results. Theoretical and experimental results also indicated that rotor unbalance and the gas film had a significant influence on nonlinear dynamic responses, although the foil structure was simplified as a uniform elastic foundation with a given value. Xu et al. [15] developed a more comprehensive GFB model that considered nonlinearity in the foil structure to predict dynamic responses in GFB-rotor systems. They calculated dynamic stiffness and damping coefficients based on a quadratic bump stiffness model derived from experimental bearing stiffness data. The simulation results of the quadratic bump stiffness model differed significantly from those of the linear bump stiffness model, indicating the significant impact of foil structure nonlinearity on the dynamic responses of GFB-rotor systems. Hoffmann et al. [16] used a link-spring model incorporating frictional contacts and bump interactions to investigate the nonlinear vibrations of GFB-rotor systems. Using experimental and numerical testing of two contrasting cases with different balanced rotors, they identified the source of subsynchronous vibrations. The simulation results showed good agreement with the measurements, indicating that interaction between the gas film and the rotor caused subsynchronous vibrations when unbalance was significant. When unbalance was minor, the subsynchronous vibrations originated from the self-excitation of the gas film.

However, the dynamic stiffness and damping coefficients were perturbation results calculated using linearized Reynolds equations when the rotor displacements were small relative to its static equilibrium position. Subsynchronous vibrations with large amplitudes were common in many tests, highlighting the limitations of the linear perturbation approach. Kim [17] investigated the dynamic responses of GFB-rotor systems using both linear and nonlinear approaches. It was found that the rotor–bearing natural frequencies predicted using the two approaches were similar. However, the onset speeds of instability differed significantly between the time domain nonlinear orbit simulations and the linear stability simulations based on the dynamic stiffness and damping coefficients. Bou et al. [18] conducted a comparison between a nonlinear time-dependent approach and the classical linear approach using dynamic stiffness and damping coefficients. The classical linear approach was found to have a valid range of relative eccentricity lower than 0.65. For high values of eccentricity, the accuracy of the results required the use of the nonlinear approach. Larsen et al. [19] also compared the two approaches for predicting the onset speed of instability. It was observed that the classical linear approach was only suitable when the uniform foil stiffness was high and the load was small. When the uniform foil stiffness was reduced or the load level was increased, there were significant discrepancies in the prediction of the onset speed of instability. The nonlinear time domain approach, however, could accurately predict the actual rotor response, which the classical linear approach was unable to achieve. Consequently, many researchers have adopted nonlinear analysis approaches. Bhore et al. [20] proposed a nonlinear time domain orbit simulation that coupled the equations of the unsteady Reynolds equation, foil deformation, and rotor motion. They conducted parametric studies on rotating speed, unbalance eccentricity, compliance, and the loss factor, revealing highly nonlinear behaviors of the disc and journal center motion. However, the foil structure was simplified as a uniform elastic foundation. Larsen et al. [21] studied the effect of unbalance and rotational speed using theoretical modeling and experimental testing. The theoretical results, solved with a mathematical approach in the time domain, showed good agreement with the experimental results. Both the theoretical and experimental results indicated that subsynchronous vibrations were primarily influenced by unbalance and rotational speed. Additionally, the correct estimation of the foil stiffness and loss factor significantly affected the accuracy of the predictions. Osmanski et al. [22] proposed a new foil model based on truss representation, considering foil mass and frictional energy dissipation at foil interfaces. They presented a nonlinear time domain model to calculate the dynamic responses of GFB-rotor systems using this foil model. The simulation results showed good agreement with the experimental results, suggesting that the natural frequencies and mode shapes were accurately captured. Bonello et al. [23] introduced a modal-based bump foil model that considered the dynamic interaction between bumps and their inertia. They used a simultaneous solution technique to integrate the rotor, gas film, and foil domains into a coupled dynamical system model. This approach accurately predicted the nonlinear behaviors of GFB-rotor systems. The imposition of pressure constraints on the gas film delayed the onset speed of instability, aligning well with experimental results. Moreover, the significant influence of gas film non-linearity on the non-linear behaviors of GFB-rotor systems was verified with measured and predicted nonlinear phenomena. The aforementioned studies demonstrate that nonlinear analysis approaches yield more accurate dynamic responses of GFB-rotor systems.

Most recently, a type of bump-type foil bearing with a thick top foil was introduced to handle heavy load conditions [24,25]. In comparison with traditional gas foil bearings (GFBs) with thin top foils, TTFBs showed significant improvements in load capacity by reducing the sagging of the top foil, as depicted in Figure 1. Wang et al. [24] developed a theoretical model for a TTFB and predicted its load capacity and dynamic force coefficients. The numerical results indicated a 23% increase in load capacity compared with a GFB, along with different dynamic force coefficients. However, the numerical results were not experimentally validated. Li et al. [25] developed a three-dimensional finite element model based on contact mechanics to investigate the load capacity of TTFBs. The numerical results, which agreed well with experimental findings, demonstrated a 100% improvement in load capacity compared with GFBs with thin top foils. The thickness of the top foil was found to have a significant impact on the degree of sagging and bearing stiffness. However, the study primarily focused on load capacity, and there is currently a lack of research on the dynamic responses of TTFB-rotor systems. This knowledge gap severely hinders the application and wider adoption of TTFBs, despite their high load capacity.

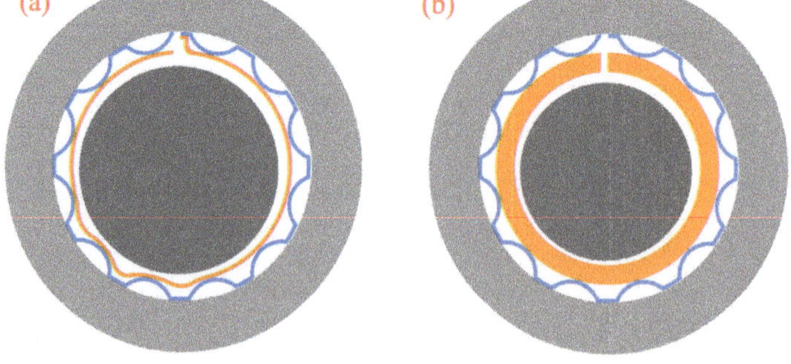

Figure 1. Schematic view of (**a**) a GFB with thin top foil and (**b**) a TTFB [25].

This paper presents an integrated nonlinear dynamic model for investigating the dynamic responses of TTFB-rotor systems, where a rigid rotor is supported by TTFBs. The model consists of a time-domain orbit simulation that couples unsteady Reynolds equations, foil deformation equations, thick top foil motion equations, and rotor motion equations. To validate the simulation results, various bump foil stiffness scenarios are tested against data obtained from a symmetrical test rig. The analysis of dynamic responses uses fast Fourier transform (FFT), waterfall plots, orbit simulations, and Poincaré maps. The simulation considers the impact of parameters such as nominal clearance, static load, and unbalance on the nonlinear responses of the TTFB-rotor system. By carefully selecting these parameters, a deeper understanding of the TTFB-rotor system is obtained. Furthermore, the obtained results and discussions are of importance to promote the application of TTFBs in high-speed and high-performance rotating machinery systems.

2. Nonlinear Numerical Prediction Method

The nonlinear model of the GFB-rotor system is typically divided into three components: the shaft part, which is based on rotor motion equations; the gas film part, which is based on unsteady Reynolds equations; and the bump structure part, which is based on foil deformation equations, as depicted in Figure 2a [26]. However, in the case of the TTFB-rotor system, the weight of the thick top foil needs to be considered. The thick top foil has a thickness of 1.5~2.0 mm, which is significantly thicker than the thin top foil's 0.1~0.2 mm. As a result, the thick top foil part should be added as the fourth component to the TTFB-rotor system, as shown in Figure 2b. A time-domain orbit simulation is conducted, integrating these four components, to accurately simulate the performance of the TTFB-rotor system.

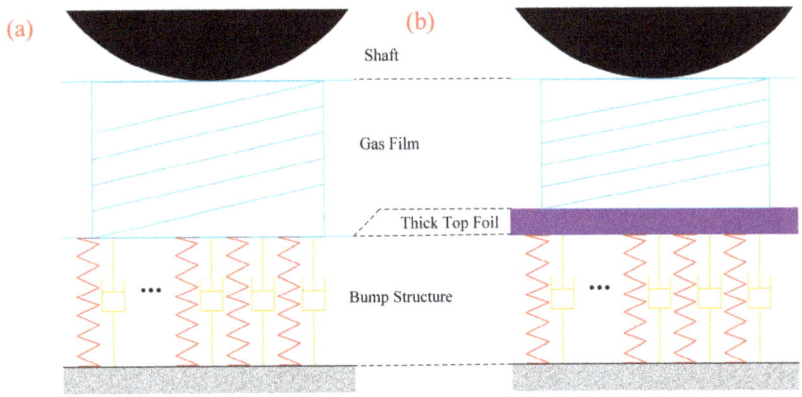

Figure 2. Schematic view of (**a**) GFB-rotor [26] and (**b**) TTFB-rotor system models.

2.1. Theoretical Model for TTFB

As shown in Figure 3, the theoretical model of a TTFB is composed of a gas film part based on unsteady Reynolds equations, a bump structure part based on foil deformation equations, and a thick top foil part based on motion equations of the thick top foil. The dimensionless transient compressible Reynolds equation that calculates the dynamic pressure and film thickness distribution of a TTFB is written as

$$\frac{\partial}{\partial \theta}\left(P_d H^3 \frac{\partial P_d}{\partial \theta}\right) + \frac{\partial}{\partial Z}\left(P_d H^3 \frac{\partial P_d}{\partial Z}\right) = \Lambda \frac{\partial}{\partial \theta}(P_d H) + 2\Lambda\gamma \frac{\partial}{\partial T}(P_d H) \quad (1)$$

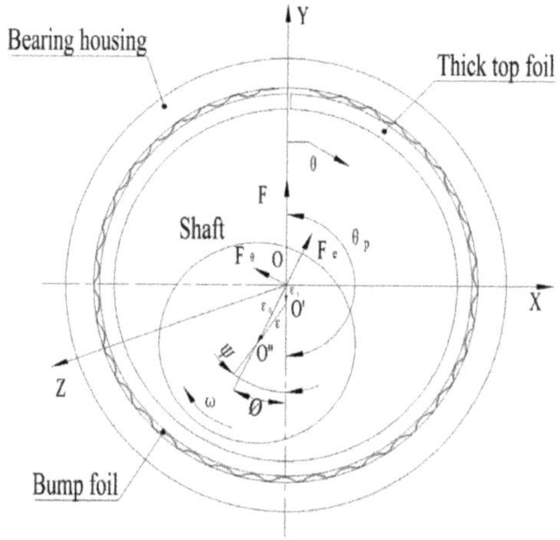

Figure 3. Coordinate system of a TTFB.

The dimensionless bearing axial width Z, pressure P_d, gas film thickness H, bearing number Λ, excitation frequency ratio γ, and time T in Equation (1) are written in Equation (2) as

$$Z=\frac{z}{R},\quad P_d=\frac{p}{p_a},\quad H=\frac{h}{C_0},\quad \Lambda=\frac{6\mu\omega}{p_a}\cdot\left(\frac{R}{C_0}\right)^2,\quad \gamma=\frac{\omega_e}{\omega},\quad T=\omega_e t \tag{2}$$

where the bearing radius R ambient pressure p_a, nominal clearance C_0, viscosity of gas μ, rotational frequency ω, excitation frequency ω_e, and time t are used in the dimensionless representation of the above-mentioned parameters.

Based on the structural characteristics of TTFBs, both the displacement of the rotor and the movement of the thick top foil dominate the eccentricity of a TTFB. Therefore, the eccentricity vector of a TTFB is the relative eccentricity vector ε obtained by subtracting the eccentricity vector of the rotor and the eccentricity vector of the thick top foil. The unsteady gas film thickness is written as

$$H=\frac{h}{C_0}=1+|\vec{\varepsilon}|\cos(\theta-\Psi) \tag{3}$$

where $|\vec{\varepsilon}|$ and Ψ are the modulus and angle of the relative eccentricity vector ε, respectively. In Equation (3), the term of the bump deflection is not included for the thick top foil is assumed as the rigid body [24].

The dimensionless transient compressible Reynolds equation is solved using the alternating-direction implicit method. When the transient gas film pressure field meets the condition of convergence, the pressure field is integrated using Simpson's one-third rule to calculate transient TTFB hydrodynamic film force in the x- and y-directions. The force is written as

$$F_{TTFB,x}=-\int_{-\frac{L}{2R}}^{\frac{L}{2R}}\int_0^{2\pi}(P_d(\theta,Z)-1)\sin\theta d\theta dZ \tag{4}$$

$$F_{TTFB,y}=-\int_{-\frac{L}{2R}}^{\frac{L}{2R}}\int_0^{2\pi}(P_d(\theta,Z)-1)\cos\theta d\theta dZ \tag{5}$$

The weight of the thick top foil is no longer small compared with that of the rotor; thus, it should be taken into account. The motion equation of the thick top foil is written as

$$m_{TF}\ddot{e}_{TF,x}+c_x\dot{e}_{TF,x}+k_x e_{TF,x}+F_{TTFB,x}=0 \tag{6}$$

$$m_{TF}\ddot{e}_{TF,y} + c_y\dot{e}_{TF,y} + k_y e_{TF,y} + F_{TTFB,y} = 0 \tag{7}$$

where $e_{TF,x}$ and $e_{TF,y}$, k_x and k_y , and c_x and c_y are the displacement, stiffness, damping of the thick top foil in the x- and y-directions, respectively.

k_x and k_y are the combined stiffness of the deformed bumps in the x- and y-directions, they can be written as

$$\begin{cases} k_x = \sum\limits_{j=1}^{n} k_j \sin^2(\phi_j - \theta_p) \\ k_y = \sum\limits_{j=1}^{n} k_j \cos^2(\phi_j - \theta_p) \end{cases} \tag{8}$$

where k_j is the stiffness of the JTH deformed bump and ϕ_j is the angle of the JTH deformed bump. In this paper, k_j is derived on the basis of the model with linear stiffness distribution [27].

c_x and c_y are the combined damping of the deformed bumps in the x- and y-directions, they can be written as

$$\begin{cases} c_x = \eta k_x/\omega \\ c_y = \eta k_y/\omega \end{cases} \tag{9}$$

where η is foil structural loss factor [20].

2.2. Theoretical Model for the Rigid Rotor

The structure of the TTFB-rotor system, as depicted in Figure 4, was investigated in this study. Unbalance was added symmetrically to the shaft, and the eddy current probe was positioned symmetrically for measurement. The shaft was divided into 20 beam elements with 80 degrees of freedom, utilizing a beam model based on the finite element method (FEM). The thrust disks and turbines mounted on the shaft were treated as beam elements and the Timoshenko beam theory was used to account for the shear effect. The nonlinear bearing forces were represented as concentrated forces at the centers of each thick top foil.

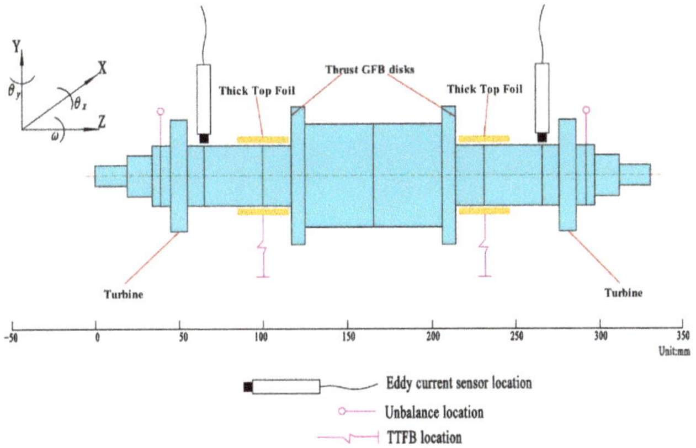

Figure 4. Schematic of the TTFB-rotor system.

When adopting the assumptions described above, the motion equation of the shaft is written as

$$M_J\ddot{e}_J + [C_J + \omega G]\dot{e}_J + K_J e_J = F_{TTFB} + F_g + F_u \tag{10}$$

where $(e_J)_i = \begin{bmatrix} x_i & y_j & \theta_{xi} & \theta_{yi} \end{bmatrix}^T$ is the displacement of each node at the shaft, x_i, y_i and θ_{xi}, θ_{yi} are the lateral and rotational displacements, respectively. M_J, C_J, G, and M_J are the mass, damping, gyroscopic, and stiffness matrixes. The damping effect of the system is considered to be related only to the nonlinear bearing force; therefore, the damping matrix C_J is set to zero [26]. F_{TTFB} is the nonlinear TTFB force vector, and F_g is the gravity force vector. F_u is the force vector induced by the unbalance, and F_u of each node at the shaft is written as

$$F_u(t) = \begin{bmatrix} F_{ux}(t) \\ F_{uy}(t) \\ M_{ux}(t) \\ M_{uy}(t) \end{bmatrix} = \omega^2 \begin{Bmatrix} m_u r_u \cos(\omega t) \\ m_u r_u \sin(\omega t) \\ 0 \\ 0 \end{Bmatrix} \tag{11}$$

where F_{ux} and F_{uy} are the unbalance force in the horizontal and vertical directions, whereas the unbalance moments M_{ux} and M_{uy} are set to zero [26]. m_u is the unbalance mass of the shaft, and r_u is the radius of the unbalance mass.

2.3. Orbit Simulation

A time-domain orbit simulation coupling unsteady Reynolds equations, foil deformation equations, thick top foil motion equations, and rotor motion equations is conducted to analyze the nonlinear dynamic responses.

Equations (6), (7) and (10) are nonlinear equations. Two kinds of numerical calculation methods, one being the explicit time history method and the other being the implicit time history method, can be adopted to solve them [28]. The Newmark method selected in this paper is an implicit time history method. The Newmark method is a method for modifying linear acceleration. By introducing two parameters α and β into the velocity and displacement expressions at time $t + \Delta t$, two basic equations of the Newmark method can be written as

$$\{\dot{u}(t+\Delta t)\} = \{\dot{u}(t)\} + [(1-\alpha)\{\ddot{u}(t)\} + \alpha\{\ddot{u}(t+\Delta t)\}] \times \Delta t \tag{12}$$

$$\{u(t+\Delta t)\} = \{u(t)\} + \{\dot{u}(t)\} \times \Delta t + \left[\left(\frac{1}{2} - \beta\right)\{\ddot{u}(t)\} + \beta\{\ddot{u}(t+\Delta t)\}\right] \times \Delta t^2 \tag{13}$$

In the Newmark method, the control parameters α and β affect the accuracy and stability of the whole algorithm. Only when $\alpha = 0.5$ and $\beta = 0.25$, the Newmark method has second-order accuracy and unconditional stability. Therefore, the control parameters α and β are assigned as 0.5 and 0.25, respectively, in research and engineering applications. The Newmark method is called the constant mean acceleration method. Equations (12) and (13) are transformed into

$$\{\dot{u}(t+\Delta t)\} = \{\dot{u}(t)\} + \frac{1}{2}(\{\ddot{u}(t)\} + \{\ddot{u}(t+\Delta t)\}) \times \Delta t \tag{14}$$

$$\{u(t+\Delta t)\} = \{u(t)\} + \{\dot{u}(t)\} \times \Delta t + \left[\frac{1}{4}(\{\ddot{u}(t)\} + \{\ddot{u}(t+\Delta t)\})\right] \times \Delta t^2 \tag{15}$$

Accordingly, the constant average acceleration method is used in this paper to calculate the trajectory of the rigid rotor and the thick top foil.

For the shaft, the dynamic response at time $n+1$ is calculated using

$$\dot{e}_J^{n+1} = \dot{e}_J^n + \frac{1}{2}\left(\ddot{e}_J^n + \ddot{e}_J^n\right) \tag{16}$$

$$e_J^{n+1} = e_J^n + \dot{e}_J^n \times \Delta t + \frac{1}{4}\left(\ddot{e}_J^n + \ddot{e}_J^n\right) \times (\Delta t)^2 \tag{17}$$

For the thick top foil, the dynamic response at time $n+1$ is calculated using

$$\hat{k} = k^n + 4\frac{m_{TF}}{(\Delta t)^2} + 2\frac{c^n}{\Delta t} \tag{18}$$

$$\hat{p}^{n+1} = -F_{film}^{n+1} + \left(4\frac{e_{TF}^{n}}{(\Delta t)^2} + 4\frac{\dot{e}_{TF}^{n}}{\Delta t} + \ddot{e}_{TF}^{n}\right)m_{TF} + \left(2\frac{e_{TF}^{n}}{\Delta t} + \dot{e}_{TF}^{n}\right)c^n \tag{19}$$

$$e_{TF}^{n+1} = \hat{p}^{n+1}/\hat{k} \tag{20}$$

$$\ddot{e}_{TF}^{n+1} = 4\frac{e_{TF}^{n+1} - e_{TF}^{n}}{(\Delta t)^2} - 4\frac{\dot{e}_{TF}^{n}}{\Delta t} - \ddot{e}_{TF}^{n} \tag{21}$$

$$\dot{e}_{TF}^{n+1} = \dot{e}_{TF}^{n} + \frac{1}{2}\left(\ddot{e}_{TF}^{n} + \ddot{e}_{TF}^{n+1}\right)\Delta t \tag{22}$$

In the TTFB-rotor system, the compressible transient Reynolds equation used to solve the gas film transient pressure distribution is a second-order partial differential equation, and the dynamic equation used to solve the rotor and thick top foil trajectory is a second-order ordinary differential equation. In the process of solving the rotor trajectory, in each time step, these two equations need to be solved simultaneously. However, in the process of solving these two equations, the results of the other side are required as the initial conditions, so the synchronous coupling solution is difficult to carry out. In order to predict the nonlinear dynamic characteristics of the TTFB-rotor system, a linear displacement prediction method is used to solve the synchronous coupling problem [28]. The linear displacement prediction method assumes that the displacement of the thick top foil and rotor changes linearly at step $n-1$, step n, and step $n+1$, and can be written as

$$\frac{e_{J(predict)}^{n+1} - e_J^n}{\Delta t} = \frac{e_J^n - e_J^{n-1}}{\Delta t} \tag{23}$$

$$\frac{e_{TF(predict)}^{n+1} - e_{TF}^n}{\Delta t} = \frac{e_{TF}^n - e_{TF}^{n-1}}{\Delta t} \tag{24}$$

According to Equations (23) and (24), based on the thick top foil and axial diameter displacement of step $n-1$ and step n, the gas film thickness of step $n+1$ can be predicted as

$$H_{predict}^{n+1} = 1 - \frac{2e_{J,x}^n - e_{J,x}^{n-1}}{C_0}cos\theta - \frac{2e_{J,y}^n - e_{J,y}^{n-1}}{C_0}sin\theta + \frac{2e_{TF,x}^n - e_{TF,x}^{n-1}}{C_0}cos\theta + \frac{2e_{TF,y}^n - e_{TF,y}^{n-1}}{C_0}sin\theta \tag{25}$$

To provide a referable analysis, 50 time steps are calculated per revolution to solve the rotor orbit. The steady-state solutions for each rotational speed are carried out to simulate the entire TTFB-rotor system speed-up test because unsteady-state solutions are very time-consuming.

3. Model Verification

To validate the method for predicting subsynchronous responses in the TTFB-rotor system, an experimental setup is designed and conducted. The TTFB used in the test is illustrated in Figure 5 and consists of a thick top foil, bump foil, bearing housing, and snap ring. The displacement of the thick top foil in the circumferential direction is restricted by inserting lugs into the grooves of the snap ring and bearing housing. In the axial direction, the displacement of the thick top foil is limited by the snap ring and the internal end face of the bearing housing. The configuration of the test rig is shown in Figure 6. A symmetric design is used for the rotor, which comprises two identical impulse turbines to ensure that both TTFBs bear the same load. A pair of gas foil thrust bearings is utilized to limit axial movement of the rotor. Four eddy current sensors are used to measure the radial displacement of the TTFBs, while a laser speed sensor tracks the rotor's speed and phase. The dimensions of the two test TTFBs are as follows: a radius of 17.56 mm and a width of 30 mm. The impulse turbine tip clearance is set at 0.3 mm, ensuring that the rotor has sufficient torque to lift off and reach a maximum speed of 60 krpm under an inlet compressed air pressure of 0.7 Mpa. All data are acquired using a dynamic signal acquisition system developed by Donghua. The shaft weight is 4 kg, resulting in a static load of 2 kg on each bearing. Both TTFBs undergo the same manufacturing process, heat treatment, and assembly. The inner diameter of the TTFBs is measured with a micrometer gauge as shown in Figure 7, and the diameter of the shaft is measured with an outside micrometer. The value of the nominal clearance is obtained by subtracting the radius of the TTFBs from the radius of the shaft. In this paper, a nominal clearance of 60 μm was tested. The thick top foil has a thickness of 1.5 mm and is coated with a 20 μm thickness of PTFE to prevent potential wear on both the TTFBs and the shaft. The rotor unbalance is reduced to a G1.0 level using a commercial dynamic balancer. Scribed lines are marked to ensure that the unbalance amount of the rotor remains unchanged after disassembly and reassembly. Table 1 provides a detailed overview of the physical and geometric parameters of the TTFB-rotor system.

Figure 5. A photo of test TTFBs.

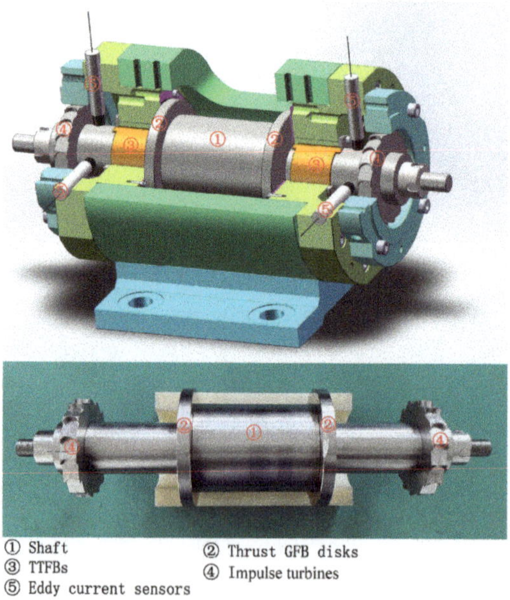

Figure 6. Schematic of the test TTFB-rotor system.

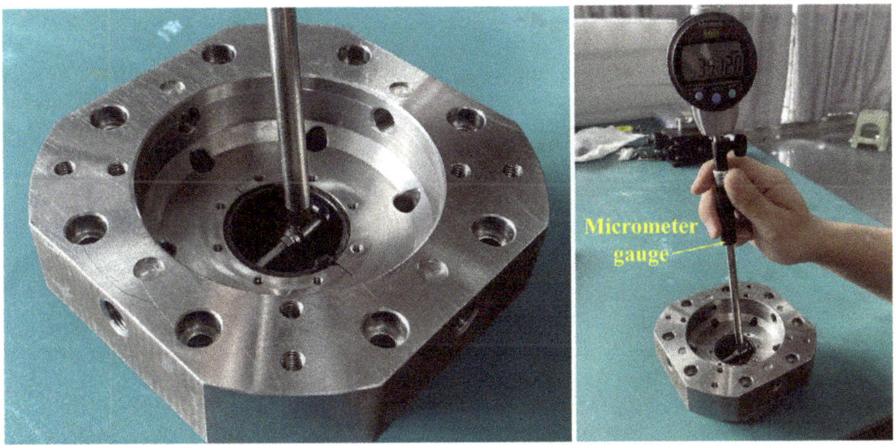

Figure 7. A photo of measuring the inner diameter of test TTFBs.

Table 1. Physical and geometric parameters of the TTFB-rotor system.

TTFB Parameters	
Bearing radius	17.56 mm
Bearing axial width	30 mm
Thick top foil thickness	1.5 mm
Thick top foil mass	0.041 kg
Bump material	3J1
Bump foil thickness	0.1 mm
Bump half length	1.25 mm
Bump height	0.51 mm
Number of bumps	38
Young's modulus	186 Gpa
Poisson's ratio	0.29
Gas viscosity	1.81×10^{-5} Pa·s
GFTB parameters	
Bearing inner radius	20 mm
Bearing outer radius	40 mm
Top foil angle (six pads)	58°
Inclined plane angle	30°
Foil thickness	0.1 mm
Rotor parameters	
Total mass	4 kg
Unbalance	4 g·mm
Radius at bearing location	17.5 mm
Total length	331 mm
Young's modulus	210 Gpa
Material density	7800 kg·m^{-3}

The maximum stable speed is achieved by accelerating the rotational speed using two symmetrically identical impulse turbines. During orbit simulations, a stable limit cycle cannot form beyond the maximum stable speed. In the experiment, when the maximum stable speed is exceeded, there is a noticeable sound of abnormal collision between the shaft and the thick top foil. To mitigate the impact of additional vibrations near the critical speed, a modal analysis of the rotor is performed, and the resulting Campbell diagram plot is presented in Figure 8. The first-order forward and backward whirling frequencies are 2247 Hz and 1999 Hz, respectively. Due to the symmetrical structure of the test rig, the vibrations on the left and right sides are similar. For research convenience, the vibration signals in the vertical direction at the right end are uniformly analyzed. Figure 9 illustrates that synchronous vibrations, marked as 1X and characterized by a frequency equal to the rotational frequency, are present. The amplitude of synchronous vibrations remains stable as the rotational speed changes, with simulation and experimental steady values of 0.88 μm and 1.21 μm, respectively. Subsynchronous vibrations, on the other hand, maintain a frequency that does not vary with the rotational speed. As the rotational speed increases, the simulation amplitude of subsynchronous vibrations also increases, and the increase becomes more pronounced as the rotational speed approaches the maximum stable speed. The experimental results exhibit a similar trend. Using ten-order zero-phase digital band pass filters [29], Figure 10 also illustrates that as the rotational speed increases, the simulation results and experimental results exhibit more pronounced subsynchronous vibrations. The predicted frequency and peak amplitude of the subsynchronous vibrations are 99 Hz and 9.12 μm, respectively, closely matching the corresponding experimental results of 102 Hz and 10.30 μm. The predicted and experimental maximum stable speeds are also close, with values of 36 krpm and 35.58 krpm, respectively. Comparing the simulation results with the experimental results, it is evident that both synchronous and subsynchronous vibrations are well-matched between the two.

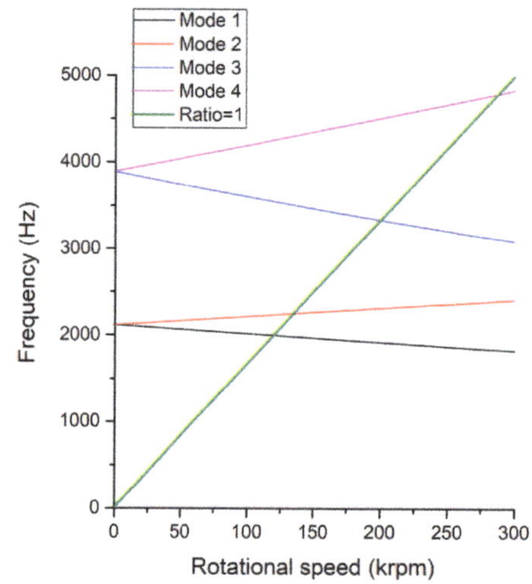

Figure 8. Campbell diagram plot of the rotor.

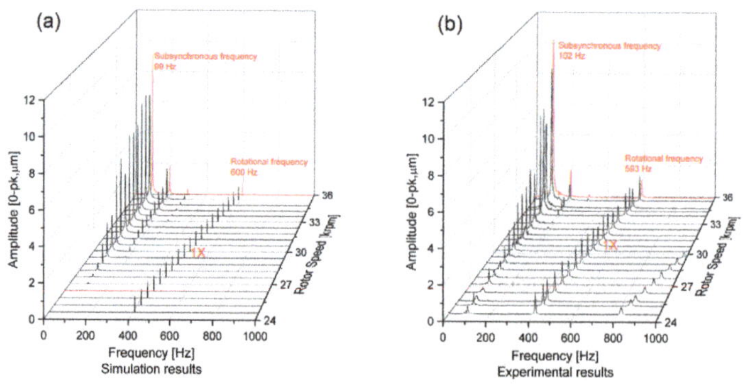

Figure 9. Waterfall plots of the TTFB-rotor system: (**a**) simulation results and (**b**) experimental results.

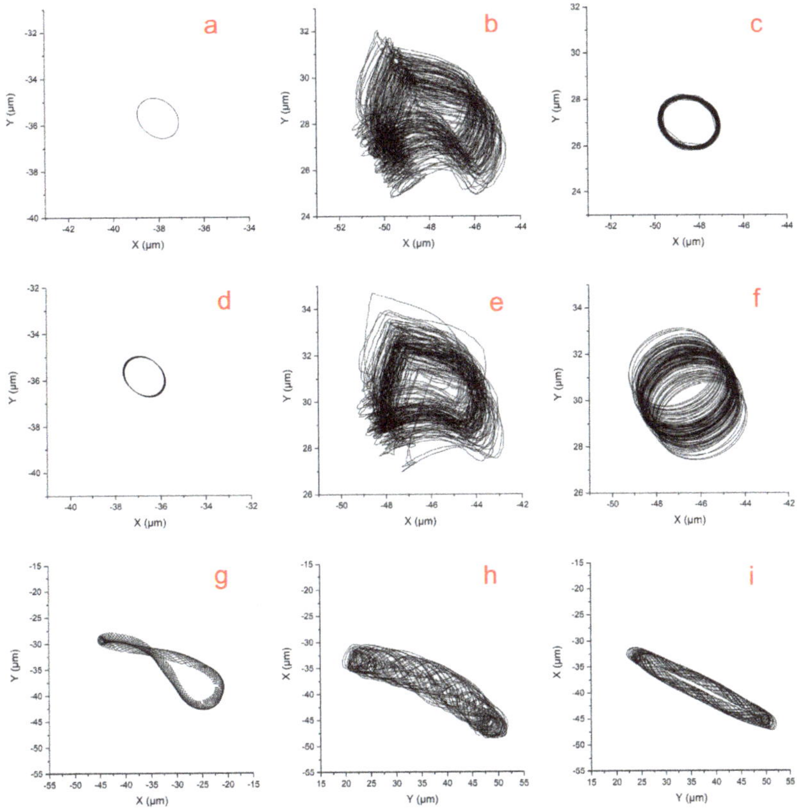

Figure 10. Orbits of rotor center at rotational speed of 24.6 krpm (**a**,**d**,**g**), 27 krpm (**b**,**e**,**h**), and 36 krpm (**c**,**f**,**i**). (**a**–**c**) Simulation results; (**d**–**f**) experimental results before filtering; and (**g**–**i**) experimental results after filtering.

To further validate the method, two additional cases were conducted with different bump foil stiffnesses. The first case involved a lower stiffness configuration, using a 0.1 mm C17200 bump foil material with a Young's modulus of 130 Gpa, resulting in 0.7 times the original stiffness as per the above model [27]. The second case utilized a higher stiffness configuration, using a 0.2 mm 3J1 alloy as the bump foil material with 8 times the original stiffness. Figure 11 illustrates waterfall plots depicting the simulation and experimental vertical vibrations at the right end. Upon comparing these plots, it is evident that even with variations in bump foil stiffness, the simulation results for synchronous and subsynchronous vibrations align well with the experimental results. The simulation's maximum stable speed and subsynchronous frequency also exhibit good agreement with the experimental findings. Thus, we can conclude that the time-domain orbit simulation accurately predicts the dynamic responses of TTFB-rotor systems.

The impact of bump foil stiffness on the dynamic responses of the TTFB-rotor system can also be inferred from these cases. It is observed that increasing the bump foil stiffness leads to a higher subsynchronous frequency. However, this observation contradicts the results of previous studies on GFB-rotor systems, which suggested a positive correlation between the subsynchronous frequency and the square root of bump foil stiffness [26]. This discrepancy implies that the thick top foil also plays a role in influencing the nonlinear responses of the TTFB-rotor system. Regarding the maximum stable speed, the simulation indicates that the maximum stable rotor speed for the lower bump stiffness bearing is 34.2 krpm, while the experimental result is 33.42 krpm. Similarly, for the original bearing, the simulation and experimental results correspond to 36 krpm and 35.58 krpm, respectively. Finally, for the higher bump stiffness bearing, the simulation and experimental results show 39 krpm and 39.12 krpm. These findings clearly demonstrate that as the bump foil stiffness increases, the stable speed of the TTFB-rotor system also increases. Consequently, a TTFB-rotor system with a higher bump foil stiffness exhibits enhanced stability.

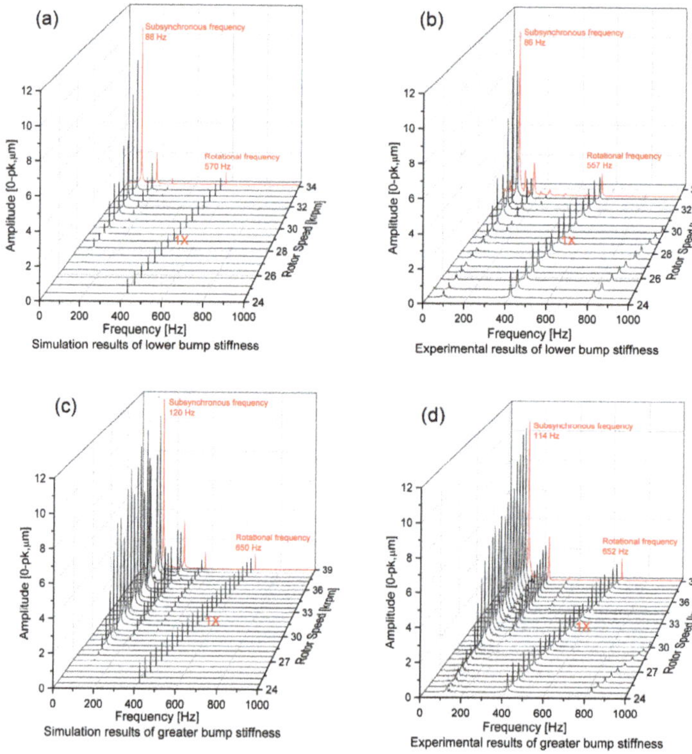

Figure 11. Waterfall plots of the TTFB-rotor system with different stiffness. (**a**) Simulation results of lower bump stiffness; (**b**) experimental results of lower bump stiffness; (**c**) simulation results of greater bump stiffness; and (**d**) experimental results of greater bump stiffness.

4. Parameter Studies and Discussion

In this section, we conduct parameter studies and discussions to examine the contributions of the gas film and shaft in the TTFB-rotor system. We investigate three key parameters: nominal clearance, static load, and unbalance. Through the use of waterfall plots, orbit simulations, Poincaré maps, and FFT plots, we present visual representations of the results obtained with varying parameters. These analyses reveal the respective influences of each parameter on the system's behavior.

4.1. Effects of Nominal Clearance

The gas film serves as a barrier between the bearing surface and the shaft, and it is responsible for generating dynamic pressure. The alteration of the nominal clearance directly affects the pressure distribution within the gas film. As a result, the nominal clearance exerts a significant influence on the stability of the TTFB-rotor system.

In this section, the system model considers the nominal clearance values of 40 μm, 50 μm, 70 μm, and 80 μm for each TTFB. The interval between the values is 10 μm, while all other conditions remain unchanged as previously described. Figure 12 illustrates the waterfall plots depicting the simulation responses for different nominal clearances. Additionally, in the previous section, the nominal clearance of 60 μm was studied and its corresponding waterfall plot is displayed in Figure 9a. Comparing Figures 12 and 9a reveals that when the nominal clearance is less than 70 μm, the subsynchronous frequency decreases as the nominal clearance increases. However, once the nominal clearance exceeds 70 μm, the subsynchronous frequency remains constant, indicating that the pressure distribution in the high-pressure region of the gas film stabilizes beyond a specific nominal clearance. As a result, the subsynchronous frequency no longer varies. Regarding the maximum stable speed, as the nominal clearance increases from 40 μm to 80 μm, the maximum stable speed of the TTFB-rotor system increases from 19.2 krpm to 63.6 krpm. This implies that subsynchronous vibrations can be suppressed by increasing the nominal clearance. Moreover, greater nominal clearance enhances the stability of the TTFB-rotor system.

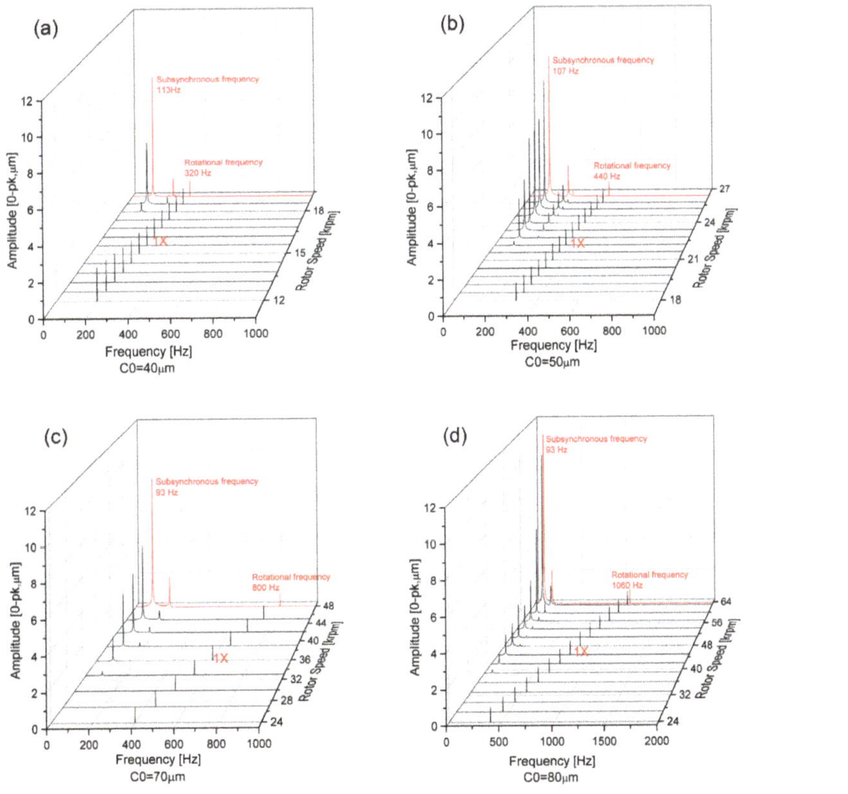

Figure 12. Waterfall plots of the TTFB-rotor system with various nominal clearances. (**a**) 40 μm; (**b**) 50 μm; (**c**) 70 μm; and (**d**) 80 μm.

The influencing mechanism of the nominal clearance on stability lies in its ability to alter the bearing attitude angle. As the nominal clearance increases, the film thickness in the high-pressure area decreases under the same static load. This results in an enlarged eccentricity ratio of the shaft and a reduction in the bearing attitude angle. The value of the bearing attitude angle represents the proportion of the tangential component of force acting on the shaft, with bearings featuring larger bearing attitude angles having smaller tangential component portions. The tangential component of force on the shaft typically injects energy into the rotor system, which is detrimental to its stable operation [30]. Consequently, increasing the nominal clearance improves the stability of the TTFB-rotor system.

4.2. Effects of Static Load

In order to further validate the influence of the bearing attitude angle on the stability of the TTFB-rotor system, this section examines the effects of static load. By increasing the static load, the eccentricity ratio of the shaft can be amplified and the bearing attitude angle can be reduced for all foil bearings [31].

In this section, the system model utilizes static loads of 1.2 kg, 1.6 kg, 2.4 kg, and 2.8 kg for each TTFB. The interval between the static loads is 0.4 kg. As the static loads are changed by extra gravity coefficients, the structure of the rotor is not changed [26]. The other conditions remain unchanged as described previously. Figure 13 illustrates waterfall plots of simulation responses corresponding to different static loads. Additionally, the static load of 2.0 kg, examined in the previous section, with its corresponding waterfall plot is shown in Figure 9a. Combining Figures 9a and 12, it becomes apparent that as the bearing static load increases from 1.2 kg to 2.8 kg, the maximum stable speed of the TTFB-rotor system rises from 19.2 krpm to 54.6 krpm, accompanied by a slight enhancement in the subsynchronous frequency at the maximum stable speed. Figure 14 showcases the orbit simulations, Poincaré maps, and FFT plots of vertical vibrations for different static loads at a rotational speed of 27 krpm. For a static load of 1.6 kg, the orbit of the rotor center becomes complex, and the Poincaré map exhibits multiple distinct points forming two large regions. The rotor motion appears quasi-periodic, with significant subsynchronous frequency amplitude present in the FFT plot. As the static load increases, the amplitude of the subsynchronous frequency and the area enclosed by the distinct points in the Poincaré map gradually decrease. Beyond a static load of 2.0 kg, the quasiperiodic rotor motion transitions to a period-1 state, and the orbit of the rotor center becomes a simple circle. The FFT plot displays a single peak, indicating the elimination of the subsynchronous frequency. These results confirm that the reduction in the bearing attitude angle achieved by adding static load can suppress subsynchronous vibration and improve the stability of the TTFB-rotor system.

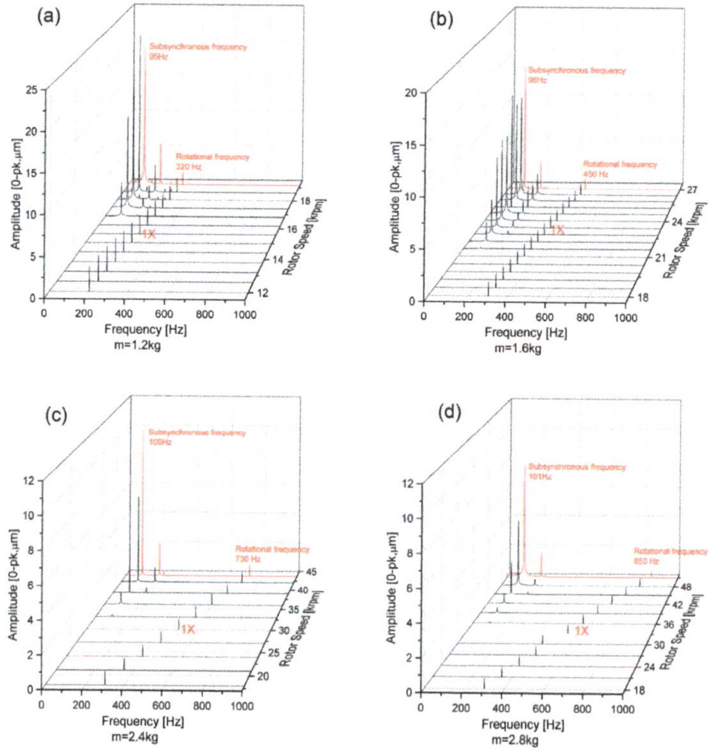

Figure 13. Waterfall plots of the TTFB-rotor system with various static loads. (**a**) 1.2 kg; (**b**) 1.6 kg; (**c**) 2.4 kg; and (**d**) 2.8 kg.

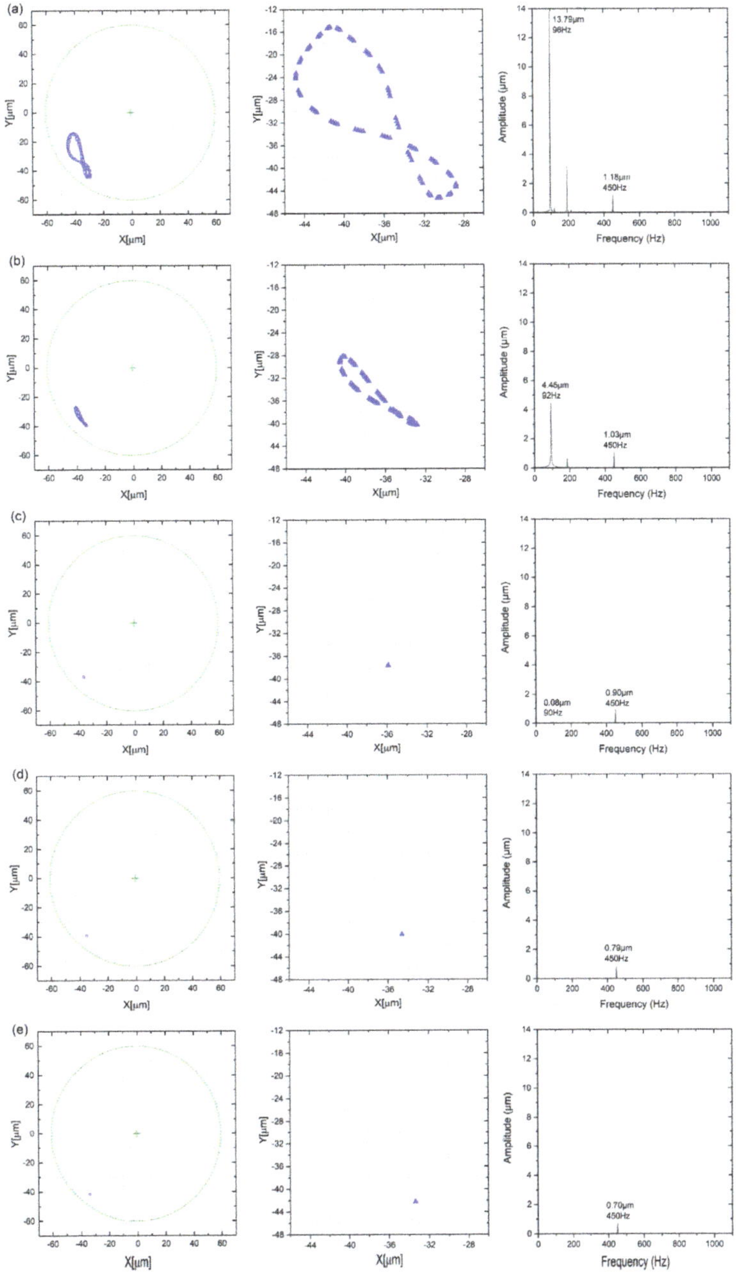

Figure 14. Orbit simulations, Poincaré maps, and FFT plots of vertical vibrations for different static loads at a rotational speed of 27 krpm. (**a**) m = 1.6 kg; (**b**) m = 1.8 kg; (**c**) m = 2.0 kg; (**d**) m = 2.2 kg; and (**e**) m = 2.4 kg.

4.3. Effects of Unbalance

Unbalance is a critical parameter that significantly influences the subsynchronous vibrations of GFB-rotor systems [20,21,32]. Therefore, this section discusses the effects of unbalance on TTFB-rotor systems. The unbalance values are set at 0 g·mm, 1 g·mm, 3 g·mm, and 4 g·mm with an interval equaling half of the base unbalance. The base unbalance, which is 2 g·mm, is addressed in the simulation validation section. To prevent additional load caused by the phase difference in the unbalance, the two unbalanced masses are positioned axially in a symmetric manner, with identical angles.

Figure 15 illustrates the waterfall plots of simulation responses with various static loads. It can be observed that, similar to Figure 9a, all the systems have a maximum stable speed of 36 krpm, and their subsynchronous vibration amplitudes exhibit the same variation with speed. On the other hand, Figure 16 presents the orbit simulations, Poincaré maps, and FFT plots of vertical vibrations for different unbalances at a rotational speed of 36 krpm. The orbits of the rotor center are highly intricate for all cases, mostly showing quasiperiodic motion with multiple distinct points in the Poincaré maps. Notably, when the unbalance is 3 g·mm, the orbit displays an interesting shape, transitioning the rotor center motion to a period-6 pattern, as the subsynchronous frequency precisely equals 1/6 of the rotational frequency. Additionally, in the FFT plots, the synchronous vibration amplitude increases with unbalance yet remains significantly smaller than the subsynchronous frequency amplitude. As depicted in Figure 17, there is a strong positive correlation (correlation coefficient of 0.9998) between the synchronous vibration amplitude and the unbalance. These findings imply that unbalance influences synchronous vibration but has minimal impact on subsynchronous vibration. This is in contrast with previous studies on GFB-rotor systems [20,21,32]. Consequently, for engineering applications utilizing TTFBs, extreme balancing of the rotor is unnecessary and of little significance.

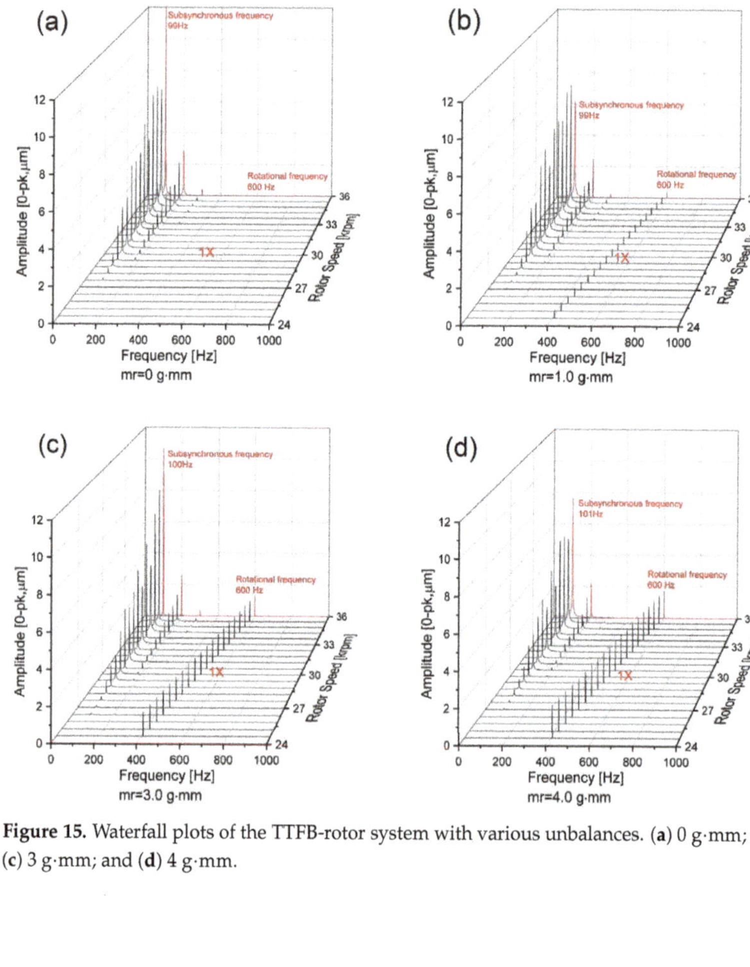

Figure 15. Waterfall plots of the TTFB-rotor system with various unbalances. (**a**) 0 g·mm; (**b**) 1 g·mm; (**c**) 3 g·mm; and (**d**) 4 g·mm.

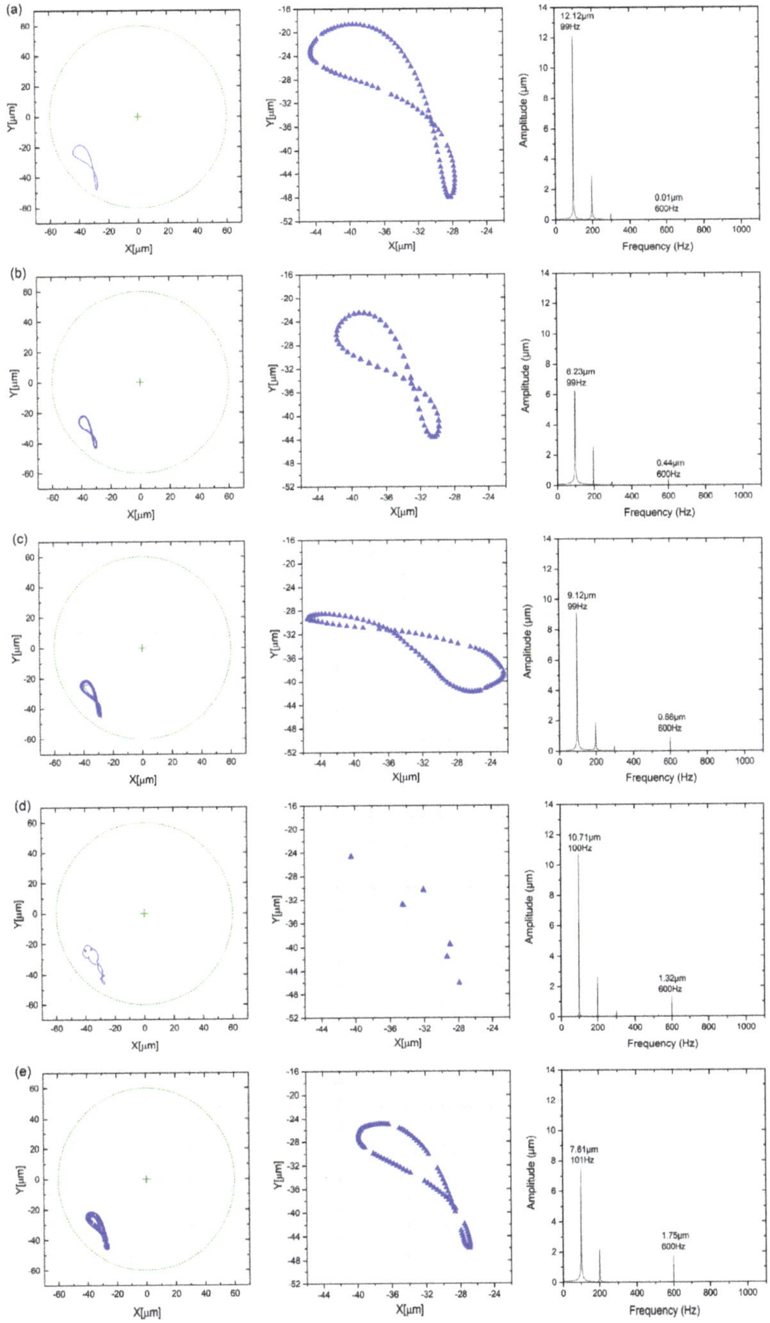

Figure 16. Orbit simulations, Poincaré maps, and FFT plots of vertical vibrations for different unbalances of the maximum stable speed. (**a**) mr = 0 g·mm; (**b**) mr = 1 g·mm; (**c**) mr = 2 g·mm; (**d**) mr = 3 g·mm; and (**e**) mr = 4 g·mm.

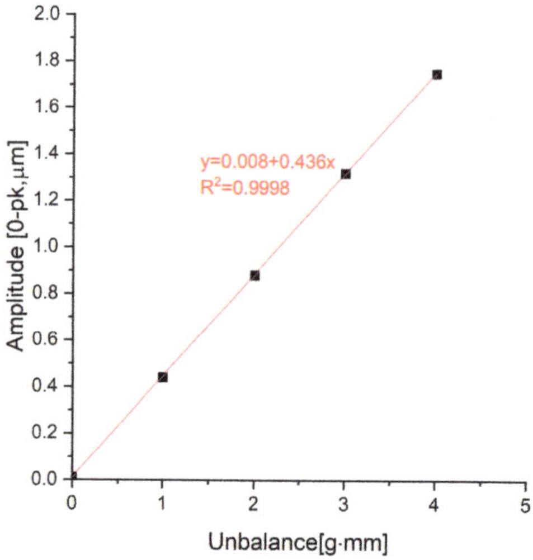

Figure 17. Synchronous vibration amplitude of various unbalances.

5. Conclusions

In this study, we developed a nonlinear dynamic model for TTFB-rotor systems by combining unsteady Reynolds equations, foil deformation equations, thick top foil motion equations, and rotor motion equations. This comprehensive model enables us to predict the dynamic responses of the TTFB-rotor system using time-domain orbit simulations, which accurately determine the rotor's orbital path.

To validate the accuracy of our model, we constructed a symmetrical test rig that effectively minimizes errors caused by varying bearing loads. We conducted calculations and tests using three different types of TTFBs with varying bump foil stiffness. The resulting waterfall plots of vibrations, both calculated and tested, indicated that higher bump foil stiffness effectively suppresses subsynchronous vibrations. Furthermore, by comparing these simulation results with the corresponding experimental results, we observed a strong correlation in vibration patterns, validating the reliability of our model.

Based on our new model, we conducted separate studies on the effects of nominal clearance, static load, and unbalance on the TTFB-rotor system. It was found that increasing the nominal clearance or the static load, both of which reduce the bearing attitude angle, enhances the maximum stable speed of the TTFB-rotor system. As a result, for TTFB-rotor systems, other related parameters reducing the bearing attitude angle contribute to improved stability. In terms of unbalance, it was observed that an increase in unbalance has a strong positive correlation with the amplitude of synchronous vibration. However, unbalance has minimal impact on subsynchronous vibrations, and all TTFB-rotor systems exhibit the same maximum stable speed. This study not only demonstrates an advanced tool for predicting the dynamic responses of TTFB-rotor systems but also paves the way for future developments of TTFB in high-speed rotating machinery systems.

Author Contributions: Conceptualization, B.H. and A.H.; methodology, B.H.; software, B.H. and X.Y.; validation, B.H. and R.W.; formal analysis, B.H.; investigation, B.H.; data curation, B.H.; writing—original draft preparation, B.H.; writing—review and editing, X.Y. and Z.W.; resources, Q.N. and Z.L.; visualization, B.H.; supervision, A.H.; funding acquisition, A.H. All authors have read and agreed to the published version of the manuscript.

Funding: This research was funded by the National Science and Technology Major Project of China, grant number J2019-V-0017-0112.

Data Availability Statement: The data presented in this study are available on reasonable request from the corresponding author.

Conflicts of Interest: Author Qifeng Ni was employed by the company Ningbo Hudu Energy Technology Co., Ltd. The remaining authors declare that the research was conducted in the absence of any commercial or financial relationships that could be construed as a potential conflict of interest.

References

1. DellaCorte, C.; Radil, K.C.; Bruckner, R.J.; Howard, S.A. Design, Fabrication, and Performance of Open Source Generation I and II Compliant Hydrodynamic Gas Foil Bearings. *Tribol. Trans.* **2008**, *51*, 254–264. [CrossRef]
2. Somaya, K.; Yamashita, T.; Yoshimoto, S. Experimental and Numerical Investigation of the High-Speed Instability of Aerodynamic Foil Journal Bearings for Micro Turbomachinery. In Proceedings of the ASME/STLE 2012 International Joint Tribology Conference, Denver, CO, USA, 7–10 October 2012; pp. 243–245.
3. Samanta, P.; Murmu, N.C.; Khonsari, M.M. The evolution of foil bearing technology. *Tribol. Int.* **2019**, *135*, 305–323. [CrossRef]
4. Heshmat, H.; Walowit, J.A.; Pinkus, O. Analysis of Gas-Lubricated Foil Journal Bearings. *J. Lubr. Technol.* **1983**, *105*, 647–655. [CrossRef]
5. DellaCorte, C. Oil-Free shaft support system rotordynamics: Past, present and future challenges and opportunities. *Mech. Syst. Signal Process.* **2012**, *29*, 67–76. [CrossRef]
6. Xiong, L.Y.; Wu, G.; Hou, Y.; Liu, L.Q.; Ling, M.F.; Chen, C.Z. Development of aerodynamic foil journal bearings for a high speed cryogenic turboexpander. *Cryogenics* **1997**, *37*, 221–230. [CrossRef]
7. Hou, Y.; Zhu, Z.H.; Chen, C.Z. Comparative test on two kinds of new compliant foil bearing for small cryogenic turbo-expander. *Cryogenics* **2004**, *44*, 69–72. [CrossRef]
8. Lee, Y.-B.; Kwak, Y.-S.; Chung, J.T.; Sim, K. Microturbocharger with Air Foil Bearings for a 100-W Class Micro Power System and Improvement of Rotordynamic Performance. *Tribol. Trans.* **2011**, *54*, 939–948. [CrossRef]
9. DellaCorte, C.; Bruckner, R.J. Remaining Technical Challenges and Future Plans for Oil-Free Turbomachinery. *J. Eng. Gas Turbines Power* **2010**, *133*, 042502. [CrossRef]
10. Guan, H.-Q.; Feng, K.; Yu, K.; Cao, Y.-L.; Wu, Y.-H. Nonlinear dynamic responses of a rigid rotor supported by active bump-type foil bearings. *Nonlinear Dyn.* **2020**, *100*, 2241–2264. [CrossRef]
11. Ku, C.P.R.; Heshmat, H. Structural Stiffness and Coulomb Damping in Compliant Foil Journal Bearings: Theoretical Considerations. *Tribol. Trans.* **1994**, *37*, 525–533. [CrossRef]
12. Lee, Y.B.; Kim, T.H.; Kim, C.H.; Lee, N.S.; Choi, D.H. Unbalance Response of a Super-Critical Rotor Supported by Foil Bearings—Comparison with Test Results©. *Tribol. Trans.* **2004**, *47*, 54–60. [CrossRef]
13. Vleugels, P.; Waumans, T.; Peirs, J.; Al-Bender, F.; Reynaerts, D. High-speed bearings for micro gas turbines: Stability analysis of foil bearings. *J. Micromech. Microeng.* **2006**, *16*, S282–S289. [CrossRef]
14. Kim, T.H.; Andrés, L.S. Effects of a Mechanical Preload on the Dynamic Force Response of Gas Foil Bearings: Measurements and Model Predictions. *Tribol. Trans.* **2009**, *52*, 569–580. [CrossRef]
15. Fangcheng, X.; Daejong, K. Dynamic performance of foil bearings with a quadratic stiffness model. *Neurocomputing* **2016**, *216*, 666–671. [CrossRef]
16. Hoffmann, R.; Liebich, R. Characterisation and calculation of nonlinear vibrations in gas foil bearing systems—An experimental and numerical investigation. *J. Sound Vib.* **2018**, *412*, 389–409. [CrossRef]
17. Kim, D. Parametric Studies on Static and Dynamic Performance of Air Foil Bearings with Different Top Foil Geometries and Bump Stiffness Distributions. *J. Tribol.* **2007**, *129*, 354–364. [CrossRef]
18. Bou-Saïd, B.; Grau, G.; Iordanoff, I. On Nonlinear Rotor Dynamic Effects of Aerodynamic Bearings with simple flexible rotors. *J. Eng. Gas Turbines Power* **2007**, *130*, 012503. [CrossRef]
19. Larsen, J.S.; Santos, I.F.; von Osmanski, S. Stability of rigid rotors supported by air foil bearings: Comparison of two fundamental approaches. *J. Sound Vib.* **2016**, *381*, 179–191. [CrossRef]
20. Bhore, S.P.; Darpe, A.K. Nonlinear dynamics of flexible rotor supported on the gas foil journal bearings. *J. Sound Vib.* **2013**, *332*, 5135–5150. [CrossRef]
21. Larsen, J.S.; Santos, I.F. On the nonlinear steady-state response of rigid rotors supported by air foil bearings—Theory and experiments. *J. Sound Vib.* **2015**, *346*, 284–297. [CrossRef]
22. von Osmanski, S.; Larsen, J.S.; Santos, I.F. A fully coupled air foil bearing model considering friction—Theory & experiment. *J. Sound Vib.* **2017**, *400*, 660–679. [CrossRef]
23. Bonello, P.; Hassan, M.F.B. An experimental and theoretical analysis of a foil-air bearing rotor system. *J. Sound Vib.* **2018**, *413*, 395–420. [CrossRef]
24. Wang, R.; Hou, A.; Li, Z. Numerical investigation of gas journal foil bearing performance with thick top foil. *J. Aerosp. Power* **2020**, *35*, 2123–2135. (In Chinese) [CrossRef]
25. Li, J.; Li, C.; Du, J. Investigations of the Bump Foil Bearing With Thick Top Foil Based on Contact Mechanics. *J. Tribol.* **2023**, *145*, 074601. [CrossRef]
26. Guo, Z.; Feng, K.; Liu, T.; Lyu, P.; Zhang, T. Nonlinear dynamic analysis of rigid rotor supported by gas foil bearings: Effects of gas film and foil structure on subsynchronous vibrations. *Mech. Syst. Signal Process.* **2018**, *107*, 549–566. [CrossRef]
27. Iordanoff, I. Analysis of an Aerodynamic Compliant Foil Thrust Bearing: Method for a Rapid Design. *J. Tribol.* **1999**, *121*, 816–822. [CrossRef]

28. Zhang, G.-h.; Sun, Y.; Liu, Z.-s.; Zhang, M.; Yan, J.-j. Dynamic characteristics of self-acting gas bearing–flexible rotor coupling system based on the forecasting orbit method. *Nonlinear Dyn.* **2012**, *69*, 341–355. [CrossRef]
29. Oppenheim, A.V.; Schafer, R.W.; Buck, J.R. *Discrete-Time Signal Processing*, 2nd ed.; Prentice-Hall, Inc.: Upper Saddle River, NJ, USA, 1999; pp. 237–240.
30. Edgar, J.; Gunter, J. *Dynamic Stability of Rotor-Bearing Systems*; National Aeronautics and Space Administration: Cleveland, OH, USA, 1966; Volume NASA SP-113, p. 228.
31. Lee, D.-H.; Kim, Y.-C.; Kim, K.-W. The effect of Coulomb friction on the static performance of foil journal bearings. *Tribol. Int.* **2010**, *43*, 1065–1072. [CrossRef]
32. Andrés, L.S.; Kim, T.H. Forced nonlinear response of gas foil bearing supported rotors. *Tribol. Int.* **2008**, *41*, 704–715. [CrossRef]

lubricants

MDPI

Article

Lubrication Performance of Misaligned Journal Bearings with Flexible Structure under Shock Load Conditions

Sung-Ho Hong [1,*] and Woo-Ju Jeon [2]

1 Department of Mechanical System Engineering, Dongguk University-WISE Campus, Gyeongju-si 38066, Republic of Korea
2 Department of Mechanical Engineering, Korea Advanced Institute of Science and Technology (KAIST), Daejeon 34141, Republic of Korea; wooju.jeon@samsung.com
* Correspondence: hongsh@dongguk.ac.kr; Tel.: +82-54-770-2211

Abstract: Bearings might be damaged due to shock loads caused by disturbances, in addition to static loads. In this study, a flexible structure was applied to enhance the lubrication characteristics of misaligned journal bearings subjected to impact loads. When an impact load is added to the bearing, a misaligned journal bearing has a high possibility of metal-to-metal contact. It might also lead to failure. Misalignment can occur at any time during bearing operation. A flexible structure is applied to the end of the bearing as a way to improve lubrication performance in a system where impact loads might be applied. The bearing's lubrication performance was numerically assessed under unsteady-state conditions. An elastohydrodynamic lubrication analysis was conducted, taking into account elastic deformation. The lubrication characteristics of misaligned journal bearings were compared with the dimensionless minimum film thickness. The flexible structure and elastic modulus of the bearing were investigated so that it could support the load without contact according to the change in the maximum magnitude of the impact load. When subjected to oil film pressure, this flexible structure underwent elastic deformation, resulting in enlargement of the oil film. A misaligned journal bearing with a suitable flexible structure provided stable lubrication without metal-to-metal contact, even under shock load conditions. The flexible structure was incorporated into the high-load-bearing region of the journal bearing as a groove. Therefore, the application of a flexible structure in misaligned journal bearings can effectively enhance lubrication performance in misaligned conditions and under shock loads.

Keywords: elastohydrodynamic lubrication (EHL); flexible structure; impact load; minimum film thickness; misaligned journal bearing

Citation: Hong, S.-H.; Jeon, W.-J. Lubrication Performance of Misaligned Journal Bearings with Flexible Structure under Shock Load Conditions. *Lubricants* **2023**, *11*, 500. https://doi.org/10.3390/lubricants11120500

Received: 6 September 2023
Revised: 7 November 2023
Accepted: 21 November 2023
Published: 25 November 2023

1. Introduction

Hydrodynamic lubrication generates oil pressure through the relative movement of mating surfaces and creates a lubricant film that completely separates the two contact surfaces [1]. As the shaft rotates, the supporting force from the lubricant film within the bearing and keeps the journal bearing separated [2–4]. However, metal-to-metal contact can occur in journal bearings for various reasons, leading to wear and failure.

Bearings often encounter misalignment of the journal axis, leading to an uneven distribution of film thickness for bearing clearance along the axial direction. This irregular distribution is primarily caused by factors such as asymmetric loads and errors during installation [5–7]. In addition, during operation, a bearing is bound to encounter diverse forms of disturbances arising from shock loads. The impact of these shock loads will inevitably influence both the bearing's performance during operation and its overall service life [8]. Moreover, shock can induce rapid and early failure of high-speed bearings [9,10].

Even a minor misalignment angle between shafts and bearings resulting from manufacturing errors, non-central loads or shaft deformation can change the distribution of the oil film. As a consequence of this misalignment, the oil film pressure, load carrying capacity and frictional force of the bearing undergo considerable variations [11,12]. Furthermore, inadequate operational quality can induce undesirable shaft vibrations [13].

Related studies can be largely divided into numerical studies, experimental studies [14] and studies [15–17] that combine numerical and experimental research. Most of them are numerical studies that perform CFD (computational fluid dynamics) analysis based on the FSI (fluid–structure interaction) technique [18], THD (thermohydrodynamic) analysis [19–21], EHD (elastohydrodynamic) analysis [22,23] and TEHD (thermoelastohydrodynamic) analysis [5,24]. Various methods have been proposed to enhance the lubrication performance of misaligned journal bearings. In terms of bearing material, a ZA-27 alloy [25] and a carbon-fiber phenolic composite [26] can be applied to improve the lubrication characteristics. To improve the lubrication characteristics, micropolar fluid [7] can be applied as lubricant in terms of lubricant, and a profile [7,27] or flexible structure [23,28–30] can be applied in terms of design. The grooved area experiences elastic deformation as a result of the oil film pressure exerted on the surface of the journal bearing. This can increase oil film thickness, effectively avoiding direct metal contact resulting from misalignment of the bearing. In a static load condition, when the groove is applied to the end of journal bearing, a larger oil film can be obtained in terms of the minimum film thickness than when the groove is not applied. However, when grooves of inappropriate geometries are applied, the minimum oil film thickness is smaller than when grooves are not applied [23]. Thus, to improve the lubricating properties of misaligned journal bearings under static load conditions, it is necessary to apply an appropriate flexible structure according to the operating conditions.

Several studies have been conducted on the shock behavior of different types of bearings, including oil bearings [31–34], air bearings [35,36] and water-lubricated bearings [37]. One study found that as the amplitude of random shock load increases, the average service life of sliding bearings decreases [31]. Another investigation revealed that rotor stability and vibration amplitude were favorable when foil bearings supported the rotor under various test conditions [35]. Furthermore, it has been reported that the use of double bladder structures can enhance the stability of water-lubricated bearings when subjected to half-sine shock load [37]. A study was conducted using a chamfer and profile to improve the lubrication characteristics of misaligned journal bearings under conditions where no shock load was applied [38] Moreover, lobe-type profiles or pockets were applied to the bearings to improve the lubrication characteristics under impact load conditions [37–40]. A study was conducted using a chamfer and profile to improve the lubrication characteristics of misaligned journal bearings under conditions where no shock load was applied [41–43].

The chamfer and profile were applied to minimize the vibration amplitude and to enhance the turbine stability limits. Research is also being conducted on methods to appropriately describe impact loads [44–46].

Some studies have been conducted to improve the lubrication characteristics of misaligned journal bearings. There are also studies on the lubrication characteristics of bearings against impact loads. However, studies applying a flexible structure when an impact load is applied to a misaligned journal bearing for improving lubrication characteristics have not been reported yet. Moreover, there are few analyses conducted under large-impact-load conditions. Thus, the objective of this study was to evaluate the lubrication performance of a groove-type flexible structure for preventing metal-to-metal contact under large-shock-load conditions, where the maximum value of the impact load was approximately three times the static load. In addition, the effect of an elastic modulus according to impact load condition within the range of the elastic modulus of bearing steel was evaluated. A numerical study was conducted on the design of an appropriate flexible structure that could prevent metal-to-metal contact under the shock load condition of misaligned journal bearings.

2. Numerical Model and Method

EHL analysis was performed to estimate the lubrication performance of the misaligned journal bearing under the given condition of an impact load. A flexible structure was applied to prevent metal-to-metal contact due to impact. EHL analysis was performed to consider elastic deformation due to oil film pressure. In this research, lubrication performance was evaluated over time using commercial software, COMSOL Version 6.0, capable of multi-physics analysis. This analysis software is based on the finite element method. It discretizes non-linear governing equations into algebraic equations. The numerical analysis involved the utilization of three key modules. The first module, known as the hydrodynamic bearing module, was utilized to analyze hydrodynamic lubrication. The second module, the solid mechanics module, was employed to study the elastic deformation of the bearing. Finally, the third module, called the solid–bearing coupling module, enabled smooth integration and interaction between the hydrodynamic bearing and solid mechanics modules [23]. The hydrodynamics bearing formula and the solid mechanics formula were combined into one matrix, and the calculation was performed using a fully coupled method. The linear elastic material model was applied to the flexible area (groove) where deformation occurs, and the rigid material model was applied to other parts. The geometric part to which the rigid body model was applied was given the condition of being fully constrained (deformation = 0). A rigid material model was applied to the rotation axis, and full constraints were applied in the axial direction (z-axis).

Misaligned journal bearings can lead to metal-to-metal contact at the end of the bearing due to various causes, as shown in Figure 1a. To prevent contact between metals, a flexible structure was applied, as represented by the black dotted line in Figure 1b. The flexible structure had a ring-type groove that could secure larger oil film thickness through elastic deformation caused by oil film pressure at the end of the bearing.

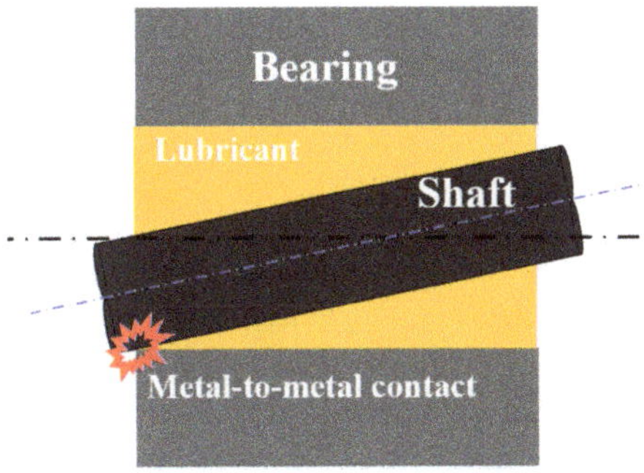

(a)

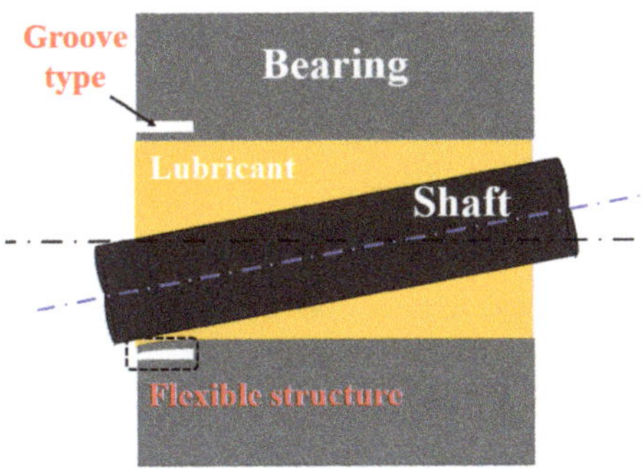

(b)

Figure 1. Misaligned journal bearing: (**a**) metal-to-metal contact; (**b**) application of a flexible structure.

Figure 2 presents a schematic of a journal bearing with a flexible structure. The rigid shaft rotates about the z-axis with an angular velocity ω. To the center of the shaft, at $z = l/2$, a load w is exerted in the direction of θ_w. The flexible structure is located in the dotted regions of Figure 2b. It is applied at the end of the bearing using a groove to induce elastic deformation. The dimensions of the flexible structure are determined by its inner end thickness (d), outer end thickness (a) and length (l_f). The shape of the flexible structure in the axial direction is determined by γ ($=d/a$), which can result in either a rectangular (for $\gamma = 1$) or a tapered (for $\gamma > 1$) shape, as shown in Figure 2c. In a previous analysis of the static load condition, lubrication performance was better when the groove shape was rectangular than when it was tapered [23]. Based on previous results, only cases where the shape of the groove was rectangular were considered in this study. The non-grooved section of the bearing surface remains rigid due to its large thickness, preventing any elastic deformation. The configurations of the journal bearing and the flexible structure are depicted by dimensionless parameters, as specified in Equation (1):

$$A = \frac{a}{r},\ L = \frac{l}{r},\ L_f = \frac{l_f}{l},\ \beta = \frac{c}{r},\ \gamma = \frac{d}{a} \tag{1}$$

The Reynolds equation is employed to solve the oil film pressure (p) under unsteady-state conditions, as demonstrated in Equation (2):

$$\frac{1}{r^2}\frac{\partial}{\partial\theta}\left(h^3\frac{\partial p}{\partial\theta}\right) + \frac{\partial}{\partial z}\left(h^3\frac{\partial p}{\partial z}\right) = 6\eta\omega\frac{\partial h}{\partial\theta} + 12\eta\frac{\partial h}{\partial t} \tag{2}$$

The dimensionless form of equation (2) becomes:

$$\frac{\partial}{\partial\theta}\left(H^3\frac{\partial P}{\partial\theta}\right) + \frac{\partial}{\partial Z}\left(H^3\frac{\partial P}{\partial Z}\right) = \frac{\partial H}{\partial\theta} + 2\frac{\partial H}{\partial T} \tag{3}$$

where p_a is the atmospheric pressure, and the dimensionless variables are:

$$H = \frac{h}{c},\ P = \frac{c^2(p - p_a)}{6r^2\eta\omega},\ X = \frac{x}{r},\ Y = \frac{y}{r},\ Z = \frac{z}{r},\ T = \omega t \tag{4}$$

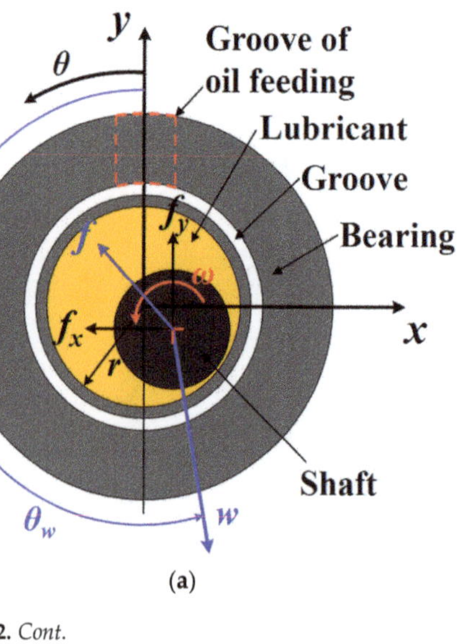

(a)

Figure 2. *Cont.*

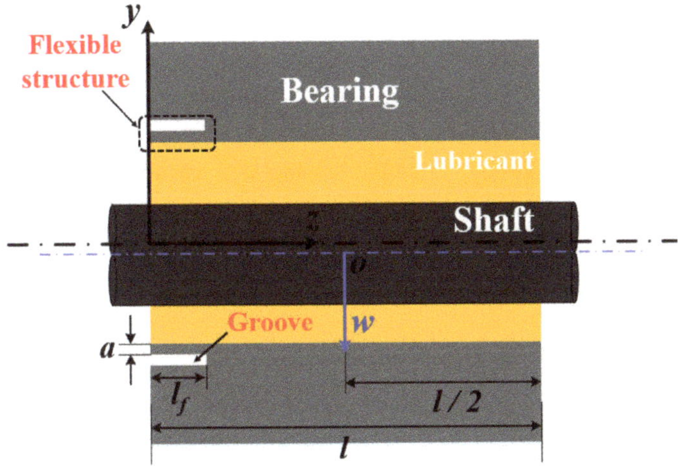

(**b**)

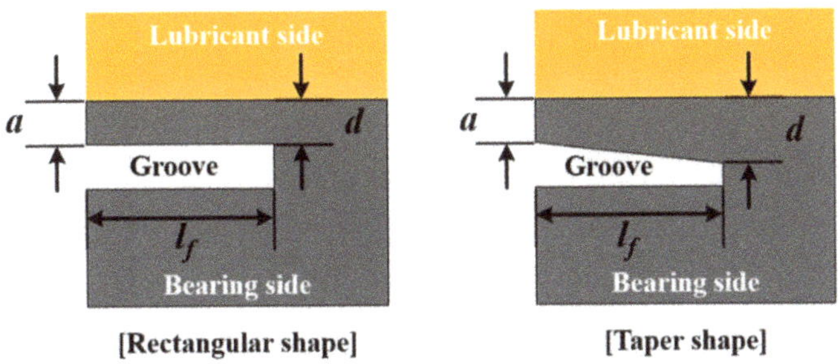

(**c**)

Figure 2. Schematic of journal bearing with flexible structure: (**a**) *x*-*y* plane; (**b**) *y*-*z* plane; (**c**) shapes of the flexible structure.

To determine the oil film pressure in Equation (2), it is essential to determine the oil film thickness (h) in the space between the bearing and shaft. Figure 3 illustrates the shaft's motion, which involves eccentric and tilted movements. The midpoint of the shaft at the location where z equals $l/2$ is designated as O, while O_1 and O_2 represent the midpoints of the shaft at the ends of the bearing. In Figure 3a, two circles depict cross-sections of the shaft projected onto the x-y plane, with O_1 and O_2 serving as their centers. The eccentricity e represents the distance on the x-y plane between the center of the bearing and O. The angle ψ on the x-y plane indicates the altitude angle between the direction of the load and the straight line passing through O and the center of the bearing. The tilting orientation of the shaft aligns with θ_w, and the tilting amount e' is equivalent to half the distance between O_1 and O_2 on the x-y plane.

Under the aligned condition with zero tilt, the shaft rotates within the bearing. The thickness of the oil film between the shaft and bearing is computed using Equation (5) [23].

$$h = c + e \cdot cos\{\theta - (\theta_W - \pi + \psi)\} + e'\left(1 - \frac{2}{l}z\right)cos\{\theta - (\theta_W - \pi)\} + h_e \tag{5}$$

The non-dimensionalized representation of Equation (5) can be expressed as follows:

$$H = 1 + \varepsilon \cdot cos\{\theta - (\theta_W - \pi + \psi)\} + \varepsilon'\left(1 - 2\frac{Z}{L}\right)cos\{\theta - (\theta_W - \pi)\} + H_e \tag{6}$$

where

$$H_e = \frac{h_e}{c}, \; \varepsilon = \frac{e}{c}, \; \varepsilon' = \frac{e'}{c} \tag{7}$$

The parameters ε and ε' featured in Equation (7) represent the shaft's eccentricity and tilting ratios, respectively. Additionally, in order to calculate Equation (5), obtaining the alteration in oil film thickness due to the elastic deformation of the flexible structure, denoted as h_e, is essential. The assessment of the elastic deformation of the bearing is conducted through the solid mechanics module. The obtained numerical results are integrated with the hydrodynamic bearing module for hydrodynamic lubrication analysis. Figure 4 displays a hexahedral mesh of the finite element model [23]. The preference for a hexahedral mesh in numerical calculations arises from its superior resolution and faster processing speed compared to the tetrahedral mesh [47]. In the analysis, 100 elements were utilized in the circumferential direction and 53 elements in the axial direction, with the exception of the groove region, where 108 elements were used in the circumferential direction. As a result, the total element count was 44,425.

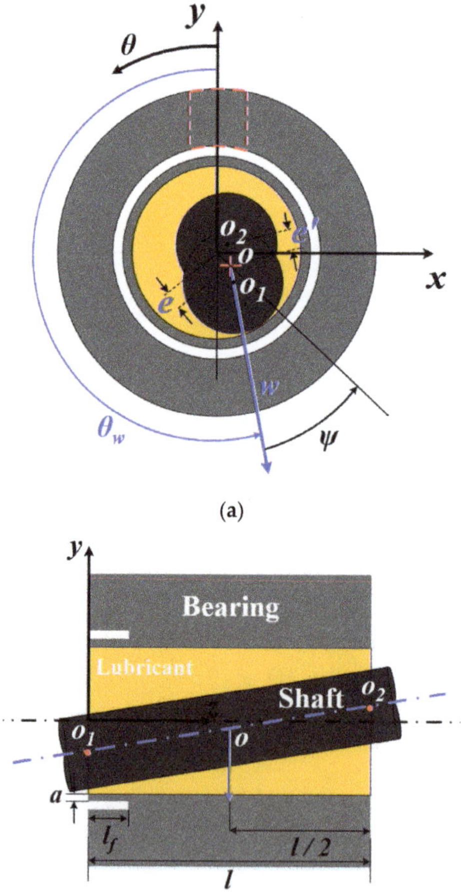

Figure 3. The shaft in the bearing under eccentric and tilted motion conditions: (**a**) *x*-*y* plane; (**b**) *y*-*z* plane.

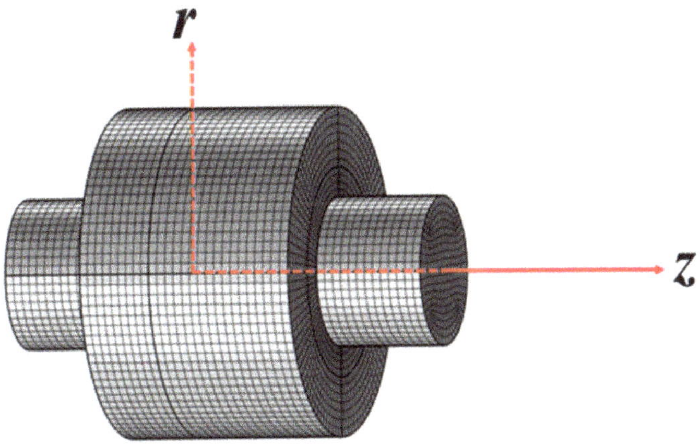

Figure 4. Finite element model of journal bearing.

The boundary conditions for solving Equations (2) and (8) are specified as follows:

- Boundary condition for the oil film fracture zone: $\frac{\partial p}{\partial n} = 0,\ p = p_a$.
- Pressure in the cavitation region:

$$p = p_a \text{ where } p < p_a. \tag{8}$$

- Pressure at the bearing ends and oil feeding groove $p(0,\ z) = p_b,\ p(\theta,\ 0) = p_b,\ p(\theta,\ l) = p_b$.
- Displacement at the inner end of the flexible structure: $u\left(x,\ y,\ l_f\right) = 0$.

The direction of vector n, as indicated in Equation (8), is perpendicular to the boundary line of the oil film fracture. Additionally, periodic conditions are applied for both pressure and displacement in the ring-shaped bearing. Equation (8) can be expressed in a dimensionless form, given by Equation (9) [23]:

- Boundary condition for the oil film fracture zone: $\frac{\partial P}{\partial n} = 0,\ P = 0$.
- Pressure in the cavitation region:

$$P = 0 \text{ where } P < 0. \tag{9}$$

- Pressure at the oil feeding groove and the bearing ends:

$$P(0,\ Z) = P_b,\ P(\theta,\ 0) = P_b,\ P(\theta,\ L) = P_b.$$

- Deformation at the inner end of the flexible structure: $U\left(X,\ Y,\ L_f\right) = 0$.

The oil film force, represented as f_o, arises from the oil film pressure. It is utilized to determine the shaft's motion through comparison with the applied load. Figure 2a illustrates the visualization of the oil film force exerted on the shaft. The components of the oil film force in the x and y directions, denoted as f_{ox} and f_{oy}, respectively, were computed using Equations (10) and (11). The calculation of the oil film force was achieved through Equation (12) [23]:

$$f_{ox} = \int_0^l \int_0^{2\pi} p\ r\ sin\theta d\theta dz \tag{10}$$

$$f_{oy} = -\int_0^l \int_0^{2\pi} p\ r\ cos\theta d\theta dz \tag{11}$$

$$f_o = \sqrt{{f_{ox}}^2 + {f_{oy}}^2} \tag{12}$$

Equations (13)–(15) are dimensionless forms of Equations (10)–(12).

$$F_{OX} = \int_0^L \int_0^{2\pi} P\, sin\theta d\theta dz \tag{13}$$

$$F_{OY} = -\int_0^L \int_0^{2\pi} P\, cos\theta d\theta dz \tag{14}$$

$$F_O = \sqrt{F_{OX}{}^2 + F_{OY}{}^2} \tag{15}$$

$$F_{OX} = \frac{c^2 f_{ox}}{6r^4\eta\omega},\ F_{OY} = \frac{c^2 f_{oy}}{6r^4\eta\omega},\ F_O = \frac{c^2 f_o}{6r^4\eta\omega} \tag{16}$$

The load used in the analysis is shown in Figure 5. The analysis was performed under the condition that the impact load was added to the static load. In Figure 5, T_1 and T_3 are time intervals in which only the dimensionless static load W acts. T_2 is the time interval in which the impact load acts together with the static load. The impact load is assumed to be in the form of a wave with attenuation. The analysis was performed for cases where the maximum value of the total load given was 1.5 times (W_{max} = 1.5 W), 2 times (W_{max} = 2 W) or 3 times (W_{max} = 3 W) the static load. In the case of actual misaligned journal bearings, shocks and impulses are applied in various directions in addition to the vertical load. In this study, in order to evaluate the lubrication characteristics of the flexible structure under impact load conditions, it was simply assumed that the impact load acts in the vertical direction.

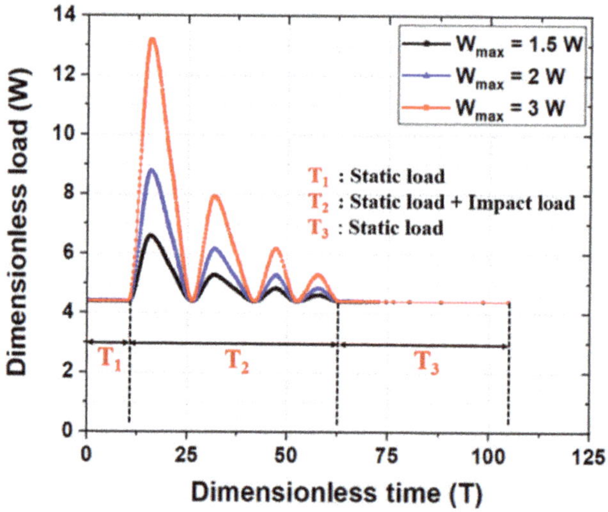

Figure 5. Dimensionless load conditions with dimensionless time.

The dynamic viscosity and density of the lubricant used in the analysis were 0.02083 Pa·s and 904 kg/m^3, respectively. It is difficult to specify information about the actual size and lubricant due to company security reasons since this is an interpretation of journal bearings applied to the home appliances of electronic company. Therefore, a dimensionless analysis was performed.

3. Numerical Results and Discussion

In a previous study [23], EHL analysis [48–53] was performed for a misaligned journal bearing with a flexible structure under a static load condition. The flexible structure of the misaligned journal bearing effectively increased the minimum film thickness. This was because the oil film thickness increased via elastic deformation in the flexible structure at the end of the bearing, with the lubrication performance of the bearings improved, as shown in Figure 6. From the overall dimensionless displacement in Figure 6, it can be confirmed that there is appropriate deformation due to oil film pressure in the area where the flexible structure was applied. In addition, a rectangular-shaped flexible structure was more effective in improving lubrication characteristics than a taper-shaped one [23]. Therefore, in this study, lubrication characteristics under impact load conditions were evaluated for misaligned journal bearings to which a rectangular-shaped flexible structure was applied. In addition, when metal-to-metal contact occurred in misaligned journal bearings, the elastic modulus of the bearing steel was changed, and the lubrication characteristics were compared with those of the existing bearing steel.

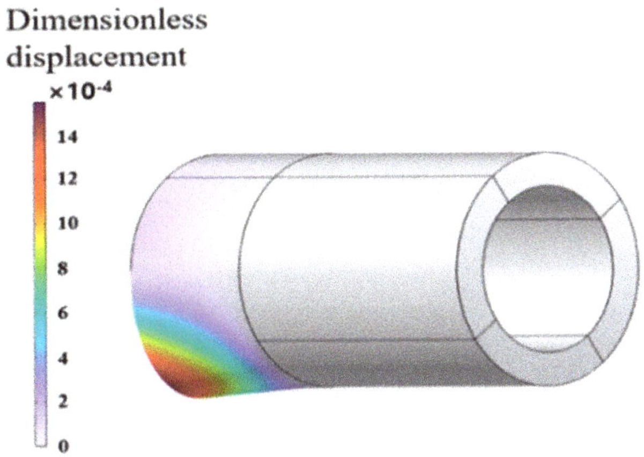

Figure 6. Dimensionless displacement of overall misalignment journal with flexible structure.

The lubrication characteristics following a change in the flexible structure geometry and a change in the tilting ratio (ε') were examined under three load conditions in which the impact load and the static load acted together. Changes in the dimensionless thickness (A) and length (L_f) were used to vary the geometry of the flexible structure. The dimensionless thickness ratio (γ) of the flexible structure was constant at 1 because it was found that the flexible structure with a rectangular shape was more effective than the taper-shaped one in improving the lubrication characteristics [23].

The dimensionless minimum oil film thickness due to changes in the dimensionless thickness (A) of the flexible structure was investigated over time, as shown in Figure 7. The lubrication characteristics were compared under three load conditions and two elastic modulus conditions with other specifications, as shown in Table 1. When A was increased, the dimensionless minimum film thickness decreased overall, resulting in poor lubrication characteristics. When A was 0.6 and 0.8, the difference in the dimensionless minimum oil film thickness was very small. However, when A exceeded 0.8, the advantage of the flexible structure did not appear under impact load conditions. This was because when A exceeded a certain value, the elastic deformation due to oil film pressure was rapidly reduced. In other words, when A exceeded a certain value, it was difficult to secure sufficient oil film thickness through elastic deformation. As the maximum load due to impact load increased, the lubrication properties deteriorated. Additionally, when the dimensionless elastic modulus (E^*) was reduced by about 32% to 1.5×10^4, the overall lubrication characteristics were improved.

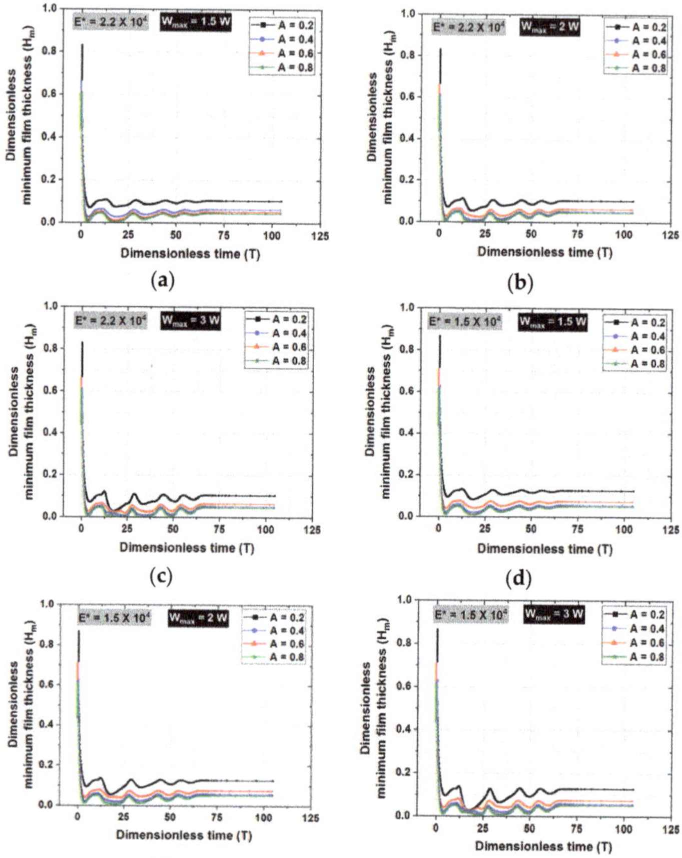

Figure 7. Dimensionless minimum film thickness with A: (**a**) $E^* = 2.2 \times 10^4$, $W_{max} = 1.5\ W$; (**b**) $E^* = 2.2 \times 10^4$, $W_{max} = 2\ W$; (**c**) $E^* = 2.2 \times 10^4$, $W_{max} = 3\ W$; (**d**) $E^* = 1.5 \times 10^4$, $W_{max} = 1.5\ W$; (**e**) $E^* = 1.5 \times 10^4$, $W_{max} = 2\ W$; (**f**) $E^* = 1.5 \times 10^4$, $W_{max} = 3\ W$.

Table 1. Specification of the analysis model (variation in A, W and E^*).

Parameter	Value	Parameter	Value
A	0.2, 0.4, 0.6, 0.8	W	$W_{max} = 1.5\ W, 2\ W, 3\ W$
E^*	2.2×10^4, 1.5×10^4	β	10^{-3}
L	3.0	γ	1.0
L_f	1/3	ε'	0.2
P_b	0.3		

The smallest dimensionless minimum film thickness within the dimensionless time region of Figure 7 is shown in Figure 8. In the analysis results under static load conditions, the lubrication characteristics were good even when A was 1.4. It is almost unrealistic to apply conditions where A is 0.8 or more. The actual applicable range will be 0.4 or less. However, this analysis was performed to show that the lubrication characteristics were improved by the flexible structure even when the total load was about three times the static load.

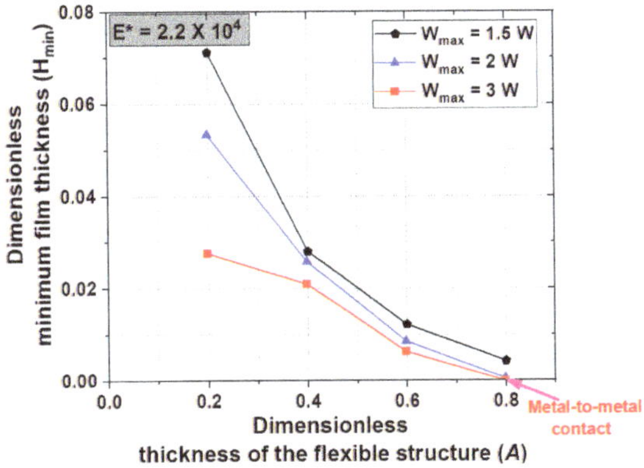

(a)

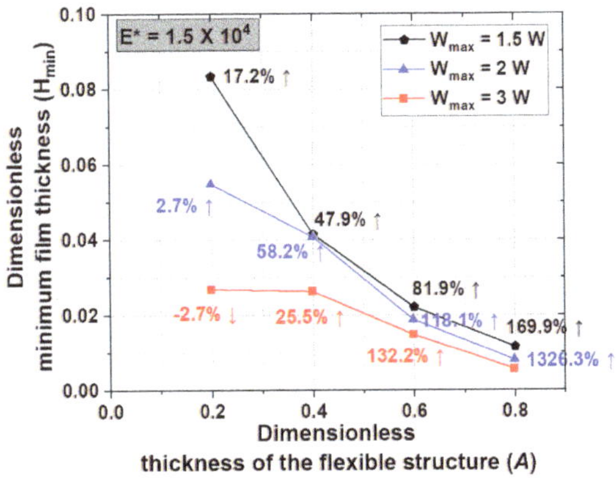

(b)

Figure 8. Dimensionless minimum film thickness with A and W_{max}: (**a**) $E^* = 2.2 \times 10^4$; (**b**) $E^* = 1.5 \times 10^4$.

The analysis was performed to show that the lubrication characteristics were improved by the flexible structure even when the total load was about three times the static load. However, as shown in Figure 8a, when the maximum load was 3 times and A was 0.8 times the static load, metal-to-metal contact occurred. Similar to the results shown in Figure 7, it was confirmed that when A was increased, the dimensionless minimum film thickness decreased. Moreover, when A was 0.8, metal-to-metal contact did not occur when the maximum load was 1.5 and 2 times the static load. However, they had a very small dimensionless minimum film thickness, leading to poor lubrication characteristics. To prevent contact between metals, the lubrication characteristics were investigated by changing the dimensionless elastic modulus of the bearing. By reducing the elastic modulus, deformation due to oil film pressure was made easier in the flexible structure. Figure 8a,b show the results when the dimensionless elastic modulus was 2.2×10^4 and 1.5×10^4, respectively. When the dimensionless elastic modulus is reduced by about 32%, the change in lubrication characteristics is expressed as a percentage, as shown Figure 8b. The percentage is expressed as a percentage of the difference in the dimensionless minimum film thickness at the two elastic moduli based on the dimensionless minimum film thickness in the case of the existing dimensionless elastic modulus ($E^* = 2.2 \times 10^4$). The lubrication characteristics were generally improved when the dimensionless elastic modulus was 1.5×10^4. In particular, looking at the case where A was 0.8, when contact occurred or the dimensionless minimum oil film thickness was very small, it was confirmed that no contact occurred and the lubrication characteristics were greatly improved. Figure 9 shows the dimensionless minimum film thickness and dimensionless maximum displacement with variation in the dimensionless time for four cases among the results in Figure 8. The point when the dimensionless minimum film thickness was the smallest is indicated by a blue dotted line. The point when the maximum dimensionless displacement was the greatest is indicated by a red dotted line. When the dimensionless elastic modulus was 2.2×10^4 and A was 0.2, the point when the dimensionless minimum film thickness was the smallest and the point when the dimensionless maximum displacement was the greatest were almost similar in the entire dimensionless time region. However, when A was 0.6, these two points in dimensionless time, when the dimensionless minimum oil film thickness was the smallest and when the dimensionless maximum deformation was the greatest, were clearly different. This pattern of results was similar even when the dimensionless elastic modulus was 1.5×10^4. Depending on the change in A, the results varied depending on the reactivity of the flexible structure. In other words, when A was small, the deformation of the flexible structure was immediate due to the oil film pressure. However, when A was large, it was believed that the influence of the deformation in the flexible structure had a slight delay. Figures 10 and 11 show dimensionless film thickness distribution and oil pressure distribution, respectively, at the point where the smallest dimensionless minimum film thickness occurs in the time region. At this time, A was 0.4, the maximum load was 1.5 times, 2 times and 3 times the static load, and the dimensionless elastic moduli were 2.2×10^4 and 1.5×10^4. Figure 10 shows a similar pattern of dimensionless film thickness distribution. However, in Figure 10f, where the maximum load was 3 times the static load and the dimensionless elastic modulus was small, at 1.5×10^4, it shows a different form. In this case, the regions where the dimensionless film thickness is large and small are relatively large. That is, under conditions where the dimensionless elastic modulus is small and the maximum load is large, the hydrodynamic pressure increases, and elastic deformation due to the generated oil film pressure is easy. In Figure 11, under the condition that the non-dimensional elastic modulus was the same, when the maximum load is increased, the hydrodynamic oil pressure in the area where oil film pressure occurred is relatively increased.

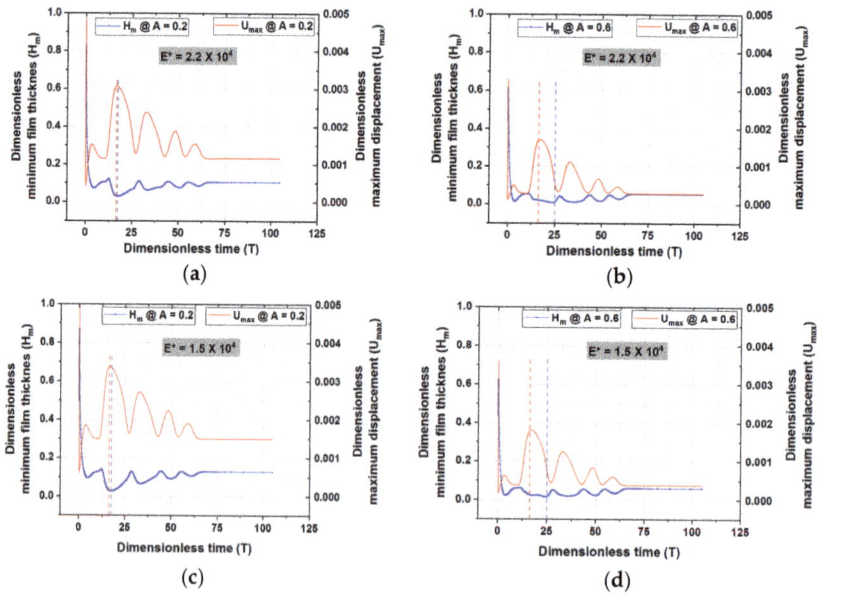

Figure 9. Relationship between dimensionless minimum film thickness and dimensionless maximum displacement with variation in dimensionless time: **(a)** $A = 0.2$, $E^* = 2.2 \times 10^4$; **(b)** $A = 0.6$, $E^* = 2.2 \times 10^4$; **(c)** $A = 0.2$, $E^* = 1.5 \times 10^4$; **(d)** $A = 0.6$, $E^* = 1.5 \times 10^4$.

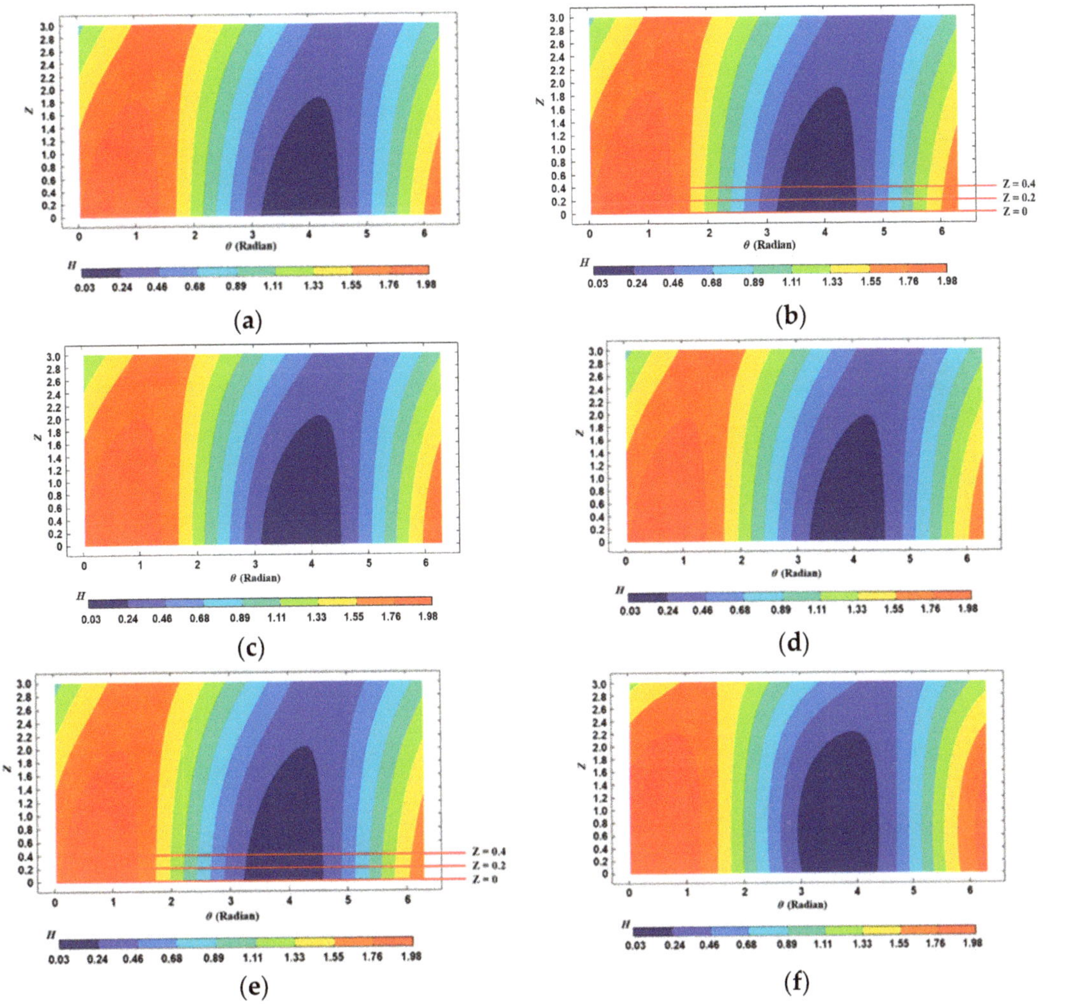

Figure 10. Dimensionless film thickness distribution with E^* and W_{max}: (**a**) $E^* = 2.2 \times 10^4$, $W_{max} = 1.5\ W$; (**b**) $E^* = 2.2 \times 10^4$, $W_{max} = 2\ W$; (**c**) $E^* = 2.2 \times 10^4$, $W_{max} = 3\ W$; (**d**) $E^* = 1.5 \times 10^4$, $W_{max} = 1.5\ W$; (**e**) $E^* = 1.5 \times 10^4$, $W_{max} = 2\ W$; (**f**) $E^* = 1.5 \times 10^4$, $W_{max} = 3\ W$.

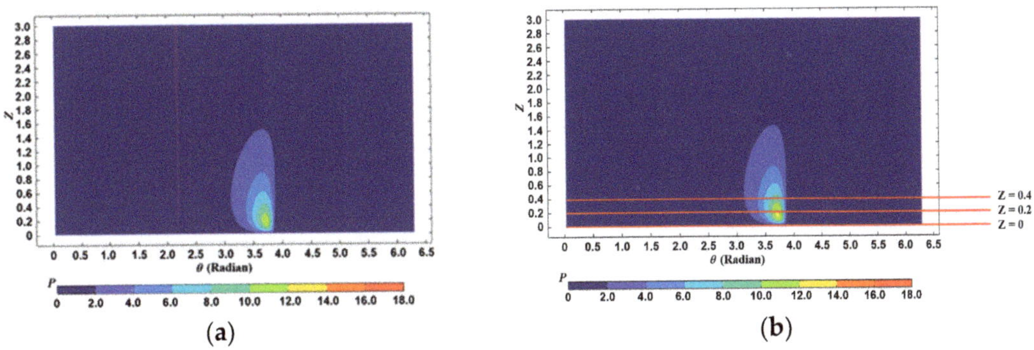

Figure 11. *Cont.*

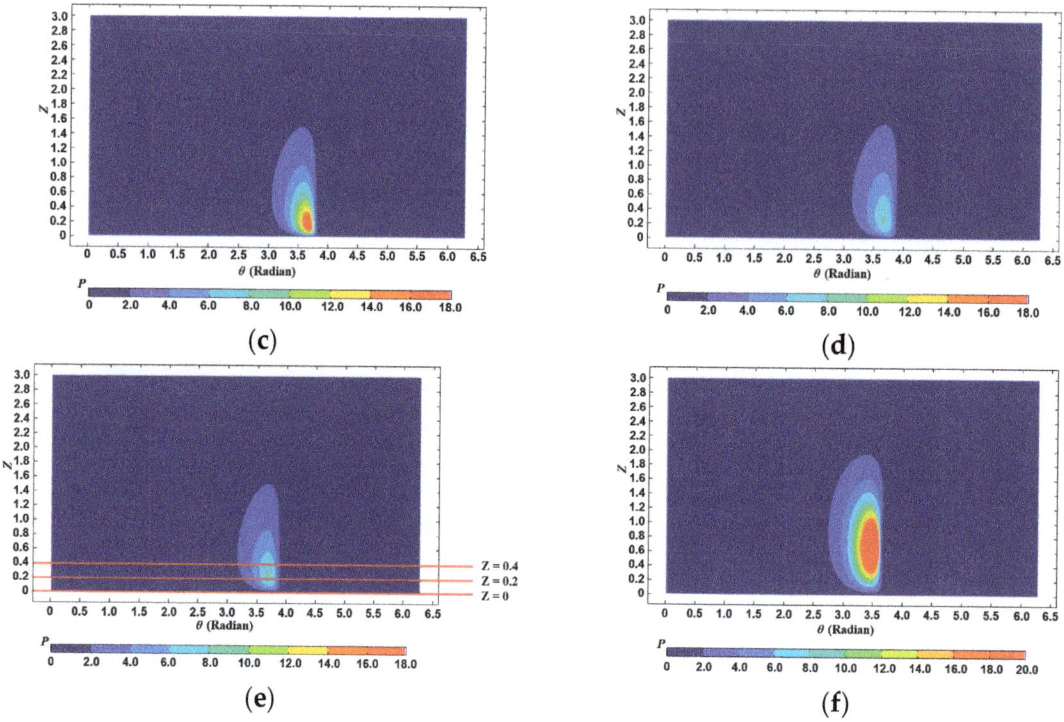

Figure 11. Dimensionless pressure distribution with E^* and W_{max}: (**a**) $E^* = 2.2 \times 10^4$, $W_{max} = 1.5\ W$; (**b**) $E^* = 2.2 \times 10^4$, $W_{max} = 2\ W$; (**c**) $E^* = 2.2 \times 10^4$, $W_{max} = 3\ W$; (**d**) $E^* = 1.5 \times 10^4$, $W_{max} = 1.5\ W$; (**e**) $E^* = 1.5 \times 10^4$, $W_{max} = 2\ W$; (**f**) $E^* = 1.5 \times 10^4$, $W_{max} = 3\ W$.

Figure 12 shows the dimensionless film thickness and dimensionless pressure in the circumferential direction for the three red lines shown in Figures 10b,e and 11b,e. Figure 12a,b show the results when the maximum load is twice the static load and the dimensionless elastic moduli are 2.2×10^4 and 1.5×10^4, respectively. Compared to the static load analysis results [23], it was found that the location where the maximum load occurred moved toward the edge of the bearing. Moreover, after the dimensionless maximum pressure occurred, a dimensionless minimum film thickness occurred, that is, the two points mentioned above did not coincide. The displacement distribution was investigated at the point where the minimum film thickness occurred. Figure 13 shows the dimensionless displacement at the point where the smallest dimensionless minimum film thickness occurs in the time region. At this time, A was 0.4, the maximum load was 1.5 times, 2 times and 3 times the static load, and the dimensionless elastic modulus was 2.2×10^4 or 1.5×10^4. As the dimensionless maximum load increased, the area where deformation occurred widened, with its size increasing overall. Moreover, when the dimensionless elastic modulus was small, more deformation occurred. In other words, the lubrication characteristics were improved as the oil film thickness was relatively large due to a larger amount of elastic deformation.

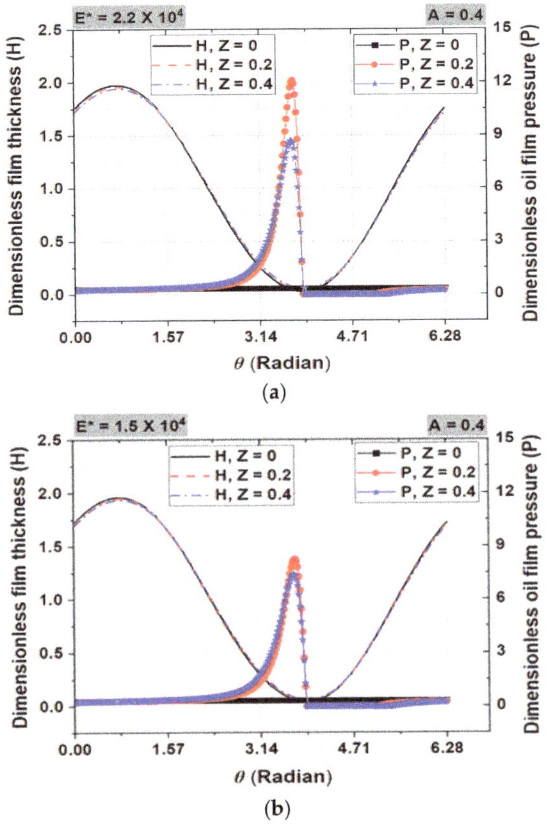

Figure 12. Circumferential distribution of dimensionless film thickness and pressure: (**a**) A = 0.4, W_{max} = 2 W, E^* = 2.2 × 10^4; (**b**) A = 0.4, W_{max} = 2 W, E^* = 1.5 × 10^4.

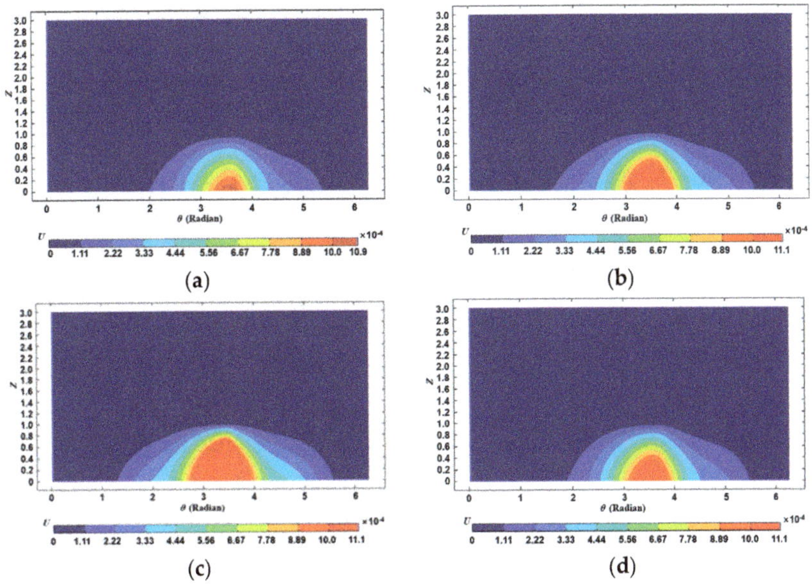

Figure 13. *Cont.*

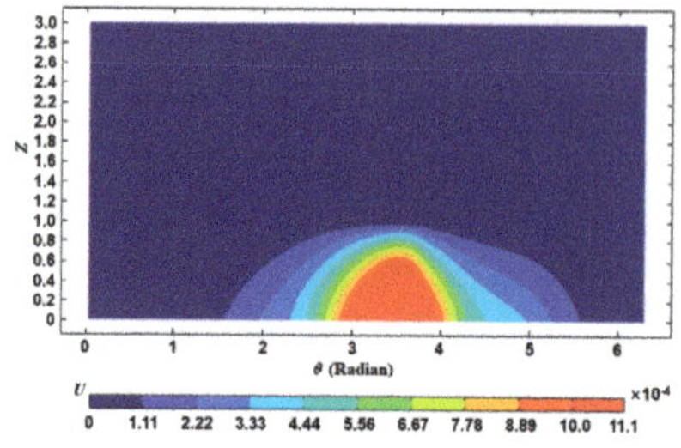

(e)

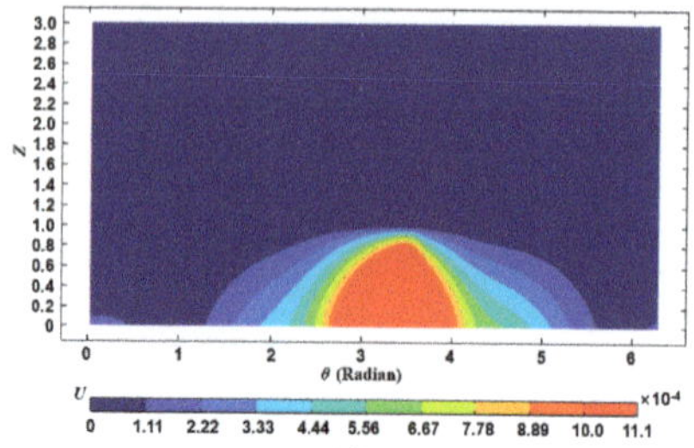

(f)

Figure 13. Dimensionless displacement distribution with E^* and W_{max}: (**a**) $E^* = 2.2 \times 10^4$, $W_{max} = 1.5\ W$; (**b**) $E^* = 2.2 \times 10^4$, $W_{max} = 2\ W$; (**c**) $E^* = 2.2 \times 10^4$, $W_{max} = 3\ W$; (**d**) $E^* = 1.5 \times 10^4$, $W_{max} = 1.5\ W$; (**e**) $E^* = 1.5 \times 10^4$, $W_{max} = 2\ W$; (**f**) $E^* = 1.5 \times 10^4$, $W_{max} = 3\ W$.

In this investigation, we performed changes in the dimensionless minimum film thickness in relation to variations in the dimensionless length (L_f) of the flexible structure, as shown in Figure 14. The lubrication characteristics were compared under three load conditions and two elastic modulus conditions with other specifications shown in Table 2. In a previous study [23], there was no metal-to-metal contact when L_f was 0.2 under the condition of a dimensionless static load, W. However, metal-to-metal contact occurred under the condition of an impact load being applied along with the static load. When L_f was 0.2, the flexible structure covered a very small area, 20% of the bearing's axial length. It is believed that it is difficult to secure a sufficient oil film due to deformation under conditions where impact load is also applied. Thus, numerical analysis was performed only for L_f between 0.3 and 0.6. Figure 14a,b show the results obtained for dimensionless elastic moduli of 2.2×10^4 and 1.5×10^4, respectively. When the dimensionless elastic modulus is reduced by about 32%, these changes in dimensionless minimum film thickness are expressed as a percentage, as shown in Figure 14b. This percentage quantifies the difference in dimensionless minimum film thickness between the two elastic moduli, based on the dimensionless minimum film thickness when the dimensionless elastic modulus is 2.2×10^4. When the dimensionless elastic modulus was 2.2×10^4, the variance in the dimensionless minimum film thickness was relatively small when the maximum load was 1.5 and 2 times the static load. However, when the maximum load was 3 times the static load, the difference in the dimensionless minimum film thickness became significantly smaller, as shown in Figure 14a. In Figure 14b, it can be seen that if the dimensionless elastic modulus was changed to a small value, the lubrication characteristics were improved in terms of minimum film thickness. In particular, when L_f was small, the improvement in the lubricating characteristics was greater due to changes in the dimensionless elastic modulus than when it was large. Moreover, when L_f increased under the same load condition, the percentage generally decreased. That is, when L_f increased and the elastic modulus decreased, the increase rate of the minimum film thickness was not large; thus, the improvement in the lubricating characteristics was not distinct.

Table 2. Specification of the analysis model (variation in L_f, W and E^*).

Parameter	Value	Parameter	Value
A	0.4	W	$W_{max} = 1.5\ W, 2\ W, 3\ W$
E^*	$2.2 \times 10^4, 1.5 \times 10^4$	β	10^{-3}
L	3.0	γ	1.0
L_f	0.3, 0.4, 0.5, 0.6	ε'	0.2
P_b	0.3		

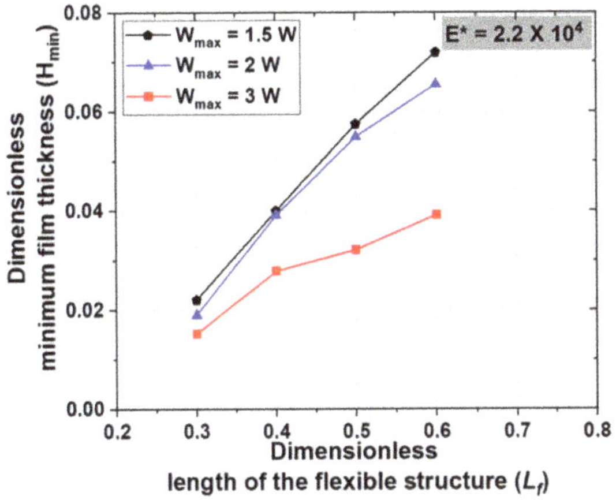

(a)

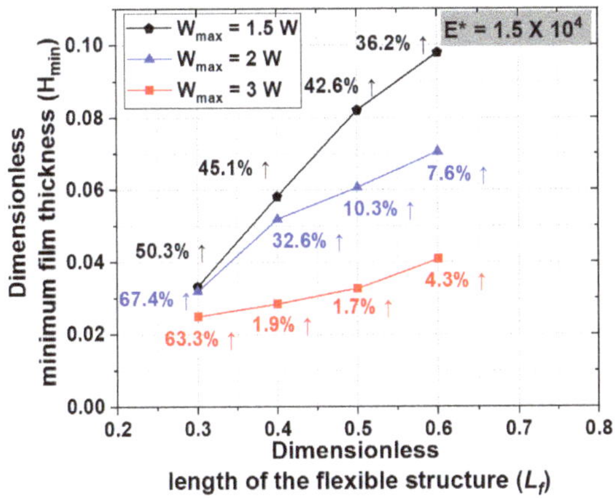

(b)

Figure 14. Dimensionless minimum film thickness with L_f and W_{max}: (**a**) $E^* = 2.2 \times 10^4$; (**b**) $E^* = 1.5 \times 10^4$.

Figure 15 illustrates the changes in the non-dimensional minimum film thickness with changes in the tilting ratio (ε'). Furthermore, we compared lubrication characteristics across three different load conditions and two elastic modulus conditions, with the specifications and details outlined in Table 3. When the dimensionless elastic modulus was 2.2×10^4 under the three load conditions, the non-dimensional minimum film thickness varied with changes in the tilting ratio, as shown in Figure 15a. As the tilting ratio was increased, the dimensionless minimum film thickness decreased. This was because as the tilting ratio was increased, misalignment became more severe, resulting in unstable driving conditions that made it difficult to secure a sufficient oil film. In previous studies [23] where a static load was applied, an oil film was secured and the lubrication characteristics were good, even at a tilting ratio of 0.4. However, under conditions where an impact load was also applied, metal-to-metal contact occurred when the tilting rate exceeded 0.2. Thus, we tried to change the dimensionless elastic modulus in order to improve the lubrication characteristics. When the dimensionless elastic modulus is reduced by about 32%, these variations in dimensionless minimum film thickness are expressed as a percentage, as shown in Figure 15b. These percentages were obtained in the same manner as previously mentioned. Due to change in the elastic modulus, the increase rate of the minimum oil film thickness at the tilting rate was smaller than the increase rate of the minimum film thickness with the same dimensionless thickness and length of the flexible structure. If the tilting ratio was large, the increase rate of the dimensionless minimum film thickness was large upon changing the dimensionless elastic modulus. However, when the tilting ratio was small, the improvement upon changing the dimensionless elastic modulus was negligible. Overall, it was ineffective in terms of improving the lubrication characteristics by facilitating elastic deformation by reducing the non-dimensional elastic coefficient with the change in tilting rate.

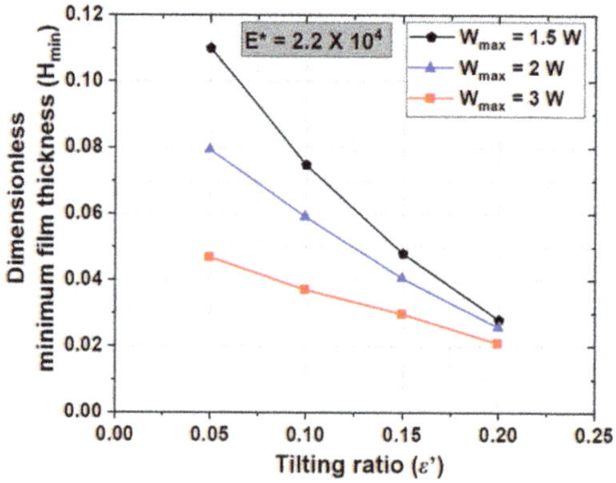

(a)

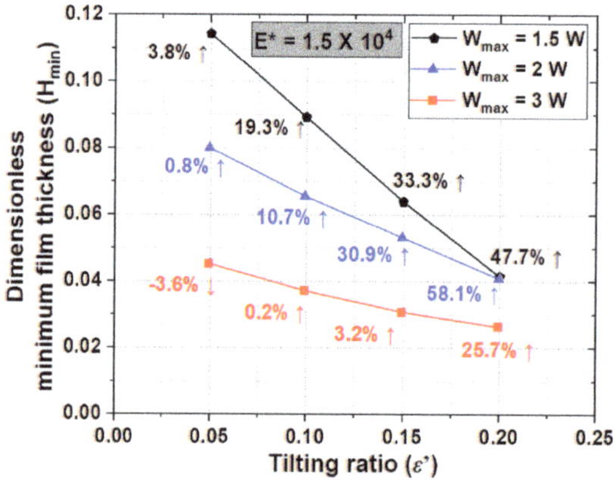

(b)

Figure 15. Dimensionless minimum film thickness with ε' and $W_{\max}$: (**a**) $E^* = 2.2 \times 10^4$; (**b**) $E^* = 1.5 \times 10^4$.

Table 3. Specification of the analysis model (variation in ε').

Parameter	Value	Parameter	Value
A	0.4	W	$W_{\max} = 1.5\ W, 2\ W, 3\ W$
E^*	2.2×10^4, 1.5×10^4	β	10^{-3}
L	3.0	γ	1.0
L_f	1/3	ε'	0.05, 0.1, 0.15, 0.2
P_b	0.3		

4. Conclusions

This study demonstrates the utilization of a flexible structure in enhancing the lubrication performance of misaligned journal bearings under shock load conditions. Shock loads applied for various reasons can cause wear and failure due to contact in misaligned journal bearings. To improve this phenomenon, a flexible structure that could facilitate elastic deformation was applied to the end of the journal bearing, and EHL analysis was performed on misaligned journal bearings. The impact load of the journal bearing was assumed to be in the form of a wave. It was applied in addition to the static load. Three conditions were used, in which the overall maximum load was applied at 1.5, 2 and 3 times the static load. In addition, in order to improve the conditions of poor lubrication, such as metal-to-metal contact, the elastic modulus was reduced within the range of the elastic modulus of bearing steel currently used, and the lubrication characteristics were compared with those of the existing elastic modulus. The assessment of lubrication performance involved a comparison of the minimum film thickness of the journal bearing equipped with a flexible structure using different dimensionless thicknesses, lengths and tilting ratios. The dimensionless thickness ratio (γ) for the flexible structure remained constant at 1, since it was determined that the rectangular-shaped flexible structure outperformed the tapered one in terms of enhancing lubrication properties in previous research. The application of the flexible structure of misaligned journal bearings is very effective in improving lubrication characteristics under impact load conditions. This is because even if an impact load is added to the static load, a sufficient oil film thickness is secured due to elastic deformation in the flexible structure. A misaligned journal bearing with a flexible structure could prevent metal-to-metal contact until a maximum load of approximately three times the rated load is reached. When contact occurs or the lubrication performance is poor in a misaligned journal bearing with a flexible structure, the lubrication performance is significantly improved by changing the modulus of elasticity to be slightly lower. Finally, the numerical results showed that the application of a flexible structure to misaligned journal bearings improved the lubrication characteristics under impact load conditions. However, from an empirical perspective, additional research is needed to apply experimental verification to actual systems such as compressors and hydraulic pumps.

Author Contributions: Conceptualization, S.-H.H. and W.-J.J.; literature review and formal analysis, S.-H.H.; writing-original draft preparation, methodology, S.-H.H.; review and editing, S.-H.H.; funding acquisition, S.-H.H. All authors have read and agreed to the published version of the manuscript.

Funding: This work was supported by Korea Hydro & Nuclear Power Co. (2023).

Data Availability Statement: Not applicable.

Acknowledgments: This work was supported by a Korea Institute of Energy Technology Evaluation and Planning (KETEP) grant funded by the Korean government (MOTIE) (No. 20214000000010).

Conflicts of Interest: The authors declare no conflict of interest.

Nomenclature

A	Dimensionless thickness of the flexible structure (=a/r)
E	Modulus of elasticity (GPa)
E^*	Dimensionless modulus of elasticity (=$c^2E/(6r^2\eta\omega)$)
F	Dimensionless force (=$c^2f/(6r^4\eta\omega)$)
F_O	Dimensionless oil film force (=$c^2f_o/(6r^4\eta\omega)$)
F_X	Dimensionless force in the X direction
F_Y	Dimensionless force in the Y direction
F_Z	Dimensionless force in the Z direction
F_{OX}	Dimensionless oil film force in the X direction
F_{OY}	Dimensionless oil film force in the Y direction
H	Dimensionless oil film thickness (=h/c)

H_e	Dimensionless oil film thickness variation by elastic deformation ($=h_e/c$)
H_m	Dimensionless minimum film thickness ($=h_m/c$)
L	Ratio of length to radius of the bearing ($=l/r$)
L_f	Dimensionless length of the flexible structure ($=l_f/l$)
O	Center of the shaft at the middle of the bearing
O_1, O_2	Center of the shaft at both ends of the bearing
P	Dimensionless oil film pressure ($=c^2(p-p_a)/(6r^2\eta\omega)$)
P_b	Dimensionless pressure at the bearing ends and oil feeding groove ($=c^2(p_b-p_a)/(6r^2\eta\omega)$)
T	Dimensionless time ($=\omega t$)
T_1, T_3	Time interval in which static load acts
T_2	Time interval in which static and impact loads acts together
U	Dimensionless displacement in the element ($=u/r$)
W	Dimensionless load acting on the shaft ($=c^2w/(6r^4\eta\omega)$)
X, Y, Z	Dimensionless rectangular coordinate system ($X=x/r$, $Y=y/r$, $Z=z/r$)
a	Thickness at the outer end of the flexible structure (mm)
c	Clearance (μm)
d	Thickness at the inner end of the flexible structure (mm)
e	Tilting amount of the shaft (μm)
e'	Tilting amount of the shaft (μm)
f	Force (N)
f_o	Oil film force (N)
f_x	Force in the x direction (N)
f_y	Force in the y direction (N)
f_z	Force in the z direction (N)
f_{ox}	Oil film force in the x direction (N)
f_{oy}	Oil film force in the y direction (N)
h	Film thickness (μm)
h_e	Change in film thickness due to elastic deformation (μm)
h_m	Minimum film thickness (μm)
l	Length of the bearing (mm)
l_f	Length of the flexible structure (mm)
p	Oil film pressure (MPa)
p_a	Atmospheric pressure (MPa)
p_b	Pressure at bearing ends and oil feeding groove (MPa)
r	Radius of the bearing (mm)
t	Time (seconds)
u	Displacement in the element (mm)
w	Load acting on the shaft (N)
x, y, z	Rectangular coordinate system (mm)
β	Ratio of clearance to bearing radius ($=c/r$)
ε	Eccentricity ratio ($=e/c$)
ε'	Tilting ratio ($=e'/c$)
γ	Ratio of thickness at both ends of the flexible structure ($=d/a$)
η	Absolute viscosity of the lubricant (Pa·s)
v	Poisson's ratio
θ	Cylindrical coordinate (rad)
θ_w	Direction of the load in the cylindrical coordinate system (rad)
ω	Angular velocity (rad/s)

References

1. Zakharov, S.M. Hydodynamic lubrication research: Current situation and future prospects. *J. Frict. Wear* **2010**, *31*, 56–67. [CrossRef]
2. Huang, Q.; Yan, X. Impact factors on lubricant performance of stern bearing with misaligned angle induced by transverse vibration of shaft. *Ocean Eng.* **2020**, *216*, 108051. [CrossRef]
3. Cabrera, D.L.; Wolley, N.H.; Allanson, D.R.; Tridimas, Y.D. Film pressure distribution in water-lubricated rubber journal bearing. *IMechE* **2005**, *219*, 125–132. [CrossRef]
4. Wang, N.; Meng, Q.; Wang, P.; Geng, T.; Yuan, X. Experimental research on film pressure distribution of water-lubricated rubber bearing with multiaxial grooves. *J. Fluids Eng.* **2013**, *135*, 084501. [CrossRef]

5. Zhang, Z.S.; Dai, X.D.; Zhang, Z.N.; Xie, Y.B. Thermoelastohydrodynamic behavior of misaligned plain journal bearings. *Proc. Inst. Mech. Eng. Part C J. Mech. Eng. Sci.* **2013**, *227*, 2582–2599. [CrossRef]
6. Zheng, L.; Zhu, H.; Zhu, J.; Deng, Y. Effects of oil film thickness and viscosity on the performance of misaligned journal bearings with couple stress lubricants. *Tribol. Int.* **2020**, *146*, 106229. [CrossRef]
7. Das, S.; Guha, S.K.; Chattopadhyay, A.K. On the steady-state performance of misaligned hydrodynamic journal bearings lubricated with micropolar fluids. *Tribol. Int.* **2002**, *35*, 201–210. [CrossRef]
8. Liang, P.; Li, X.; Guo, F.; Cao, Y.; Zhang, X.; Jiang, F. Influence of sea wave shock on transient start-up performance of water-lubricated bearing. *Tribol. Int.* **2022**, *167*, 107332. [CrossRef]
9. Wang, C.; Zhang, C.; Gu, L.; Bi, M.; Hou, P.; Zheng, D.; Wang, L. Analysis on surface damage of M50 steel at impact-sliding contacts. *Tribol. Int.* **2020**, *150*, 106384. [CrossRef]
10. Wang, C.; Zang, C.; Bao, M.; Gu, L.; Wang, L.; Wei, X.; Li, W. Scratch damage on guiding surfaces of high-speed bearings under oil cut-off condition. *Eng. Fail. Anal.* **2022**, *140*, 106624. [CrossRef]
11. Xiang, G.; Han, Y.; Wang, J.; Xiao, K.; Li, J. A transient hydrodynamic lubrication comparative analysis for misaligned micro-grooved bearing considering axial reciprocating movement of shaft. *Tribol. Int.* **2019**, *132*, 11–23. [CrossRef]
12. Wang, Y.; Shi, X.; Zhang, I. Experimental and numerical study on water-lubricated rubber bearings. *Ind. Lubric. Tribol.* **2014**, *66*, 282–288. [CrossRef]
13. Murawski, L. Shaft line alignment analysis taking ship construction flexibility and deformations into consideration. *Mar. Struct.* **2005**, *18*, 62–84. [CrossRef]
14. Bouyer, J.; Fillon, M. An experimental analysis of misalignment effects on hydrohynamic plain journal bearing performances. *J. Tribol.* **2002**, *124*, 313–319. [CrossRef]
15. Sun, J.; Changlin, G. Hydrodyamic lubrication analysis of journal bearing considering misalignment caused by shaft deformation. *Tribol. Int.* **2004**, *37*, 841–848. [CrossRef]
16. Sun, J.; Gui, C.; Li, Z.; Li, Z. Influence of journal misalignment caused by shaft deformation under rotational load on performance of journal bearing. *Proc. Inst. Mech. Eng. Part J J. Eng.* **2005**, *219*, 275–283. [CrossRef]
17. Sun, J.; Gui, C.; Li, Z. An experimental study of journal bearing lubrication effected by journal misalignment as a result of shaft deformation under load. *J. Tribol.* **2005**, *127*, 813–819. [CrossRef]
18. Li, Q.; Liu, S.I.; Pan, X.H.; Zheng, S.Y. A new method for the studying the 3D transient flow of misaligned journal bearings in flexible rotor-bearing systems. *J. Zhejiang Univ. Sci. A* **2012**, *13*, 647–664. [CrossRef]
19. Pierre, I.; Bouyer, J.; Fillon, M. Thermohydrodynamic behavior of misaligned plain journal bearings: Theoretical and experimental approaches. *Tribol. Trans.* **2004**, *47*, 594–604. [CrossRef]
20. Xu, G.; Zhou, J.; Geng, H.; Lu, M.; Yang, I.; Yu, L. Research on the static and dynamic characteristics of misaligned journal bearing considering the turbulent and thermohydrodynamic effects. *J. Tribol.* **2015**, *137*, 024504. [CrossRef]
21. Li, B.; Sun, J.; Zhu, S.; Fu, Y.; Zhao, X.; Wang, H.; Teng, Q.; Ren, Y.; Li, Y.; Zhu, G. Thermohydrodynamic lubrication analysis of misaligned journal bearing considering the axial movement of journal. *Tribol. Int.* **2019**, *135*, 397–407. [CrossRef]
22. Nikolakopoulos, P.G.; Papadopoulos, C.; Kaiktsis, I. Elastohydrodynamic analysis and Pareto optimization of intact, worn and misaligned journal bearings. *Meccanica* **2011**, *46*, 577–588. [CrossRef]
23. Jeon, W.J.; Hong, S.H. A new type of misaligned journal bearing with flexible structure. *Lubricants* **2023**, *11*, 256. [CrossRef]
24. Thomsen, K.; Klit, P. Improvement of journal bearing operation at heavy misalignment using bearing flexibility and compliant liners. *Proc. Inst. Mech. Eng. Part J J. Eng. Tribol.* **2012**, *226*, 651–660. [CrossRef]
25. Sharma, S.C.; Girish, B.M.; Kamath, R.; Satish, B.M. Graphite particles reinforced ZA-27 alloy composite materials for journal applications. *Wear* **1998**, *219*, 162–168. [CrossRef]
26. Kim, S.S.; Park, D.C.; Lee, D.G. Characteristics of carbon fiber phenolic composite for journal bearing materials. *Compos. Struct.* **2004**, *66*, 359–366. [CrossRef]
27. Bouyer, J.; Fillon, M. Improvement of the THD performance of a misaligned plain journal bearing. *J. Tribol.* **2003**, *125*, 334–342. [CrossRef]
28. Hattori, H. EHL analysis of a journal bearing for rotary compressors under dynamic loading: Effect of flexible structure at bearing end. *Trans. Jap. Soc. Mech. Eng.* **1998**, *64*, 3171–3178. [CrossRef]
29. Ito, Y.; Hattori, H.; Miura, K. Mixed lubrication analysis of vane sliding surface in rotary compressor mechanisms. *Tribol. Online* **2009**, *4*, 96–102. [CrossRef]
30. Ikeda, A.; Satodate, K.; Aoki, T.; Miura, K. Development of the rotary compressor which is high performance in the wide range, and has high refrigerating capacity. In Proceedings of the JSRAE Annual Conference, Tokyo, Japan, 12–14 September 2012.
31. Tao, Y.; Zhao, J.; Feng, S.Z. A reliability assessment model for journal bearing based on natural degradation and random shocks. *J. Mech. Sci. Technol.* **2020**, *34*, 4641–4648. [CrossRef]
32. Lee, J.; Jang, G.; Jung, K. Optimal design of fluid dynamic bearings to develop a robust disk-spindle system in a hard disk drive utilizing model analysis. *Microsyst. Technol.* **2013**, *19*, 1495–1504. [CrossRef]
33. Jang, G.H.; Yoon, J.W. Nonlinear dynamic analysis of a hydrodynamic journal bearing considering the effect of a rotating or stationary herringbone groove. *J. Tribol.* **2002**, *124*, 297–304. [CrossRef]
34. Park, K.Y.; Jang, G.H. Dynamic of a hard disk drive spindle system due to its structural design variables and the design variables of fluid dynamic bearings. *IEEE Trans. Magn.* **2009**, *45*, 5135–5140. [CrossRef]

35. Walton, J.F.; Heshmat, H.; Tomaszewski, M.J. Testing of s small turbocharger/turbojet sized simulator rotor supported on foil bearings. *J. Eng. Gas Turbines Power* **2004**, *130*, 67–73. [CrossRef]
36. Zhang, Q.D.; Shan, X.C. Dynamic characteristics of micro air bearings for microsystems. *Microsyst. Technol.* **2008**, *14*, 229–234. [CrossRef]
37. Ren, T.M.; Feng, M. Anti-shock characterictics of water lubricated bearing for fuel cell vehicle air compressor. *Tribol. Int.* **2017**, *107*, 56–64. [CrossRef]
38. Jamali, H.U.; Sultan, H.S.; Abdullah, O.I.; Al-Tamimi, A.N.J.; Abbud, L.H.; Ruggiero, A.; Al-Dujaili, Z.A. Effect of chamfer form and parameters on the characteristics of finite length journal bearing under impact load. *Lubricants* **2023**, *11*, 73. [CrossRef]
39. Jamali, H.U.; Sultan, H.S.; Abdullah, O.I.; Al-Tamimi, A.N.J.; Abbud, L.H.; Ruggiero, A.; Al-Dujaili, Z.A. Analysis of the performance of chamfered finite-length journal bearings under dynamic loads. *Mathematics* **2023**, *11*, 587. [CrossRef]
40. Pai, R.; Majumdar, B.C. Stability of submerged four-lobe oil journal bearings under dynamic load. *Wear* **1992**, *154*, 95–108. [CrossRef]
41. Strzelecki, S. Operating characteristics of heavy loaded cylindrical journal bearing with variable axial profile. *Mater. Res.* **2005**, *8*, 481–486. [CrossRef]
42. Nacy, S.M. Effect of chamfering on side-leakage flow rate of journal bearings. *Wear* **1997**, *212*, 95–102. [CrossRef]
43. Nicoletti, R. Optimization of journal bearing profile for higher dynamic stability limits. *J. Tribol.* **2012**, *135*, 011702. [CrossRef]
44. Ghosh, M.K.; Satish, M.R. Stability of multillobe hybrid bearing with short sills—Part II. *Tribol. Int.* **2003**, *36*, 633–636. [CrossRef]
45. Jiang, L.; Feng, Q.; Coit, D.W. Modeling zoned shock effects on the stochastic degradation in dependent failure processes. *IIE Trans.* **2015**, *47*, 460–470. [CrossRef]
46. Song, S.I.; Coit, D.W.; Feng, Q.M. Reliability analysis of multi-component series systems subject to hard and soft failures with dependent shock effects. *IIE Trans.* **2016**, *48*, 720–735. [CrossRef]
47. Hong, S.H.; Kim, K.W. A new type groove for hydraulic spool valve. *Tribol. Int.* **2016**, *103*, 629–640. [CrossRef]
48. Oh, K.P.; Huebner, K.H. Solution of the elastohydrodynamic finite journal bearing problem. *ASME Lubr. Technol.* **1973**, *95*, 342–351. [CrossRef]
49. Kalin, M.; Kus, M. New strategy for reducing the EHL friction in steel contacts using additive-formed oleophobic boundary film. *Friction* **2021**, *9*, 1346–1360. [CrossRef]
50. Wang, Y.; Wang, Q.J.; Lin, C. A mixed-EHL analysis of effects of misalignments and elastic deformations on the performance of a coupled journal-thrust bearing system. *Tribol. Int.* **2006**, *39*, 281–289. [CrossRef]
51. Cai, J.; Xiang, G.; Li, S.; Guo, J.; Wang, J.; Chen, S.; Yang, T. Mathematical modeling for nonlinear dynamic mixed friction behaviorsof novel coupled bearing lubricated with low viscosity fluid. *Phys. Fluids* **2022**, *34*, 093612. [CrossRef]
52. Zhou, Y.; Wang, Y.; Zhao, J. Influence on journal bearing considering wall-slip in EHL. *IOP Conf. Ser. Mater.* **2018**, *394*, 042042. [CrossRef]
53. Xiang, G.; Yang, T.; Guo, J.; Wang, J.; Liu, B.; Chen, S. Optimization transient wear and contact performances of water-lubricated bearings under fluid-solid-thermal coupling condition using profile modification. *Wear* **2022**, *502–503*, 204379. [CrossRef]

lubricants MDPI

Article

Study on Thermal Characteristics of Angular Contact Ball Bearings Considering Roundness Error

Yongjian Yu [1,2,3,*], Ruixiang Ma [1], Yujun Xue [1,3] and Yonggang Liu [1]

1 School of Mechatronics Engineering, Henan University of Science and Technology, Luoyang 471003, China; xiang1037388414@163.com (R.M.); xue_yujun@163.com (Y.X.); martin_lyg@163.com (Y.L.)
2 Longmen Laboratory, Luoyang 471000, China
3 Henan Key Laboratory for Machinery Design and Transmission System, Henan University of Science and Technology, Luoyang 471003, China
* Correspondence: yyjhkd@126.com

Abstract: To develop an angular contact ball bearing with low power consumption, a heat generation calculation model for angular contact ball bearings has been established based on bearing quasi dynamics, elastohydrodynamic lubrication theory, heat transfer theory, and Kirchhoff's law of energy conservation, considering the effects of roundness error, bearing preload, centrifugal effect, and thermal expansion. The correctness of the model is verified through experiments. The influence of different operating conditions and roundness errors on the thermal characteristics of angular contact ball bearings is analyzed. The results of the calculation indicate that when the roundness error order is equal to the number of balls $n/2 \pm 2$ (where $n = 1, 2, 3, \ldots$), the overall heat generation of the bearing is lower than that without considering the roundness error. When the roundness error order is equal to $(2n-1)/4 \pm 2$ (where $n = 1, 2, 3, \ldots$), the overall heat generation of the bearing is higher than that without considering the roundness error. At the same rotating speed, the overall heat generation fluctuates as the roundness error order changes, and the trend becomes more pronounced as the rotating speed increases. The maximum overall heat generation is achieved when the roundness error order equals $(2n-1)/4$ times (where $n = 1, 2, 3, \ldots$) the number of balls. When the roundness error order is equal to $n/2$ times the number of balls (where $n = 1, 2, 3, \ldots$), the bearing's overall heat generation is minimal. The variation in the total heat generated by the bearing is directly proportional to the amplitude of the roundness error. With the increase in roundness error harmonic order, the bearing integral heat generation shows a periodic change, and the change period has a mapping relationship with the number of balls.

Keywords: angular contact ball bearing; quasi dynamics; thermal network method; roundness error

Citation: Yu, Y.; Ma, R.; Xue, Y.; Liu, Y. Study on Thermal Characteristics of Angular Contact Ball Bearings Considering Roundness Error. *Lubricants* **2024**, *12*, 43. https://doi.org/10.3390/lubricants12020043

Received: 23 November 2023
Revised: 25 January 2024
Accepted: 28 January 2024
Published: 3 February 2024

1. Introduction

Angular contact ball bearings are used in wide applications, such as precision machine tools, aerospace, robotics, instrumentation, advanced rail transit, and other domains. During processing and production, the roundness error on the components can lead to uneven loading of the rolling elements, changes in contact angle, increased friction, rapid increase in internal heat generation, thermal expansion of the bearing, and changes in structural dimensions, which in turn cause changes in the friction and wear characteristics of the bearing, affecting the working accuracy of the bearing. A significant temperature rise can further cause the tempering and softening of bearing contact surface materials, leading to fatigue failure of bearings and ultimately resulting in bearing scrapping.

Currently, numerous studies have been reported on the heat generation characteristics and roundness error of bearings. In terms of bearing heat generation, Palmgren et al. [1] conducted tests on different types and sizes of bearings and analyzed the data to obtain an empirical formula for the bearing friction moment. Based on Palmgren's method, Harris et al. [2] revised the empirical formula of the bearing friction moment on the basis of experiments and proposed a local heating calculation method considering six elements of rolling bearing friction. Rami Kerrouche et al. [3] started from the root cause of the friction power loss in high-speed cylindrical roller bearings, introduced a detailed calculation method for friction power loss, namely the local method, calculated the total power loss of the bearing, and then established a thermal network model of the bearing with the friction power loss as a boundary condition, solving for the temperature of each component of the bearing. Pouly et al. [4,5] investigated the power consumption and heat generation mechanisms of high-speed rolling element bearings. They employed a thermal network approach to estimate temperatures at different locations within thrust angular ball bearings. The study revealed the impact of power loss distribution, including sliding friction and oil evaporation losses, on temperature. Furthermore, it emphasized the crucial role of the oil-air mixture in this context. Zhang et al. [6] established a ball load balance model, optimizing the thermal grid model of high-speed spindle bearings. The model was validated using the Newton–Raphson method, with results showing good agreement with experimental values, enhancing the accuracy of predicting bearing operational precision and lifespan. Tarawneh et al. [7] studied the temperature distribution of bearing parts, generation of friction heat, and surface temperature of the bearing housing by the finite element method. Hong Y. et al. [8] employed finite element simulation software to investigate the temperature field of deep groove ball bearings. Using Hertz's elastic contact theory as a basis, an elliptical contact region sliding model was employed to compute the differential frictional heat generated between the ball and raceway. Zheng et al. [9] considered the influence of the contact angle of angular contact ball bearings on the thermal expansion and deformation of the bearings. They established a comprehensive thermal grid model for a pair of front bearings in a high-speed main spindle and its surrounding environment to predict the temperature rise of the bearings. Actual tests were conducted on the temperature variation of the bearings, and the test results were compared with the corresponding numerical solutions. The study results indicated that the established model could relatively accurately predict the temperature changes in the bearings. Nicolas et al. [10] employed a thermal network method to analyze the temperature field of sliding bearings, successfully overcoming challenges such as the exponential dependence of viscosity on temperature. The obtained solutions closely aligned with the experimental and numerical results from other researchers, providing a valuable tool for evaluating the operational status of industrial machines under limited time and computing resources. Dong et al. [11] proposed a transient bearing temperature field prediction method by combining the thermal network method and finite element method. They analyzed the relationship between time step, computational efficiency, and calculation results. By comparing with traditional static thermal analysis results, the study validated that this simulation method exhibited higher accuracy. Lei et al. [12] analyzed the temperature rise characteristics of angular contact ball bearings lubricated under the ring based on the dynamics, calculated the heat generation by the local method, and finally calculated the temperature of each part by the thermal network method. The finite element software ANSYS was used for comparison and verification.

In terms of bearing roundness error, Hassan E. Rasheed [13] conducted a theoretical study on the influence of circumferential, axial, and combined surface waviness on the performance of bearings. It was demonstrated that combined waviness can enhance the load-carrying capacity and friction characteristics when circumferential and axial waviness numbers are kept below approximately 9 and 2, respectively. An increase in waviness amplitude resulted in more pronounced changes in load-carrying capacity and friction variables. Rodionov et al. [14] examined the connection between the machining error of ball bearing raceways, ball raceway surfaces, and bearing friction torque. They arrived at the conclusion that the shape error of the raceway surface is the primary factor responsible for bearing friction torque fluctuations. Cui L. [15] developed a quasi-dynamic analysis model for rolling bearings. This model was used to analyze the vibration performance of high-speed ball bearings and high-speed roller bearings and to determine the relationship between the waviness, structural parameters, and bearing vibration frequency of the inner and outer rings, as well as the rolling bodies of rolling bearings. The conclusion drawn was that the waviness of rolling bearings can alter the vibration characteristics of the system, and the vibration frequencies induced by the waviness of the outer ring, inner ring, and rolling elements are functions of the waviness order, rotor, and cage rotation frequencies. Zheng H. et al. [16] developed a dynamic model for a cylindrical roller bearing system using the Newton–Euler equation and subsequently examined the impact of outer ring waviness on the bearing's dynamic characteristics. The results indicated that a strong vibration can be generated by the bearing system when the wave number of the outer ring waviness is equal to or multiplied by the number of rollers. Yang et al. [17] studied the static characteristics of bearing capacity, attitude angle, end leakage flow, and friction coefficient under different waviness parameters through numerical simulation. Their results indicated that the position of the phase angle determines whether the waviness of the bearing bush may deteriorate or enhance the bearing system. Liu et al. [18] established a TDDE (time-dependent displacement excitation) model considering the coupling error of roundness and waviness. They analyzed the influence of roundness error and waviness amplitude on the vibration of a single-row angular contact ball bearing, providing valuable guidance for its fault diagnosis. Deng et al. [19] theoretically analyzed the correlation between the ripple amplitude, harmonic order of the working contact surface, and the friction torque of angular contact ball bearings with low-friction torque operating at a speed of 5000 r/min and an axial load of 20 N. The study was experimentally validated. The research results indicated that the friction torque of angular contact ball bearings fluctuates either significantly or minimally when the harmonic order of the inner and outer raceways follows certain relationships. Gao et al. [20] conducted experiments to investigate the relationship between channel waviness and friction torque in ball bearings under a speed of 1 r/min and an axial load of 1 N. The results suggested that, under certain conditions, the amplitude of waviness on the outer channel surface only affects the maximum fluctuation of friction torque without altering the frequency of the maximum fluctuation. The excitation frequency of friction torque fluctuation increases with the harmonic order of waviness on the outer raceway surface. Liu et al. [21] presented a model for calculating bearing friction torque, which takes into account the roundness error of rolling elements in needle roller bearings. They then compared the calculation results with those obtained by Palmgren and SKF companies to validate the feasibility of this approach. The model provides a more accurate method for predicting the friction torque of needle roller bearings with roundness error compared to previous empirical methods.

Building on the aforementioned research, the primary approach for determining bearing heat generation involves employing quasi-static, dynamic, or finite element methods. This typically includes locally calculating heat generation or determining local friction torque conversion. However, the quasi-static model proves unsuitable for high-speed bearings. The dynamic model and finite element method, while accurate, incur lengthy calculation times and high costs, rendering them less practical for engineering applications. Consequently, the calculation of local friction torque based on the quasi-dynamic model has been adopted to ascertain the overall heat generation of the bearing.

Currently, research on heat generation resulting from bearing friction considering roundness error is predominantly conducted under conditions of low speed and light load, and it does not account for the thermal expansion of components and the impact of changes in lubricant parameters on bearing heat generation. In light of this, this research is founded on the theories of bearing quasi dynamics, elastohydrodynamic lubrication, heat transfer, and Kirchhoff's law of energy conservation. An angular contact ball bearing heat generation calculation model has been established, which takes into account the impact of roundness error, bearing preload, centrifugal effect, and thermal coupling. The analysis aims to investigate the influence of roundness error in the bearing channel on the bearing's overall heat generation, and experiments are carried out to verify the rationality of the model. It provides a theoretical basis for analyzing the thermal characteristics of diagonal contact ball bearings with roundness error.

2. Calculation Method for Steady-State Heat Generation of Angular Contact Ball Bearings Considering Roundness Error

A thorough consideration is conducted to take into account factors such as roundness error, the force between the rolling element and the inner and outer rings, the cage, the drag force, resistance, and friction caused by the lubricating oil, as well as the axial displacement generated during pre-tightening, based on Hertz contact theory and elastohydrodynamic lubrication theory. Numerical calculation methods are employed to solve the diverse mechanical and kinematic parameters of the bearing under steady-state conditions. These parameters are then substituted into the friction torque calculation formula to determine the overall heat generation of the bearing. Lastly, the overall heat generation is substituted into the heat network method to calculate the temperature of each part of the bearing, the expansion generated by the components, and the changes in the viscosity characteristics of the lubricating oil. The quasi-dynamic calculation model for angular contact ball bearings is revised, and iterative solutions are executed to determine the bearing mechanics parameters, kinematics, heat generation, and temperature of each component. The fundamental concept is illustrated in Figure 1.

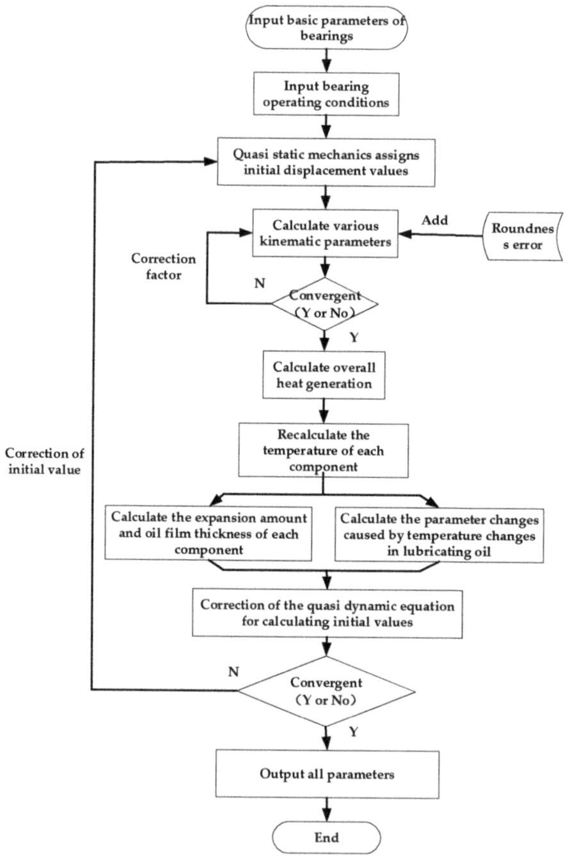

Figure 1. Flow chart of the calculation model for angular contact ball bearings considering roundness error.

2.1. Quasi Dynamics of Angular Contact Ball Bearings Considering Roundness Error

In order to conveniently represent the motion states and acting forces of various components in bearing motion, two coordinate systems are established on angular contact ball bearings, as shown in Figure 2, namely the inertial coordinate system and the rolling element coordinate system. The inertial coordinate system is fixed, with the coordinate origin O chosen at the centroid of the bearing. The X-axis coincides with the direction of the bearing axis, the Y-axis represents the radial direction of the bearing, and the Z-axis direction is determined by the right-hand screw rule. The inner and outer ring speeds of the bearing, the revolution angular velocity of the rolling elements, the revolution angular velocity of the cage, the axial displacement of the bearing active race relative to the stationary race, and the centroid displacement of the cage are all measured in this coordinate system. The rolling element coordinate system is a dynamic coordinate system, with the coordinate origin at the center of the ball. The direction of the x_b-axis always coincides with the X-axis of the inertial coordinate system, and the direction of the y_b-axis is perpendicular to the X-axis of the inertial coordinate system and points towards the outer ring of the bearing, determined by the right-hand screw rule for the z_b-axis direction. The self-rotation angular velocity of the ball and the relative displacement between the ball and the cage pocket hole are measured in this coordinate system.

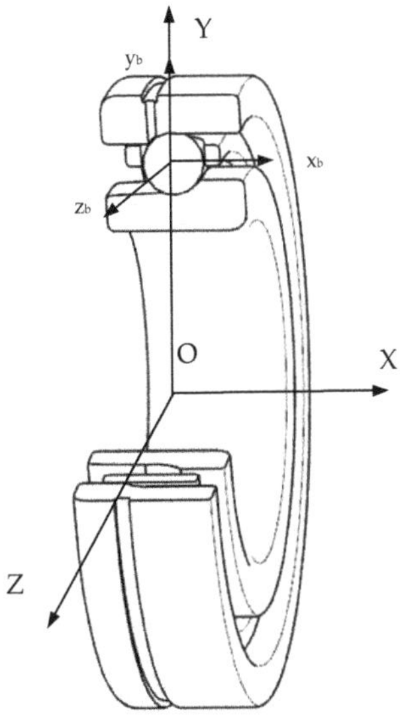

Figure 2. Angular contact ball bearing coordinate system.

2.1.1. Channel Roundness Error of Angular Contact Ball Bearings

As a result of machining errors, there exists a common roundness deviation in the grooves of the inner and outer rings of angular contact ball bearings, which is illustrated in Figure 3. The machining accuracy of the balls is generally higher than that of the inner and outer rings, and this article will not focus on analyzing it.

The roundness error on the normal line of the groove of angular contact ball bearings is represented by Fourier series, and the polar coordinate equation of the groove roundness error is given by:

$$\Delta S_{\mathrm{i}j}(\theta) = \sum_{n=2}^{\infty} \left(A_n \cos\left(n\theta_j + \varphi_{\mathrm{i}}\right)\right) \tag{1}$$

$$\Delta S_{\mathrm{e}j}(\theta) = \sum_{m=2}^{\infty} \left(A_m \cos\left(m\theta_j + \varphi_{\mathrm{e}}\right)\right) \tag{2}$$

where θ_j is the position angle of the point of the ball on the channel in the radial plane; n and m are the order of the roundness error of the inner and outer rings, respectively; A_n and A_m are the amplitude of the roundness error of the inner and outer rings, respectively; and φ_i and φ_e are the initial phase angles of the inner and outer rings, respectively.

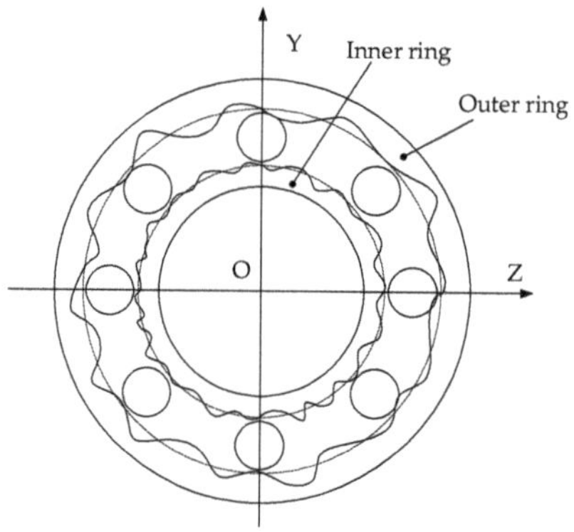

Figure 3. A model of an angular contact ball bearing with roundness error.

Due to the existence of channel roundness error, the curvature radii of the inner and outer channels can be expressed as follows:

$$r_{\text{i}j} = r_\text{i} + \Delta S_{\text{i}j}(\theta_{\text{i}j}) \tag{3}$$

$$r_{\text{e}j} = r_\text{e} + \Delta S_{\text{e}j}(\theta_{\text{e}j}) \tag{4}$$

where r_i and r_e are the radii of the curvature of the channels in the inner and outer rings; $r_{\text{i}j}$ and $r_{\text{e}j}$ are the curvature radii of the inner and outer ring channels after considering the roundness error.

The coefficients of the curvature radii of the channels in the inner and outer rings can be expressed as follows:

$$f_\text{i} = r_{\text{i}j}/D_\text{w} \tag{5}$$

$$f_\text{e} = r_{\text{e}j}/D_\text{w} \tag{6}$$

where D_w is the diameter of the ball bearing.

As the curvature radius coefficient of the channel changes, the elastic deformation of the bearing surface changes as follows:

$$\delta_{\text{i}j} = \sqrt{(A_\text{x} - x_1)^2 + (A_\text{y} - y_1)^2} - (f_\text{i} - 0.5)D_\text{w} \tag{7}$$

$$\delta_{\text{e}j} = \sqrt{x_1{}^2 + y_1{}^2} - (f_\text{e} - 0.5)D_\text{w} \tag{8}$$

where A_x and A_y are the distances of the center of the curvature of the groove of the inner and outer rings of the bearing in the coordinate system in the X- and Y-directions. x_1 and y_1 are the distances between the center of the ball and the center of the curvature of the outer channel in the coordinate system in the X- and Y-directions.

2.1.2. Quasi-Dynamic Model

The quasi dynamics of angular contact ball bearings are founded on Hertzian contact theory and the theory of elastohydrodynamic lubrication. The presence of roundness error has a significant impact on the curvature radii of the inner and outer raceways in these bearings, which in turn causes modifications to crucial parameters such as normal contact load (Q), steady-state contact angle (α), oil film thickness, and hydrodynamic friction force

(F). Therefore, establishing a quasi-dynamic equilibrium equation system must take into account the roundness error. For detailed meanings of the parameters in the formula, please refer to reference [15].

(1) The equilibrium equation of the ball

Figure 4 shows the force on the ball in the plane, from which the equilibrium equation of the ball can be obtained as follows:

$$\sum F_x = 0$$

$$\begin{gathered} Q_{ij}\sin\alpha_{ij} - Q_{ej}\sin\alpha_{ej} + F_{Tsij}\cos\alpha_{ij} - F_{Tsej}\cos\alpha_{ej} - F_{Rsij}\cos\alpha_{ij} + F_{Rsej}\cos\alpha_{ej} \\ + F_{Hsij}\cos\alpha_{ij} - F_{Hsej}\cos\alpha_{ej} + P_{Sgj} + P_{Rgj} = 0 \end{gathered} \tag{9}$$

$$\sum F_y = 0$$

$$\begin{gathered} Q_{ij}\cos\alpha_{ij} - Q_{ej}\cos\alpha_{ej} - F_{Tsij}\sin\alpha_{ij} + F_{Tsej}\sin\alpha_{ej} + F_{Rsij}\sin\alpha_{ij} - F_{Rsej}\sin\alpha_{ej} \\ - F_{Hsij}\sin\alpha_{ij} + F_{Hsej}\cos\alpha_{ej} - P_{Ssj} - P_{Rsj} = 0 \end{gathered} \tag{10}$$

$$\sum F_z = 0$$

$$F_{Tgej} - F_{Tgij} - F_{Rgej} + F_{Rgij} + F_{Hgej} - F_{Hgij} + Q_{cj} - F_{cbj} - F_{dj} = 0 \tag{11}$$

$$\sum M_x = 0$$

$$(F_{Tgej} - F_{Rgej})\frac{D_w}{2}\cos\alpha_{ej} + (F_{Tgij} - F_{Rgij})\frac{D_w}{2}\cos\alpha_{ij} - (P_{Ssj} + P_{Rsj})\frac{D_W}{2} - J_x\omega'_{xj} = 0 \tag{12}$$

$$\sum M_y = 0$$

$$(F_{Rgej} - F_{Tgej})\frac{D_w}{2}\sin\alpha_{ej} + (F_{Rgij} - F_{Tgij})\frac{D_w}{2}\sin\alpha_{ij} - (P_{Sgj} + P_{Rgj})\frac{D_w}{2} - J_y\omega'_{yj} + G_{yj} = 0 \tag{13}$$

$$\sum M_z = 0$$

$$(F_{Tsej} - F_{Rsej})\frac{D_w}{2} + (F_{Tsij} - F_{Rsij})\frac{D_W}{2} - J_z\omega'_{zj} - G_{zj} = 0 \tag{14}$$

where the subscript i represents the inner ring of the bearing; the subscript e represents the outer ring of the bearing; the subscript s denotes the long axis direction of the contact area between the ball and the inner and outer rings; the subscript g denotes the short axis direction; the subscript j represents the sorting of the rolling element, the same below; Q is the normal contact load of the contact area; α represents the working contact angle; F_T represents the oil film drag force; F_R is fluid friction; Q_c represents the force between the ball and the retainer pocket; F_{cb} is the centrifugal force of the ball; F_d is the oil mist resistance generated by the mixing of oil and gas on the ball; J is the moment of inertia of the ball; and G is the gyroscopic moment of the ball.

(2) Cage frame balance equation

Figure 5 shows the force situation of the cage frame, from which the equilibrium equation of the cage can be obtained as follows:

$$\sum_{j=1}^{Z}\left[Q_{cj}\sin\phi_j + (P_{Ssj} - P_{Rsj})\cos\phi_j\right] + F_{cy} - G_c = 0 \tag{15}$$

$$\sum_{j=1}^{Z}\left[-Q_{cj}\cos\phi_j + (P_{Ssj} - P_{Rsj})\sin\phi_j\right] + F_{cz} = 0 \tag{16}$$

$$\frac{1}{2}\sum_{j=1}^{Z}\left(-Q_{cj}\cdot d_{\rm m}\right)+\frac{1}{2}\sum_{j=1}^{Z}D_{\rm cp}\left(P_{Ssj}-P_{Rsj}\right)+M_{\rm cx}=0 \tag{17}$$

where P_S is the sliding friction force of the contact area between the ball and the retainer pocket; P_R is the rolling friction force at the contact area between the ball and the retainer pocket; and ϕ_j is the angular position of the rolling element, the same below.

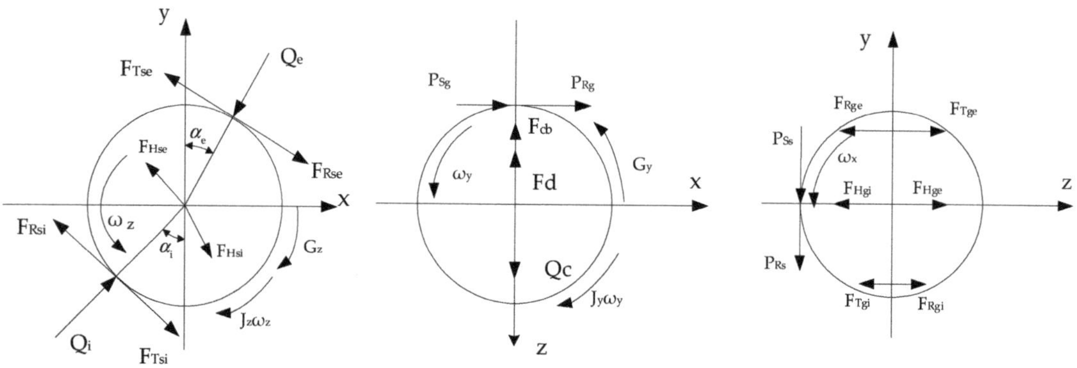

Figure 4. Balance diagram of ball force.

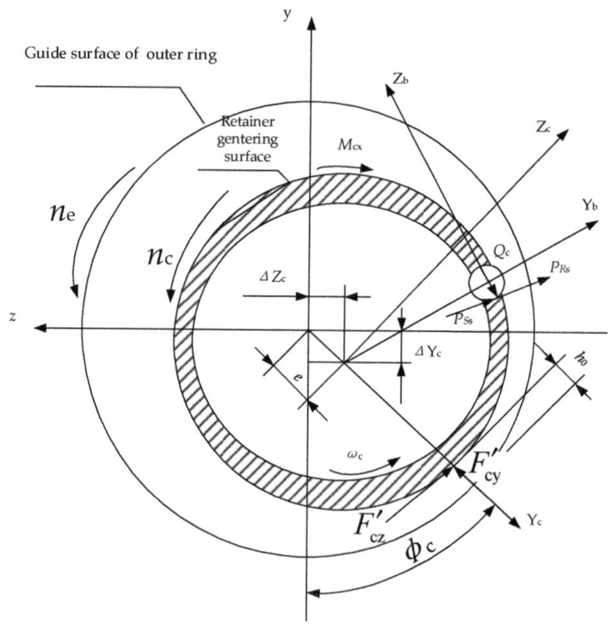

Figure 5. Cage frame force diagram.

(3) Inner ring balance equation

According to the mechanical relationship between the ball and the inner ring, the balance equation of the bearing inner ring can be obtained as follows:

$$F_X-\sum_{j=1}^{Z}\left(Q_{ij}\sin\alpha_{ij}+F_{Tsij}\cos\alpha_{ij}-F_{Rsij}\cos\alpha_{ij}\right)=0 \tag{18}$$

$$F_Y - \sum_{j=1}^{Z} (Q_{ij} \cos \alpha_{ij} - F_{Tsij} \sin \alpha_{ij} + F_{Rsij} \sin \alpha_{ij}) \cos \phi_j = 0 \tag{19}$$

$$F_Z - \sum_{j=1}^{Z} (Q_{ij} \cos \alpha_{ij} - F_{Tsij} \sin \alpha_{ij} + F_{Rsij} \sin \alpha_{ij}) \sin \phi_j = 0 \tag{20}$$

$$\begin{aligned} M_Y - \sum_{j=1}^{Z} [r_{i0}(Q_{ij} \sin \alpha_{ij} + F_{Tsij} \cos \alpha_{ij} - F_{Rsij} \cos \alpha_{ij}) - f_i D_w F_{Tsij} \cos \alpha_{ij} \\ + f_i D_w F_{Rsij} \cos \alpha_{ij}] \sin \phi_j = 0 \end{aligned} \tag{21}$$

$$\begin{aligned} M_Z - \sum_{j=1}^{Z} [r_{i0}(Q_{ij} \sin \alpha_{ij} + F_{Tsij} \cos \alpha_{ij} - F_{Rsij} \cos \alpha_{ij}) - f_i D_w F_{Tsij} \cos \alpha_{ij} \\ + f_i D_w F_{Rsij} \cos \alpha_{ij}] \cos \phi_j = 0 \end{aligned} \tag{22}$$

where r_{i0} refers to the radius of the curvature of the center track of the inner channel; F refers to the external force borne by the bearing in the inertial coordinate system; and M is the external moment around the inertial coordinate system.

The model comprises a total of 6Z + 8 equations and is computed with the assistance of DVE-C++. Due to the significant difference in the order of magnitude between displacement and velocity in the solution variable, the sequence of first division and then total is employed for solution. The Powell optimization algorithm and the traditional Newton iteration method are combined to perfectly solve the problem of the equation set that cannot be solved due to the irreversible partial derivative matrix. Meanwhile, to accelerate the solution speed and improve solution accuracy, the golden section search method [22] is employed to find the minimum value of the function within the interval.

2.2. Calculation of Friction Torque

The problem of friction torque in angular contact ball bearings is highly complex. The bearing's processing technology, selected materials, structural dimensions, geometric accuracy, lubrication conditions, and working load all have an impact on it. The bearing friction torque is divided into six parts based on its characteristics: friction torque caused by elastic hysteresis, differential sliding friction torque of the rolling element, friction torque caused by spin sliding of the rolling element, friction torque between the rolling element and the cage pocket, friction torque between the cage and the guide ring, and viscous friction torque of the lubricating oil. For detailed meanings of the parameters in the formula, please refer to reference [23]. The mechanical and kinematic parameters obtained from the quasic dynamic calculation model of the angular contact ball bearing are utilized to calculate the friction torque, resulting the overall heat generation.

2.2.1. Friction Torque Caused by Elastic Hysteresis

The properties of material lag lead to friction torque when rolling the rolling element on the inner and outer ring raceway. The expression of friction torque caused by elastic lag is as follows:

$$M_1 = \frac{d_m}{4}(1 - \gamma^2) \times \left(\sum_{z=1}^{z} (\Phi_i) + \sum_{z=1}^{z} (\Phi_e) \right) \times s \tag{23}$$

where d_m is the diameter of the bearing pitch circle; γ is the dimensionless parameter, which can be calculated by $\gamma = \frac{D_w}{d_m} \cos(\alpha_0)$; α_0 is the initial bearing contact angle; and s is the elastic hysteresis coefficient, generally 0.007.

2.2.2. Lubricating Oil Viscous Friction Torque

The characteristics of lubricating oil viscosity lead to the friction torque generated by the rolling body in the process of movement, which is expressed as follows:

$$M_2 = 6.35 \times \frac{1}{a^*_{\mathrm{i(e)}}} \times S_{\mathrm{r}} \times \frac{d_{\mathrm{m}}}{2} \times \sum\left[\left(\frac{h_{\mathrm{i}} + h_{\mathrm{e}}}{2}\right)(a_{\mathrm{i}} + a_{\mathrm{e}})\right] \tag{24}$$

where $a^*_{\mathrm{i(e)}}$ is the viscosity coefficient of the lubricating oil; S_{r} is the lubrication sufficient coefficient, an optional oil film lubrication coefficient; and h_{e} and h_{i} are the oil film thickness at the center of the inner and outer ring contact zone.

2.2.3. Friction Torque Caused by Differential Sliding

The friction torque caused by the inconsistency of the linear velocity of the ball and ferrule at the contact point during the motion of the angular contact ball bearing is expressed as follows:

$$M_3 = \frac{d_{\mathrm{m}}}{2D_{\mathrm{w}}}(1-\gamma^2) \times \left(\sum_{z=1}^{z} M_{De} + \sum_{z=1}^{z} M_{\mathrm{Di}}\right) \times f_{\mathrm{s}} \tag{25}$$

where f_{s} is the sliding friction coefficient, which can be taken as the sliding friction coefficient of the lubricating oil in the contact area.

2.2.4. Friction Torque Caused by Spin Slip

In angular contact ball bearings, the friction torque caused by the spin sliding of the ball around the normal direction of the contact zone is expressed as follows:

$$M_4 = \frac{3}{8} \times f_{\mathrm{s}} \times \sum_{z=1}^{z}\left(\Gamma_{\mathrm{i(e)}} a_{\mathrm{i(e)}} Q_{\mathrm{i(e)}} \sin \alpha_{\mathrm{i(e)}}\right) \tag{26}$$

where $\alpha_{\mathrm{i(e)}}$ is the working contact angle of the inner and outer rings of the bearing.

2.2.5. Friction Torque Caused by Friction between Rolling Element and Cage

The friction torque between the cage and the guide ring is mainly generated by the hydrodynamic friction force of the lubricating oil, and it is expressed as follows:

$$M_5 = \frac{d_{\mathrm{m}}}{4}(1-\gamma^2) \times \sin(\alpha_0 + \arctan(\frac{2D_w \sin(\alpha_0)}{d_{\mathrm{m}}(1-\gamma)})) \times G_c \times f_{\mathrm{c}} \tag{27}$$

where G_c is the cage weight, and f_{c} is the sliding friction coefficient between the rolling element and the cage pocket.

2.2.6. Friction Torque Caused by Friction between Cage and Guide Ring

The friction torque between the cage and the guide ring is mainly generated by the hydrodynamic friction force of lubricating oil, and it is expressed as follows:

$$M_6 = 1.38 \times Et - 7G_c \times f_{\mathrm{c}} \times \omega^2_{\mathrm{ci(e)}} \times e_{\mathrm{c}} \times D_{\mathrm{yi(e)}} \times (1-\gamma^2) \times 10^{-3} \tag{28}$$

where $\omega_{\mathrm{ci(e)}}$ is the angular velocity of the guide surface relative to the cage; e_{c} is the relative eccentricity of the center of the cage: and $D_{\mathrm{yi(e)}}$ is the diameter of the guard edge of the guide ring.

2.2.7. Total Friction Torque

The bearing overall friction torque (M_0) is expressed as follows:

$$M_0 = M_1 + M_2 + M_3 + M_4 + M_5 + M_6 \tag{29}$$

The expression of bearing total heat generation (Q) is as follows:

$$Q = 1.05 \times 10^{-4} N M_0 \tag{30}$$

where N is the rotational speed.

2.3. Temperature Calculation by Thermal Network Method

Using the theory of heat transfer and Kirchhoff's law of energy conservation, the thermal network method is employed to determine the temperature. As illustrated in Figure 6, the thermal network grid method is employed to establish a temperature field model, formulate temperature node solution equations, and compute the bearing temperature. In Figure 7, the contact points between the bearing pedestal, inner ring, outer ring, main shaft, ball, inner ring and ball, and outer ring have been simplified as temperature nodes.

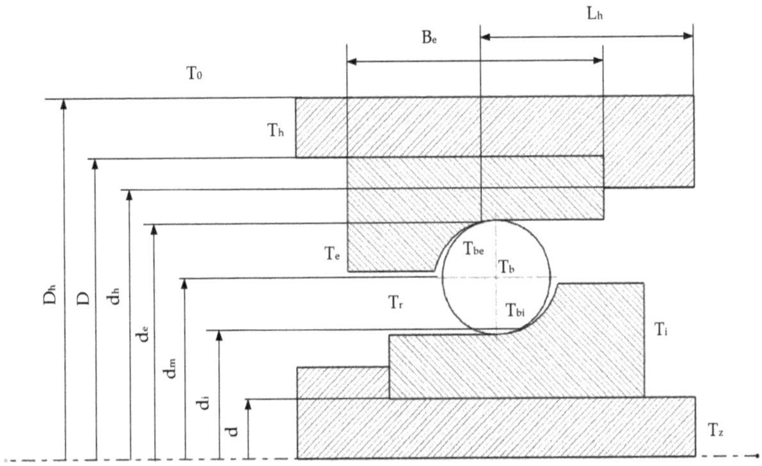

Figure 6. Schematic diagram of bearing temperature nodes.

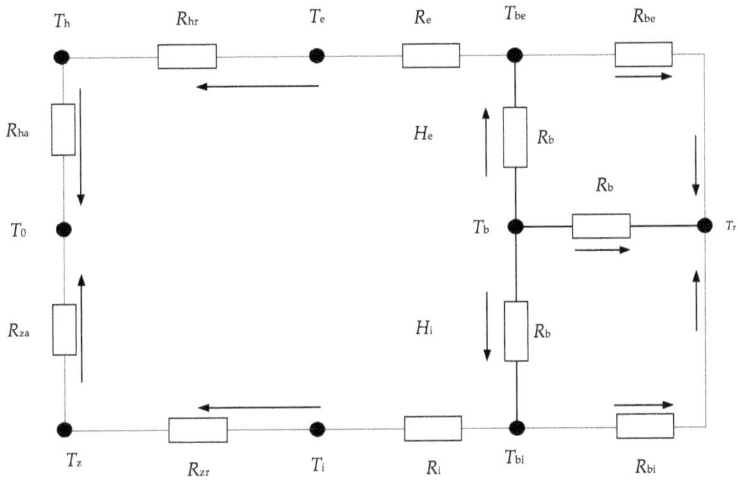

Figure 7. Schematic diagram of bearing heat transfer network.

Based on Burton and Steph's viewpoint, the friction heat generated by the bearing is equally distributed to the rolling elements and the ferrules, and the heat transfer equations are expressed as follows:

$$\frac{T_z - T_0}{R_{za}} + \frac{T_z - T_i}{R_{zr}} = 0 \tag{31}$$

$$\frac{T_i - T_z}{R_{zr}} + \frac{T_i - T_{bi}}{R_i} = 0 \tag{32}$$

$$\frac{T_{bi} - T_i}{R_i} + \frac{T_{bi} - T_r}{R_{ir}} + \frac{T_{bi} - T_b}{R_b} = \frac{H_i}{2} \tag{33}$$

$$\frac{T_b - T_r}{R_{br}} + \frac{T_b - T_{be}}{R_b} + \frac{T_b - T_{bi}}{R_b} = \frac{H_i + H_e}{2} \tag{34}$$

$$\frac{T_{be} - T_e}{R_e} + \frac{T_{be} - T_r}{R_{er}} + \frac{T_{be} - T_b}{R_b} = \frac{H_e}{2} \tag{35}$$

$$\frac{T_e - T_h}{R_{hr}} + \frac{T_e - T_{be}}{R_e} = 0 \tag{36}$$

$$\frac{T_h - T_0}{R_{ha}} + \frac{T_h - T_e}{R_{hr}} = 0 \tag{37}$$

where T_0 and T_r respectively represent the ambient temperature and lubricating oil inlet temperature.

The literature [24] contains information on heat transfer resistance. The convection heat transfer coefficient can be divided into three components: the forced convection heat transfer coefficient between the bearing seat and the air, between the main shaft and the air, and between the lubricating oil in the bearing, as shown below.

The coefficient of thermal convection between the bearing pedestal and the air can be expressed as follows:

$$k_h = 9.7\,\mathrm{W}/(\mathrm{m}^2 \cdot \mathrm{K}) \tag{38}$$

The coefficient of thermal convection between the main shaft and the air can be expressed as follows:

$$k_z = 9.7 + 5.33(25\pi N/6000)^{0.8} \tag{39}$$

The average coefficient of forced convection of lubricating oil in rolling contact bearing can be approximated as follows [25]:

$$k_a = 0.0986\left\{\frac{N}{v}\left[1 \pm \frac{D_w \cos(a_0)}{d_m}\right]\right\}^{0.5} \lambda P_r^{\frac{1}{3}} \tag{40}$$

where λ represents the fluid thermal coefficient, and P_r denotes the Prandtl number.

3. Comparative Verification

To validate the model, Figure 8 illustrates the test shaft system of the testing machine. Figure 9 illustrates its working principle. The loading mechanism of the test machine is capable of simultaneously applying axial and radial loads, with a maximum capacity of 1000 N, and the motor can reach a maximum speed of 36,000 r/min. The critical component of the test machine is the test bearing shaft system, with the ability to install four sets of bearings in each test. The bearings are divided into two end bearings and a central supporting bearing. Radial loads are applied to the supporting bearing and transmitted through the test shaft to the end bearings. Simultaneously, axial loads are applied to the front-end bearing and transmitted through the test bearing to the rear-end bearing. The test bearing is an H7006C ceramic ball bearing, featuring an accuracy class of P4. The parameters are presented in Table 1. The lubricating oil used is 4106 aviation lubricating oil, as illustrated in Table 2. The method of lubrication employed is oil injection lubrication. The bearing's motion form involves the fixation of the outer ring and the rotation of the inner ring. During numerical simulation of the model, the external channel roundness error order is 2, the internal channel roundness error order is 3, and the internal and external channel roundness error amplitude is 0.3 μm.

Figure 8. Bearing testing machine.

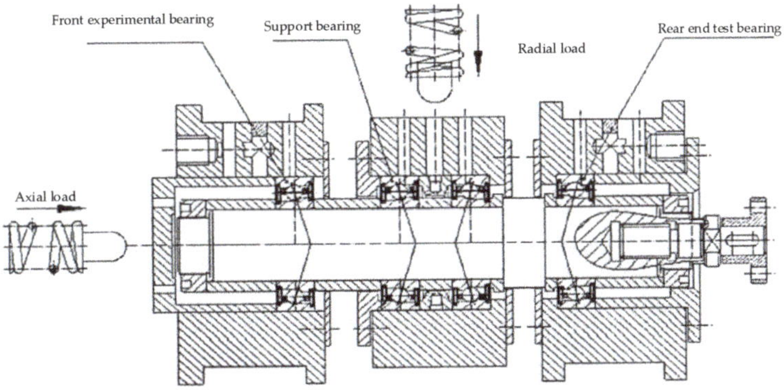

Figure 9. Schematic diagram of working principle of testing machine.

Table 1. H7006C Parameters of ceramic ball bearing.

Parameter	Value
Outside diameter (mm)	55
Inner diameter (mm)	30
Pitch diameter (mm)	42.85
Number of balls	18
Initial contact angle (°)	16
Spherical diameter (mm)	5.5
Width (mm)	13
Inner race groove diameter (mm)	39.927
Outer race groove diameter (mm)	48.079
Cage pocket hole diameter (mm)	5.89
Clearance (mm)	0.15
Cage width (mm)	8.8
Guide face diameter (mm)	46.07
Ceramic thermal conductivity (W/(m·K))	16.7

Table 2. Performance parameters of lubricating oil at normal temperature.

Parameter	Value
Dynamic viscosity (Pa*s)	0.055
Coefficient of viscous pressure ($\times 10^{-8}$ Pa^{-1})	1.85
Viscosity-temperature coefficient (°C^{-1})	0.0315
Coefficient of heat transfer (W/(m·K))	0.0966

The working condition is presented in Table 3. The experiment on bearing temperature measurement is conducted under five different working conditions. At an ambient temperature of 25 °C, the steps for temperature measurement are as follows: Firstly, the electric spindle is mounted onto the test bench, and the oil and gas apparatus, as well as the power supply, are connected. Subsequently, the data for axial and radial loading are transmitted through a pressure sensor to the control system. Next, a temperature sensor is inserted into the outer ring of the front-end bearing, and the collected temperature data are transferred to the computer control system. Figure 10 illustrates the interface of the test data acquisition instrument.

Table 3. Test conditions.

Condition	1	2	3	4	5
Axial load F_a (N)	300	500	200	400	600
Radial load F_r (N)	200	400	100	300	500
Inner ring speed (r/min)	6000	6000	10,000	10,000	10,000
Elapsed time (min)	10	10	10	10	10

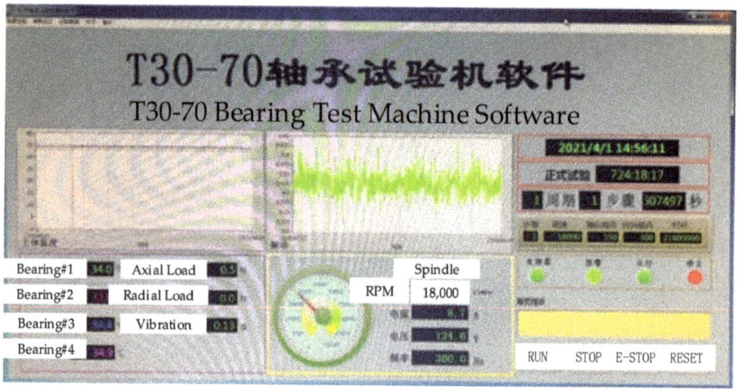

Figure 10. Result acquisition instrument.

For a single operating condition, the temperature is collected every 10 min of operation, and if the value fluctuation is minimal, it is considered to have reached a steady state. The results obtained thus far are compared with those of the numerical simulation.

Figure 11 shows the comparison between the results of the bearing test and the calculation results. The figure reveals that the calculation results of the model established in this paper are essentially consistent with the test results. The maximum relative error occurs under condition 3, where the test temperature is 33.5 °C and the calculation model results are 31.7 °C. The error is 5.37%, which confirms the validity of the model in this paper.

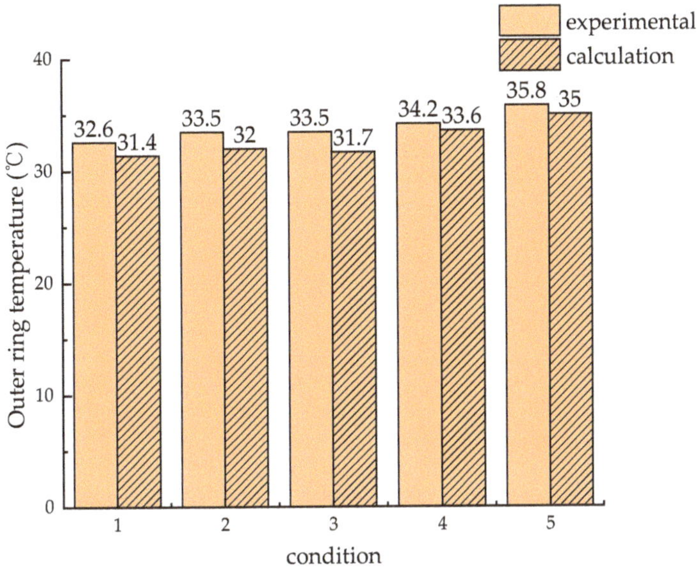

Figure 11. Result comparison chart.

The calculation results generally tend to be lower than the test results, as the oil inlet temperature in the established bearing steady-state heat generation model is a fixed value, failing to reflect the temperature change that occurs during the circulation of lubricating oil.

4. Calculation Results and Analysis

Using angular contact ball bearing 7008C as the research subject, its rated static load is 7.7 KN, the rated dynamic load is 11.5 KN, and the maximum speed is 40,000 r/min. Its primary structural parameters are displayed in Table 4. The material parameters are presented in Table 5. Ambient temperature is 20 °C. The lubricating oil selected is 4106 aviation power oil, and the oil injection lubrication method is adopted. The form of bearing movement is such that the outer ring remains stationary while the inner ring rotates, with the oil inlet temperature set at 20 °C.

Table 4. 7008C bearing parameters.

Parameter	Value
Outside diameter (mm)	68
Inner diameter (mm)	40
Pitch diameter (mm)	53.98
Number of balls	20
Initial contact angle (°)	18
Spherical diameter (mm)	6.35
Width (mm)	15
Inner race groove curvature radius (mm)	3.49
Outer race groove curvature radius (mm)	3.3
Cage pocket diameter (mm)	6.73
Cage pocket clearance (mm)	0.19
Cage width (mm)	10
Guide face diameter (mm)	58.08

Table 5. Material parameters.

Parameter	Value
Bearing Steel Modulus of Elasticity (N/m^2)	2.04×10^{11}
Holder elastic modulus (Polyamide, N/m^2)	2.32×10^{9}
Bearing Steel Density (kg/m^3)	7805.6
Cage density (kg/m^3)	1120
Poisson's ratio of bearing steel	0.3
Cage Poisson's ratio	0.34
Heat conductivity coefficient of bearing steel (W/(m*K))	40.1
Thermal expansion coefficient of bearing steel (1/°C)	0.1224×10^{-4}

4.1. *Influence of Bearing Working Conditions on Overall Heat Generation of Bearings*

Figure 12 shows the relationship between the rotational speed of the bearing and the total heat generation of the bearing under different outer ring groove roundness error orders when the axial load is F_a = 500 N and the outer ring groove circularity error amplitude is 0.5 μm. From the graph, it can be seen that the overall heat generation of the bearing linearly increases with the increase in speed. As the speed increases, the centrifugal force on the ball increases, resulting in an increase in sliding friction between the ball and the inner and outer contact areas, and a linear increase in overall heat generation.

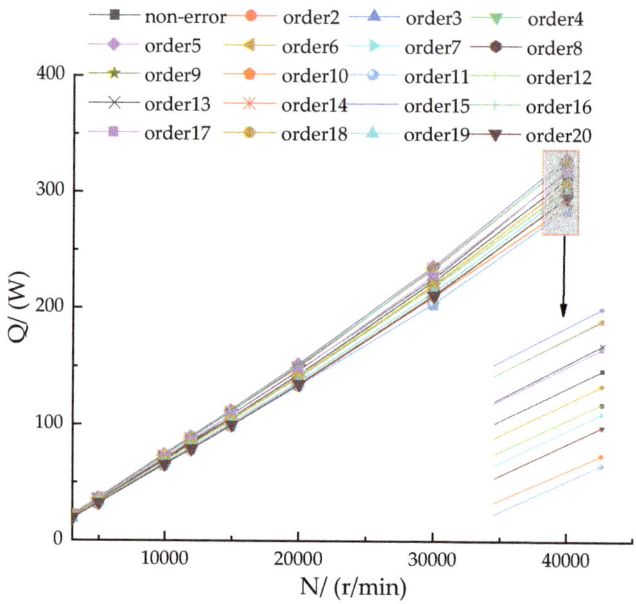

Figure 12. Relationship between rotating speed and bearing heating under different roundness error orders.

Figure 13 illustrates the correlation between axial load and total bearing heat generation of the angular contact ball bearing under various outer channel roundness error orders, with the rotating speed set at 10,000 r/min and the outer ring channel roundness error amplitude measured at 0.5 μm. Figure 12 indicates that there is a gradual increase in overall bearing heat generation as the axial load increases. The increase in axial load inhibits the rotation of the bearing and also leads to an increase in the normal contact force between the bearing rolling elements and the inner and outer rings. This causes an increase in the friction moment in the contact area, resulting in an overall increase in heat generation. When the order of the outer channel roundness error is equal to the number of balls $n/2 \pm 2$ (where $n = 1, 2, 3, \ldots$), the contact point between the ball and the outer ring is near the peak or valley, resulting in a small contact area and a low average normal load on the ball. The overall friction torque is correspondingly reduced, so the overall heat generation of the bearing is lower than when the roundness error is not considered. When the order of the outer channel roundness error is equal to the number of balls $(2n-1)/4 \pm 2$ (where $n = 1, 2, 3, \ldots$), the contact area between the rolling elements and the inner and outer channels is large, the average normal load on the balls is large, and the friction torque in the contact area increases accordingly, resulting in a higher overall heat generation of the bearing compared to the case without roundness error. Under the same axial load, the overall heat generation of the bearing fluctuates significantly as the order of the roundness error increases.

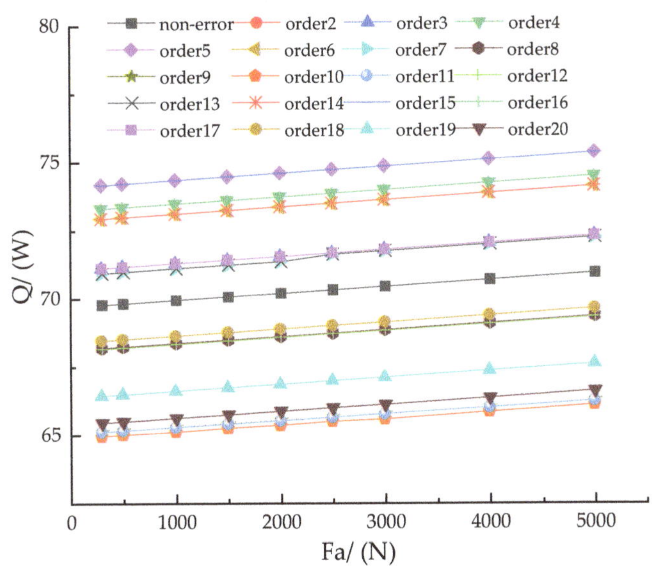

Figure 13. Relationship between axial load and bearing heat generation under different roundness error orders.

4.2. Influence of Roundness Error Order on the Whole Heat Generation of Bearings

Figure 14 shows that at a rotational speed of 10,000 r/min, the amplitude of the circularity error in the outer ring groove is 0.5 μm. The figure also illustrates the relationship between the circularity error of the outer ring groove and heat generation under various axial loads. As can be seen from the figure, as the order of the roundness error increases, the overall heat generation of the bearing fluctuates. When the order of the roundness error is equal to the number of balls $(2n-1)/4$ times (where $n = 1, 2, 3, \ldots$), the contact area between the ball and the raceway is maximized, and the spin friction power consumption generated between the ball and the inner and outer raceways is maximized, resulting in maximum overall friction torque and maximum overall heat generation of the bearing. When the order of the roundness error is equal to $n/2$ times the number of balls (where $n = 1, 2, 3, \ldots$), the contact point between the ball and the outer raceway is at the peak or valley, at which point the contact area between the ball and the raceway, the deformation of the Hertz contact area, and the normal load on the ball are minimized, resulting in minimum friction torque and minimum overall heat generation of the bearing.

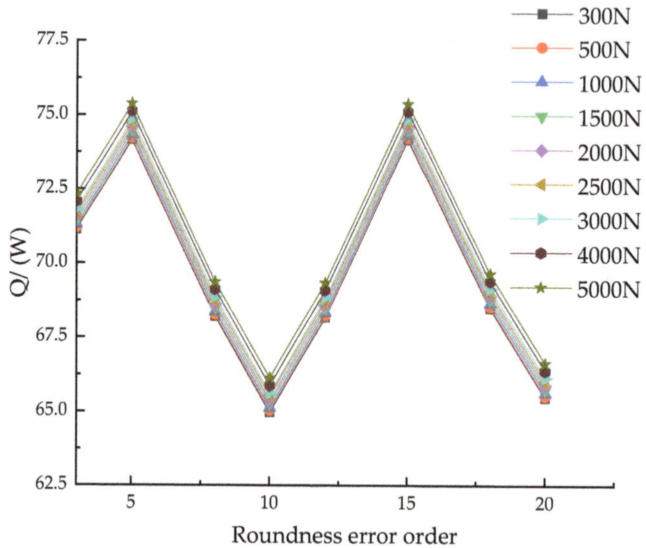

Figure 14. Relationship between roundness error and heat generation under different axial loads.

Figure 15 illustrates the relationship between the roundness error of the outer ring channel and heat generation under different rotational speeds, with an axial load of F_a = 500 N and outer ring channel roundness error amplitude of 0.5 μm. The figure illustrates that as the roundness error of the outer ring channel increases, it causes irregular changes in the contact area, resulting in fluctuations in the contact pressure and friction force, and consequently in heat generation. The faster the rotating speed, the greater the change in contact pressure and friction in the contact area, and the more pronounced the trend becomes.

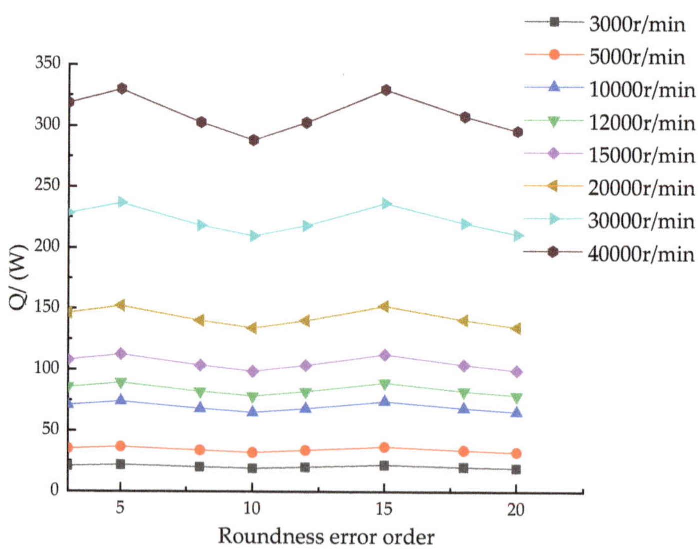

Figure 15. Relationship between roundness error and heat generation under different rotating speeds.

In Figure 16, the relationship between the amplitude of the outer ring groove roundness error and heat generation is shown when the axial load is F_a = 500 N and the rotational speed is 10,000 r/min. As can be seen from the figure, the overall heat generation fluctuation of the bearing under steady state increases with the increase in the roundness error amplitude, and its fluctuation is proportional to the amplitude. The increase in the amplitude of the roundness error will lead to an increase in the unevenness of the radial clearance. When the magnitude of the roundness error increases, the non-uniformity also increases, resulting in a decrease in the radial clearance of the rolling elements. With the increase in the roundness error amplitude, the contact deformation also increases. As the roundness error order rises, the normal contact load exhibits fluctuating changes, and the intensity of these fluctuations is proportional to the roundness error amplitude. This is the primary factor causing fluctuations in heat generation. At specific orders, although the roundness error amplitude is particularly large, the overall force on the bearing is uniform, and the normal load is relatively small. Therefore, the overall heat generation is smaller. The amplitude of the circularity error selected in this article is intended to fully demonstrate its regularity. In practical work, excessive roundness error amplitude can compress the radial clearance, causing the bearing to fail to operate normally, and should be kept within a reasonable range. This phenomenon needs to be considered in the design and selection of bearings to ensure that the bearings can operate normally and maintain thermal balance during operation.

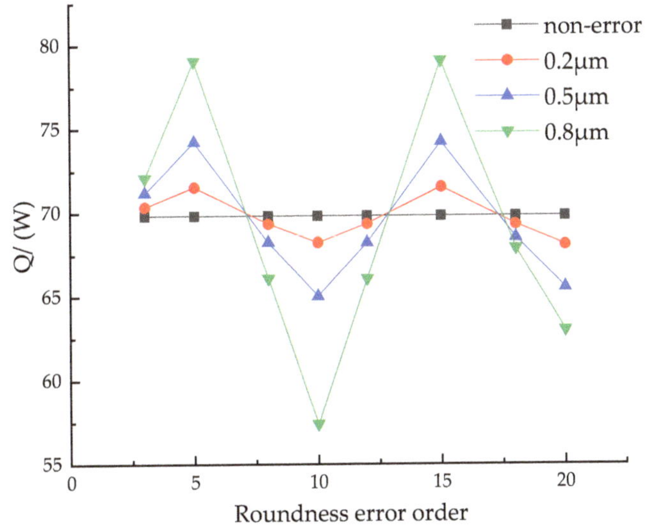

Figure 16. Relationship between roundness error amplitude of the outer groove roundness error and heat generation.

Figure 17 shows the relationship between the roundness error and heat generation of different numbers of balls when the axial load is F_a = 500 N and the rotational speed is 10,000 r/min. As can be seen from the figure, the overall heat generation of the bearing exhibits periodic changes with the increase in harmonic order of the outer ring groove roundness error. Under the same working conditions and roundness error, an increase in the number of balls will lead to an increase in the amount of heat generated, as the increased number of balls will result in an increase in the contact area between them, thereby generating more friction.

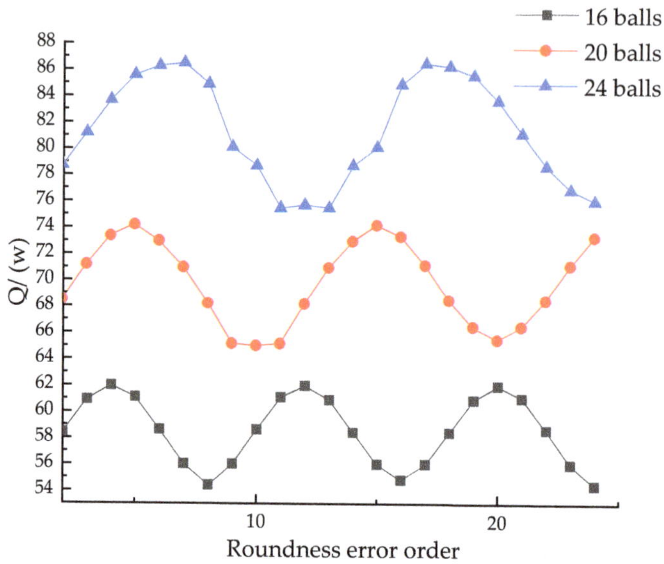

Figure 17. Relationship between roundness error, the numbers of balls and heat generation.

5. Conclusions

In this study, a thermal calculation model for angular contact ball bearings has been established based on the quasi-dynamic theory of bearings, the theory of elastohydrodynamic lubrication, the theory of heat conduction, and Kirchhoff's law of energy conservation. The model considers the effects of roundness error, bearing preload, centrifugal effects, and thermal coupling. The influence of bearing raceway roundness error on the overall heat generation of the bearing has been analyzed. The following are the conclusions of this study:

(1) As the speed and axial load increase, the overall heat generation linearly increases, and the increase in speed produces a more significant increase in heat generation compared to the axial load. When the order of the roundness error is equal to the number of balls $n/2 \pm 2$ (where n = 1, 2, 3, ...), the overall heat generation of the bearing is lower than when the roundness error is not considered. When the order of the roundness error is equal to the number of balls $(2n-1)/4 \pm 2$ (where n = 1, 2, 3, ...), the overall heat generation of the bearing is higher than that without roundness error.

(2) When the order of the roundness error is equal to the number of balls $(2n-1)/4$ times (where n = 1, 2, 3, ...), the overall heat generation of the bearing is maximum. When the order of the roundness error is equal to the number of balls $n/2$ times (where n = 1, 2, 3, ...), the overall heat generation of the bearing is minimum.

(3) As the order of the roundness error increases, the overall heat generation fluctuates. The faster the speed, the more obvious the trend of fluctuation. Under the same order of roundness error, the load has little effect on the overall heat generation, which increases with the increase in bearing speed.

(4) The overall heat generation fluctuation of the bearing under steady state increases with the increase in the roundness error amplitude, and its fluctuation is proportional to the amplitude. The overall heat generation of the bearing exhibits periodic changes with the increase in the harmonic order of the roundness error, and the change period is mapped to the number of balls.

Author Contributions: Conceptualization, Y.Y.; methodology, Y.Y.; software, Y.Y. and R.M.; validation, Y.Y., Y.X. and Y.L.; formal analysis, Y.Y. and R.M.; investigation, Y.Y. and R.M.; resources, Y.Y., Y.X. and Y.L.; data curation, Y.Y. and R.M.; writing—original draft preparation, Y.Y. and R.M.; writing—review and editing, Y.Y., R.M. and Y.X.; visualization, Y.Y., R.M., Y.X. and Y.L.; project administration, Y.Y., Y.X. and Y.L. All authors have read and agreed to the published version of the manuscript.

Funding: This work was supported by the Ministry of Industry and Information Technology Special Projects (Grant No. TC220H05V) and also supported by Key Research Projects of Higher Education Institutions of Henan Province (Grant No. 22A460018).

Institutional Review Board Statement: Not applicable.

Informed Consent Statement: Not applicable.

Data Availability Statement: Data are contained within the article.

Conflicts of Interest: The authors declare no conflicts of interest.

References

1. Palmgren, A. *Ball and Roller Bearing Engineering*; Svenska Kullager Fabriken (SKF) Industries Incorporation: Philadelphia, PA, USA, 1959.
2. Harris, T.A.; Kotzalas, M.N. *Advanced Concepts of Bearing Technology Rolling Bearing Analysis*; CRC Press, Taylor Francis Group: New York, NY, USA, 2006.
3. Kerrouche, R.; Dadouche, A.; Mamou, M.; Boukraa, S. Power Loss Estimation and Thermal Analysis of an Aero-Engine Cylindrical Roller Bearing. *Tribol. Trans.* **2021**, *64*, 1079–1094. [CrossRef]
4. Pouly, F.; Changenet, C.; Ville, F.; Velex, P.; Damiens, B. Power loss predictions in high-speed rolling element bearings using thermal networks. *Tribol. Trans.* **2010**, *53*, 957–967. [CrossRef]
5. Pouly, F.; Changenet, C.; Ville, F. Investigations on the Power Losses and Thermal Behavior of Rolling Element Bearings. *J. Eng. Tribol.* **2010**, *224*, 925–933.
6. Zheng, D.; Chen, W.; Li, M. An optimized thermal network model to estimate thermal per formances on a pair of angular contact ball bearings under oil-air lubrication. *Appl. Therm. Eng.* **2018**, *131*, 328–339.
7. Tarawneh, C.M.; Fuentes, A.A.; Kypuros, J.A.; Navarro, L.A.; Vaipan, A.G.; Wilson, B.M. Thermal Modeling of a Railroad Tapered-Roller Bearing Using Finite Element Analysis. *J. Therm. Sci. Eng. Appl.* **2012**, *4*, 031002. [CrossRef]
8. Hong, Y.; Kou, W. Dynamic Simulation of the Slide Friction Heat Generation of Deep Groove Ball Bearing Based on Hertz Theory. *Equip. Manuf. Technol.* **2014**, *12*, 39–41. (In Chinese)
9. Zheng, D.; Chen, W. Effect of structure and assembly constraints on temperature of high-speed angular contact ball bearings with thermal network method. *Mech. Syst. Signal Process.* **2020**, *145*, 106929. [CrossRef]
10. Nicolás, J.A.M.; de León Hijes, F.C.G.; Alhama, F. Solution of temperature fields in hydrodynamics bearings by the numerical network method. *Tribol. Int.* **2007**, *40*, 139–145. [CrossRef]
11. Dong, Y.; Ma, Y.; Qiu, M.; Chen, F.; He, K. Analysis and experimental research of transient temperature rise characteristics of high-speed cylindrical roller bearing. *Sci. Rep.* **2024**, *14*, 711. [CrossRef] [PubMed]
12. Lei, J.; Su, B.; Zhang, S.; Yang, H.; Cui, Y. Dynamics-Based Thermal Analysis of High-Speed Angular Contact Ball Bearings with Under-Race Lubrication. *Machines* **2023**, *11*, 691. [CrossRef]
13. Hassan, E. Rasheed, Effect of surface waviness on the hydrodynamic lubrication of a plain cylindrical sliding element bearing. *Wear* **1998**, *223*, 1–6.
14. Rodionov, E.M. *Moment Originating from Errors in the Form or Rolling Surfaces of a Ball Bearing*; Technical Report FTD-HT-66-374; Foreign Technology Div.: Wright-Patterson AFB, OH, USA, 1966.
15. Cui, L. Research on Dynamic Performances of High-Speed Rolling Bearing and Rotor System of Aeroenging. Ph.D. Dissertation, Harbin Institute of Technology, Harbin, China, 2008. (In Chinese).
16. Zheng, H.; Wang, F.; Ji, B. The Study for Effect of Outer Race Waviness on the Dynamic Behavior of Cylindrical Roller Bearing. *Mach. Des. Manuf.* **2014**, *12*, 82–84+88. (In Chinese)
17. Yang, M.; Lu, H.; Zhang, X.; Duan, M.; Bao, L.; Wang, B.; Wu, W. Influence of Surface Waviness of Journal and Bearing Bush on the Static Characteristics of Hydrodynamic Bearing. *Processes* **2021**, *9*, 110. [CrossRef]
18. Liu, J.; Pang, R.; Xu, Y.; Ding, S.; He, Q. Vibration analysis of a single row angular contact ball bearing with the coupling errors including the surface roundness and waviness. *Sci. China Technol. Sci.* **2020**, *63*, 943–952. [CrossRef]
19. Deng, S.; Li, X.; Wang, J.; Wang, Y.; Teng, H. Analysis on the friction torque fluctuation of angular contact ball bearings. *J. Mech. Eng.* **2011**, *23*, 104–112. (In Chinese) [CrossRef]
20. Gao, Y.; Deng, S.; Zheng, C.; Liang, B.; Li, J. Experimental study on effects of groove waviness on ball bearing friction torque. *Bearing* **2009**, *10*, 41–44. (In Chinese)

21. Liu, J.; Yan, Z.; Shao, Y. An investigation for the friction torque of a needle roller bearing with the roundness error. *Mech. Mach. Theory* **2018**, *121*, 259–272. [CrossRef]
22. Burton, R.A.; Staph, H.E. Thermally activated seizure of angular contact bearings. *ASLE Trans.* **1967**, *10*, 408–417. [CrossRef]
23. Deng, S.; Li, X.; Wang, J. Friction al Torque Characteristic of Angular Contact Ball Bearings. *J. Mech. Eng.* **2011**, *47*, 114–120. (In Chinese) [CrossRef]
24. Ai, S.; Wang, W.; Wang, Y.; Zhao, Z. Temperature rise of double-row tapered roller bearings analyzed with the thermal network method. *Tribol. Int.* **2015**, *87*, 11–22. [CrossRef]
25. Crecelius, W.J.; Pirvics, J. *Computer Program Operation Manual on SHABERTH: A Computer Program for the Analysis of the Steady State and Transient Thermal Performance of Shaft-Bearing Systems*; U.S. Air Force: Washington, DC, USA, 1976.

MDPI
St. Alban-Anlage 66
4052 Basel
Switzerland
www.mdpi.com

Lubricants Editorial Office
E-mail: lubricants@mdpi.com
www.mdpi.com/journal/lubricants

www.ingramcontent.com/pod-product-compliance
Lightning Source LLC
Chambersburg PA
CBHW061759210926
51850CB00006B/595
* 9 7 8 3 7 2 5 8 1 1 2 9 8 *